MILLER'S

GUIDE TO THE DISSECTION OF THE DOG

THIRD EDITION

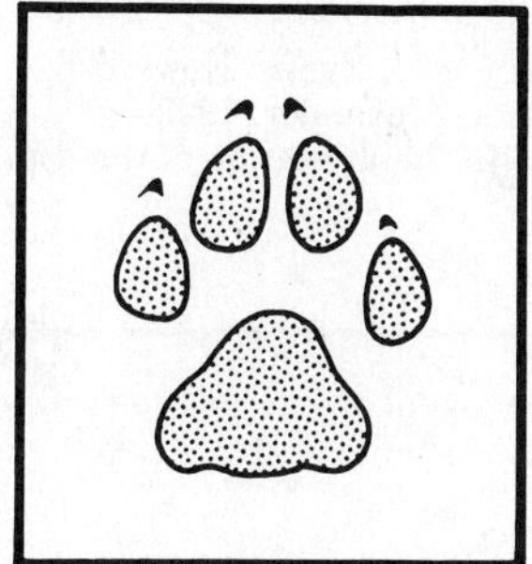

HOWARD E. EVANS, Ph.D.

Professor Emeritus of Veterinary and Comparative Anatomy
New York State College of Veterinary Medicine
Cornell University, Ithaca, New York

ALEXANDER de LAHUNTA, D.V.M., Ph.D.

Professor and Chairman of Veterinary Anatomy
New York State College of Veterinary Medicine
Cornell University, Ithaca, New York

W. B. SAUNDERS COMPANY
Harcourt Brace Jovanovich, Inc.
Philadelphia ○ London ○ Toronto ○ Montreal ○ Sydney ○ Tokyo

W. B. SAUNDERS COMPANY
Harcourt Brace Jovanovich, Inc.

The Curtis Center
Independence Square West
Philadelphia, PA 19106

Library of Congress Cataloging-in-Publication Data

Miller, Malcolm Eugene.

[Guide to the dissection of the dog]
Miller's guide to the dissection of the dog.—3rd ed./Howard
E. Evans, Alexander de Lahunta.
p. cm.
Bibliography: p.
Includes index.
ISBN 0-7216-2323-9
1. Dogs—Anatomy. 2. Dissection. I. Evans, Howard E.,
1922– II. de Lahunta, Alexander, 1932–
III. Title. IV. Title: Guide to the dissection of the dog.
QL813.D64M54 1988
636.7'0891—dc19 87-37470
CIP

Listed here is the latest translated edition of this book together with the language of the translation and the publisher.

Spanish (1st Edition)—Nueva Editorial Interamericana, S.A. de C.V., Mexico City, Mexico

Japanese (2nd Edition)—Gashuka, Tokyo

Editor: John Dyson
Developmental Editor: Linda Mills
Designer: Karen O'Keefe
Production Manager: Bill Preston
Manuscript Editor: Mark Crowe
Illustration Coordinator: Brett MacNaughton
Indexer: Ella Shapiro

Miller's Guide to the Dissection of the Dog ISBN 0-7216-2323-9

Last digit is the print number: 9 8 7 6 5 4 3

PREFACE

Dr. Malcolm E. Miller, D.V.M., Ph.D., was Professor and Head of the Department of Anatomy in the New York State College of Veterinary Medicine at Cornell University. He taught veterinary anatomy from 1935 until his death in 1960. His *Guide to the Dissection of the Dog*, first published in 1947 by Edwards Brothers Press of Michigan, was revised with the help of his wife, Mary, through a third edition in 1952. *Miller's Guide to the Dissection of the Dog* by Evans and de Lahunta was first published in 1971, translated into Spanish and revised in 1973. A second edition, published in 1980, was translated into Japanese. This third edition alters some dissection procedures, updates anatomical nomenclature and adds new descriptive material and illustrations. A new feature is the inclusion of directions for palpation of the live dog at appropriate times during the dissection. Of the 233 figures, 33 have not appeared in the guide previously.

We are grateful to the many students and colleagues in various institutions for their helpful suggestions. Most of the illustrations were prepared by Marion Newson and Pat Barrow, former medical illustrators in the Department of Anatomy. There are additional drawings by William Hamilton and our present illustrator, Michael Simmons. The preparation of osteological and preserved materials was the responsibility of Reinhold Dembinski.

The purpose of this guide is to facilitate a thorough dissection of the dog in order to learn basic mammalian structure. We have attempted to emphasize what we believe is essential anatomical knowledge for the veterinary curriculum. The descriptions are based upon the dissection of embalmed, arterially injected adult dogs. The anatomical terms used are those that appear in Nomina Anatomica Veterinaria (third edition, 1983). A more detailed consideration of the structures dissected can be found in the second edition of *Miller's Anatomy of the Dog* by Evans and Christensen, published in 1979 by W. B. Saunders Company, Philadelphia.

HOWARD E. EVANS
ALEXANDER DE LAHUNTA

Ithaca, N.Y.

PREFACE

CONTENTS

LIST OF ILLUSTRATIONS

INTRODUCTION

Anatomy is the study of structure. Physiology is the study of function. Structure and function are inseparable as the foundation of the science and art of medicine. One must know the parts before one can appreciate how they work. **Gross anatomy,** the study of structures that can be dissected and observed with the unaided eye or with a hand lens, forms the subject matter of this guide.

The anatomy of one part in relation to other parts of the body is **topographic anatomy.** The practical application of such knowledge in the diagnosis and treatment of pathological conditions is **applied anatomy.** The study of structures too small to be seen without a microscope is **microscopic anatomy.** Examination of structure in even greater detail is possible with an electron microscope and constitutes **ultrastructural anatomy.** When an animal becomes diseased or its organs function improperly, its deviation from the normal is studied as **morbid anatomy** or **pathology.** The study of the development of the individual from the fertilized oocyte to birth is **embryology;** from the zygote to the adult, **developmental anatomy. Teratology** is the study of abnormal development.

MEDICAL ETYMOLOGY AND ANATOMICAL NOMENCLATURE

The student of anatomy is confronted with an array of unfamiliar terms and names of anatomical structures. A better understanding of the language of anatomy helps make its study more intelligible and interesting. For the publication of scientific papers and communication with colleagues, the mastery of anatomical terminology is a necessity. To assure knowledge of basic anatomical terms, a medical dictionary should be kept readily accessible and consulted frequently. It is very important to learn the spelling, pronunciation and meaning of all new terms encountered. Vertebrate structures are numerous, and in many instances common names are not available or are so vague as to be meaningless. One soon realizes why it is desirable to have an international glossary of terms that can be understood by scientists in all countries. The learning of a medical vocabulary can be aided by the mastery of Greek and Latin roots and affixes.

Our present medical vocabulary has a history dating back more than 2000 years and reflects the influences of the world's languages. The early writings in anatomy and medicine were almost entirely in Latin, and as a consequence the majority of anatomical terms stem from this classical language. Latin terms are commonly translated into the vernacular of the person using them. Thus the Latin *hepar* becomes the English *liver,* the French *foie,* the Spanish *higado* and the German *Leber.*

Although anatomical terminology has been rather uniform, differences in terms have arisen between the different fields and different countries. In 1895 a group consisting mainly of German anatomists proposed a standard list of terms from those currently in use throughout the world. This list, known as the Basle Nomina Anatomica (BNA), was not adopted internationally but forms the basis for the present Nomina Anatomica (NA), which was adopted by the International Congress of Anatomists in Paris in 1955. Of the 5640 standard terms, more than 80 per cent are continued from the BNA. The latest revision of Nomina Anatomica is the fifth edition, published by Williams & Wilkins in 1983.

The guiding principles as stated in the Nomina Anatomica are to make no changes for purely pedantic or etymological reasons; to keep terms short and simple; to give similar names to closely related structures; to discard eponyms; to use terms with some informative or descriptive value; to arrange differentiating adjectives as opposites (for example: superficialis and profundus); to adopt single terms as a rule and to allow official alternatives only as exceptions; to resist naming every minute structure ever discovered or described; and to use Latin for all terms. Considerable thought and time have gone into anatomical terminology, and committees are actively engaged in improving it. The International Committee on Veterinary Anatomical Nomenclature, appointed by the World Association of Veterinary Anatomists, published **Nomina Anatomica Veterinaria** of domestic animals in 1968. These Nomina, as revised in the third edition of 1983, serve as the basis for the nomenclature used in this guide.

DIRECTIONAL TERMS

An understanding of the following planes, positions and directions relative to the animal body or its parts is necessary in order to follow the procedures for dissection (Fig. 1).

PLANE: a surface, real or imaginary, along which any two points can be connected by a straight line.

Median plane: divides the head, body or limb longitudinally into equal right and left halves.

Sagittal plane: passes through the head, body or limb parallel to the median plane.

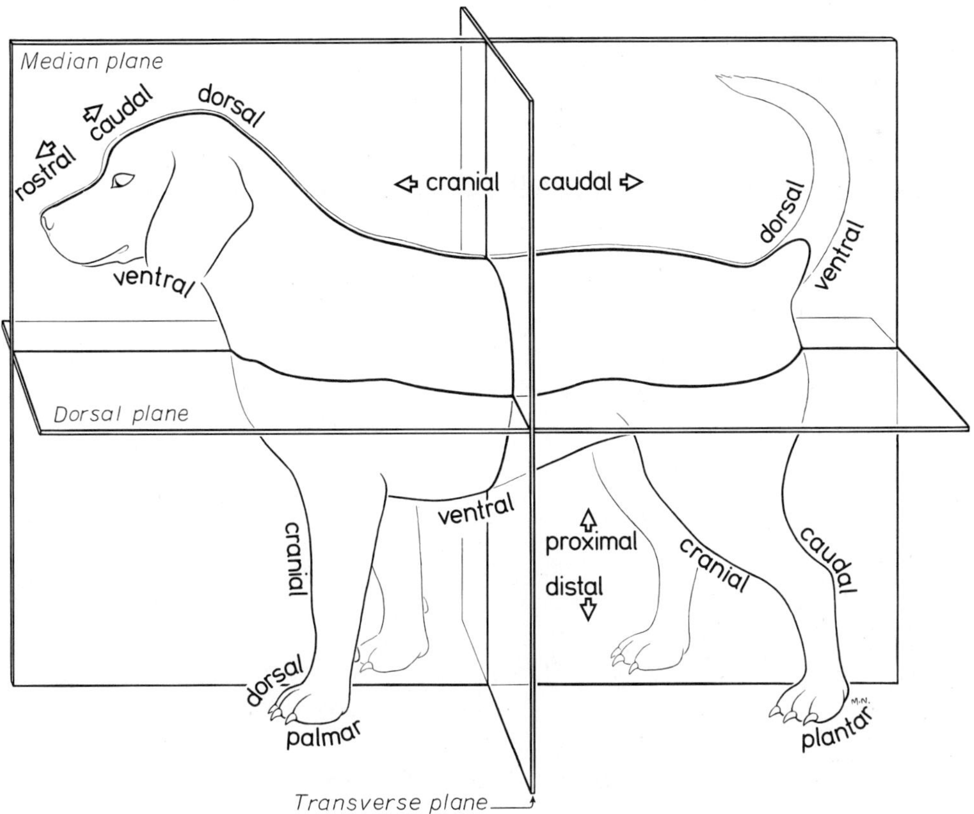

FIGURE 1. Directional terms.

Transverse plane: cuts across the head, body or limb at a right angle to its long axis or across the long axis of an organ or a part.

Dorsal plane: runs at right angles to the median and transverse planes and thus divides the body or head into dorsal and ventral portions.

DORSAL: toward or relatively near the back and corresponding surface of the head, neck, and tail; on the limbs it applies to the upper or front surface of the carpus, tarsus, metapodium and digits (opposite to the side with the pads).

VENTRAL: toward or relatively near the belly and the corresponding surface of the head, neck, thorax, and tail. This term is never used for the limbs.

MEDIAL: toward or relatively near the median plane.

LATERAL: away from or relatively farther from the median plane.

CRANIAL: toward or relatively near the head; on the limbs it applies proximal to the carpus and tarsus. When referring to the head it is replaced by the term rostral.

ROSTRAL: toward or relatively near the nose; applies to the head only.

CAUDAL: toward or relatively near the tail; on the limbs it applies proximal to the carpus and tarsus. Also used in reference to the head.

The adjectives for directional terms may be modified to serve as adverbs by replacing the ending “al” with the ending “ally,” as in *dorsally*.

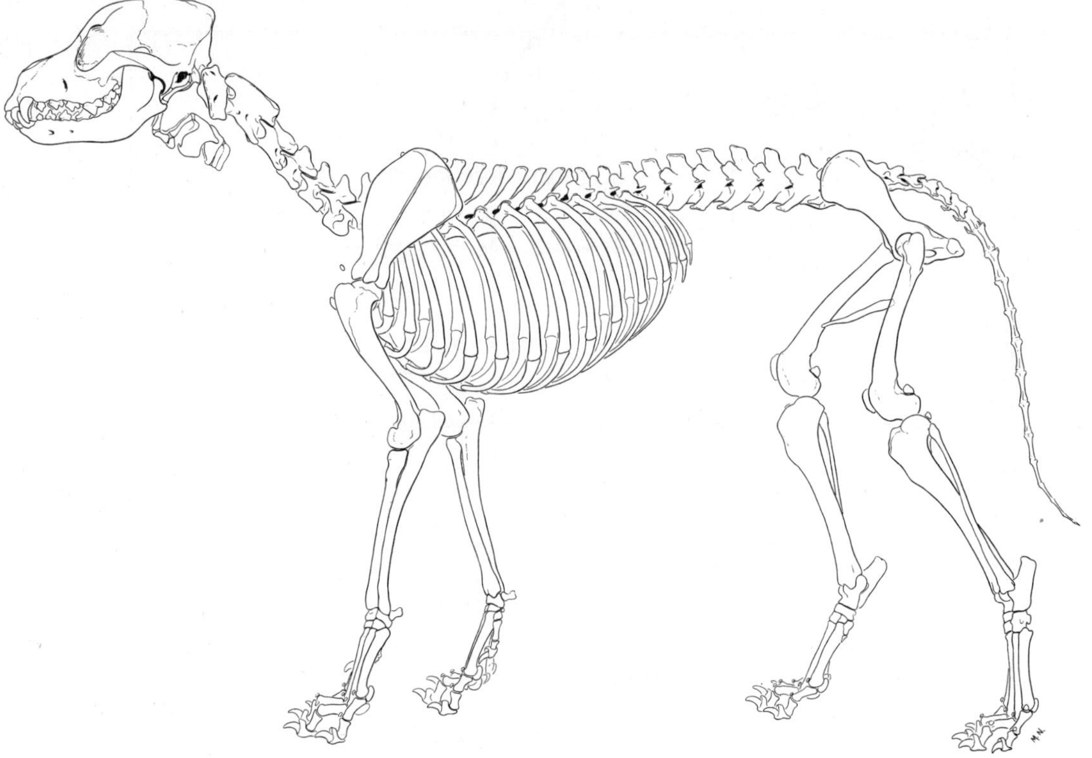

FIGURE 2. Skeleton of male dog.

Certain terms whose meanings are more restricted are used in the description of organs and appendages.

INTERNAL OR INNER: close to, or in the direction of, the center of an organ, body cavity or structure.

EXTERNAL OR OUTER: away from the center of an organ or structure.

SUPERFICIAL: relatively near the surface of the body or the surface of a solid organ.

DEEP: relatively near the center of the body or the center of a solid organ.

PROXIMAL: relatively near the main mass or origin; in the limbs and tail, the attached end.

DISTAL: away from the main mass or origin; in the limbs and tail, the free end.

RADIAL: on that side of the forearm in which the radius is located.

ULNAR: on that side of the forearm in which the ulna is located.

TIBIAL AND FIBULAR: on the corresponding sides of the leg, the tibial side being medial, and fibular lateral.

In the dog and similar species the paw is that part of the thoracic or pelvic limb distal to the radius and ulna or to the tibia and fibula. The hand (manus) and foot (pes) of man are homologous with the forepaw and hind paw respectively.

PALMAR: the aspect of the forepaw on which the pads are located—the surface that contacts the ground in the standing animal.

PLANTAR: the aspect of the hind paw on which the pads are located—the surface that contacts the ground in the standing animal. The opposite surface of both forepaw and hind paw is known as the dorsal surface.

AXIS: the central line of the body or any of its parts.

AXIAL, ABAXIAL: of, pertaining to or relative to the axis. In referring to the digits, the functional axis of the limb passes between the third and fourth digits. The axial surface of the digit faces the axis, the abaxial surface faces away from the axis.

The following terms apply to the various basic movements of the parts of the body.

FLEXION: the movement of one bone in relation to another in such a manner that the angle formed at their joint is reduced: The limb is retracted or folded; the digit is bent; the back is arched dorsally.

EXTENSION: the movement of one bone upon another is such that the angle formed at their joint increases: The limb reaches out or is extended; the digit is straightened; the back is straightened. Extension beyond 180 degrees is overextension.

ABDUCTION: the movement of a part away from the median plane.

ADDUCTION: the movement toward the median plane.

CIRCUMDUCTION: the movement of a part when outlining the surface of a cone (for example, the arm extended drawing a circle).

ROTATION: the movement of a part around its long axis (for example, the action of the radius when using a screwdriver). The direction of rotation of a limb or segment of a limb on its long axis is designated by the direction of movement of its cranial or dorsal surface (example: In medial rotation of the arm the crest of the greater tubercle is turned medially).

SUPINATION: lateral rotation of the appendage so that the palmar or plantar surface of the paw faces medially or dorsally.

PRONATION: medial rotation of the appendage from the supine position so that the palmar or plantar surface is facing ventrally.

On radiographs the view is described in relation to the direction of penetration by the X ray: from point of entrance to point of exit of the body part prior to striking the film. Oblique views are described with combined terms (example: A view of the carpus with the x-ray tube perpendicular to the dorsal surface and the film on the palmar surface would be a dorsopalmar view. If the x-ray tube was turned so that it pointed towards the dorsomedial surface of the carpus and the film was on the palmarolateral surface, the view would be dorsomedial-palmarolateral oblique).

DISSECTION

The dog provided as a specimen for dissection has been humanely prepared by injection of pentobarbital for anesthesia via the cephalic vein and by exsanguination through a cannula inserted in the common carotid artery. This procedure allows the pumping action of the heart to empty the blood vessels prior to the injection of embalming fluid consisting of 8 per cent formalin and 2 per cent phenol in aqueous solution. It is injected under 5 lb. of pressure over a period of approximately 30 minutes. After embalming, the arteries are injected with red latex, also through the common carotid artery, from a 50 cc. hand syringe. The prepared specimen represents a considerable investment of materials and labor and should be cared for accordingly. A well-kept specimen facilitates dissection and study throughout the term. Gauze, 2 per cent phenoxyethanol, 1 per cent phenol or other antifungal agent and plastic sheeting are provided to moisten and wrap the paws and head and to cover the entire specimen between dissection periods. Refrigeration is helpful for storage but not essential.

There are certain principles and procedures that are generally accepted as aids in the learning of anatomy. The purpose of the dissection is to gain a clear understanding of the normal structures of the body and an appreciation for individual variation.

THE SKELETAL AND MUSCULAR SYSTEMS

Before dissection of muscles is explained, the bones of that region will be described. A thorough understanding of the relationships of muscles and bones will facilitate learning the muscular attachments and functions.

The **appendicular skeleton** includes the bones of the thoracic girdle and forelimbs and the pelvic girdle and hind limbs (Fig. 3).

BONES OF THE THORACIC LIMB

The **thoracic girdle** consists of paired scapulae and clavicles. The scapula is large, whereas the clavicle is often reduced or absent. The dog's **clavicle** (see Figs. 11, 13) is a small oval plate located cranial to the shoulder in the brachiocephalicus muscle. The clavicle is one of the first bones to show a center of ossification in the fetal dog, but in the adult it is partly cartilaginous. It is frequently visible in dorsoventral radiographs of the trunk, medial to the shoulder.

Thoracic Limb (forelimb)		**Pelvic Limb (hind limb)**	
Thoracic girdle	Scapula Clavicle	Pelvic girdle	Ilium Ischium Pubis
Arm or Brachium	Humerus	Thigh	Femur
Forearm or Antebrachium	Radius Ulna	Leg or Crus	Tibia Fibula
Forepaw or Manus	Carpal bones Metacarpal bones Phalanges	Hind paw or Pes	Tarsal bones Metatarsal bones Phalanges

FIGURE 3. Topography of appendicular skeleton.

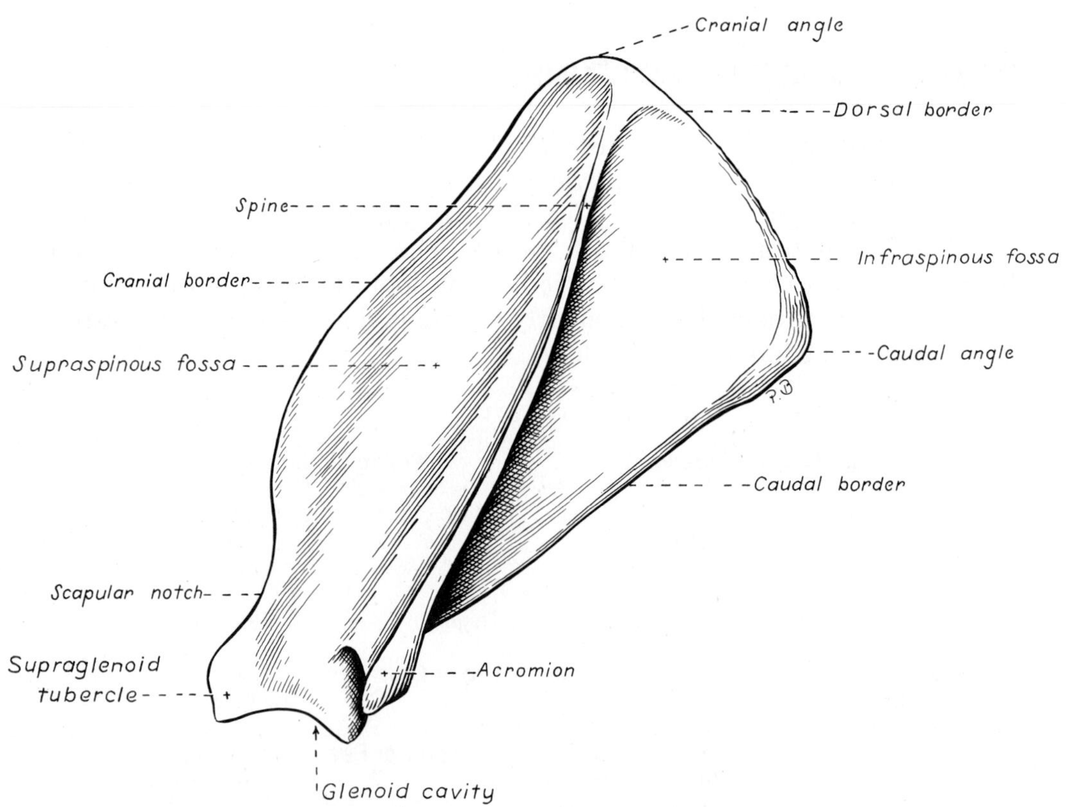

FIGURE 4. Left scapula, lateral surface.

Scapula

The scapula (Figs. 4, 5), a flat, roughly triangular bone, possesses two surfaces, three borders and three angles. The ventral angle is the distal or articular end which forms the **glenoid cavity,** and the constricted part that unites with the expanded blade is referred to as the **neck.**

The lateral surface (Fig. 4) of the scapula is divided into two nearly equal fossae by a shelf of bone, the **spine** of the scapula. The spine is the most prominent feature of the bone. It begins at the dorsal border as a thick, low ridge and becomes thinner and wider toward the neck. In all breeds the free border is slightly thickened, and in some it is everted caudally. The distal end is a truncated process, the **acromion,** where part of the deltoideus muscle arises. On a continuation of the spine proximally the omotransversarius attaches. The remaining part of the spine provides a place for insertion of the trapezius and for origin of that part of the deltoideus that does not arise from the acromion.

The **supraspinous fossa** is the entire surface cranial to the spine of the scapula. The supraspinatus arises from all but the distal part of this fossa.

The **infraspinous fossa,** caudal to the spine, is triangular, with the apex at the neck. The infraspinatus arises from the infraspinous fossa.

The medial or costal surface has two areas. A small proximal and cranial rectangular area, the **serrated face,** serves as insertion for the powerful serratus ventralis muscle. The large remaining part of the costal surface is the **subscapular fossa,** which is nearly flat and usually presents three straight muscular lines that converge distally. The subscapularis arises from the whole subscapular fossa.

The **cranial border** of the scapula is thin. Near the ventral angle the border is concave as it enters into the formation of the neck. The notch thus formed is the **scapular notch.** The dorsal end of the cranial border thickens and, without definite demarcation at the cranial angle, is continuous with the dorsal border.

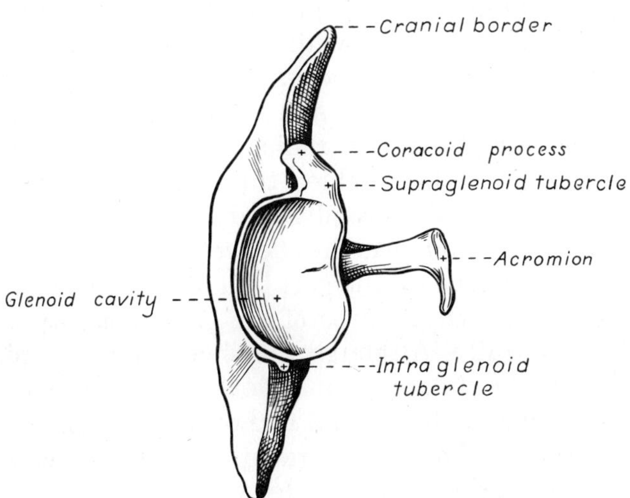

FIGURE 5. Left scapula, ventral angle.

The **dorsal border** extends from the cranial to the caudal angles. In life it is capped by a narrow band of cartilage, but in the dried specimen the cartilage is destroyed by ordinary preparation methods. The rhomboideus attaches to this border.

Just proximal to the ventral angle, the thick **caudal border** bears the **infraglenoid tubercle,** from which arise the teres minor and the long head of the triceps. The middle third of the caudal border of the scapula is broad and smooth; part of the subscapularis and long head of the triceps arise from it. Somewhat less than a third of the dorsal segment of the caudal border is thick and gives rise to the teres major.

The **ventral angle** forms the expanded distal end of the scapula. The adjacent constricted part, the neck, is the segment of the scapula ventral to the spine and dorsal to the expanded part of the bone that forms the glenoid cavity. Clinically, the ventral angle is by far the most important part of the scapula, since it enters into the formation of the shoulder. The glenoid cavity articulates with the head of the humerus. Observe the shallowness of the cavity.

The **supraglenoid tubercle** is an eminence at the cranial part of the glenoid cavity. The tubercle shows a slight medial inclination on which a small tubercle, the **coracoid process,** can be distinguished. The coracobrachialis arises from the coracoid process, while the biceps brachii arises from the supraglenoid tubercle.

Live Dog

Palpate the borders of the scapula, the spine, acromion and supraglenoid tubercle.

Humerus

The humerus (Fig. 6) is located in the arm, or brachium. This bone enters into the formation of both the shoulder and elbow. The shoulder is formed by the articulation of the scapula and humerus, and the elbow by the articulation of radius and ulna with each other and with the humerus. The proximal extremity of the humerus includes the head, neck and the greater and lesser tubercles. The distal extremity, the condyle, includes the trochlea, capitulum, and the radial and olecranon fossae, which communicate proximal to the trochlea through the supratrochlear foramen. The medial and lateral epicondyles are situated on the sides of the condyle. The **body** of the humerus lies between the two extremities.

The **head** of the humerus is the part that articulates with the scapula. It presents more than twice the area of the glenoid cavity of the scapula and is elongated sagittally. Although the shoulder is a typical ball-and-socket joint, it normally undergoes only flexion and extension. The **intertubular groove** begins at the cranial end of the articular area. It lodges the tendon of origin of the biceps brachii and is deflected toward the median plane by the **greater tubercle,** which forms the craniolateral part of the proximal extremity. The greater tubercle is convex at its

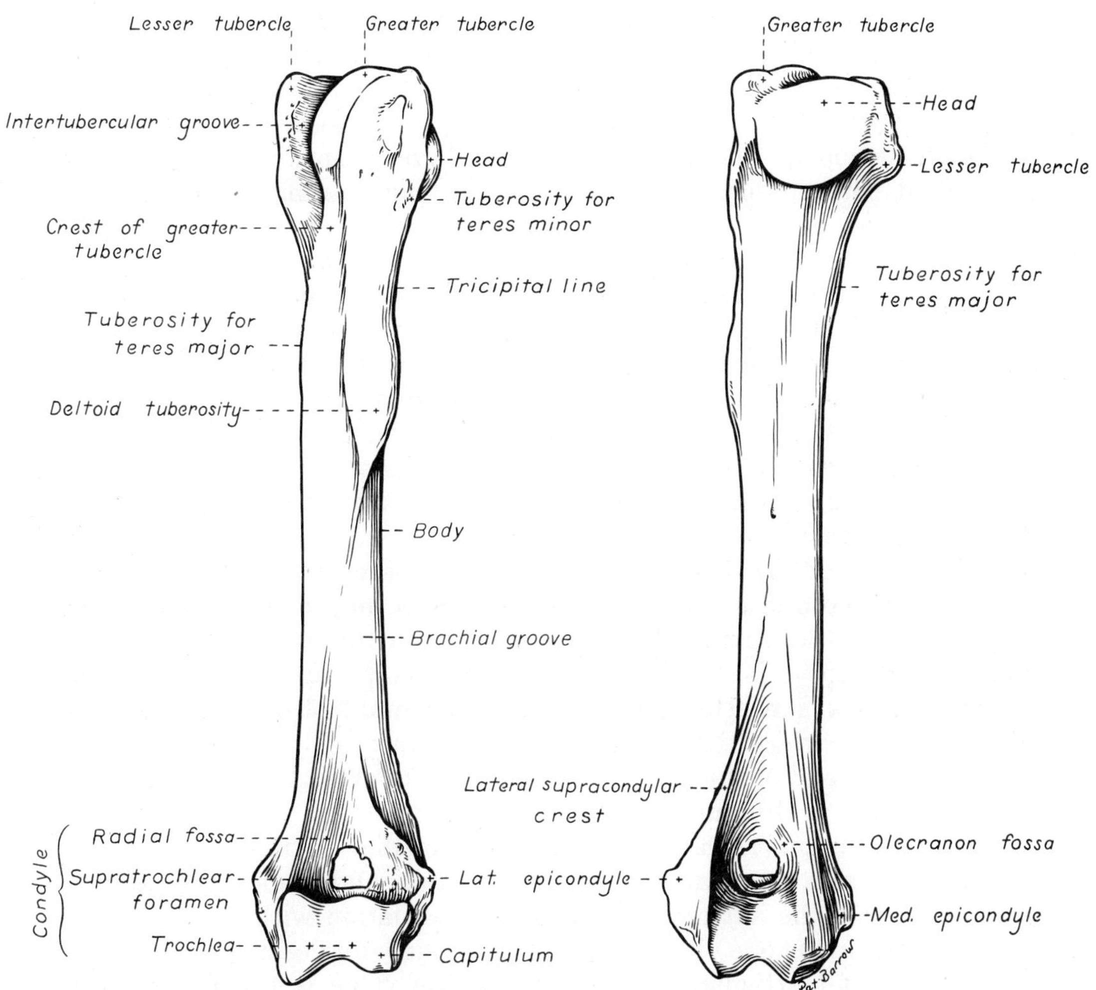

FIGURE 6. Left humerus, cranial and caudal views.

summit and, in most breeds, higher than the head. It receives the insertions of the supraspinatus and the infraspinatus and part of the deep pectoral. Between the head of the humerus and the greater tubercle are several foramina for the transmission of vessels. The infraspinatus is inserted on the smooth facet on the lateral side of the greater tubercle. The **lesser tubercle** lies on the medial side of the proximal extremity of the humerus, caudal to the intertubercular groove. It is not as high or as large as the greater tubercle. The subscapularis attaches to its proximal border. The neck of the humerus is not distinct except caudally. It is the line along which the head and parts of the tubercles have fused with the body.

The **cranial surface** of the humerus is distinct in the middle third of the body, where it furnishes attachment for the brachiocephalicus and part of the pectorals. Distally it fades but may be considered to continue to the medial lip of the trochlea. On the proximal third of the cranial border are two ridges. They continue to the cranial and caudal parts of the greater tubercle. The ridge that extends proximally in a craniomedial direction is the **crest of the greater tubercle** and is also the cranial border of the bone. This forms part of the area of insertion of the pectorals.

The ridge extending to the caudal part of the greater tubercle is on the **lateral surface** of the humerus. Distally it is thickened to form the **deltoid tuberosity.** The deltoideus inserts here. From this tuberosity to the caudal part of the greater tubercle the ridge forms the prominent **tricipital line.** The lateral head of the triceps arises from this line. The teres minor inserts on and adjacent to the proximal extremity of the tricipital line. The smooth **brachialis groove** is on the lateral surface of the body. The brachialis, which originates in the proximal part of the groove, spirals around the bone in the groove so that distally it lies on the craniolateral surface. Distal to this groove is the thick **lateral epicondylar crest.** The extensor carpi radialis and part of the anconeus attach here. The crest extends distally to the lateral epicondyle.

The **caudal surface** is smooth and rounded transversely and ends in the deep olecranon fossa.

The **crest of the lesser tubercle** crosses the proximal end of the **medial surface** and ends distally at the **teres major tuberosity.** The teres major and latissimus dorsi are inserted on this tuberosity. Caudal and proximal to this the medial head of the triceps arises and the coracobrachialis is inserted. Approximately the middle third of the medial surface is free of muscular attachment and is smooth.

The distal end of the humerus, including its articular areas and the adjacent fossae, is the **humeral condyle.** The articular surface is divided unevenly by a low ridge. The large area medial to the ridge is the **trochlea,** which articulates with both radius and ulna and extends proximally into the adjacent fossae. The articulation with the trochlear notch of the ulna is one of the most stable hinge joints (ginglymus) in the body. The small articular area lateral to the ridge is the **capitulum,** which articulates only with the head of the radius.

The **lateral epicondyle** is smaller than the medial one and occupies the enlarged distolateral end of the humerus proximal to the capitulum. It gives origin to the common digital extensor, lateral digital extensor, ulnaris lateralis and supinator. The lateral ligament of the elbow also attaches here. The lateral epicondylar crest extends proximally from this epicondyle.

The **medial epicondyle** is the enlarged distomedial end of the humerus proximal to the trochlea. Its caudal projection deepens the olecranon fossa. The anconeus arises from this projection. The elevated portion of the medial epicondyle serves as origin for flexor carpi radialis, flexor carpi ulnaris, pronator teres and the superficial and deep digital flexor muscles. The medial ligament of the elbow also attaches here.

The **olecranon fossa** is a deep excavation of the caudal part of the humeral condyle. It receives the anconeal process of the ulna during extension of the elbow. On the cranial surface of the humeral condyle is the **radial fossa,** which communicates with the olecranon fossa by an opening, the **supratrochlear foramen.** No soft structures pass through this foramen.

Live Dog

Flex and extend the shoulder. Palpate the greater tubercle of the humerus. Follow its cranial part to the crest of the greater tubercle and cranial border of the humerus. Palpate the groove medial to the greater tubercle and its crest. Palpate the deltoid tuberosity on the lateral surface of the body and the epicondyles on the condyle of the humerus. Note the width of the condyle at the elbow.

Because of the size of the medial epicondyle most elbow joint separations—luxation—caused by injury result in the humerus being displaced medial to the ulna. Note that replacement would require flexion of the elbow to allow the anconeal process of the ulna to pass over the crest of the lateral epicondyle.

Radius

The radius and ulna are the bones of the antebrachium, or forearm. It is important to know that they cross each other obliquely so that the proximal end of the ulna is medial and the distal end lateral to the radius. The radius (Fig. 7), the shorter of the two bones of the forearm, articulates proximally with the humerus and distally with the carpus. It also articu-

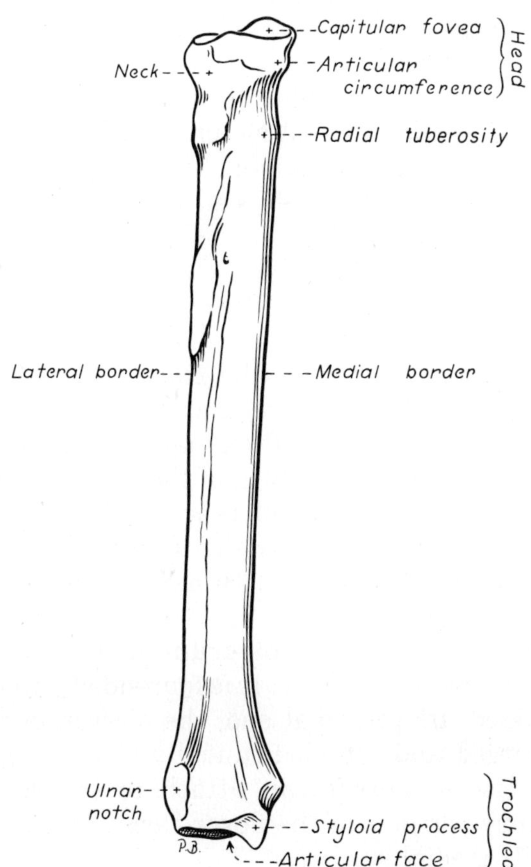

FIGURE 7. Left radius, caudal view.

lates with the ulna, proximally by its caudal surface and distally near its lateral border.

The proximal extremity consists of head, neck and tuberosity. The **head** of the radius, like the whole bone, is widest medial to lateral. It forms proximally an oval, depressed articular surface, the **fovea capitis,** which articulates with the capitulum and part of the trochlea of the humerus. The small **radial tuberosity** lies distal to the neck on the medial border of the bone. The biceps brachii inserts in part on this tubercle.

The **body** of the radius is compressed so that it possesses cranial and caudal surfaces and medial and lateral borders. It is slightly convex cranially. At the carpal end the body blends without sharp demarcation with the enlarged distal extremity. The caudal surface of the radius is roughened and slightly concave. It has a ligamentous attachment to the ulna. Distally it broadens and becomes the expanded caudal surface of the distal extremity. The cranial surface of the radius, convex transversely, is relatively smooth throughout.

The distal extremity of the radius is the **trochlea.** Its carpal articular surface is concave. On the lateral surface of the distal extremity is the **ulnar notch,** a slightly concave area with a facet for articulation with the ulna. The medial surface of the distal extremity ends in a rounded projection, the **styloid process.** The medial ligament of the carpus attaches proximal to the styloid process. The cranial surface of the distal extremity presents three distinct grooves. The most medial groove, small, short and oblique, contains the tendon of the abductor pollicis longus. The middle and longest groove, extending proximally on the shaft of the radius, is for the extensor carpi radialis. The most lateral of the grooves on this surface is wide and of variable distinctness. It contains the tendon of the common digital extensor.

Ulna

The ulna (Fig. 8) is located in the caudal part of the forearm. It exceeds the radius in length and is irregular in shape and generally tapers from its proximal to its distal end. Proximally the ulna is medial to the radius and articulates with the trochlea of the humerus by the **trochlear notch** and with the articular circumference of the radius by the **radial notch.** This forms the elbow. Distally the ulna is lateral and articulates with the radius medially and with the ulnar and accessory carpal bones distally.

The proximal extremity is the **olecranon.** It serves as a lever arm for the extensor muscles of the elbow. It is four-sided, laterally compressed and medially inclined. Its proximal end, the olecranon tuber, is grooved cranially and enlarged and rounded caudally. The triceps brachii, anconeus and tensor fasciae antebrachii attach to the caudal part of the olecranon. The ulnar portions of the flexor carpi ulnaris and deep digital flexor arise from its medial surface.

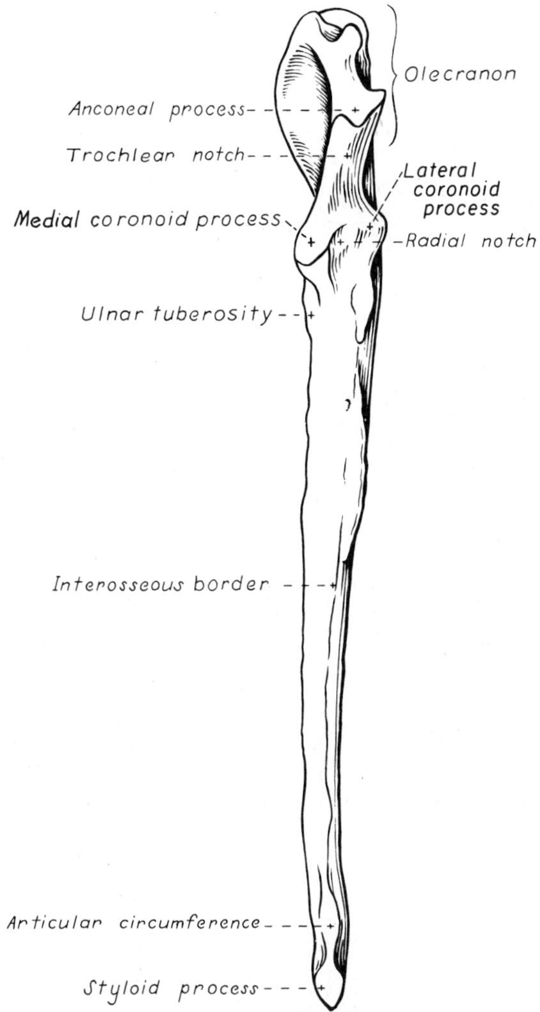

FIGURE 8. Left ulna, cranial view.

The **trochlear notch** is a smooth, vertical, half-moon–shaped concavity facing cranially. The whole trochlear notch articulates with the trochlea of the humerus. At its proximal end a sharp-edged, slightly hooked **anconeal process** fits into the olecranon fossa of the humerus when the elbow is extended. At the distal end of the notch are the **medial** and **lateral coronoid processes,** which articulate with the humerus and radius. The medial coronoid process is larger.

The **body** of the ulna is three-sided in its middle third; proximal to this the bone is compressed laterally, while the distal third gradually loses its borders, becomes irregular and is continued by the pointed distal extremity. The **ulnar tuberosity** is a small, elongated eminence on the medial surface of the bone at its proximal end, just distal to the medial coronoid process. The biceps brachii and the brachialis insert on this eminence. The **interosseous border** is distinct, rough and irregular, especially at the junction of the proximal and middle thirds of the bone,

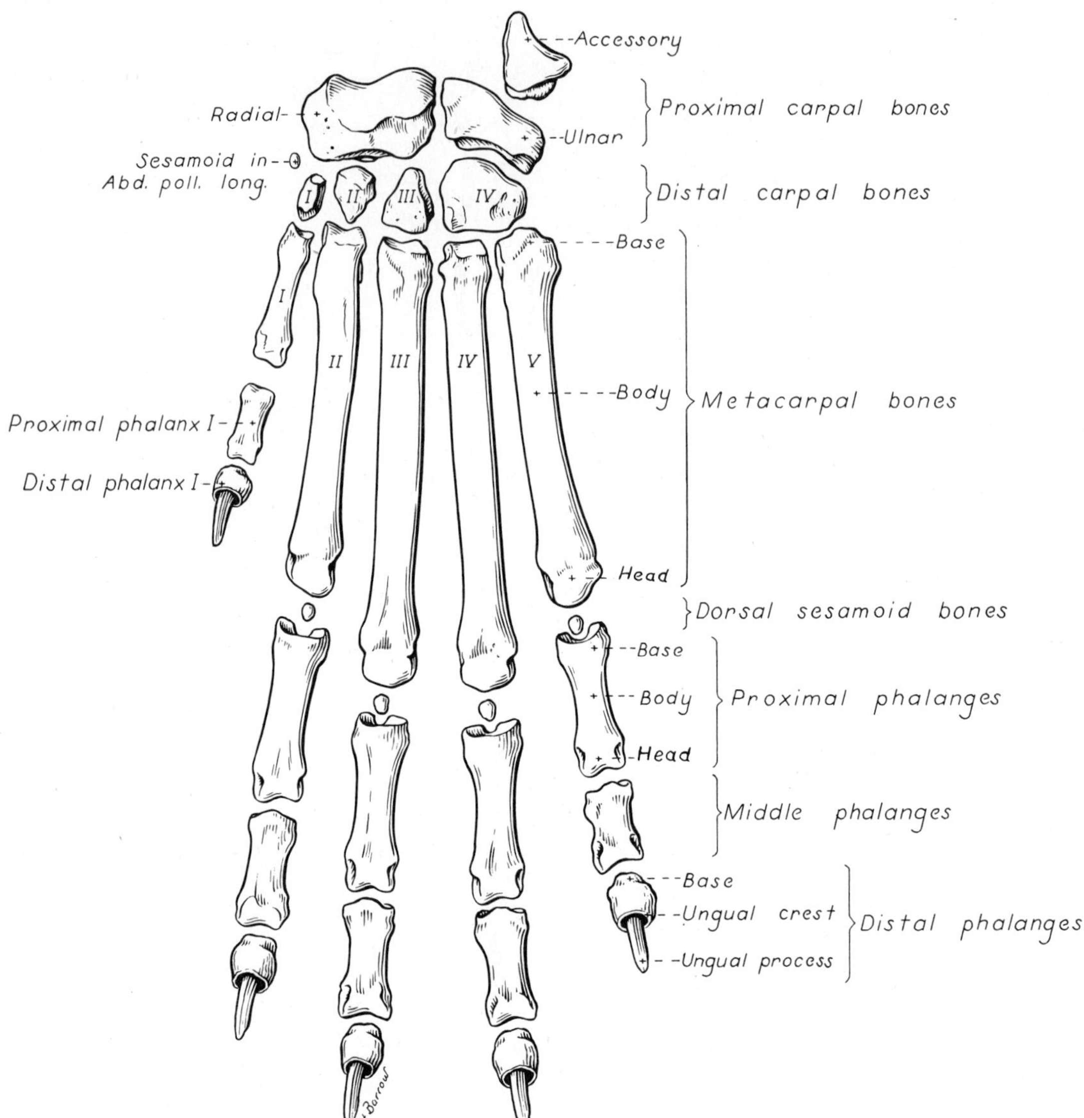

FIGURE 9. Bones of left forepaw, dorsal view.

where a large, expansive but low eminence is found. This eminence indicates the place of articulation with the radius by means of a heavy ligament. Frequently, a vascular groove medial to the crest marks the position of the caudal interosseous artery in life. This groove is most conspicuous in the middle third of the ulna. The body shows a distinct caudal concavity.

The distal extremity of the ulna is the head with its prominent **styloid process.** A part of this process articulates with the ulnar and accessory carpal bones. The head articulates medially with the radius.

Carpal Bones

The term **carpus** (Figs. 9, 10) is used to designate that part of the extremity between the forearm and metacarpus that includes all the soft

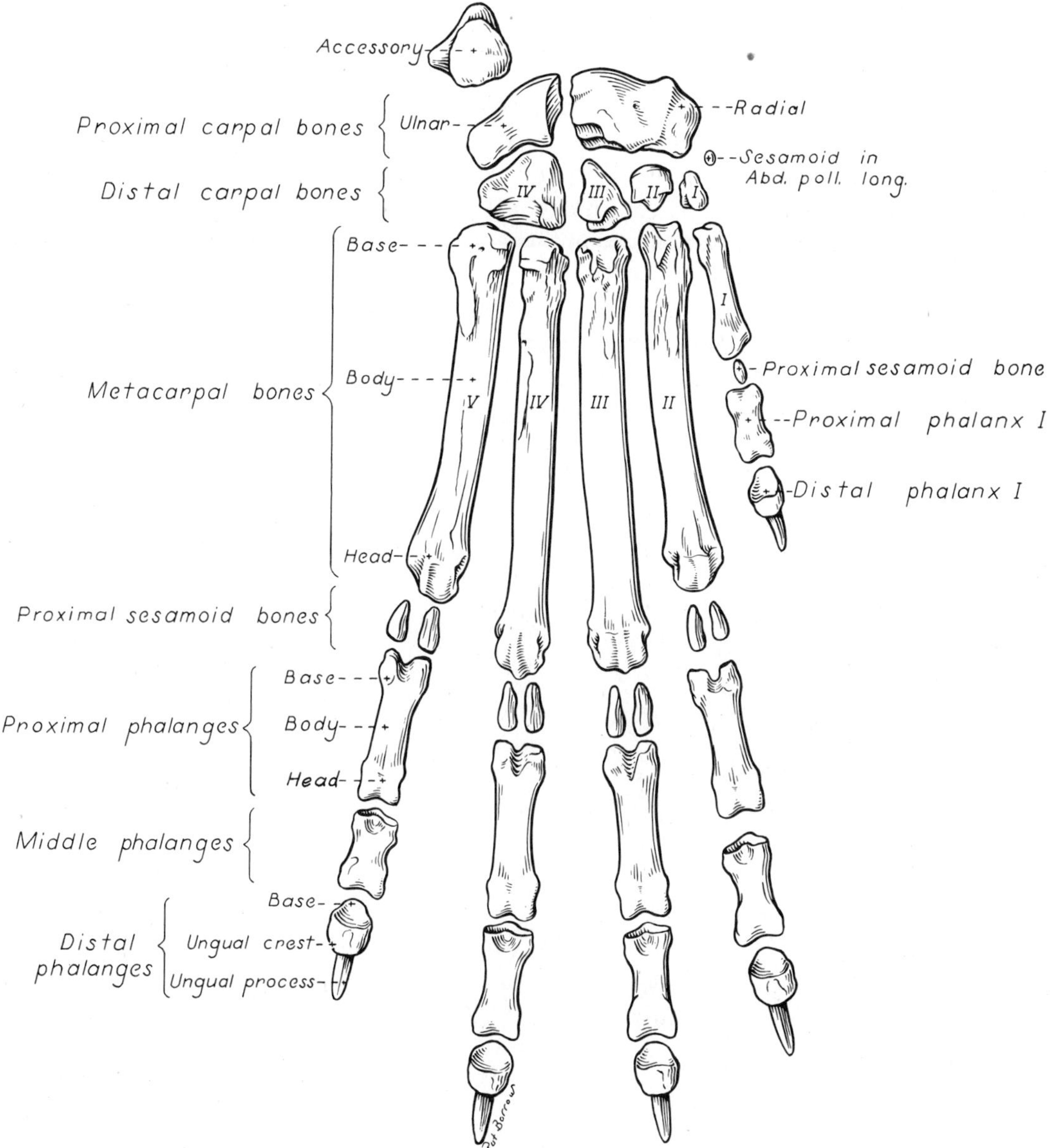

FIGURE 10. Bones of left forepaw, palmar view.

structures as well as the bones. The carpus includes seven small, irregular bones arranged into two rows. These are most conveniently studied on radiographs. The proximal row contains three bones. The largest of these, the **radial carpal,** is on the medial side and articulates proximally with the radius. The **ulnar carpal** is the lateral member of the proximal row. Its palmar portion projects distally palmar and lateral to the fourth carpal bone. The **accessory carpal,** the palmar member, is a short rod of bone that articulates with the styloid process of the ulna and the ulnar carpal bone and serves as a lever arm for some of the flexor muscles of the carpus. The distal row consists of four bones numbered from the medial to the lateral side. From the smallest on the medial side, these are the first, second, third and fourth carpal bones. The fourth carpal bone is the largest and articulates with the base of the fourth and fifth metacarpals.

Metacarpal Bones

The **metacarpus** (Figs. 9, 10) contains five bones. The metacarpal bones are long bones in miniature, possessing a slender **body,** or shaft, and large extremities. The proximal extremity is the **base,** the distal one the **head.** The metacarpals, like the carpals and digits, are numbered from medial to lateral. Proximally all articulate principally with the corresponding carpal bones, except the fifth, which articulates with the fourth carpal. Distally all articulate with the corresponding proximal phalanges. The interosseous muscles largely fill the intermetacarpal spaces palmar to the metacarpal bones.

The first metacarpal bone is atypical. It is a vestigial structure, but unlike the first metatarsal bone in the hind paw, it is constantly present.

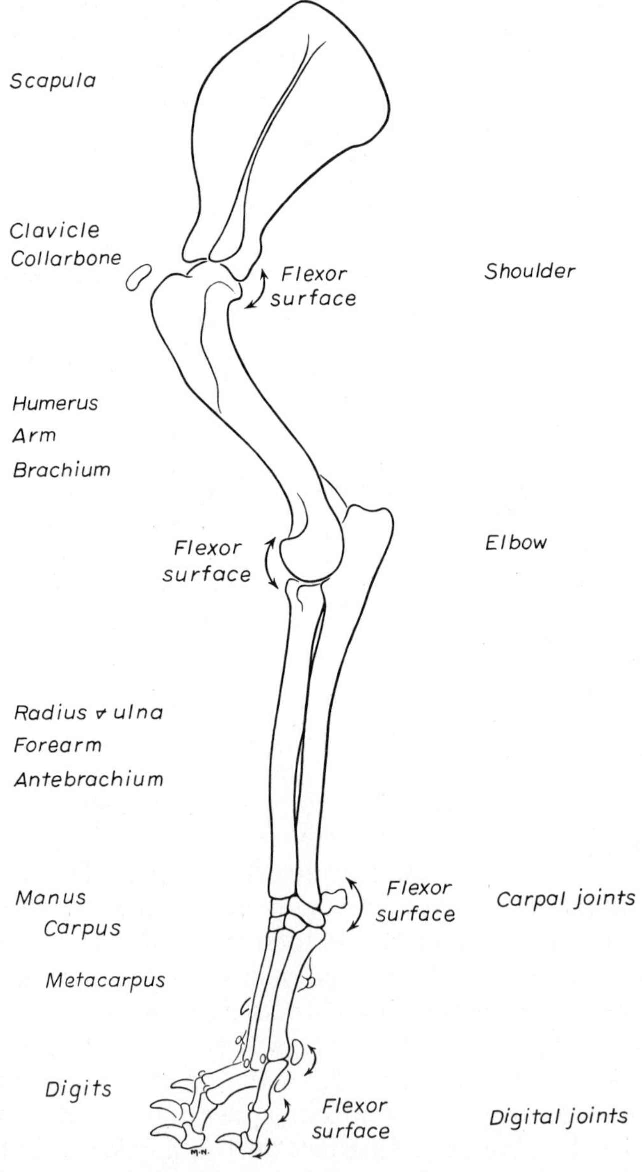

FIGURE 11. Left thoracic limb skeleton and flexor surfaces of joints.

Phalanges

In the forepaw there are three phalanges (Figs. 9, 10) for each of the four main digits; the first digit, or pollex, a dewclaw, has two phalanges. Each phalanx has a proximal **base** and distal **head.**

On the distal or third phalanx a tubercle projects proximally from the palmar border of the base, upon which the deep digital flexor is inserted. A thin shelf of bone, the **ungual crest,** overlaps the claw and forms a band of bone around the root of the claw. The rounded dorsal part of the base is the **extensor process,** on which the common digital muscle is inserted. The **ungual process** is a curved conical extension of the distal phalanx into the claw.

Two **proximal sesamoid bones** are located in the interosseous tendons on the palmar surface of each metacarpophalangeal joint (digits two to five). Four small **dorsal sesamoid bones** (none for the first digit) are imbedded in the extensor tendons as they pass over the metacarpophalangeal joints (Figs. 9, 10).

Live Dog

Flex and extend the elbow. Palpate the olecranon tubercle proximally, the head of the radius laterally and the medial coronoid process of the ulna medially deep to the muscles. Note that the combined bones at the elbow are not as wide as the humeral condyle. Flex and extend the antebrachiocarpal joint and palpate the styloid process of the radius medially and ulna laterally.

Flex and extend the carpus. Note that the major motion is at the radiocarpal joint. Palpate the accessory carpal. Palpate the metacarpals and phalanges and flex and extend the metacarpophalangeal and interphalangeal joints. Note the relationship of the metacarpal pad to the metacarpophalangeal joints and the digital pads to the distal interphalangeal joints.

MUSCLES OF THE THORACIC LIMB

Superficial Structures of the Thorax

General study of the ventral surface of the trunk should be made before the dissection of the thoracic region is begun. Find the **umbilicus.** This is represented by a scar that may be either flat or slightly raised and is located on the midventral line, one-third to one-fourth of the distance from the xiphoid cartilage to the scrotum or vulva. In most dogs it is hidden by hair. The umbilicus is irregularly oval and may be from a few millimeters to a centimeter in length. The umbilicus serves as a landmark in abdominal surgery. Notice that the hair over a large area around the umbilicus slants toward it, thus forming a **vortex.**

Pick up a fold of skin, or **common integument,** which consists of an outer thin epithelium, the **epidermis,** and an underlying thicker layer of connective tissue, the **dermis.** Skin thickness varies on different parts of the body depending upon the extent of the dermis. Notice that the skin is thickest in the neck region, thinner over the sternum and thinnest on the ventral surface of the abdomen; also notice that the skin of the dorsum of the neck and thorax is loosely attached.

The mammae vary in number from 8 to 12, but 10 is average. They are situated in two rows, usually opposite each other. The number is most often reduced in the smaller breeds.

When 10 glands are present, the cranial four are the **thoracic mammae,** the following four the **abdominal mammae,** and the caudal two the **inguinal mammae.** When the abdominal and inguinal mammae are maximally developed, the glandular tissue in each row appears to form a continuous mass. It should be observed that the mammae lie in the areolar connective tissue. The cranial pair of thoracic mammae are smaller than the other pair. Each mamma has a papilla that is partly hairless and contains about 12 openings, but these vary and are difficult to see if the animal is not lactating.

The costal cartilages of the tenth, eleventh and twelfth ribs unite with each other to form the **costal arch.** Palpate this arch and the caudal border and free end of the last or thirteenth rib. This rib does not attach to the costal arch and is therefore a floating rib.

Make a midventral incision *through the skin only* from the cranial end of the neck to the umbilicus (Fig. 12). From the umbilicus extend a transverse incision to the middorsal line on the **left side.** From a point on the midventral incision directly opposite the arm extend a transverse incision to the left elbow. Make a complete circular incision through the skin around the elbow. Extend a third transverse incision from the cranial end of the midventral incision to the middorsal line on the left side. This should pass just caudal to the ear. Carefully reflect the skin of the thorax and neck to the middorsal line. The skin will be intimately fused with the thin underlying **cutaneous** muscle over the neck, thorax and abdomen. The latter should be left on the specimen as far as is possible.

The subcutaneous tissue that now confronts the dissector is composed of areolar tissue and fascia. **Areolar tissue** is loose, irregularly arranged

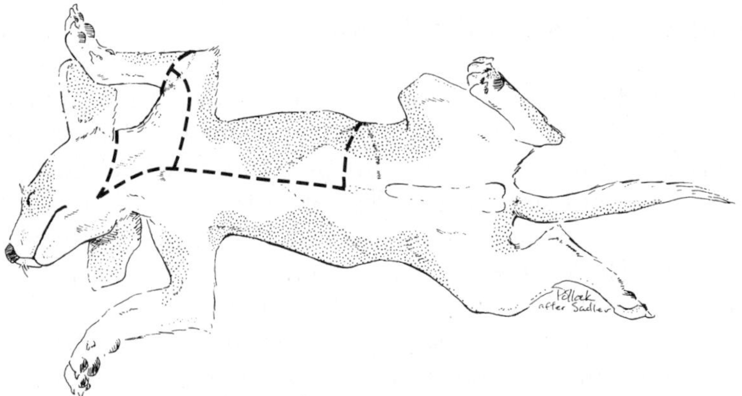

FIGURE 12. Dissection position and first skin incisions.

connective tissue that often contains fat. **Fascia** is a denser, more regularly arranged thin layer of connective tissue. It is more fibrous, and it envelops the body beneath the skin and encloses individual muscles or groups of muscles. The **superficial fascia** is deep to the areolar tissue, forming the deep portion of the subcutaneous tissue that covers the entire body. It blends with the **deep fascia,** which is more firmly attached to the muscle that it encloses. These are not always easily distinguished from each other. In your specimen the areolar tissue is often distended with embalming fluid. Subcutaneous injections are made into this tissue. When the fatty areolar tissue and fascia with vessels and nerves are removed from a muscle, the muscle is said to be cleaned.

All muscles have attachments. In most instances the more proximal attachment, the part that moves the least, is the **origin.** The **insertion** is the more distal attachment, or the part that moves the most. The origin is usually a direct attachment of the muscle cells to the bone. The insertion often is by a tendon or aponeurosis extending from the muscle cells to the bone. A **tendon** consists of dense, regularly arranged fibrous connective tissue organized into a small well-defined bundle. An **aponeurosis** has the same consistency as a tendon but the fibrous tissue is arranged as a thin sheet of tissue. A **ligament** is fibrous connective tissue between bones, although the term is used for a variety of fibrous connections between organs or between an organ and the body wall.

Read before you dissect! In many instances during the study of muscles a description of a specific muscle will be given before the instructions for dissection. At no time should muscles be removed or even transected without instructions. In each instance clean the exposed surface of the muscle being described, isolate its borders and verify its origin and insertion. If the muscle is to be transected, free it from underlying structures first.

The **cutaneus trunci** (see Fig. 16) is a thin sheet of muscle that covers most of the dorsal, lateral and ventral walls of the thorax and abdomen. It is more closely applied to the skin than to underlying structures and is often reflected with the skin before being observed. Like all cutaneous muscles, it is developed in the superficial fascia of the thorax and abdomen. Behind the shoulder the fibers sweep obliquely toward the axilla; farther caudally they are principally longitudinal and arise from the superficial fascia over the rump.

The attachments of the cutaneus trunci are the superficial fascia of the trunk and the skin. The muscle sends a fasciculus to the medial side of the forelimb; caudal and ventral to this the fibers fray out over the deep pectoral muscle. The cutaneus trunci twitches the skin. It is innervated by the lateral thoracic nerve. In the male dog there is a distinct development of this muscle adjacent to the ventral midline caudal to the xiphoid. This is the **preputial muscle,** which passes caudally and radiates into the prepuce, forming an arch with the muscle of the opposite side. It functions to support the cranial end of the prepuce during the nonerect state and to pull the prepuce back over the glans penis after erection.

Sever the axillary and ventral attachments of the cutaneus trunci and reflect it dorsally. Caution: Beneath the thin cutaneus trunci is the thicker latissimus dorsi muscle, which should be left in place on the lateral side of the trunk.

Live Dog

Grasp the skin in several areas and note the variation in thickness. Note where it is especially loose and suitable for subcutaneous injections of fluids. Pinch the skin over the side of the thorax and observe the skin wrinkling that occurs because of reflex activity of the cutaneus trunci muscle.

Extrinsic Muscles of the Thoracic Limb and Related Structures

The extrinsic muscles of the thoracic limb are those that attach the limb to the axial skeleton; the intrinsic muscles extend between the bones that compose the limb itself.

In the ventral thoracic region are the superficial and deep pectoral muscles, which extend between the sternum and the arm. Thoroughly clean these muscles. In thin specimens this will require little dissecting. In pregnant or lactating bitches it will require reflecting the two thoracic mammae caudally, while in fat specimens the forelimb will probably have to be manipulated so that the borders of the muscles are clearly discernible before cleaning. Always clean the extremities of a muscle as well as the middle part. Actually see and feel the attachments. Visualize the muscle's position and action on the skeleton and attempt to palpate it on a live dog.

Extrinsic Muscles of the Thoracic Limb:
- Superficial pectoral
- Deep pectoral
- Brachiocephalicus
- Omotransversarius
- Trapezius
- Rhomboideus
- Latissimus dorsi
- Serratus ventralis

1. The two **superficial pectoral muscles** (Figs. 13–15) lie under the skin between the cranial part of the sternum and the humerus. Their caudal border is thin; their cranial border is thick and rounded and forms the caudal border of a triangle at the base of the neck. The smaller **descending pectoral** is superficial to the transverse pectoral, which it obliquely crosses from its origin on the first sternebra to its insertion on the crest of the greater tubercle of the humerus. The **transverse pectoral**

arises from the first two or three sternebrae and inserts over a longer distance on the crest of the greater tubercle of the humerus. It is related on its deep surface to the deep pectoral muscle (ascending pectoral). At their insertions these muscles lie between the brachiocephalicus cranially and the biceps brachii and humerus caudally. Clean both of these superficial pectoral muscles. Transect them 1 centimeter from the sternum, and reflect them toward the humerus.

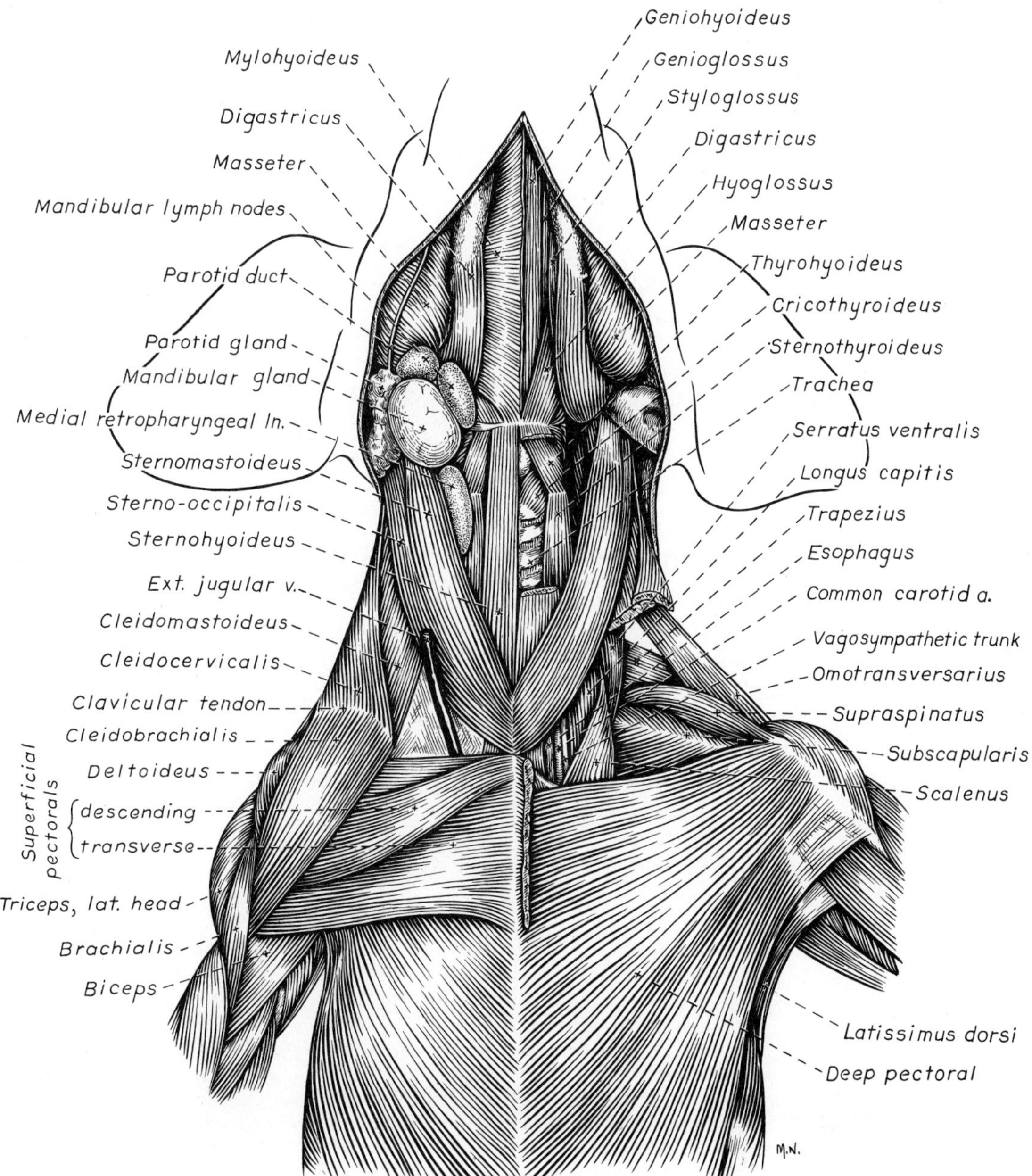

FIGURE 13. Superficial muscles of neck and thorax, ventral view.

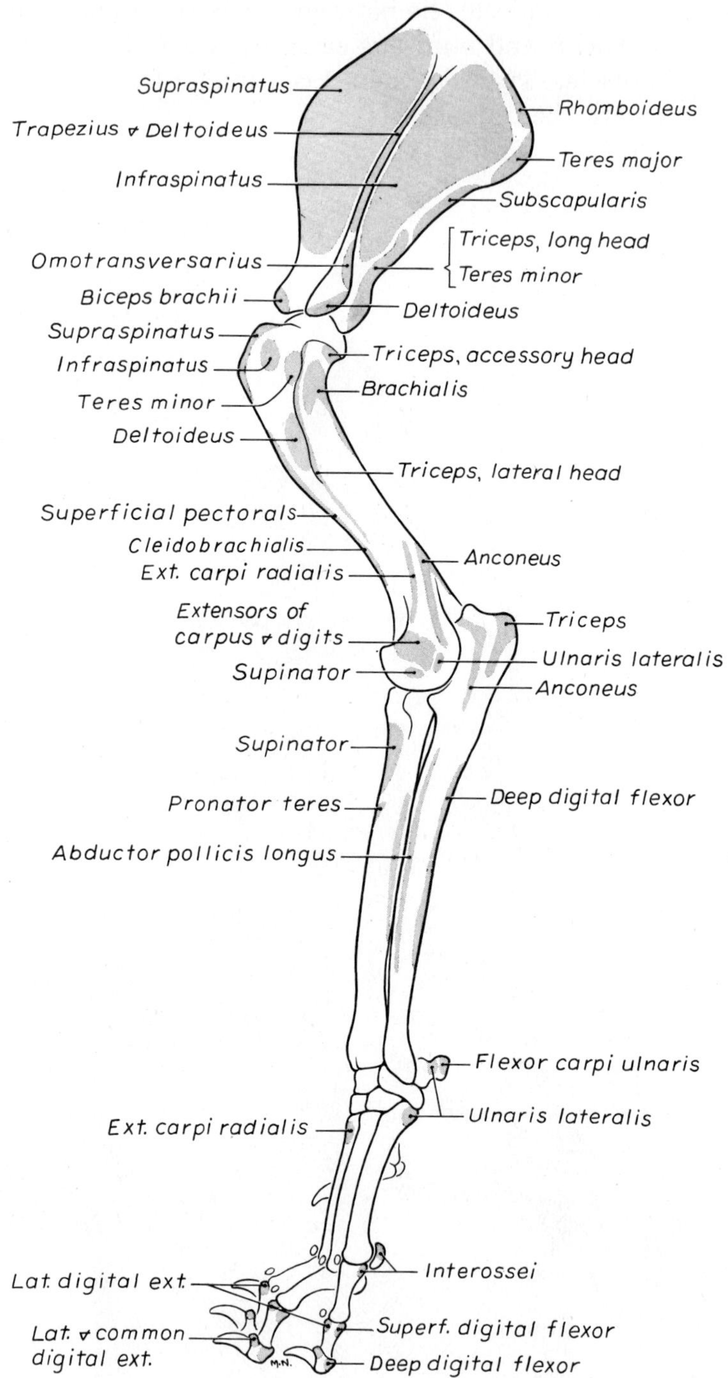

FIGURE 14. Left thoracic limb skeleton, lateral view of muscle attachments.

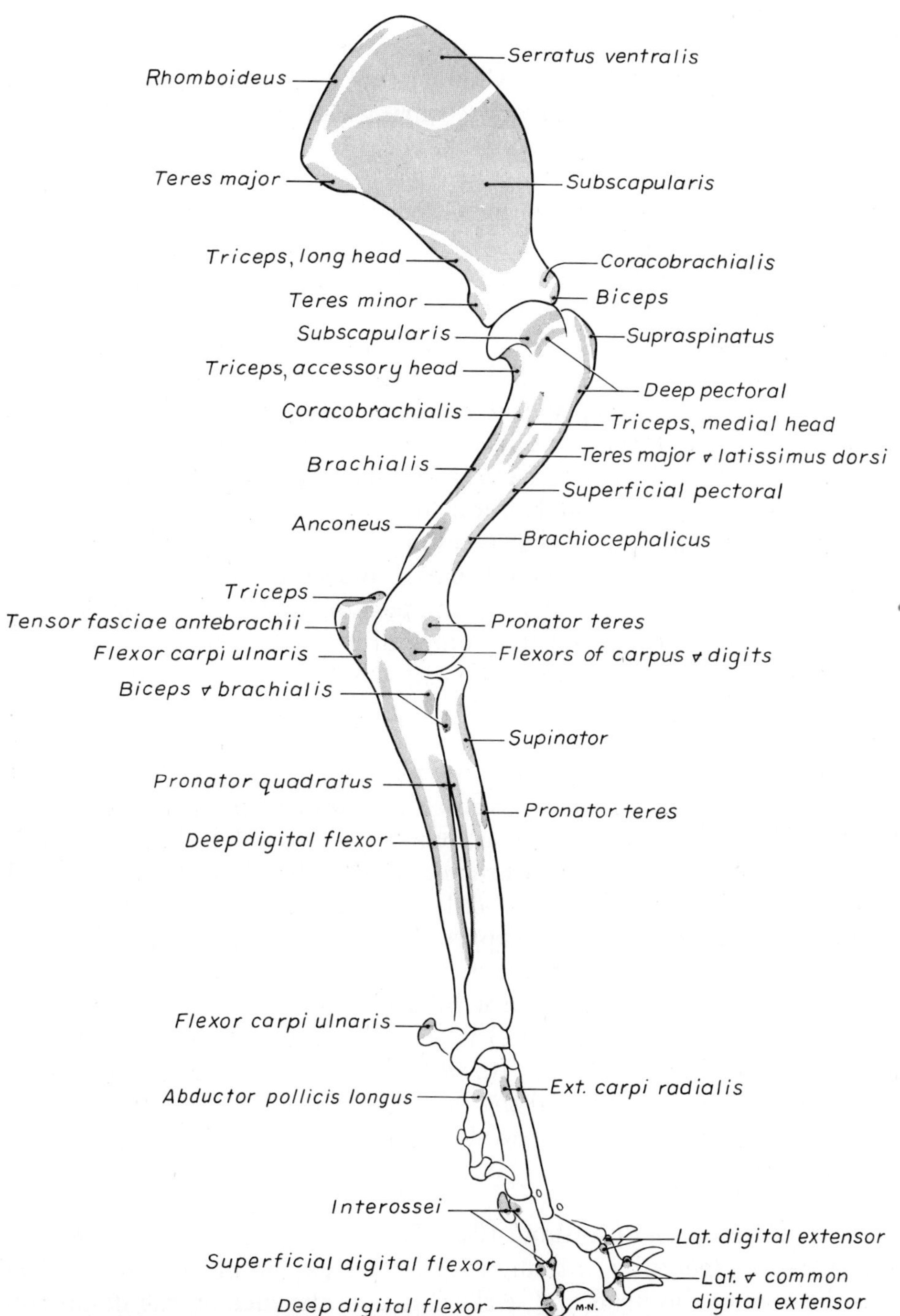

FIGURE 15. Left thoracic limb skeleton, medial view of muscle attachments.

As muscle attachments are being cleaned, examine the skeleton parts involved.

ORIGIN: The first two sternebrae and usually a part of the third; the fibrous raphe between fellow muscles.

INSERTION: The whole crest of the greater tubercle of the humerus.

ACTION: To adduct the limb when it is not bearing weight or to prevent the limb from being abducted when bearing weight.

INNERVATION: Cranial pectoral nerves (C7, C8).

2. The **deep pectoral muscle** extends from the sternum to the humerus and is larger and longer than the superficial pectoral muscles. It lies largely under the skin and the thoracic mammae. The papilla of the caudal thoracic mamma usually lies at the caudal border of the muscle. Only the cranial part is covered by the superficial pectoral muscles. An abdominal slip of this muscle is often present on the caudolateral border. Transect the deep pectoral muscle 2 centimeters from, and parallel to, the sternum, and clean the distal part to its insertion.

ORIGIN: The ventral part of the sternum and the fibrous raphe between fellow muscles; the deep abdominal fascia in the region of the xiphoid cartilage (the caudal end of the sternum).

INSERTION: The major portion partly muscular, partly tendinous on the lesser tubercle of the humerus; an aponeurosis to the greater tubercle and its crest; the caudal part to the medial brachial fascia.

ACTION: To pull the trunk cranially when the limb is advanced and fixed; to extend the shoulder; to draw the limb caudally when it is not supporting weight.

INNERVATION: Caudal pectoral nerves (C8, T1).

The **superficial fascia of the neck** is continued on the head as the superficial fascia of the various regions of the head. Caudally it becomes continuous with the superficial brachial and pectoral fasciae. Some of the fascia is also continued into the axillary space. Notice that the external jugular vein is completely wrapped by it. Save this vein for future orientation. The cutaneous muscles of the neck are completely enveloped by the fascia. Only the platysma will be dissected.

The **platysma** (see Fig. 192) is the best developed of the cutaneous muscles of the neck and head. Its fibers sweep cranioventrally over the dorsal part of the neck toward the lateral surface of the face. The muscle may have been reflected with the skin. Its insertion will be seen when the head is dissected.

3. The **brachiocephalicus** (Figs. 13–16) of the dog is a compound muscle developmentally, although it appears as one muscle that extends from the arm to the head and neck. One end attaches on the distal third of the humerus, where it lies between the biceps brachii medially and the brachialis laterally. Proximally on the humerus it partly covers the pectoral muscles at their insertions and lies craniomedial to the deltoid muscle. It crosses the cranial surface of the shoulder, divides into two parts and obliquely traverses the neck. At the shoulder a faint line crosses

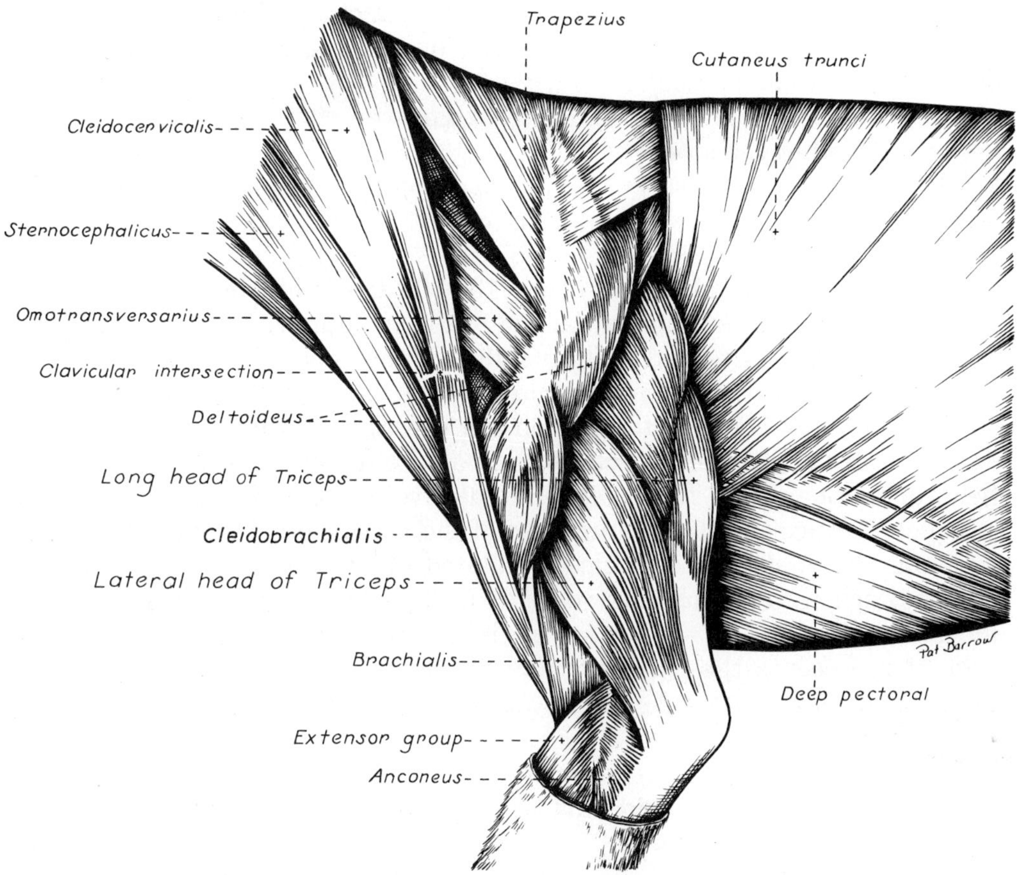

FIGURE 16. Superficial muscles of left shoulder and arm.

the muscle. This is the edge of a fibrous plate, the **clavicular intersection,** on the deep surface of which the vestigial **clavicle** (collar bone) is connected. Although the clavicle has lost its functional significance in the dog, it is still considered the origin of the components of the brachiocephalicus muscle. Thus the muscle distal to the clavicular tendon that attaches to the arm is the **cleidobrachialis,** the muscle that extends from the clavicular tendon to the dorsum of the neck is the **cleidocervicalis** and beneath this is the **cleidomastoideus,** which attaches to the skull. The cleidocervicalis is bounded caudally by the trapezius and cranially by the sternocephalicus.

Transect the cleidocervicalis to expose the full extent of the cleidomastoideus. Note that the cleidomastoideus runs toward the head deep to the sternocephalicus. Transect the cleidomastoid muscle and search for the clavicle by inserting your finger on the medial side of the clavicular intersection.

ATTACHMENTS: All attachments are movable, but the clavicle or clavicular intersection is considered as the origin for purposes of naming the muscles. The cleidobrachialis attaches to the distal end of the cranial border of the humerus. There is also a significant fascial tie into the axilla. The cleidocervicalis attaches to the cranial half of the middorsal fibrous raphe and sometimes to the nuchal crest of the occipital bone. The cleidomastoideus attaches to the mastoid part of the temporal bone with the sternomastoideus muscle.

ACTION: To advance the limb; to extend the shoulder and draw the neck and head to the side.

INNERVATION: Accessory nerve, and ventral branches of cervical nerves.

4. The **sternocephalicus** (Figs. 13, 16) arises on the sternum and inserts on the head. At the cranial end of the sternum the muscle is thick, rounded and closely united with its fellow of the opposite side. Even after the main parts of the paired muscle diverge, there may be considerable crossing of fibers between the two on the ventral surface of the neck. The dorsal border of the sternocephalicus is adjacent to the ventral border of the brachiocephalicus. The external jugular vein crosses its lateral surface obliquely. Notice that the cranial part of the muscle divides into two parts and that the ventral part is closely related to the cleidomastoid division of the brachiocephalicus. The ventral part of the sternocephalicus, the **sternomastoideus,** is similar to the cleidomastoideus in shape and insertion. It represents the chief continuation of the sternocephalicus to the head. The thin but wide dorsal part of the sternocephalicus is the **sternooccipitalis.**

ORIGIN: The first sternebra or manubrium.

INSERTION: The mastoid part of the temporal bone and the nuchal crest of the occipital bone.

ACTION: To draw the head and neck to the side.

INNERVATION: Accessory nerve and ventral branches of cervical nerves.

Transect the left sternocephalicus close to the manubrium and reflect it. The sternohyoid and sternothyroid muscles are both covered by the deep fascia of the neck and lie dorsal to the sternocephalicus at their origin.

The **sternohyoideus** (Fig. 13) lies on the trachea. A midventral groove indicates the separation of right and left muscles.

ORIGIN: The first sternebra and the first costal cartilage.

INSERTION: The basihyoid bone.

ACTION: To pull the tongue and larynx caudally.

INNERVATION: Ventral branches of cervical nerves.

The **sternothyroideus** (Fig. 13) is covered at its origin by the sternohyoideus. The sternothyroideus inserts on the lateral surface of the thyroid cartilage. The left muscle is bounded dorsally by the esophagus and medially by the trachea. Notice that a tendinous intersection runs across the muscle 3 or 4 centimeters cranial to its origin. It is at this level that the sternohyoideus separates from the sternothyroideus.

ORIGIN: The first costal cartilage.

INSERTION: The caudolateral surface of the thyroid cartilage.

ACTION: Same as the sternohyoideus—to draw the larynx and tongue caudally.

INNERVATION: Ventral branches of cervical nerves.

5. The **omotransversarius** (Figs. 13, 14, 16) is in a deeper plane than the brachiocephalicus. It is straplike and extends from the distal end of the spine of the scapula to the atlas. It is related to the deep cervical fascia medially. Its caudal part is subcutaneous, but cranially it is covered by the cervical part of the brachiocephalicus. Transect the omotransversarius through its middle and reflect each half toward its attachment. This will expose the **superficial cervical lymph nodes** located cranial to the scapula.

ATTACHMENTS: The distal end of the spine of the scapula; cranially, the wing of the atlas.

ACTION: To advance the limb or flex the neck laterally.

INNERVATION: Accessory nerve.

The **deep fascia of the neck** is a strong wrapping that extends under the sternocephalicus, cleidomastoideus, omotransversarius and cleidocervicalis muscles. It covers the sternohyoideus and sternothyroideus ventrally and surrounds the trachea, thyroid gland, larynx and esophagus. The deep fascia that covers the common carotid artery, vagosympathetic nerve trunk, internal jugular vein and tracheal lymphatic trunk is the **carotid sheath.** Locate these structures in the carotid sheath between the omotransversarius dorsally and the sternothyroideus ventrally. The deep fascia of the neck continues dorsally and laterally to invest the deep cervical muscles.

The **supraspinous ligament** connects the dorsal aspects of all vertebral spines except the cervical vertebrae. The **nuchal ligament,** composed predominately of yellow elastic fibrous tissue, extends from the spine of the first thoracic vertebra to the spine of the axis. The **median raphe** of the neck is a longitudinal fibrous septum between right and left epaxial muscles dorsal to the nuchal ligament. It serves as the attachment for many of the cervical muscles. Observe these during the dissection of the following muscles.

6. The **trapezius** (Figs. 14, 16) is thin and triangular. It is divided into cervical and thoracic parts, separated by an aponeurosis. The muscle as a whole extends from the median raphe of the neck and the supraspinous ligament to the spine of the scapula. The cervical part is overlapped by the cleidocervicalis, while the thoracic part overlaps the latissimus dorsi. Make an arching cut through the muscle, beginning at the middle of the cranial border. Extend it over the dorsal border of the scapula and continue through the aponeurotic area to the middle of the caudal border. Reflect the muscle to its attachments.

ORIGIN: The median raphe of the neck and the supraspinous ligament from the level of the third cervical vertebra to the level of the ninth thoracic vertebra.

INSERTION: The spine of the scapula.

ACTION: To elevate and abduct the forelimb.

INNERVATION: Accessory nerve.

7. The **rhomboideus** (Figs. 14, 15, 17) lies beneath the trapezius and holds the dorsal border of the scapula close to the body. It has capital,

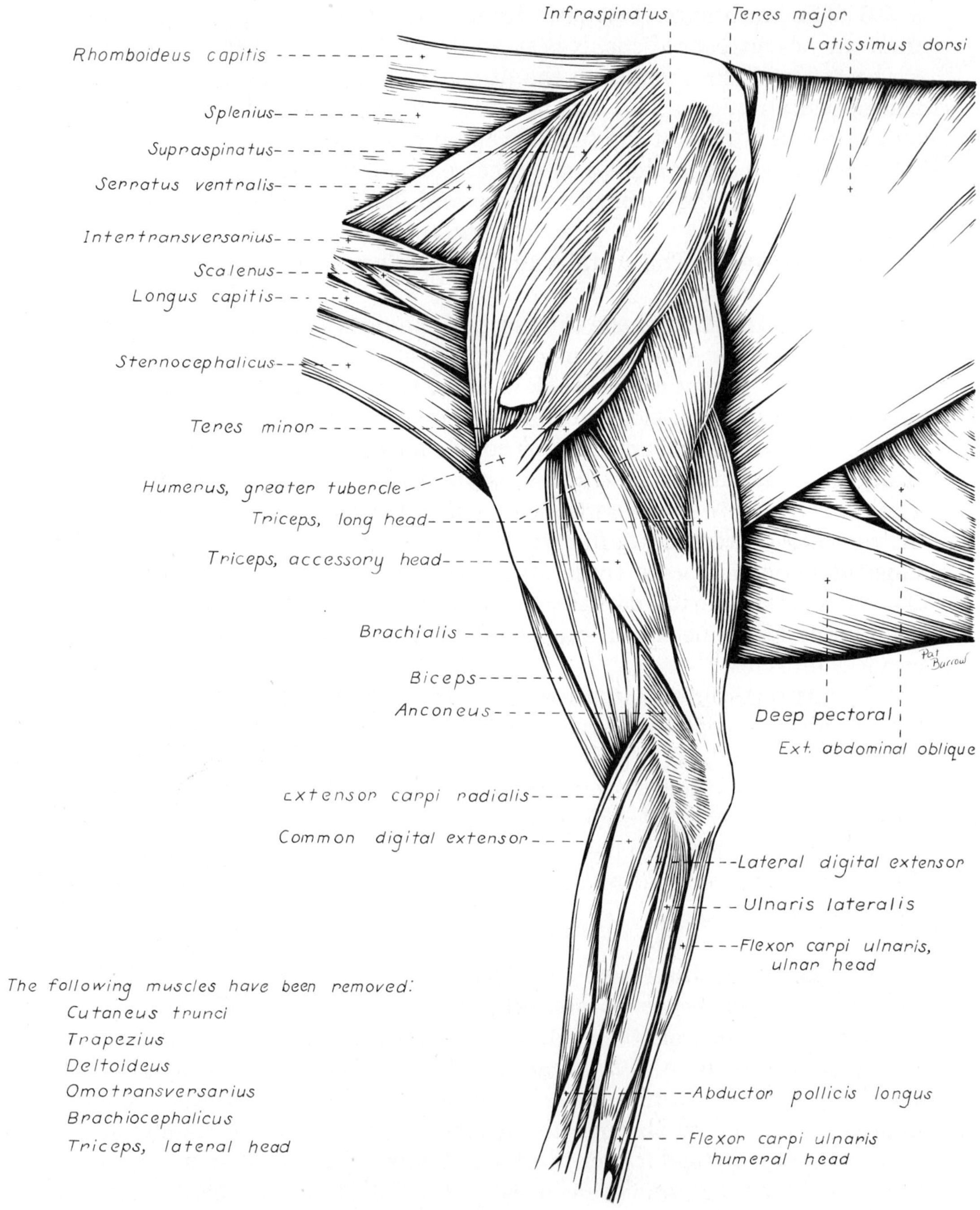

FIGURE 17. Muscles of left shoulder, arm and antebrachium.

cervical and thoracic parts. The narrow **rhomboideus capitis** attaches the cranial dorsal border of the scapula to the nuchal crest of the occipital bone. The **rhomboideus cervicis** runs from the median raphe of the neck to the dorsal border of the scapula. The **rhomboideus thoracis** is short and thick. Its caudal border is deep to the latissimus dorsi. The cervical and thoracic parts of the rhomboideus are contiguous on the dorsal border of the scapula. Transect the entire muscle a few centimeters from the scapula.

ORIGIN: The nuchal crest of the occipital bone; the median fibrous raphe of the neck; the spinous processes of the first seven thoracic vertebrae.

INSERTION: The dorsal border and adjacent surfaces of the scapula.

ACTION: To elevate the forelimb and draw the scapula against the trunk.

INNERVATION: Ventral branches of cervical and thoracic nerves.

8. The **latissimus dorsi** (Figs. 15, 17) is large and roughly triangular. It lies caudal to the scapula, where it covers most of the dorsal and some of the lateral thoracic wall. Clean its ventrocaudal border. Directly caudal to the forelimb, transect the latissimus dorsi at a right angle to its fibers.

ORIGIN: The thoracolumbar fascia from the spinous processes of the lumbar and the last seven or eight thoracic vertebrae; a muscular attachment to the last two or three ribs.

INSERTION: The teres major tuberosity of the humerus and the teres major tendon.

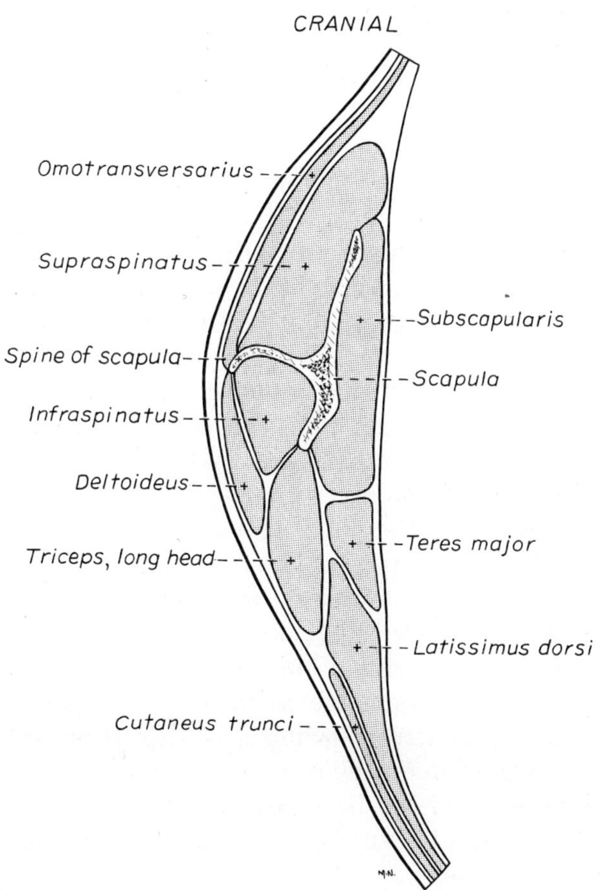

FIGURE 18. Transverse section through middle of left scapula.

ACTION: To draw the free limb caudally as in digging; to flex the shoulder.

INNERVATION: Thoracodorsal nerve (C7, C8, T1).

The **thoracolumbar fascia** is deep fascia of the trunk. It arises from the supraspinous ligament and spines of the thoracic and lumbar vertebrae, covers the muscles of the vertebrae, ribs and abdomen and fuses with the opposite fascia on the ventral midline along a median fibrous raphe, the linea alba. The thoracolumbar fascia serves as an attachment for numerous muscles. It will be dissected with the abdominal muscles.

9. The **serratus ventralis** (Figs. 15, 17) is a large, fan-shaped muscle that acts as a sling to support the body between the limbs. Abduct the forelimb. This will require severing the axillary artery and vein, the nerves contributing to the brachial plexus, and the axillary fascia. As the forelimb is progressively abducted, the attachment of the serratus ventralis will detach from the serrated face of the scapula. Since the forelimb is removed at this stage of the dissection, the detachment of the serratus ventralis may be completed. It is the only extrinsic muscle of the forelimb that has not been transected.

ORIGIN: The transverse processes of the last five cervical vertebrae and the first seven or eight ribs ventral to their middle.

INSERTION: The dorsomedial third of the scapula (serrated face).

ACTION: To support the trunk and depress the scapula.

INNERVATION: Ventral branches of cervical nerves and the long thoracic nerve (C7).

Live Dog

Stand over a dog and palpate on both sides the superficial pectoral muscles from in front of the thoracic limbs and the deep pectoral muscles from behind. Place your fingers on the sternum and grasp the deep pectoral between your fingers medially and your thumb laterally. Extend the neck and palpate the sternomastoideus, sternohyoideus and sternothyroideus muscles at their attachment to the first sternebra. Palpate the trachea and appreciate the muscles that must be separated to expose the trachea to open it—a tracheostomy.

Palpate the spines of the thoracic vertebrae. The cervical spines caudal to the axis are not palpable. Try to separate the dorsal border of the scapula from the thoracic vertebral spines. The trapezius and rhomboideus prevent this. Palpate the sides of the thorax covered by the latissimus dorsi and feel its ventral border.

Visualize the attachments of the serratus ventralis and how it keeps the limb attached to the trunk when the dog is supporting weight. Occasionally its termination on the scapula is torn by injury. This results in an abnormal elevation of the limb when the dog bears weight and the dorsal border of the scapula will protrude dorsally beside or above the level of the thoracic spines.

Intrinsic Muscles of the Thoracic Limb

Lateral: scapula and shoulder
- Deltoideus
- Infraspinatus
- Teres minor
- Supraspinatus

Caudal: arm
- Tensor fasciae antebrachii
- Triceps brachii
- Anconeus

Caudal and medial: forearm
- Flexor carpi radialis
- Superficial digital flexor
- Flexor carpi ulnaris
- Deep digital flexor
- Pronator quadratus

Medial: scapula and shoulder
- Subscapularis
- Teres major
- Coracobrachialis

Cranial: arm
- Biceps brachii
- Brachialis

Cranial and lateral: forearm
- Extensor carpi radialis
- Common digital extensor
- Lateral digital extensor
- Ulnaris lateralis
- Supinator
- Pronator teres
- Abductor pollicis longus

Lateral Muscles of the Scapula and Shoulder

1. The **deltoideus** (Figs. 14, 16, 18) is composed of two portions that fuse and act in common across the shoulder. The proximal portion arises as a wide aponeurosis from the length of the scapular spine and covers the infraspinatus. The latter muscle can be seen through this aponeurosis and should not be confused with the deltoideus. Observe the distal portion of the deltoideus, which arises from the acromion and has a fusiform shape. Both portions of the muscle fuse before they insert on the humerus. Transect the combined muscle 2 centimeters distal to the acromion and reflect the stumps. Free the proximal portion from the infraspinatus and work under the aponeurosis of origin to verify its attachment to the spine of the scapula.

ORIGIN: The spine and acromial process of the scapula.

INSERTION: The deltoid tuberosity.

ACTION: To flex the shoulder.

INNERVATION: Axillary nerve.

2. The **infraspinatus** (Figs. 14, 17, 18) is fusiform and lies principally in the infraspinous fossa. Transect the infraspinatus halfway between its extremities. Free and reflect the distal half from the scapula by scraping the fibers away from the spine and fossa with the handle of the scalpel. Reflect the distal half to its insertion on the side of the greater tubercle. This will expose a bursa between the tendon of insertion and the greater tubercle of the humerus. A bursa is a closed sac containing synovial fluid, which reduces friction.

ORIGIN: The infraspinous fossa.

INSERTION: A small, circumscribed area on the lateral side of the greater tubercle of the humerus.

ACTION: To extend or flex the joint, depending on the degree of extension or position of the joint when the muscle contracts. To abduct the shoulder and to rotate the arm laterally.

INNERVATION: Suprascapular nerve.

3. The **teres minor** (Figs. 14, 17) a small, wedge-shaped muscle is now exposed caudal to the shoulder. It is covered superficially by the deltoideus, caudally by the triceps and cranially by the infraspinatus.

ORIGIN: The infraglenoid tubercle and distal third of the caudal border of the scapula.

INSERTION: The teres minor tuberosity of the humerus.

ACTION: To flex the shoulder and rotate the arm laterally.

INNERVATION: Axillary nerve.

4. The **supraspinatus** (Figs. 14, 15, 17–19), wider and larger than the infraspinatus, is largely covered by the cervical part of the trapezius and the omotransversarius. It lies in the supraspinous fossa and extends over the cranial border of the scapula so that a part of the muscle is closely united with the subscapularis. Clean and observe the insertion on the greater tubercle of the humerus.

ORIGIN: The supraspinous fossa.

INSERTION: The greater tubercle of the humerus, by a thick tendon.

ACTION: To extend the shoulder.

INNERVATION: Suprascapular nerve.

Medial Muscles of the Scapula and Shoulder

1. The **subscapularis** (Figs. 14, 15, 18, 20) occupies the entire subscapular fossa, the boundaries of which it overlaps slightly. The

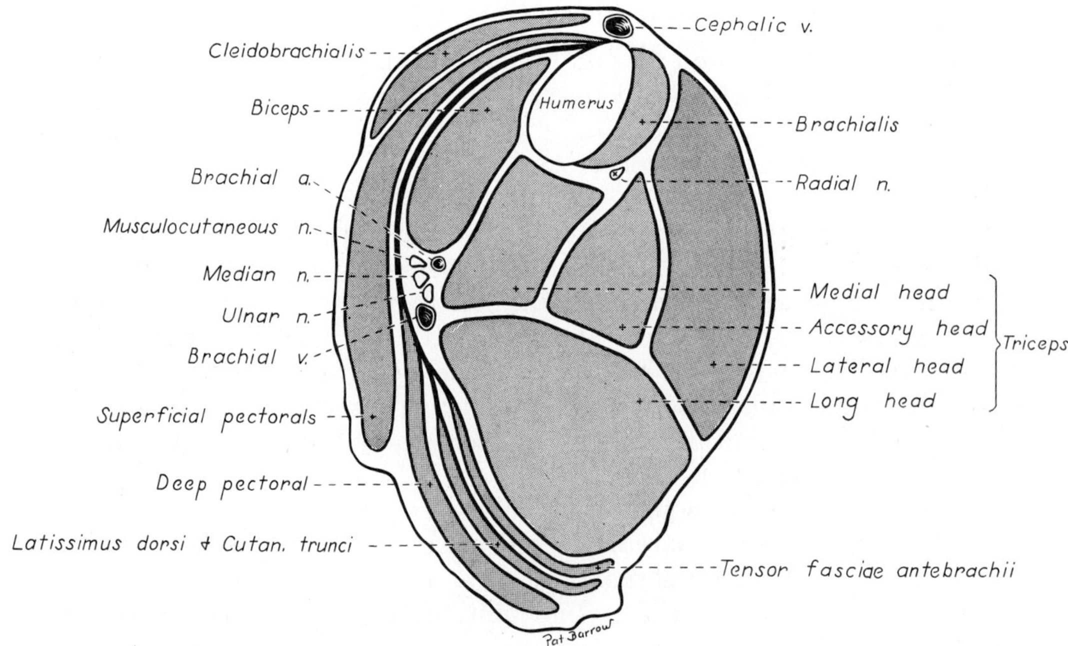

FIGURE 19. Transverse section through middle of right arm.

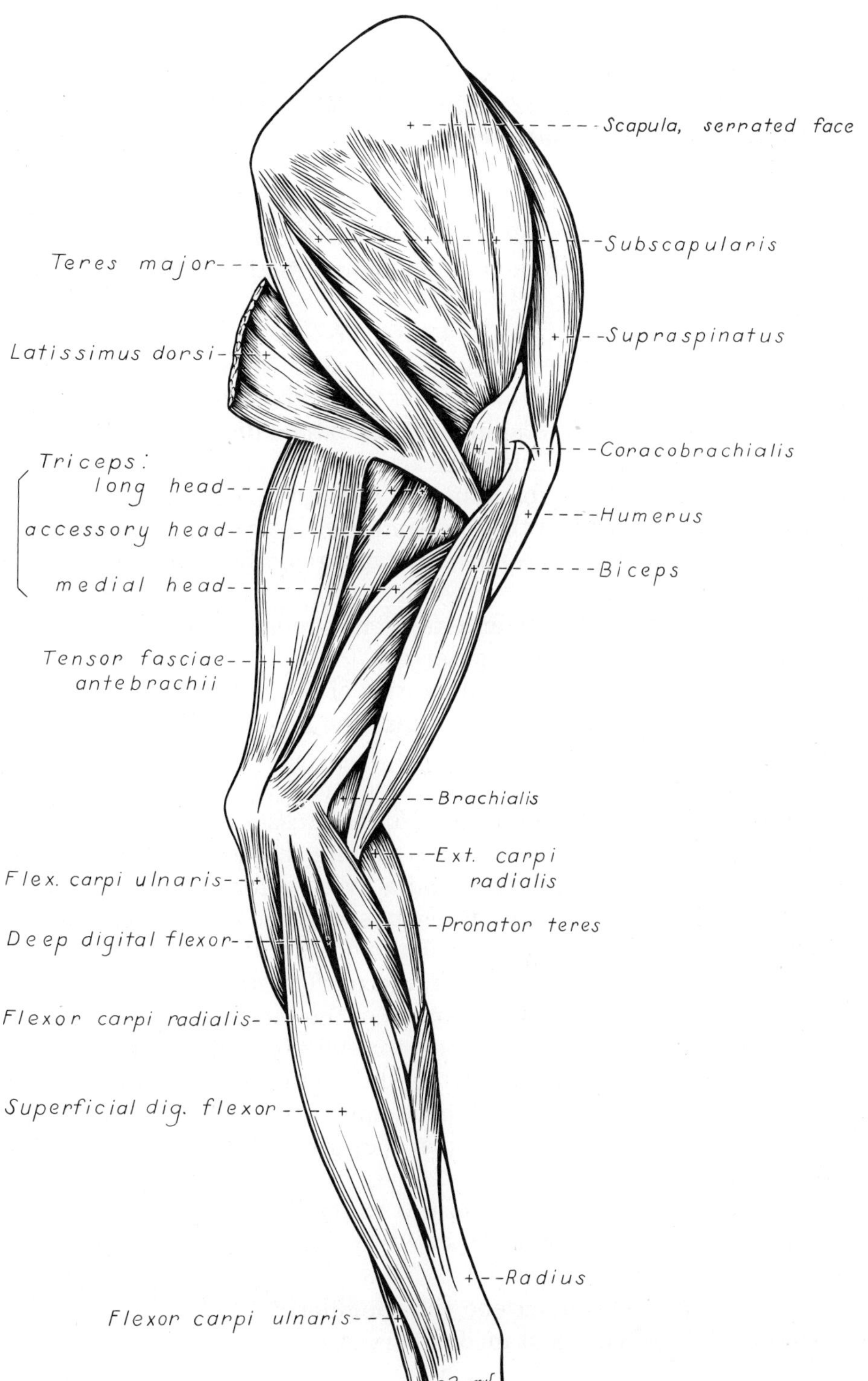

FIGURE 20. Muscles of left thoracic limb, medial view.

supraspinatus is closely associated with it cranially, while the teres major has a similar relation caudally. Clean the insertion but do not transect the muscle.

ORIGIN: The subscapular fossa.

INSERTION: The lesser tubercle of the humerus.

ACTION: To adduct and extend the shoulder.

INNERVATION: Subscapular nerves.

Although some support is provided to the shoulder by the subscapularis muscle medially and the teres minor, infraspinatus and supraspinatus muscles laterally, luxation will not occur without injury to the joint capsule and its glenohumeral ligaments (see Fig. 30).

2. The **teres major** (Figs. 14, 15, 17–20), directly caudal to the subscapularis, belies its descriptive name since it is not round but has three surfaces. Its proximal end arises from the subscapularis and the dorsal caudal border of the scapula. Fibers extend distally to attach to the tendon of insertion of the latissimus dorsi. Work between the distal half of the muscle and the subscapularis. Observe the close relationship between the teres major and the latissimus dorsi. Transect the teres major and reflect the combined insertion to expose the belly of the coracobrachialis muscle.

ORIGIN: The caudal angle and adjacent caudal border of the scapula; the caudal surface of the subscapularis.

INSERTION: The teres major tuberosity of the humerus.

ACTION: Flex the shoulder and rotate the arm medially.

INNERVATION: Axillary nerves.

3. The **coracobrachialis** (Figs. 15, 20) crosses the medial surface of the shoulder obliquely. It is a small spindle-shaped muscle that arises from the coracoid process of the scapula by a relatively long tendon that courses caudodistally across the lesser tubercle. There it crosses the tendon of insertion of the subscapularis. It is provided with a synovial sheath as it crosses the lesser tubercle of the humerus. The muscle belly is distal to the lesser tubercle. The conjoined tendon of the teres major and latissimus dorsi crosses the insertion of the coracobrachialis. Notice that the coracobrachialis tendon courses cranial to the center of the shoulder. This accounts for its action as an extensor muscle of the shoulder. Free the coracobrachialis and isolate the tendon of origin by cutting into its synovial sheath.

ORIGIN: The coracoid process of the scapula.

INSERTION: The crest of the lesser tubercle of the humerus proximal to the teres major tuberosity.

ACTION: To adduct and extend the shoulder.

INNERVATION: Musculocutaneous nerves.

Live Dog

Locate the spine of the scapula and palpate the supraspinatus cranial to it and the infraspinatus and deltoideus caudal to it. Find the acromion and palpate the acromial part of the deltoideus. Rotate the arm medially and laterally. The restriction to lateral rotation is by the teres major and subscapularis whereas the restriction to medial rotation is by the teres minor and infraspinatus. Injury that tears the termination of the latter muscles allows excessive medial rotation. In the weight-bearing phase of walking this causes the elbow to abduct. Occasionally injury to the infraspinatus results in contracture with shortening of the muscle. This causes excessive lateral rotation of this arm. The dog stands with the elbow adducted. This is exacerbated in the swing phase (protraction) where there is also a compensatory abduction of the paw. On manipulation there is increased restriction to medial rotation. This can be corrected by cutting the tendon of the contracted infraspinatus.

Caudal Muscles of the Arm (Brachium)

This group is a large, muscular mass that almost completely fills the space between the caudal border of the scapula and the olecranon. It consists of three muscles: the triceps brachii, the tensor fasciae antebrachii and the anconeus. By far the largest of these muscles is the triceps. All of the caudal muscles of the arm are extensors of the elbow.

1. The **tensor fasciae antebrachii** (Figs. 15, 19, 20) is a thin strap that extends from the latissimus dorsi to the medial fascia of the forearm and the olecranon. It lies on the long head of the triceps.

ORIGIN: The fascia covering the lateral side of the latissimus dorsi.

INSERTION: The olecranon.

ACTION: To extend the elbow.

INNERVATION: Radial nerve.

2. The **triceps brachii** in the dog consists of four heads instead of the usual three, with a common tendon to the olecranon. Only the long head arises from the scapula. The other three arise from the proximal end of the humerus.

The **long head** (Figs. 14–20) completely bridges the humerus: It arises from the caudal border of the scapula and inserts on the olecranon. It appears to have two bellies. Caudal to the shoulder palpate a groove between the long and lateral heads of the triceps. Separate these two heads along this groove. Expose the tendon of the long head and notice the bursa between it and the groove of the olecranon. Notice how the tendons of the other heads blend with that of the long head.

ORIGIN: The caudal border of the scapula.

INSERTION: The olecranon.

ACTION: To extend the elbow and flex the shoulder.

INNERVATION: Radial nerve.

The **lateral head** (Figs. 14, 16, 19) of the triceps lies distal to the long head, caudal to the acromial part of the deltoideus and lateral to the accessory head, which it covers. Transect the origin of the lateral head and reflect it to expose the underlying accessory and medial heads. This also exposes the brachialis muscle.

ORIGIN: The tricipital line of the humerus.
INSERTION: The olecranon.
ACTION: To extend the elbow.
INNERVATION: Radial nerve.

The **accessory head** (Figs. 14, 15, 17, 19, 20) lies between the lateral and medial heads.

ORIGIN: The neck of the humerus.
INSERTION: The olecranon.
ACTION: To extend the elbow.
INNERVATION: Radial nerve.

The **medial head** (Figs. 15, 19, 20) lies caudally on the humerus medial to the biceps brachii. Separate the muscle from the long head caudally and from the accessory head laterally. The long tendon of the accessory head is closely bound to its lateral surface.

ORIGIN: The crest of the lesser tubercle near the teres major tubercle.
INSERTION: The olecranon.
ACTION: To extend the elbow.
INNERVATION: Radial nerves.

3. The **anconeus** (Figs. 14–17) is a small muscle located almost completely in the olecranon fossa. Reflect the insertion of the lateral head of the triceps to uncover the lateral surface of this muscle. Notice that the most distal fibers lie in a trnsverse plane. On the lateral side cut the origin of the anconeus from the lateral supracondylar crest and epicondyle. Reflect it to expose the elbow joint capsule. Open the joint capsule to expose the elbow.

ORIGIN: The lateral epicondylar crest and the lateral and medial epicondyles of the humerus.

INSERTION: The lateral surface of the proximal end of the ulna.
ACTION: To extend the elbow.
INNERVATION: Radial nerve.

Live Dog

Palpate the triceps brachii caudal to the arm and its termination on the olecranon. Appreciate the function of this muscle in extension of the elbow to support the weight of the animal in standing and during locomotion. Avulsion of its tendon or denervation of this muscle will cause the limb to collapse when weight is placed on it.

Cranial Muscles of the Arm

1. The **biceps brachii** (Figs. 13–15, 17, 19–21) has only one head. It is a long, fusiform muscle that lies on the medial and cranial surfaces of the humerus. It completely bridges this bone as it arises on the supraglenoid tuberosity of the scapula and inserts on the proximal ends of the radius and ulna. It is covered superficially by the pectoral muscles. Clean the muscle and transect it through its middle. Reflect the proximal half

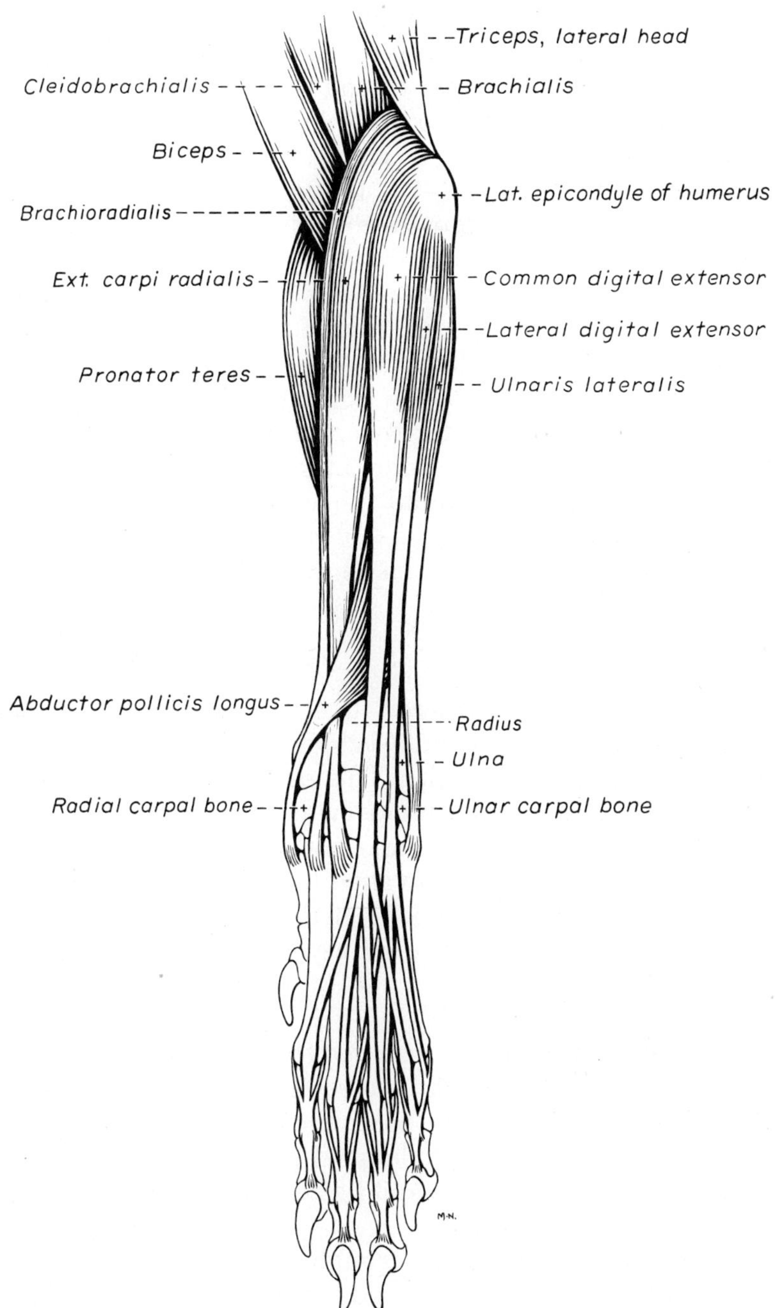

FIGURE 21. Muscles of left antebrachium, cranial view.

to its origin. This will require severing the **transverse humeral retinaculum,** a band of fibrous tissue that joins the greater and lesser tubercles and holds the tendon of origin in the intertubercular groove. An extension of the shoulder joint capsule acts as a synovial sheath for this tendon. Reflect the distal half of the biceps to the proximal end of the radius and ulna, where it meets the brachialis tendon and bifurcates. The tendons of insertion lie on the elbow joint capsule. Delay cleaning these tendons until after the pronator teres muscle has been dissected.

ORIGIN: The supraglenoid tubercle.
INSERTION: The ulnar and radial tuberosities.
ACTION: To flex the elbow and extend the shoulder.
INNERVATION: Musculocutaneous nerve.

2. The **brachialis** (Figs. 13–17, 19–21) should be studied from the lateral side. It is a long, thin muscle that lies in the brachialis groove of the humerus. From the proximal third of this groove the brachialis curves laterally and cranially as it courses distally, crosses the elbow and inserts by a terminal tendon on the medial side of the proximal end of the ulna. A large part of its lateral surface is covered by the lateral head of the triceps. Distally it runs medial to the origin of the extensor carpi radialis. Its insertion will be dissected later with the biceps insertion.

ORIGIN: The proximal third of the lateral surface of the humerus.
INSERTION: The ulnar and radial tuberosities.
ACTION: To flex the elbow.
INNERVATION: Musculocutaneous nerve.

Live Dog

Palpate the crest of the greater tubercle and feel the tendon of the biceps brachii in the intertubercular groove medial to it covered by the termination of the pectoral muscles. This part of the tendon is covered by a synovial sheath that is continuous with the shoulder joint capsule. Swelling of the tendon sheath will be felt when there is increased synovial fluid from lesions of the tendon, its sheath or the shoulder joint. Distally in the arm feel the terminal portions of the biceps brachii medially and the brachialis laterally on the cranial aspect of the elbow. Both muscles can be palpated here.

Before the rest of the skin is removed from the thoracic limb, examine the foot pads. The small pad that protrudes palmar to the carpus is the **carpal pad.** The largest in the paw, the **metacarpal pad,** is on the palmar side of the metacarpophalangeal joints and is triangular. The **digital pads** are ovoid in outline and flattened. Each is located palmar to the distal interphalangeal joint.

Make a midcaudal incision through the skin from the olecranon through the carpal and metacarpal pads to the interdigital space between digits III and IV. Reflect the skin, dissect it free from the fascia and remove it from the forelimb. The pads are closely attached to the underlying structures but may be dissected free and removed. Be careful not to cut too deeply and sever the underlying small tendons. Work distally on each of the four main digits and completely remove the skin and digital pads.

The subcutaneous tissue distal to the elbow is scanty. As in most other places in the body it connects the skin with the underlying fascia. The principal veins and cutaneous nerves lie in large part on the superficial fascia of the antebrachium. For descriptive purposes the superficial and deep fascia distal to the elbow may be divided into antebrachial, carpal, metacarpal and digital parts.

The **deep antebrachial fascia** forms a single sleeve for the muscles of the forearm on the caudal surface. Make an incision through the deep antebrachial fascia from the olecranon to the accessory carpal bone. Carefully reflect the fascia cranially on the forearm. At first the fascia is easily reflected, as it lies on the epimysium of the muscles beneath. Cranially it sends delicate septal leaves between muscles, and on reaching the radius it firmly unites with its periosteum.

Cranial and Lateral Muscles of the Forearm (Antebrachium)

The cranial and lateral antebrachial muscles are, from cranial to caudal, the pronator teres, supinator, extensor carpi radialis, common digital extensor, lateral digital extensor, ulnaris lateralis and the abductor pollicis longus. Most of these muscles arise from the lateral or extensor epicondyle of the humerus. A slender, inconstant muscle, the brachioradialis (Figs. 21, 22), arises from the lateral epicondylar crest of the humerus and passes distally and medially to insert on the distal fourth of the radius. If present the muscle is frequently removed with the skin.

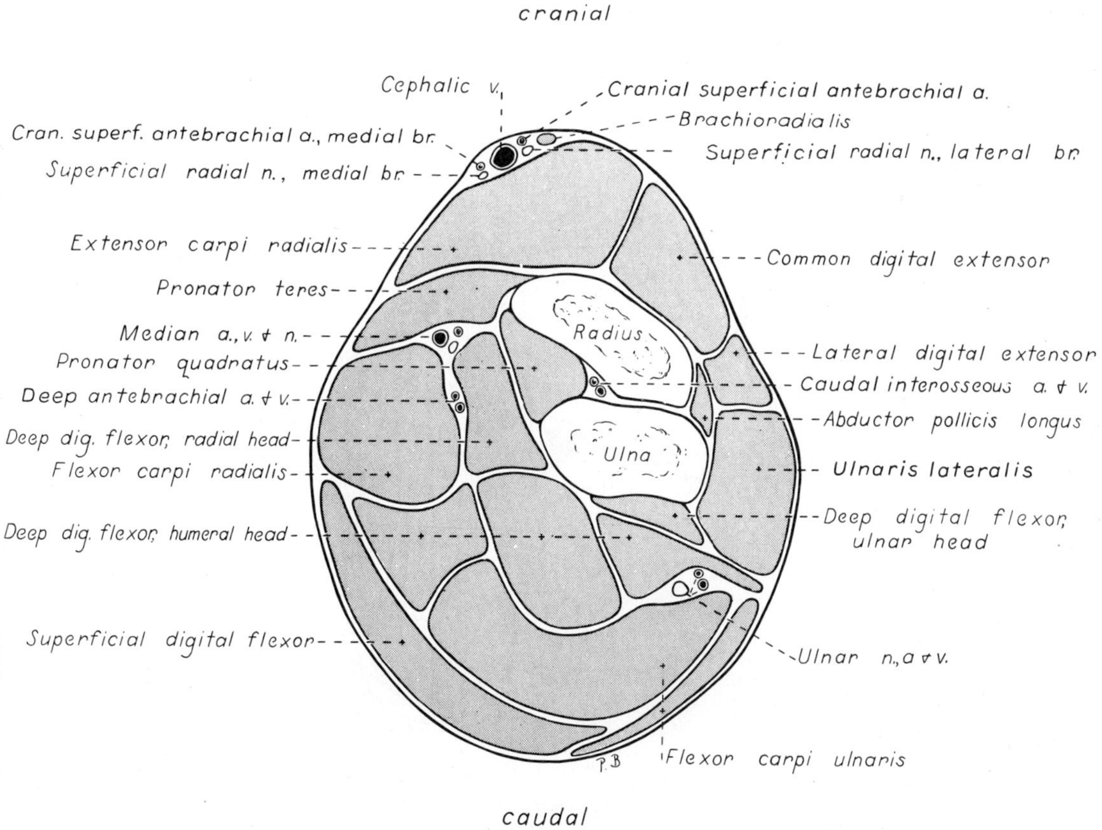

FIGURE 22. Transverse section of right forearm between proximal and middle thirds.

1. The **extensor carpi radialis** (Figs. 14, 15, 17, 21–23) is the largest of the craniolateral antebrachial muscles. It lies on the cranial surface of the radius throughout most of its course and is easily palpated in the live dog. The tendon looks single but is distinctly double throughout its distal third. These closely associated tendons run first under the tendon of the abductor pollicis longus, then in the middle groove of the radius farther distally and finally over the carpus. They are held in place by the **extensor retinaculum.** This is a transversely oriented condensation of carpal fascia that aids in holding in grooves all the tendons that cross the dorsum of the carpus. Between bundles of tendons the extensor retinaculum dips down to blend with the fibrous dorsal part of the joint capsule. Define by dissection the proximal and distal margins of the extensor retinaculum, but do not sever it along the tendons.

ORIGIN: The lateral epicondylar crest.

INSERTION: The small tuberosities on the proximal ends and dorsal surfaces of metacarpals II and III.

ACTION: To extend the carpus.

INNERVATION: Radial nerve.

2. The **common digital extensor** (Figs. 14, 15, 17, 21–23) is shaped like, and lies caudal to, the extensor carpi radialis on the lateral side. Its four individual tendons that leave the muscle are closely combined where

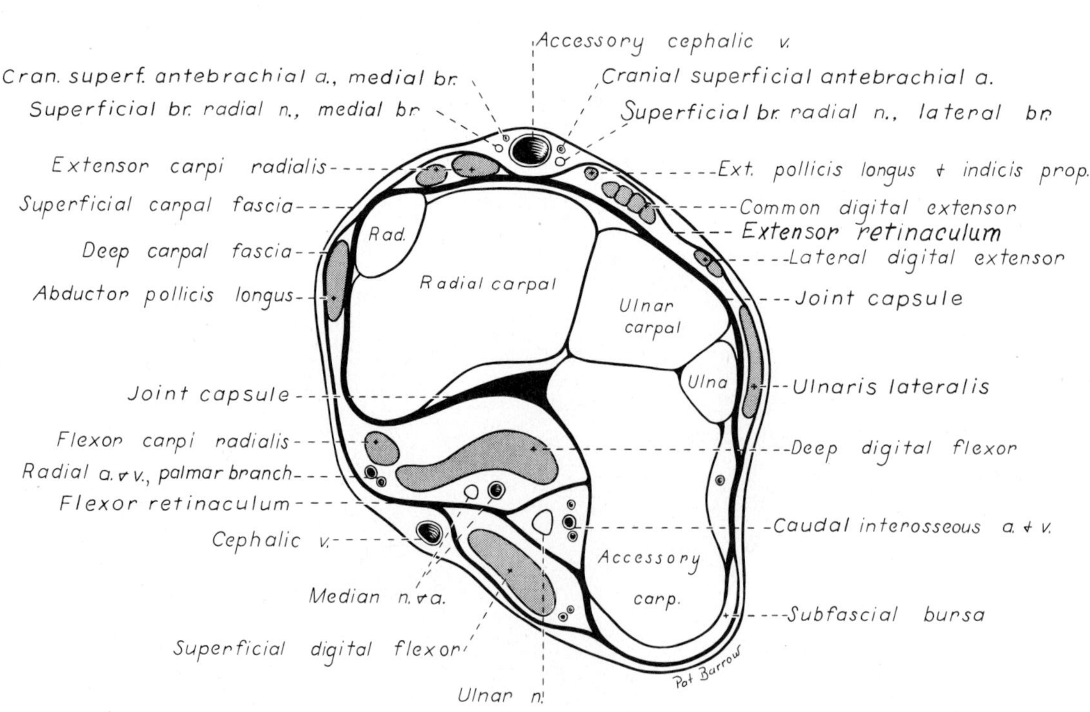

FIGURE 23. Transverse section of right carpus through accessory carpal bone.

they cross the cranial surface of the abductor pollicis longus and then the carpus; here they are held in the lateral groove of the radius by the extensor retinaculum. Distal to the ligament the four tendons diverge and each goes to the distal phalanx of one of the four main digits. Dissect the tendon of the common extensor that goes to the third or fourth digit. Free the tendon as it crosses each of the joints. It often contains a sesamoid bone at the metacarpophalangeal joint. Insertion is on the extensor process of the distal phalanx.

ORIGIN: The lateral epicondyle of the humerus.

INSERTION: The extensor processes of the distal phalanges of digits II, III, IV and V.

ACTION: To extend the joints of the four principal digits.

INNERVATION: Radial nerve.

Notice that the distal interphalangeal joint is in a marked degree of overextension. This is brought about by the elastic **dorsal ligament,** which lies on each side of the common extensor tendon. The ligament attaches proximally to the sides of the proximal end of the middle phalanx. Distally it attaches to the dorsal surface of the ungual crest of the distal phalanx (see Fig. 27). Its elasticity retracts the claw.

3. The **lateral digital extensor** (Figs. 14, 17, 21–23) is about half the size of the common digital extensor. It lies between the common digital extensor and the ulnaris lateralis. Its tendon begins at the middle of the forearm, passes under the extensor retinaculum in a groove between the radius and ulna and immediately splits into three branches. The main part of each tendon attaches to the extensor process of the distal phalanx of digits III, IV and V in common with the common digital extensor tendon.

ORIGIN: The lateral epicondyle of the humerus.

INSERTION: The proximal ends of all the phalanges of digits III, IV and V, but mainly the extensor processes of the distal phalanges of these digits.

ACTION: To extend joints of digits III, IV and V.

INNERVATION: Radial nerve.

4. The **ulnaris lateralis** (Figs. 14, 17, 21–25) is larger than the lateral digital extensor, behind which it lies. It is bounded deeply by the ulna and the large flexor group of digital muscles, which lie caudal to it. It is the only flexor to arise on the lateral epicondyle. Expose the muscle and notice the two tendons of insertion.

ORIGIN: The lateral epicondyle of the humerus.

INSERTION: The lateral aspect of the proximal end of metacarpal V and the accessory carpal bone.

ACTION: To abduct and flex the carpus.

INNERVATION: Radial nerve.

5. The **supinator** (Figs. 14, 15, 26) is short, broad and flat and obliquely placed across the lateral side of the flexor surface of the elbow

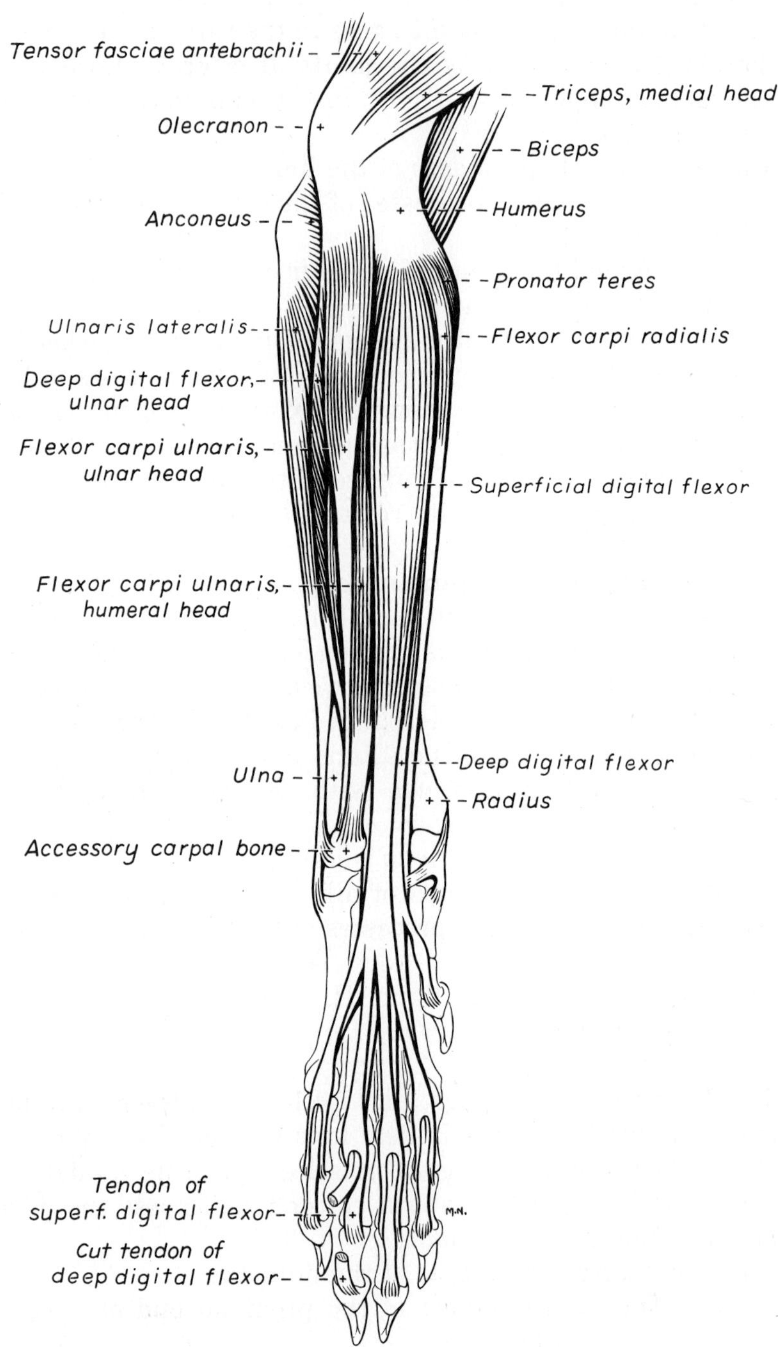

FIGURE 24. Muscles of left antebrachium, caudal view.

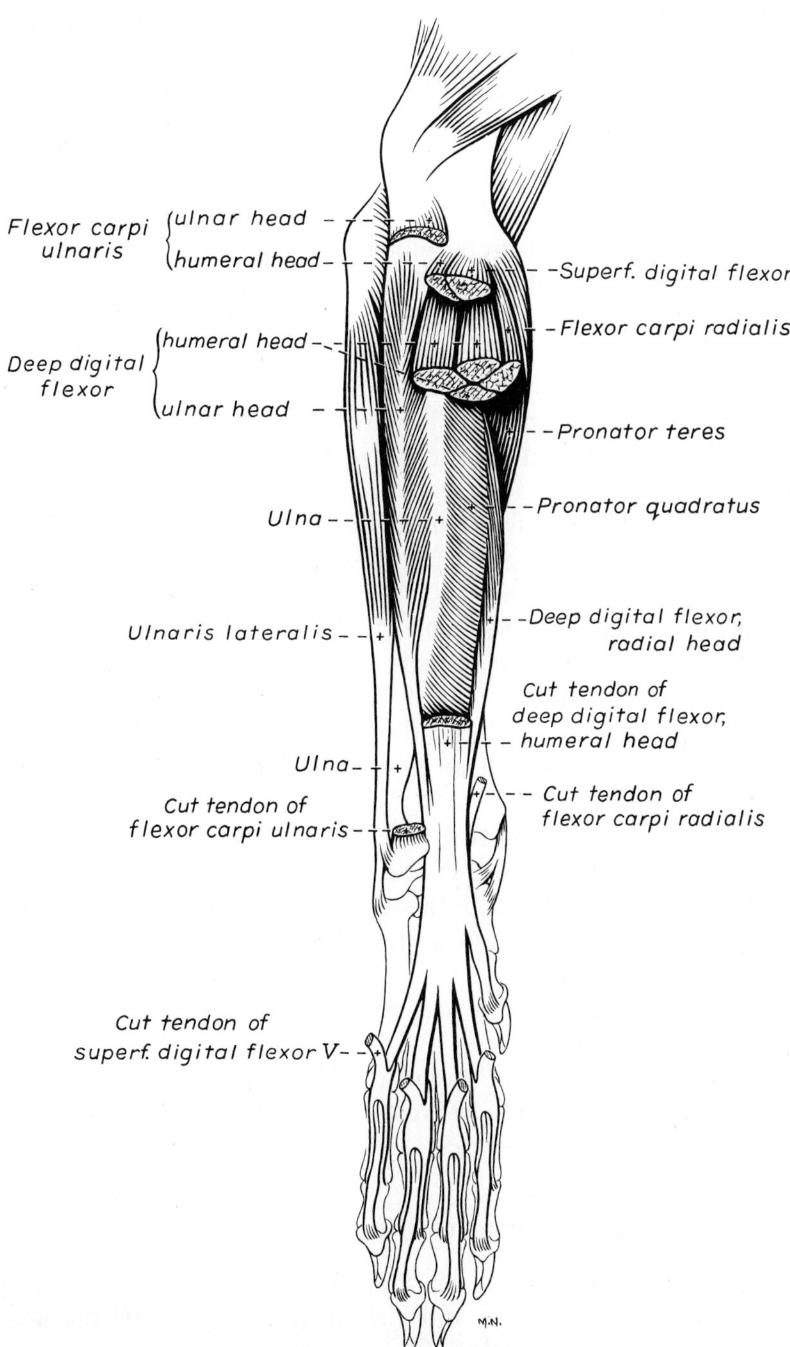

FIGURE 25. Deep muscles of left antebrachium, caudal view.

joint. It is covered superficially by the extensor carpi radialis and common digital extensors, which should be transected in the middle of their muscle bellies and reflected. The supinator lies principally on the proximal fourth of the radius.

ORIGIN: The lateral epicondyle of the humerus.

INSERTION: The cranial surface of the proximal fourth of the radius.

ACTION: To rotate the forearm laterally so that the palmar side of the paw faces medially (supination), and to flex the elbow.

INNERVATION: Radial nerve.

The **pronator teres** (Figs. 14, 15, 20–22, 24–26) extends obliquely across the medial surface of the elbow. It is round in transverse section at its origin and flat at its insertion. It lies between the extensor carpi radialis cranially and the flexor carpi radialis caudally. Displace adjacent muscles to see its origin and insertion. Transect the muscle and reflect the extremities.

ORIGIN: The medial epicondyle of the humerus.

INSERTION: The medial border of the radius between the proximal and middle thirds.

ACTION: To rotate the forearm medially so that the palmar side of the paw faces the ground (pronation); to flex the elbow.

INNERVATION: Median nerve.

Clean the tendons of insertion of the biceps and brachialis muscles, which are now exposed. The tendon of insertion of the biceps splits into two parts (see Fig. 31). The larger of the two inserts on the ulnar tuberosity and the smaller on the radial tuberosity. The terminal tendon of the brachialis inserts between these two tendons of the biceps on the ulnar tuberosity.

The **abductor pollicis longus** (Figs. 14, 15, 17, 21–23) lies primarily in the groove between the radius and ulna and is triangular. Displace the digital extensors so that the bulk of the muscle is uncovered. Clean the muscle and transect its tendon as it obliquely crosses the extensor carpi radialis.

ORIGIN: The lateral border and cranial surface of the body of the ulna; the interosseous membrane.

INSERTION: The proximal end of metacarpal I.

ACTION: To abduct the first digit or pollex.

INNERVATION: Radial nerve.

Caudal and Medial Muscles of the Forearm

The deep antebrachial fascia has been removed from this group of muscles. When dissecting the individual muscles, it will be necessary to clean their tendons of insertion. The muscles in this group include, from the radius caudally, the flexor carpi radialis, the deep digital flexor, the superficial digital flexor and, on the lateral side, the flexor carpi ulnaris.

1. The **flexor carpi radialis** (Figs. 15, 20, 22–25) lies between the pronator teres cranially and the superficial digital flexor caudally. It

covers the deep digital flexor, part of which can be seen. The flexor carpi radialis has a thick, fusiform belly, which, partly imbedded in the deep flexor, extends only to the middle of the radius. There it gives rise to a flat tendon that is augmented by fibers leaving the medial border of the radius. Clean the tendon to the point where it passes deep to a thick layer of fibrous tissue on the palmar side of the carpus, the **flexor retinaculum** (Fig. 23). Do not cut through this fibrous tissue now. A synovial sheath extends from the distal end of the radius almost to the insertion of the muscle on the second and third metacarpal bones. This will be exposed later.

ORIGIN: The medial epicondyle of the humerus and the medial border of the radius.

INSERTION: The palmar side of the proximal ends of metacarpals II and III.

ACTION: To flex the carpus.

INNERVATION: Median nerve.

2. The **superficial digital flexor** (Figs. 14, 15, 20, 22–24) lies beneath the skin and antebrachial fascia on the caudomedial side of the forearm. It covers the deep digital flexor and is fleshy almost to the carpus. Its tendon is at first single, then crosses the palmar (flexor) surface of the carpus medial to the accessory carpal bone, where it is covered by the superficial part of the flexor retinaculum, and finally divides into four tendons of nearly equal size. These insert on the proximal palmar surfaces of the middle phalanges of the four principal digits. At the metacarpophalangeal joint, each forms a collar around the deep flexor tendon that passes through it. Clean each of the individual tendons as far as the metacarpophalangeal joints. Transect the muscle at the middle of the antebrachium, and turn the distal part toward the digits. Since all parts of the superficial digital flexor tendon are similar, only that to the third digit will be dissected. The superficial and deep digital flexor tendons are held firmly in place at the metacarpophalangeal joint by the **palmar annular ligament.**

If any of the structures mentioned below are not clearly seen on the third digit, they should be verified on one or more of the other main digits. Observe that the tendon of the superficial digital flexor sheathes the deep digital flexor for a distance of more than a centimeter at the metacarpophalangeal joint; that the superficial digital flexor tendon lies on the palmar side of the deep flexor tendon at the proximal end of its encircling sheath but on the dorsal side at the distal end; and that the superficial flexor tendon with its sheath and the deep flexor tendon are in a common synovial membrane, the **digital synovial sheath** (see Fig. 27).

ORIGIN: The medial epicondyle of the humerus.

INSERTION: The palmar surface of the base (proximal end) of the middle phalanges of digits II, III, IV and V.

ACTION: To flex digits II, III, IV and V.

INNERVATION: Median nerve.

3. The **flexor carpi ulnaris** (Figs. 14, 15, 17, 20, 22, 24, 25) consists of two parts that are distinct throughout their length. The **ulnar head** arises from the caudal border of the proximal end of the ulna. It is thin and wide proximally but narrow distally. It lies between the ulnaris lateralis and superficial digital flexor. The **humeral head** is large and fleshy and lies cranial to the ulnar head, except distally, where its tendon lies caudal to it. Dissect the insertion of this muscle on the accessory carpal bone and clean its origin. A subfascial bursa is present over the tendon of insertion of the humeral head and an intertendinous bursa is found between the two tendons of insertion at the carpus.

ORIGIN: Ulnar head—the caudal border and medial surface of the olecranon; humeral head—the medial epicondyle of the humerus.

INSERTION: The accessory carpal bone.

ACTION: To flex the carpus.

INNERVATION: Ulnar nerve.

4. The **deep digital flexor** (Figs. 14, 15, 20, 22–25, 27) has three heads of origin of dissimilar size, which arise from the humerus, radius and ulna. Their bellies, along with the pronator quadratus, lie on the caudal surfaces of the radius and ulna. Transect both muscle bellies of the flexor carpi ulnaris in the middle of the antebrachium. Reflect the stumps to expose and identify the three heads of the deep digital flexor muscle. Notice that the **humeral head** of this muscle is much larger than the other two heads and has several bellies. The **ulnar head** is small and arises from the caudal border of the ulna. The **radial head** is the smallest and comes from the medial border of the radius. The tendons of all three heads fuse at the carpus to form a single tendon. This tendon is held in place in the carpal canal by the thick, fibrous **flexor retinaculum.** The **carpal canal** is formed by the accessory carpal bone laterally, the joint capsule and carpal bones dorsally and the flexor retinaculum on the palmar surface. Cut this retinaculum medially and reflect it laterally to the accessory carpal bone to expose the deep digital flexor tendon.

Distal to the carpus the deep digital flexor tendon divides into five branches. Each branch goes to the palmar surface of the base of the distal phalanx of its respective digit. There is a synovial bursa under the humeral head at the elbow and a carpal synovial sheath in the carpal canal. Digital synovial sheaths extend from above the metacarpophalangeal joints to the insertion of the tendons on the distal phalanges of all the digits. The digital synovial sheaths, except that of the first digit, are common to both the superficial and the deep flexor tendons. Dissect the deep digital flexor tendon to its insertion on the distal phalanx of the third digit. It has already been exposed with the superficial digital flexor tendon at the metacarpophalangeal joint. Note the **annular digital ligaments** that support the deep digital flexor tendon proximal and distal to the palmar surface of the proximal interphalangeal joint (see Fig. 27).

ORIGIN: Humeral head—the medial epicondyle of the humerus; ulnar head—the proximal three-fourths of the caudal border of the ulna; radial head—the middle third of the medial border of the radius.

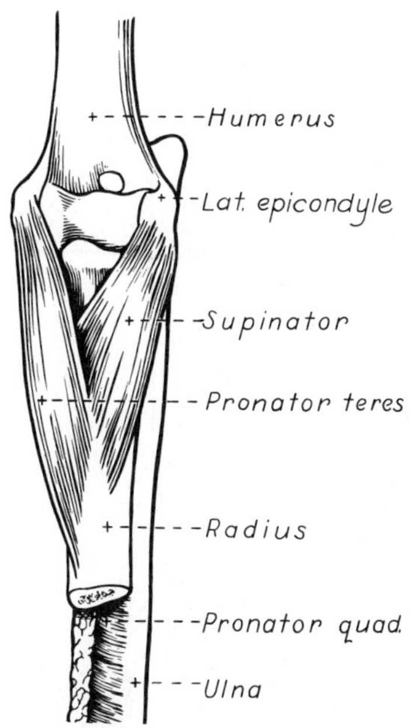

FIGURE 26. Rotators of left antebrachium.

INSERTION: The palmar surface of the base (proximal end) of the distal phalanx of each digit.

ACTION: To flex the digits.

INNERVATION: Median and ulnar nerves.

5. The **pronator quadratus** (Figs. 15, 22, 25, 26) fills in the space between the radius and ulna. Spread the flexor muscles and observe the pronator. The fibers of this muscle run transversely between the ulna and radius.

ATTACHMENTS: The apposed surfaces of the radius and ulna.

ACTION: To pronate the paw.

INNERVATION: Median nerve.

Muscles of the Forepaw

There are several special muscles of the digits. Only the **interossei** will be dissected. The four interosseous muscles are fleshy and similar in size and shape (Fig. 27). They lie deep to the deep digital flexor tendon and cover the palmar surfaces of the four main metacarpal bones. Transect the deep digital flexor tendon at the proximal end of the carpus and reflect it distally. Dissect the interosseous muscle of the third digit. Each muscle arises from the proximal end of its respective metacarpal bone and the carpal joint capsule. After a short course, each divides into two tendons, which attach to the proximal end of the proximal phalanx. Imbedded in each tendon is a sesamoid bone that lies on the palmar surface of the metacarpophalangeal joint. There are thus two **proximal sesamoids** embedded in the tendon of insertion at metacarpophalangeal joints II, III,

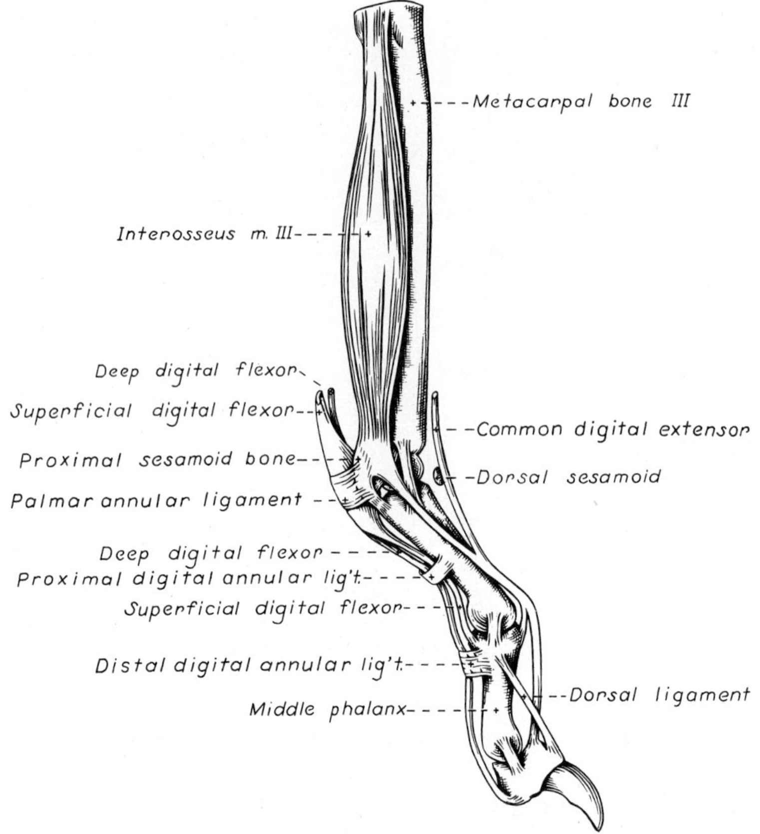

FIGURE 27. Third digit, medial view.

IV and V. A lesser tendon continues obliquely across the proximal end on each side of the proximal phalanx and joins the common extensor tendon on the dorsal surface. The interosseous muscle is a flexor of the metacarpophalangeal (fetlock) joint and maintains the joint angle when the dog bears weight on the paw. Complete exposure of the tendon of insertion of an interosseous muscle can be accomplished by separating the third and fourth digits to the level of the carpometacarpal joint. Abduct the fourth and fifth digits to observe the full length of the interosseous muscle on the third digit. Transect the muscle at the middle of the metacarpal bone and reflect the distal portion to expose the sesamoid bones in the tendon at the metacarpophalangeal joint.

Live Dog

Palpate the lateral epicondyle of the humerus and follow the extensor muscles distally in the forearm. Follow the tendons of the extensor carpi radialis and common digital extensor across the carpus. Flex the carpus and palpate the antebrachiocarpal joint on either side of these tendons where a needle puncture can be made.

Palpate the medial humeral epicondyle and the origins of the flexor muscles on it. At the carpus palpate the terminations of the flexor carpi ulnaris and ulnaris lateralis on the accessory carpal. Palpate the tendon

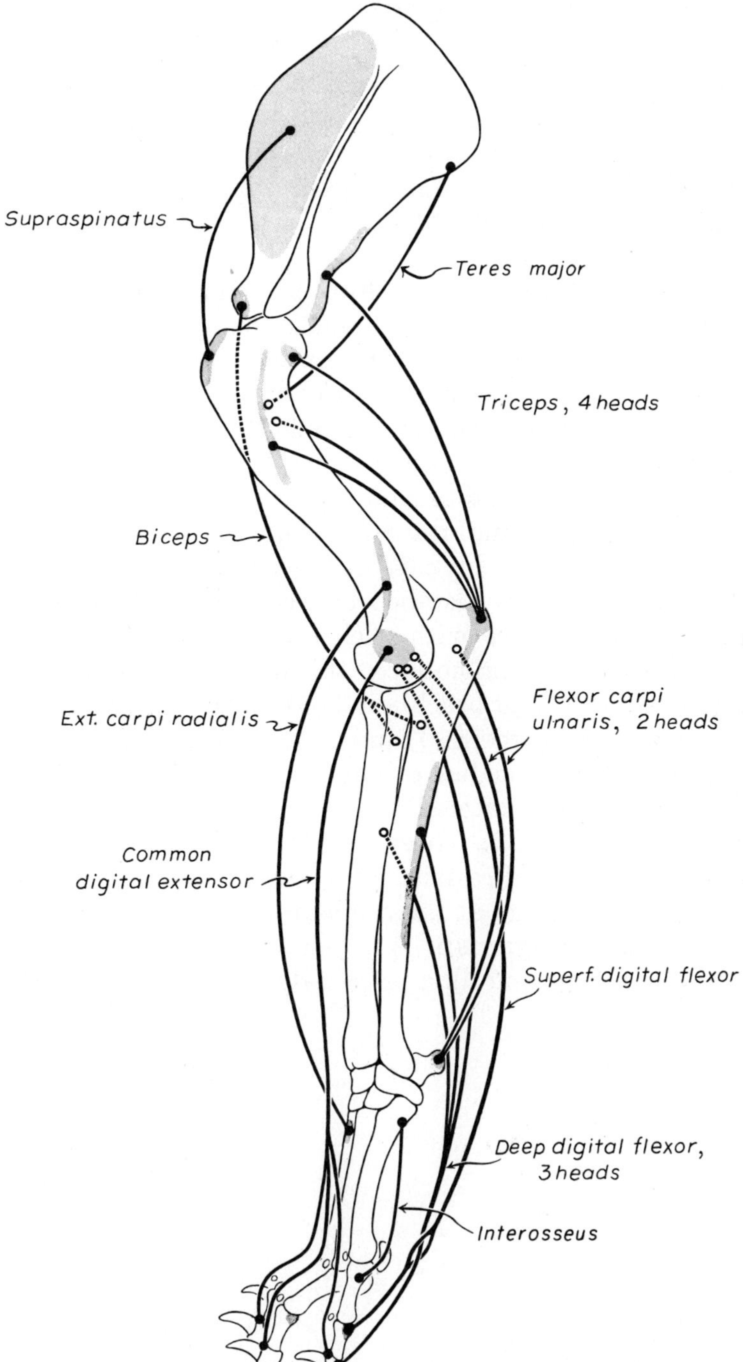

FIGURE 28. Major extensors and flexors of left thoracic limb.

of the superficial digital flexor at the carpal canal. Flex the carpus and feel the tendon loosen. Extend it and feel the tendon become taut. Flex and extend the metacarpophalangeal and interphalangeal joints and appreciate the action of the digital extensors and flexors on these joints. Laceration of the digital flexors will cause overextension of the interphalangeal joints affected.

JOINTS OF THE THORACIC LIMB

The **shoulder** (Fig. 29) is a ball-and-socket joint between the glenoid cavity of the scapula and the head of the humerus. It is capable of movements in any direction but the chief movements it undergoes are flexion and extension. The shoulder joint capsule is a loose sleeve of synovial membrane and thin fibrous tissue that unites the scapula and humerus. Thickenings on each side of the membranous joint capsule are called the **medial** and **lateral glenohumeral ligaments** (Fig. 30). There is a collagenous thickening across the tendon of origin of the biceps at the intertubercular groove; this is the **transverse humeral retinaculum.** The joint capsule surrounds the tendon of origin of the biceps brachii in the intertubercular groove forming its tendon sheath.

The **elbow** (Fig. 31) is a hinge joint formed by the condyle of the humerus, the head of the radius and the trochlear notch of the ulna. In addition, there is the proximal radio-ulnar articulation, although it is not weight bearing. The elbow joint capsule attaches to the articular margins; it extends distally a short distance between the radius and ulna. All compartments communicate with each other. The **lateral** and **medial collateral ligaments** are pronounced thickenings in the fibrous layer of the capsule. The biceps and brachialis tendons cover the distal portion of the medial collateral ligament. Transect the brachialis and reflect these insertions to expose the medial collateral ligament. On the lateral side dissect the origin of the lateral digital extensor to expose the lateral collateral ligament (Fig. 32). These taut ligaments prevent adduction or abduction of the elbow and restrict movement to the sagittal plane. Some rotational movement occurs at the radio-ulnar joint.

The **interosseous ligament** is a condensation of collagenous tissue, which unites the radius and ulna proximally.

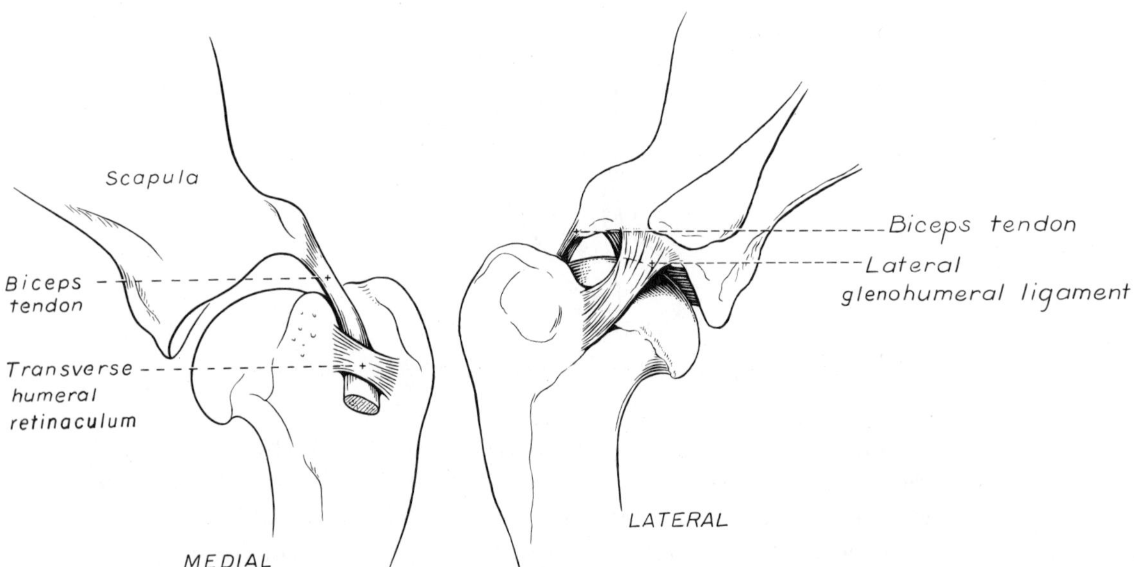

FIGURE 29. Ligaments of left shoulder, lateral and medial views.

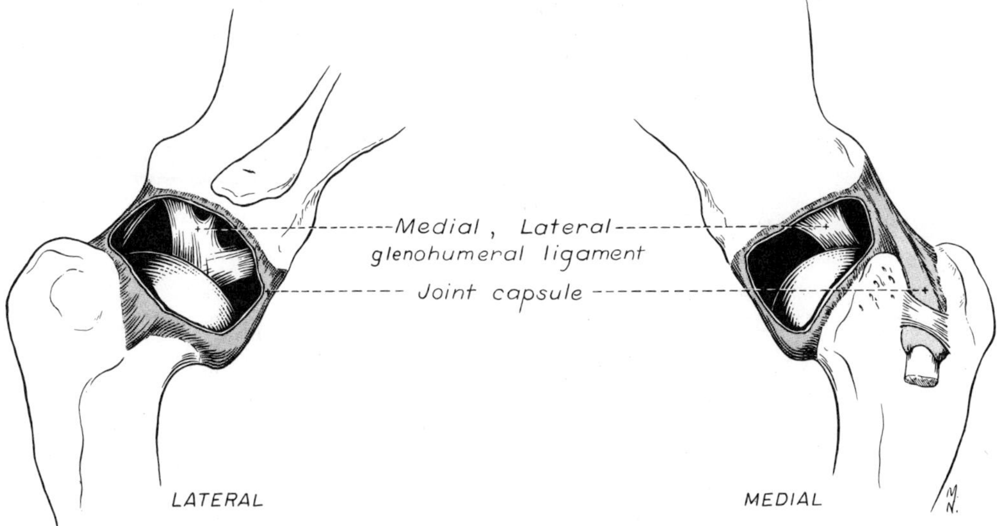

FIGURE 30. Capsule of left shoulder joint, lateral and medial views.

The carpal joint is a composite of three articular levels: 1. proximally an antebrachiocarpal joint between the radius and ulna articulating with the radial and ulnar carpal bones; 2. an accessory mediocarpal joint between the two rows of carpal bones; and 3. a carpometacarpal joint between the distal row of carpal bones and the metacarpals. The carpal joint capsule extends as a sleeve from the distal ends of the radius and ulna to the metacarpus. It attaches to the carpal bones in its course across the joint and forms separate compartments. The antebrachiocarpal joint compartment does not communicate with the middle carpal joint com-

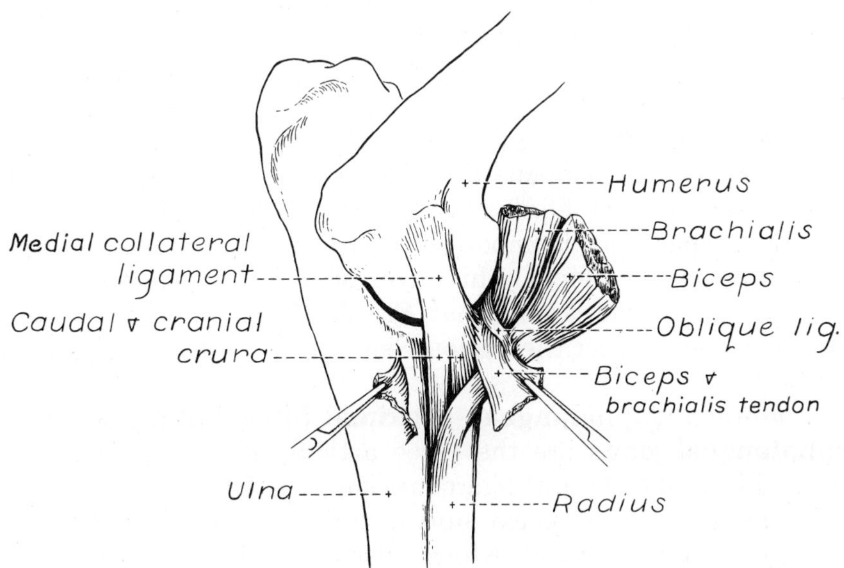

FIGURE 31. Left elbow joint, medial aspect.

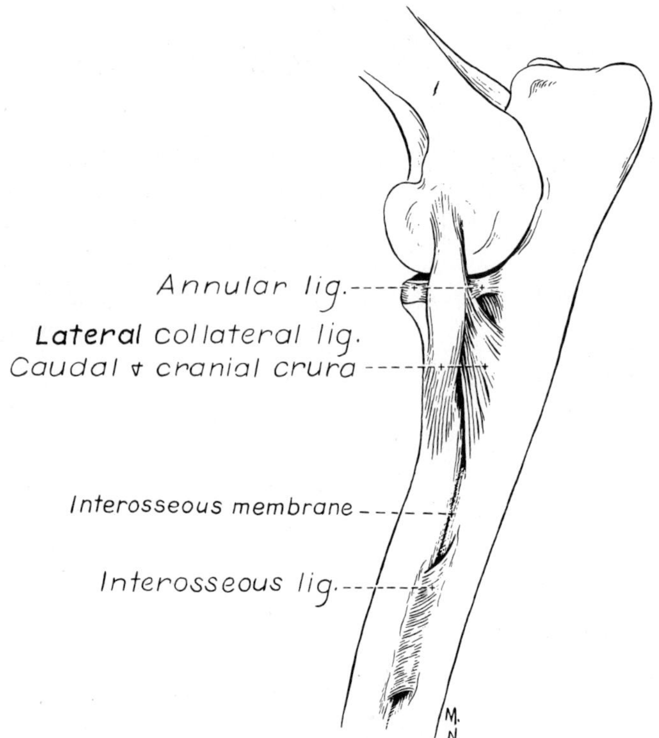

FIGURE 32. Left elbow joint, lateral aspect.

partment. The middle carpal and carpometacarpal joint compartments communicate between the distal row of carpal bones.

The joint capsule of the carpus differs from that of typical hinge joints in that both the dorsal and palmar surfaces are heavily reinforced by the fibrous layer of the joint capsule. On the dorsal surface of the joint the fibrous layer of the capsule contains grooves in which the extensor tendons lie. This layer is loose between the radius and ulna proximally and the proximal row of carpals distally, as most of the movement of the carpus takes place there. Cut through the joint capsule of this antebrachiocarpal joint to observe its components and the degree of movement of the joint. On the palmar side, the fibrous layer is thick and firmly attaches to the bones of the carpus. This is important in preventing collapse of the carpus when the limb bears weight. This layer of fibrocartilage forms the deep (dorsal) boundary of the carpal canal (Fig. 23). Numerous ligaments within the joint capsule connect the various bones of the carpus. These will not be dissected.

The **metacarpophalangeal, proximal interphalangeal** and **distal interphalangeal** joints are the three articulations of each main digit. Medial and lateral collateral ligaments support these joints.

Each metacarpophalangeal joint includes two proximal sesamoids in the tendons of the interossei, which articulate with the flexor surface of the metacarpal head.

BONES OF THE PELVIC LIMB

The **pelvic girdle,** or pelvis, of the dog consists of two hip bones, which are united at the **symphysis pelvis** midventrally and join the sacrum dorsally. Each hip bone, or **os coxae,** is formed by the fusion of three primary bones and the addition of a fourth in early life (Fig. 33). The largest and most cranial of these is the **ilium,** which articulates with the sacrum. The **ischium** is the most caudal, while the **pubis** is located ventromedial to the ilium and cranial to the large obturator foramen. The **acetabulum,** a socket, is formed where the three bones meet. It receives the head of the femur in the formation of the hip. The small **acetabular bone,** which helps form the acetabulum, is incorporated with the ilium, ischium and pubis when they fuse (about the third month).

The **pelvic canal** is short ventrally but long dorsally. Its lateral wall is composed of the ilium, ischium and pubis. Dorsolateral to the skeletal part of the wall, the pelvic canal is bounded by soft tissues. The **pelvic inlet** is limited laterally and ventrally by the arcuate line. Its dorsal boundary is the promontory of the sacrum. The **pelvic outlet** is bounded ventrally by the **ischiatic arch** (the ischiatic arch is formed by the concave caudal border of the two ischii), middorsally by the first caudal vertebra and laterally by the superficial gluteal muscle and the sacrotuberous ligament.

Os Coxae

1. The **ilium** (Figs. 33–35), a flat bone presenting two surfaces and three borders, forms the cranial half to three-fifths of the os coxae. It can be divided into a wide cranial part, which is concave laterally and is known as the **wing,** and a narrow, laterally compressed caudal part, the **body.**

The cranial border is arciform and usually roughened and is more commonly known as the **iliac crest.** It is thin but gradually increases in thickness dorsally. The angle of junction of the iliac crest with the ventral border is known as the **cranial ventral iliac spine,** which provides a place of origin for both bellies of the sartorius and a part of the tensor fasciae latae. The **tuber coxae** is composed of the cranial ventral iliac spine and the adjacent part of the ventral border of the wing of the ilium. The rest of the ventral border is concave. It ends in the **tuberosity for the rectus femoris** just cranial to the acetabulum.

The dorsal border of the ilium is broad and massive. The junction of the dorsal border with the iliac crest forms an obtuse angle that is a rounded prominence, the **cranial dorsal iliac spine.** Caudal to the cranial dorsal iliac spine is the wide but blunt **caudal dorsal iliac spine.** The two spines and intervening bone comprise the **tuber sacrale** which occupies nearly half the length of the dorsal border of the ilium. The caudal half of the dorsal border is gently concave. It forms the **greater ischiatic notch** and also helps form the ischiatic spine, which is dorsal to the acetabulum.

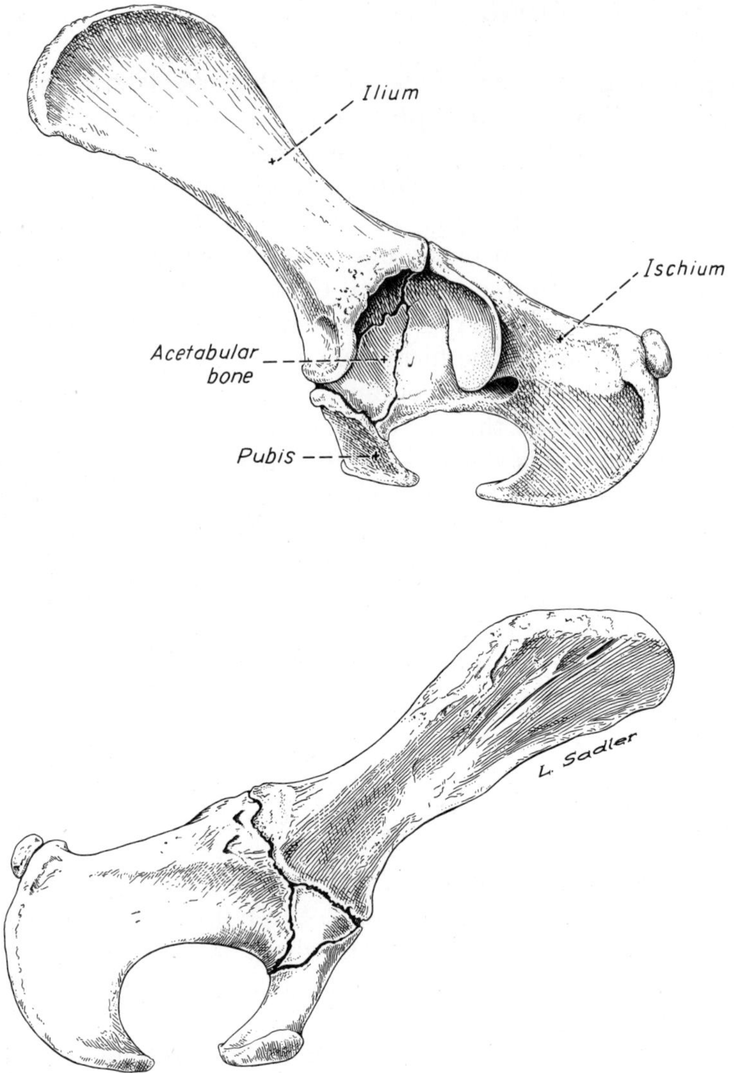

FIGURE 33. Left hip bone, 15-week-old Beagle.

The external or **gluteal surface** (Fig. 34) of the wing of the ilium is nearly flat caudally and concave cranially, where it is limited by the iliac crest. The dorsal part of this concave area is bounded by a heavy ridge. The gluteal surface is rough ventrocranially.

The internal or **sacropelvic surface** (Fig. 35) of the wing of the ilium presents a smooth, nearly flat area that provides attachment for the longissimus and the quadratus lumborum muscles. The **auricular surface** is rough and articulates with a similar surface of the sacrum, forming the sacroiliac joint. The **arcuate line** is located along the ventromedial edge of the sacropelvic surface of the body of the ilium and runs from the auricular surface to the iliopubic eminence of the pubis.

2. The **ischium** (Figs. 33–35) consists of tuberosity, body and ramus. It forms the caudal part of the os coxae and enters into the formation of the acetabulum, obturator foramen and symphysis pelvis. The **ischiatic**

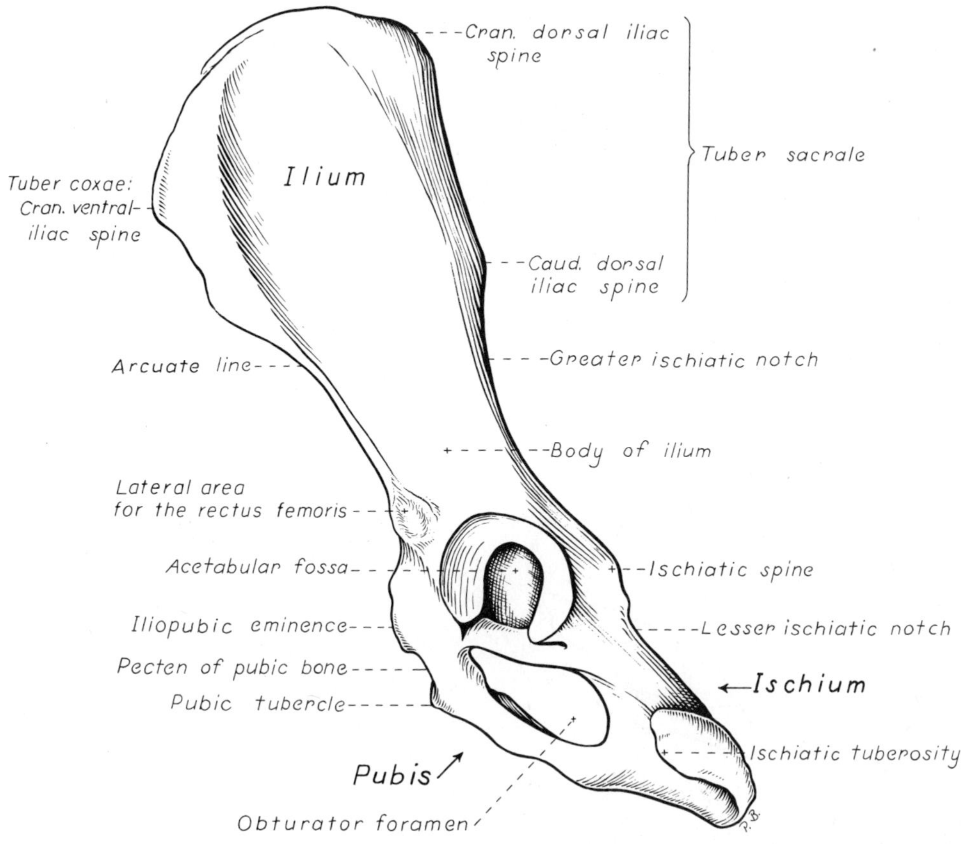

FIGURE 34. Left hip bone, lateral view.

tuberosity is the thick caudolateral margin of the bone. The lateral angle of the tuber is enlarged and hooked; it furnishes attachment for the sacrotuberous ligament. The medial angle is rounded. The ventral surface is the place of origin for the biceps femoris, semitendinosus and semimembranosus. The crus of the penis and the muscle that surrounds it also attach to the ischiatic tuberosity.

The internal obturator arises from the dorsal surface of the ischium. The ventrolateral surface of the ischium (Fig. 34) is broad and expansive caudally, where it gives rise to the quadratus femoris, the external obturator and the adductor. From the convex craniolateral part of this surface arise the gemelli.

The **ischiatic spine,** a rounded crest dorsal to the acetabulum, is continuous from the ilium along the dorsal part of the body of the ischium. It is limited caudally by a series of low ridges produced by the tendon of the internal obturator. This area is known as the **lesser ischiatic notch** (Figs. 34, 35).

The **ramus** of the ischium is thin and wide. It is bounded laterally by the obturator foramen and blends caudally with the body of the ischium. The ramus meets its fellow at the symphysis and is fused with the pubis cranially. The adductor and the external obturator arise in part from its

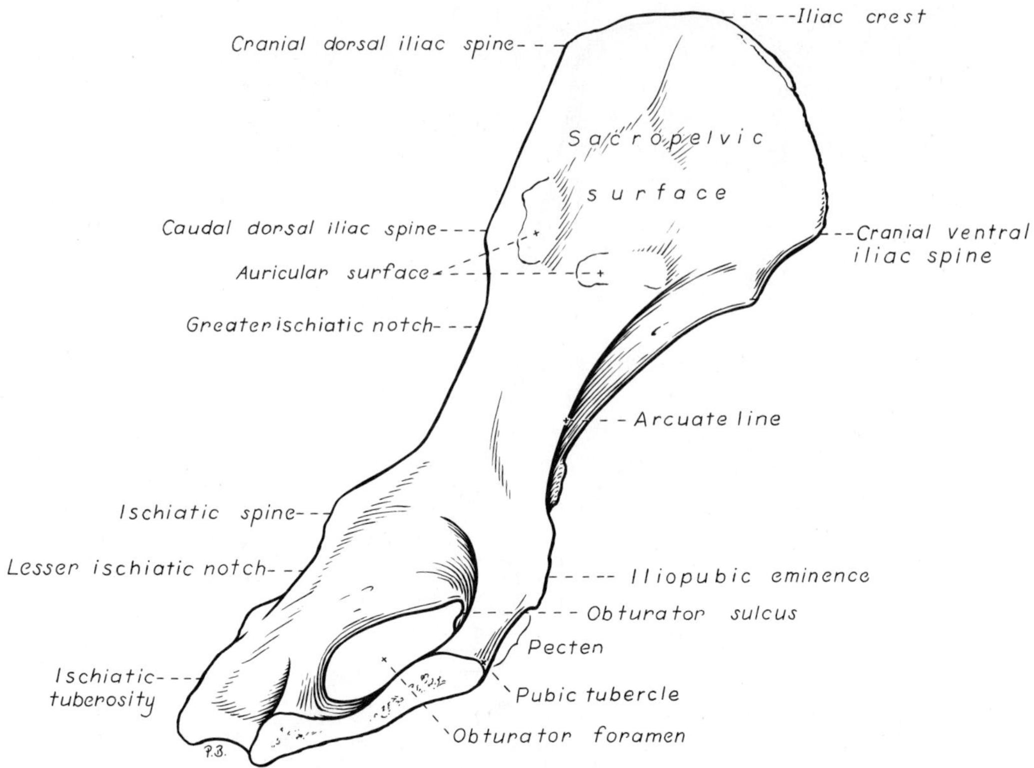

FIGURE 35. Left hip bone, medial view.

ventral surface; the internal obturator arises in part from its dorsal surface.

The **ischiatic arch** is formed by the caudal border of each ischium.

3. The **pubis** (Figs. 33–35) extends from the ilium and ischium laterally to the symphysis medially and consists of a body and two rami. The **body** is located cranial to the obturator foramen. The **cranial ramus** extends from the body to the ilium and enters into the formation of the acetabulum. The **caudal ramus** fuses with the ischium at the middle of the pelvic symphysis. The ventral surface of the pubis serves as origin for the gracilis, the adductor and the external obturator. The dorsal surface gives rise to a small part of the internal obturator and the levator ani. The **obturator sulcus,** a groove for the obturator nerve, is located at the cranial end of the obturator foramen and passes dorsally over the pelvic surface of the body of the bone. The **iliopubic eminence** projects from the cranial border of the cranial ramus of the pubis. The **pubic tubercle** projects cranially from the pubis on the midline. The roughened cranial border between the iliopubic eminence laterally and the pubic tubercle medially is the **pecten.** The pectineus arises from the iliopubic eminence. The psoas minor is inserted on the tubercle of the ilium dorsal to it.

The **acetabulum** (Figs. 33, 34) is a cavity that receives the head of the femur. Its articular surface is semilunar and is composed of parts of the ilium, ischium and pubis and, in young animals, the acetabular bone

(Fig. 33). The circumference of the articular surface is broken at the caudomedial part by the **acetabular notch.** The **acetabular fossa** is formed by the ischium and the acetabular bone. The ligament of the head of the femur attaches in this fossa. The fossa and the notch are the nonarticular parts of the acetabulum. The two sides of the notch are connected by the transverse ligament.

The **obturator foramen** is closed in life by the obturator membrane and the external and internal obturator muscles that the membrane separates.

Femur

The femur (Figs. 36, 37), or thigh bone, is the largest bone in the body. The flexor angle of the hip is about 110 degrees. The flexor angle at the stifle is from 130 to 135 degrees.

The femur is a typical long bone with a cylindrical body and two expanded extremities. The proximal extremity presents on its medial side a smooth, nearly hemispherical **head,** most of which is articular except for a small shallow fossa beginning near the middle of the head and usually extending to its caudomedial margin. This fossa is the **fovea capitis femoris,** to which the ligament of the head of the femur attaches. The head is attached to the medial part of the proximal extremity by the **neck** of the femur. The neck is distinct but short and provides attachment for the joint capsule. The **greater trochanter,** the largest eminence of the proximal extremity, is located directly lateral to the head. To it attach the middle gluteal and deep gluteal. The **trochanteric fossa** is a deep cavity medial to the greater trochanter. The gemelli and the external and internal obturators insert in this fossa. The **lesser trochanter,** a pyramidal projection at the proximal end of the medial side of the body of the femur, serves for the insertion of the iliopsoas. A ridge of bone extends from the summit of the greater trochanter to the lesser trochanter. This, the **intertrochanteric crest,** represents the caudolateral boundary of the trochanteric fossa. The quadratus femoris inserts on the crest at the level of the lesser trochanter. The **third trochanter** is poorly developed. It appears at the base of the greater trochanter as a small rough area on which the superficial gluteal inserts. The third and lesser trochanters are located in about the same transverse plane. The vastus parts of the quadriceps femoris attach to the smooth proximal part of the femur.

The **body** of the femur is slightly convex cranially. Viewed cranially, the body presents a smooth, rounded surface. The caudal surface is rough and is limited by **medial** and **lateral lips.** The lips, closest together in the middle of the body, diverge as they approach each extremity. The proximal part of the medial lip ends in the lesser trochanter, the distal part at the medial supracondylar tuberosity. The proximal part of the lateral lip ends in the third trochanter, the distal part at the lateral supracondylar tuberosity. The adductor inserts on most of the caudal rough surface, while a tendon extends from the pectineus to the distal part of the medial lip.

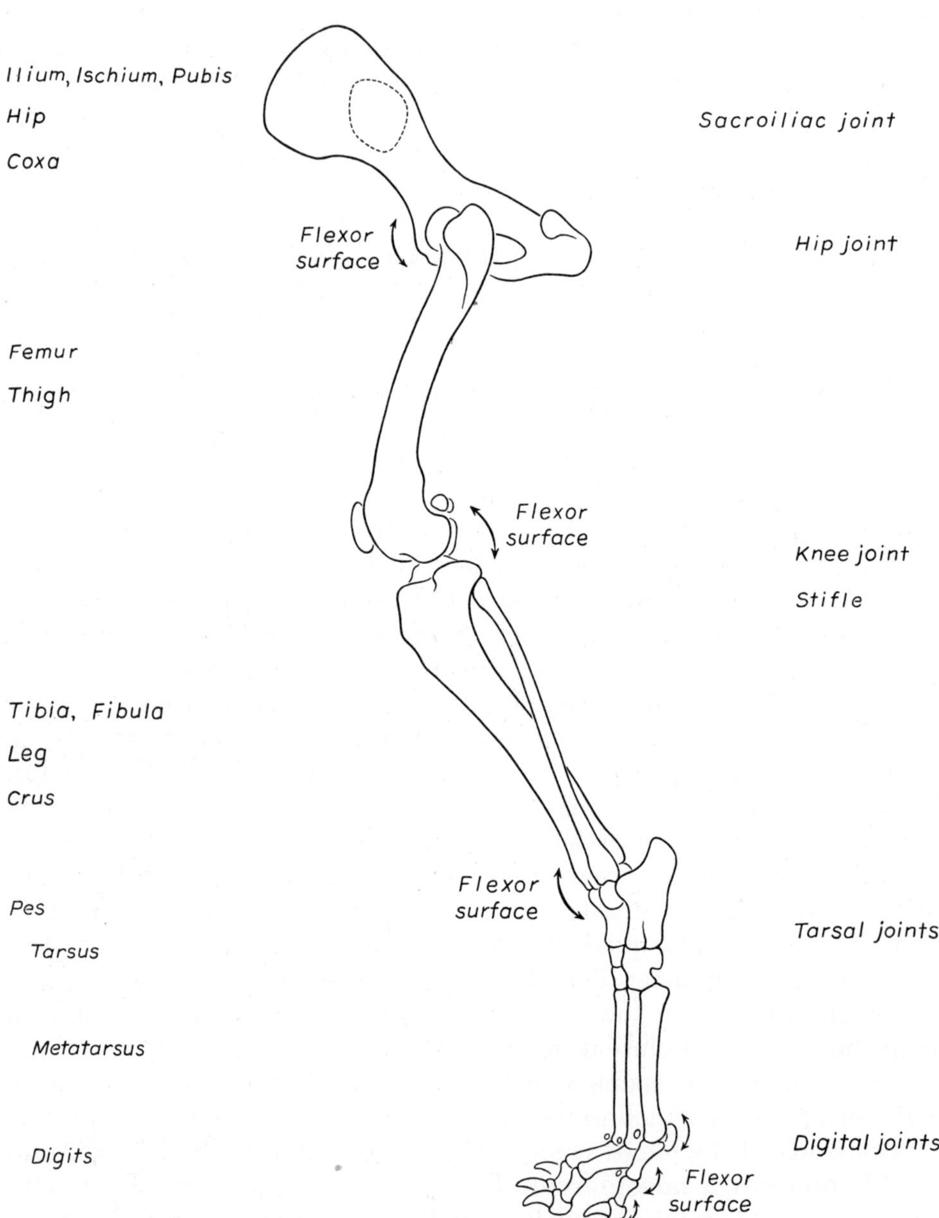

FIGURE 36. Bones of left pelvic limb.

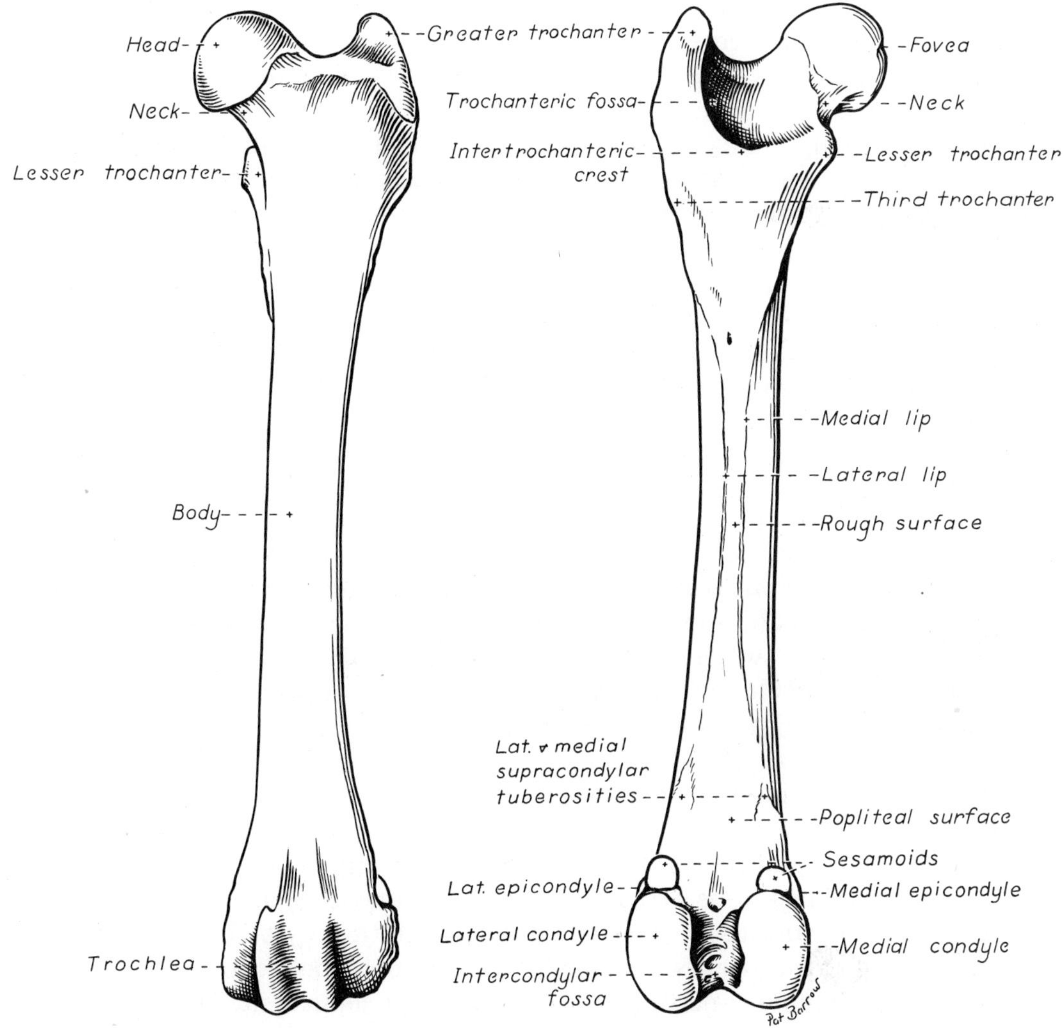

FIGURE 37a. Left femur, cranial view.

FIGURE 37b. Left femur, caudal view.

The distal extremity of the femur presents several articular surfaces. The **trochlea** is the smooth groove on the craniodistal part of the bone for articulation with the **patella.** The medial trochlear lip is usually thicker than the lateral. The patella (knee cap) is a sesamoid in the tendon of insertion of the large quadriceps femoris that extends the stifle. It aids in the protection of the tendon and the joint, but its chief purpose is redirection of the tendon of insertion of the quadriceps. The trochlea of the femur is continuous with the condyles, which articulate, both directly and through fibrocartilaginous menisci, with the tibia. The **medial** and **lateral condyles** are separated from each other by the **intercondylar fossa,** a deep wide space. The two condyles are similar in shape and surface area. Each is convex transversely and longitudinally. At the depth of the intercondylar fossa the cruciate ligaments attach. On the caudodorsal aspect of each femoral condyle is a facet upon which a sesamoid bone (fabella) rests. The medial and lateral fabellae are in the tendons of origin of the medial and lateral heads of the gastrocnemius muscles. Proximal to these sesamoid facets are the **medial** and **lateral supracondylar tuberosities** from which the gastrocnemii arise. The superficial digital flexor also arises from the lateral tuberosity. The **popliteal surface** is a large, flat, triangular area on the caudal surface of the distal extremity proximal to the condyles and intercondylar fossa. The **medial** and **lateral epicondyles** are rough areas on each side, proximal to the condyles. They serve for the attachment of the collateral ligaments of the stifle. The lateral epicondyle also gives rise to the popliteus. The small **extensor fossa** is located on the lateral epicondyle at the junction of the lateral condyle and the lateral lip of the trochlea; from it arises the long digital extensor. The semimembranosus is inserted just proximal to the medial epicondyle.

Tibia

The tibia (Fig. 38), the shin or leg bone, has a proximal articular surface that flares out transversely and is also broad craniocaudally. It is wider than the distal end of the femur, with which it articulates, and is formed largely by two condyles. The **medial condyle** is separated from the **lateral condyle** by the **intercondylar eminence.** Both condyles include the articular areas on their proximal surfaces and the adjacent nonarticular parts of the proximal extremity. The lateral condyle is particularly prominent. It possesses a facet on its lateral side for articulation with the head of the fibula and provides origin for part of the peroneus longus and cranial tibial muscles. A sesamoid bone in the tendon of origin of the popliteus articulates with the lateral condyle of the tibia. The semimembranosus is inserted on the medial condyle. Two biconcave fibrocartilages, the **menisci,** fill part of the space between the apposed condyles of the femur and tibia, making the joint congruent. The intercondylar eminence consists of two small, elongated tubercles, which form its highest part, and a shallow intermediate fossa. The **cranial intercondylar area** is a depression cranial to the eminence and in large part between the condyles. It affords attachment to the cranial parts of the menisci and the cranial cruciate ligament. The **caudal intercondylar area** occupies a place similar to that of the cranial area but caudal to the eminence. It provides attachment for the caudal parts of the menisci and the caudal cruciate ligament. The **popliteal notch** is caudal to the caudal intercondylar area and is located between the two condyles. The popliteal vessels pass through the notch. The **tibial tuberosity** is the large quadrangular process on the proximocranial surface of the tibia. The quadriceps femoris, the biceps femoris and the sartorius attach to this tuberosity by means of the patella and patellar ligament. The tibial tuberosity is continued distally by the **cranial border of the tibia.** It inclines laterally on the body. The following muscles attach wholly or in part to the cranial border of the tibia: biceps femoris, semitendinosus, gracilis and sartorius. The **extensor groove** is a small, smooth groove located at the junction of the lateral condyle and the tibial tuberosity. The long digital extensor passes through it.

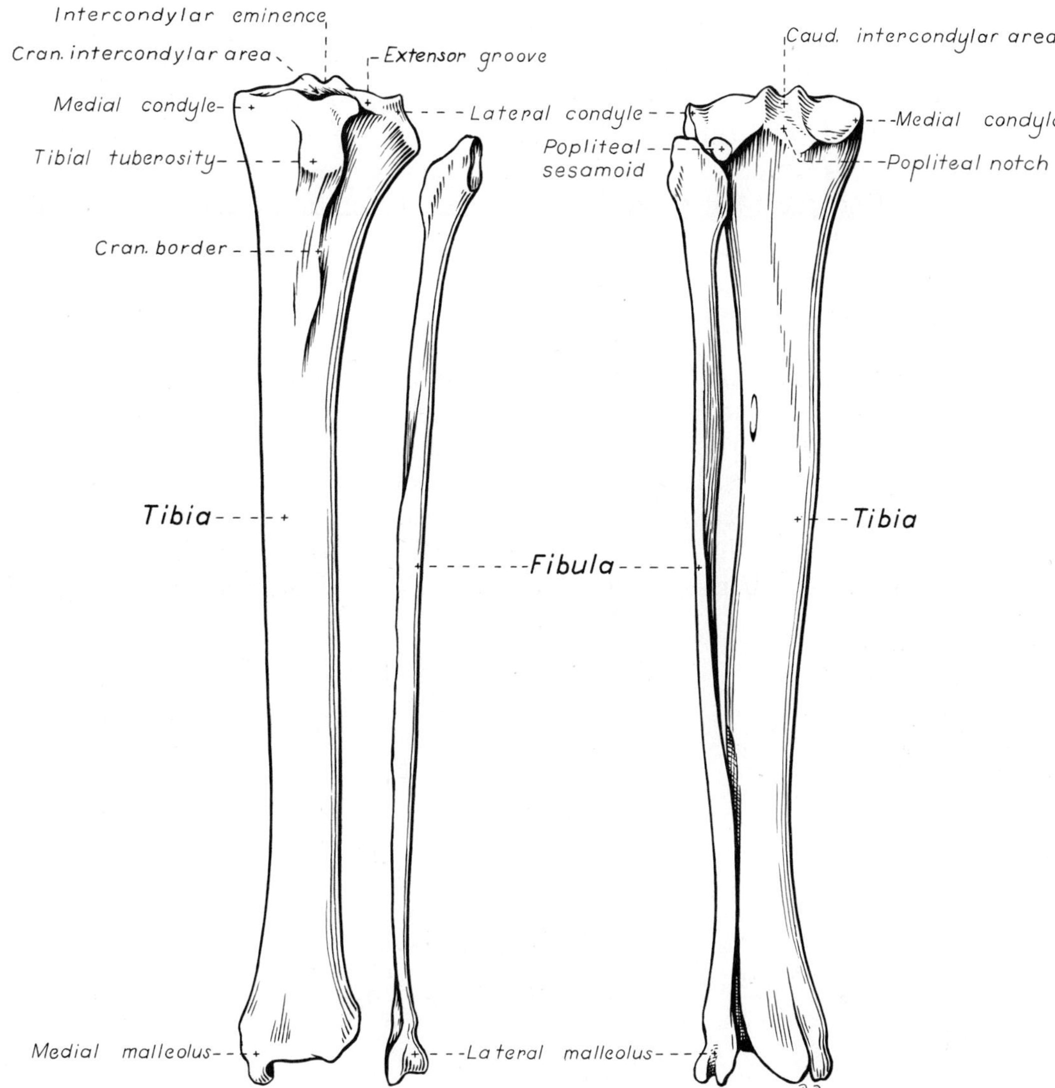

FIGURE 38a. Left tibia and fibula, cranial view.

FIGURE 38b. Articulated left tibia and fibula, caudal view.

The **body** is triangular proximally, nearly cylindrical in the middle and four-sided distally. The semitendinosus and gracilis are inserted on the proximal medial surface. The proximal third of the caudal surface serves for the insertion of the popliteus medially and for the origins of the deep digital flexor laterally.

The distal extremity of the tibia is quadrilateral in transverse section. The **tibial cochlea,** the articular surface, consists of two grooves that receive the ridges of the proximal trochlea of the talus. The medial part of the distal extremity of the tibia is the **medial malleolus.** The lateral surface of the distal extremity articulates with the fibula by a small facet. No muscles arise from the distal half of the tibia.

Fibula

The fibula (Fig. 38) has proximal and distal extremities and an intermediate body. The proximal extremity, or **head,** articulates with the lateral condyle of the tibia. The distal extremity, **lateral malleolus,** has two grooves which redirect the force of contraction. On its medial surface is a distinct facet for articulation with the distolateral surface of the tibia and with the talus.

Tarsal Bones

The **tarsus** (Fig. 39), between the metatarsus and the leg, is composed of seven tarsal bones and the related soft tissues. It is also called the hock. The bones are arranged in three irregular rows. The proximal row is composed of a long, laterally located **calcaneus** and a shorter, medially located **talus.** The tibia and fibula articulate with the talus. The calcaneus articulates with the talus and the fourth tarsal bone. The distal row consists of four bones. Three small bones, the **first, second** and **third**

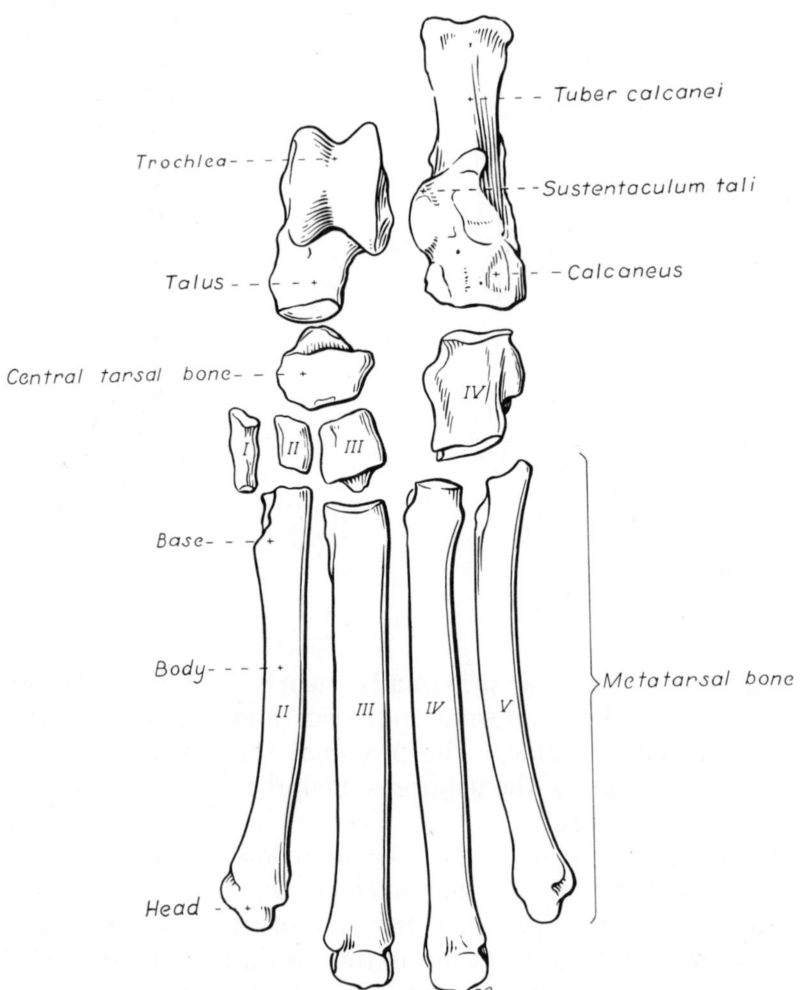

FIGURE 39. Left tarsal and metatarsal bones, dorsal view.

tarsal bones, are located side by side and are separated from the proximal row by the **central tarsal** bone. The large **fourth tarsal** bone, which completes the distal row laterally, articulates with the calcaneus proximally. The fourth tarsal is as long as the combined lengths of the third and central tarsal bones against which it lies.

The **talus** has a **trochlea** on its proximal end for articulation with the tibial cochlea. This is the tarsocrural joint where flexion and extension occur between the leg and hind paw. The **tuber calcanei** is a traction process of the calcaneus that projects proximally and caudally. The extensor muscles of the hock insert on this process via the calcanean tendon. On the medial side of the calcaneus is a bony process, the **sustentaculum tali.** The tendon of the lateral digital flexor glides over the plantar surface of this process. The fourth tarsal bone is grooved on the distal half of its lateral surface for the passage of the tendon of the peroneus longus.

Metatarsal Bones

The metatarsal bones (Fig. 39) resemble the metacarpal bones except for the first, which may be divided, rudimentary or absent.

Phalanges

The phalanges (Fig. 40) and sesamoids form the skeleton of the digit. Those of the hind paw are similar to those of the forepaw.

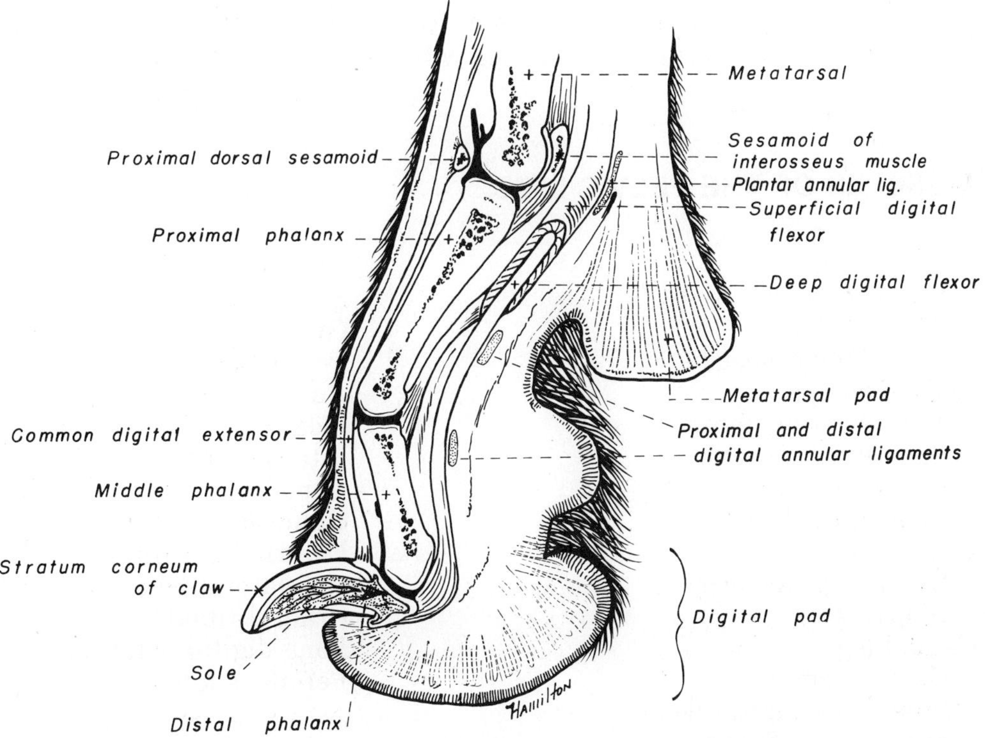

FIGURE 40. Median section of the third digit of the hind paw.

The first digit, or **hallux,** is frequently absent. When present, it is called a dewclaw and may vary from a fully developed digit articulating with a normal first metatarsal bone to a vestigial structure composed only of a terminal phalanx.

Live Dog

Palpate the iliac crests beside the sacrum at the cranial aspect of the pelvis, the tuber ischii caudolaterally and the ischiatic arch between these tuberosities. Note their symmetry. Pelvic fractures and sacroiliac luxation are common and result in palpable asymmetry.

Palpate the greater trochanter of the femur lateral to the hip. Place your thumb in the groove between the greater trochanter and the ischial tuberosity. Rotate the thigh laterally and note the displacement of your thumb by the greater trochanter. This will not occur when there is a hip luxation.

At the distal end of the femur palpate the trochlea and the edges of the patella that articulate with it. Palpate the condyles of the femur and tibia. Flex, extend and rotate the stifle to appreciate the level of this articulation. The tibial tuberosity and cranial border are palpable on the proximal tibia. Feel the body of the tibia where it is subcutaneous on the medial side.

Palpate the narrow distal tibia and fibula. Flex and extend the tarsocrural joint and feel the edges of the trochlea of the talus that participate in this joint. Palpate the tuber calcanei and note how a proximal pull on this lever will extend the joint. A fracture of this will cause the joint to overflex and the dog will walk partly plantigrade.

MUSCLES OF THE PELVIC LIMB

Caudal thigh:
- Biceps femoris
- Semitendinosus
- Semimembranosus

Lateral rump:
- Tensor fasciae latae
- Superficial gluteal
- Middle gluteal
- Deep gluteal

Cranial thigh:
- Quadriceps femoris
- Iliopsoas

Caudal leg:
- Gastrocnemius
- Superficial digital flexor
- Deep digital flexors
- Popliteus

Medial thigh:
- Sartorius
- Gracilis
- Pectineus
- Adductor

Caudal hip:
- Internal obturator
- Gemelli
- Quadratus femoris
- External obturator

Craniolateral leg:
- Cranial tibial
- Long digital extensor
- Peroneus longus

Remove the skin from the caudal part of the left half of the trunk, the rump and the thigh. Continue the midventral incision from the umbilicus to the root of the tail. In making this incision, closely circle the external genital parts and the anus. Extend an incision distally on the medial surface of the left thigh to the tarsus. Encircle the tarsus with a skin incision. First reflect the skin from the medial surface of the thigh and leg and then, starting at the tarsus, reflect the whole flap of skin from the lateral surface of the leg, stifle, thigh, rump and abdomen to the middorsal line. The cutaneus trunci may be removed with the skin since it is more intimately attached to the skin than to the underlying structures.

There are superficial and deep fasciae of the pelvic limb. They cannot always be separated, and, in general, the deep one is more dense.

The **superficial fascia of the trunk** (Fig. 41) continues dorsally on the rump as the **superficial gluteal fascia.** In obese specimens much fat is present between this fascia and the deep fascia. The cutaneus trunci arises as an aponeurosis from the superficial gluteal fascia. It may be examined in the skin just reflected and in the previous dissection. Its fibers run cranioventrally to the caudal part of the axilla. Its most ventral border lies in the fold of the flank; dorsally it is separated from its fellow by a narrow band of fascia over the loin and rump. The superficial gluteal fascia passes to the tail as the **superficial caudal fascia** and continues distally on the limb as the superficial fascia of its respective parts. Remove the fat-laden areolar tissue and the superficial fascia covering the caudal part of the trunk and rump. Do not cut into the glistening deep fascia underneath.

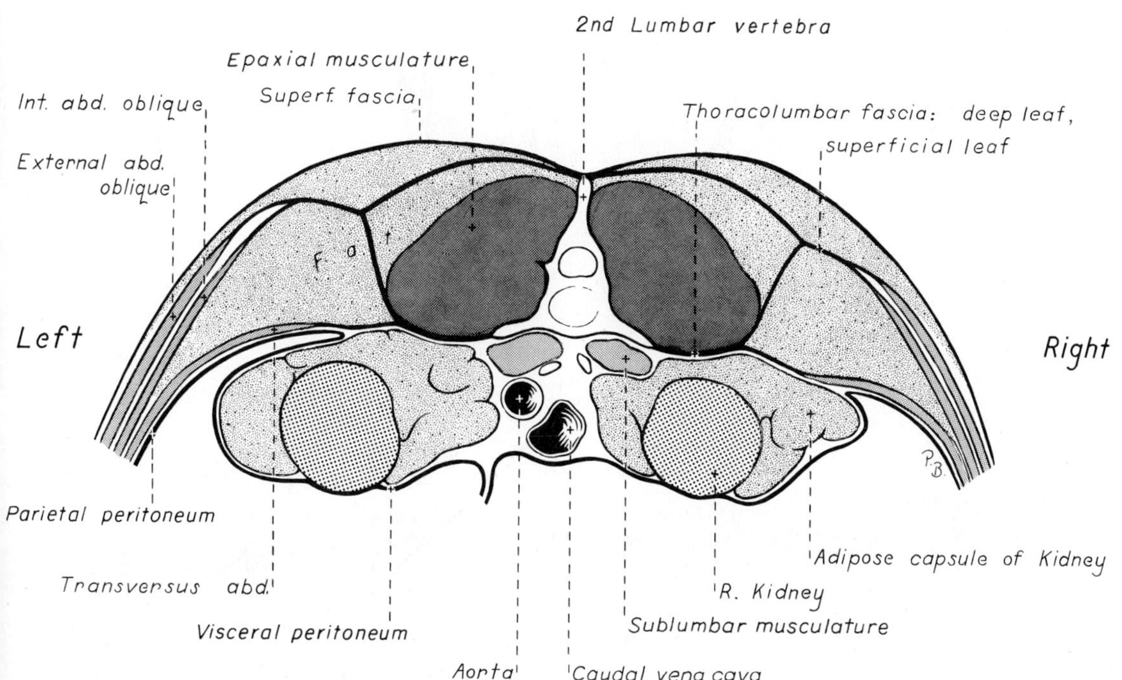

FIGURE 41. Schematic transection in the lumbar region showing fascial layers.

A thick deep fascia covers the dorsal muscles of the loin, rump and tail. On the dorsocaudal part of the trunk this is the **thoracolumbar fascia,** a region of the **deep fascia of the trunk** (Fig. 41). It is continued caudally at the iliac crest by the **deep gluteal fascia.** This distinct glistening fascia covers the muscles of the rump and serves in part as origin for the middle and superficial gluteals. The deep gluteal fascia is continued caudally on the tail by the **deep caudal fascia.** The latter follows the irregularities of the caudal muscles and closely binds these to the vertebrae of the tail. Distally the deep gluteal fascia blends with the fascia of the thigh, where it becomes the **medial** and **lateral femoral fasciae.** The medial fascia is thin, whereas the lateral femoral fascia, the **fascia lata,** is thick and serves as an aponeurotic insertion for thigh

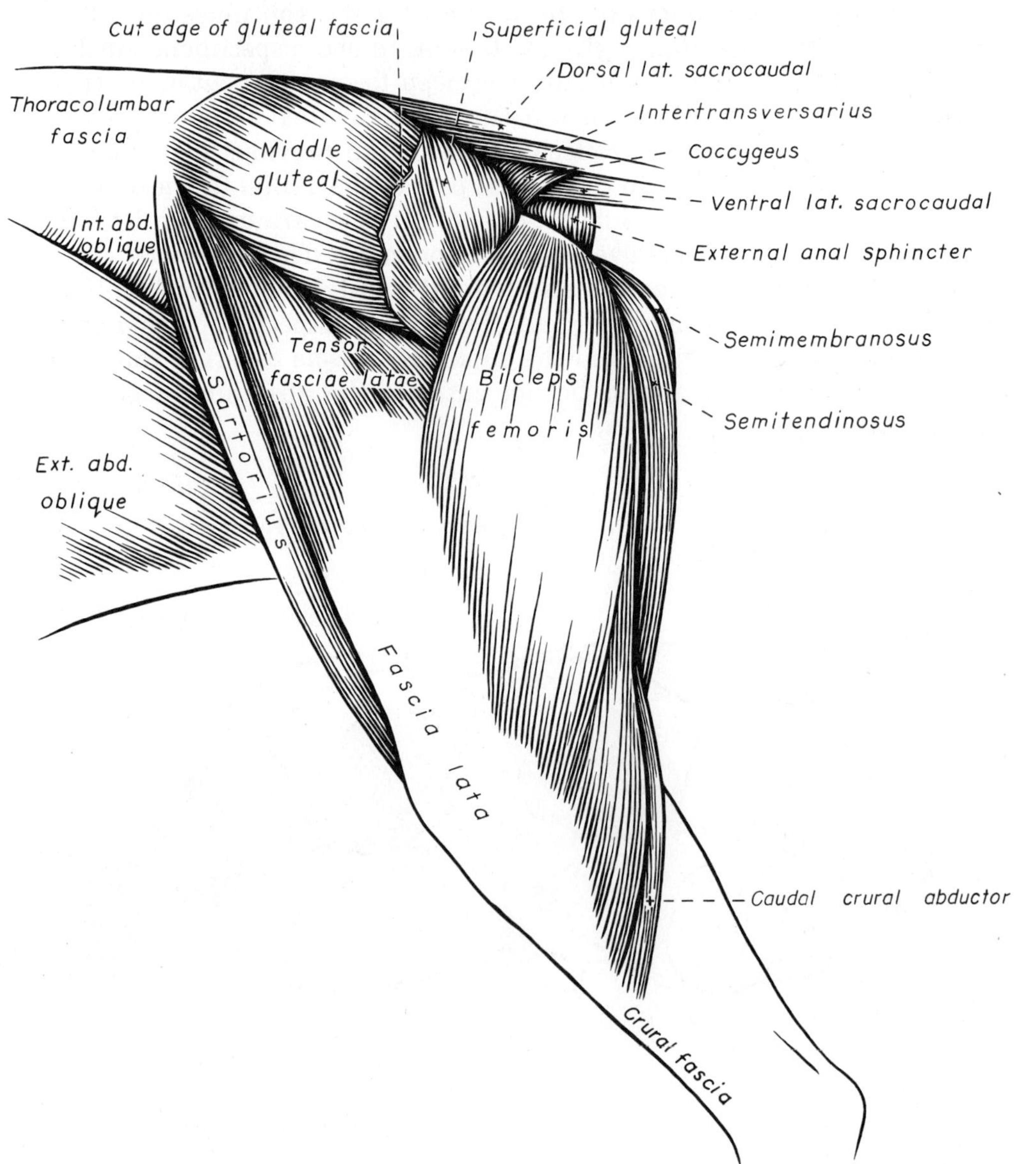

FIGURE 42. Superficial muscles of left pelvic limb, lateral view.

muscles. The femoral fascia is continued on the leg as the **crural fascia.** Clean but do not cut any of the deep fascia until instructed to do so.

Caudal Muscles of the Thigh

This group consists of three primary muscles: biceps femoris, laterally; semitendinosus, caudally; and semimembranosus, medially. (A slender caudal crural abductor muscle is closely associated with the mediocaudal surface of the biceps [Figs. 42, 44].)

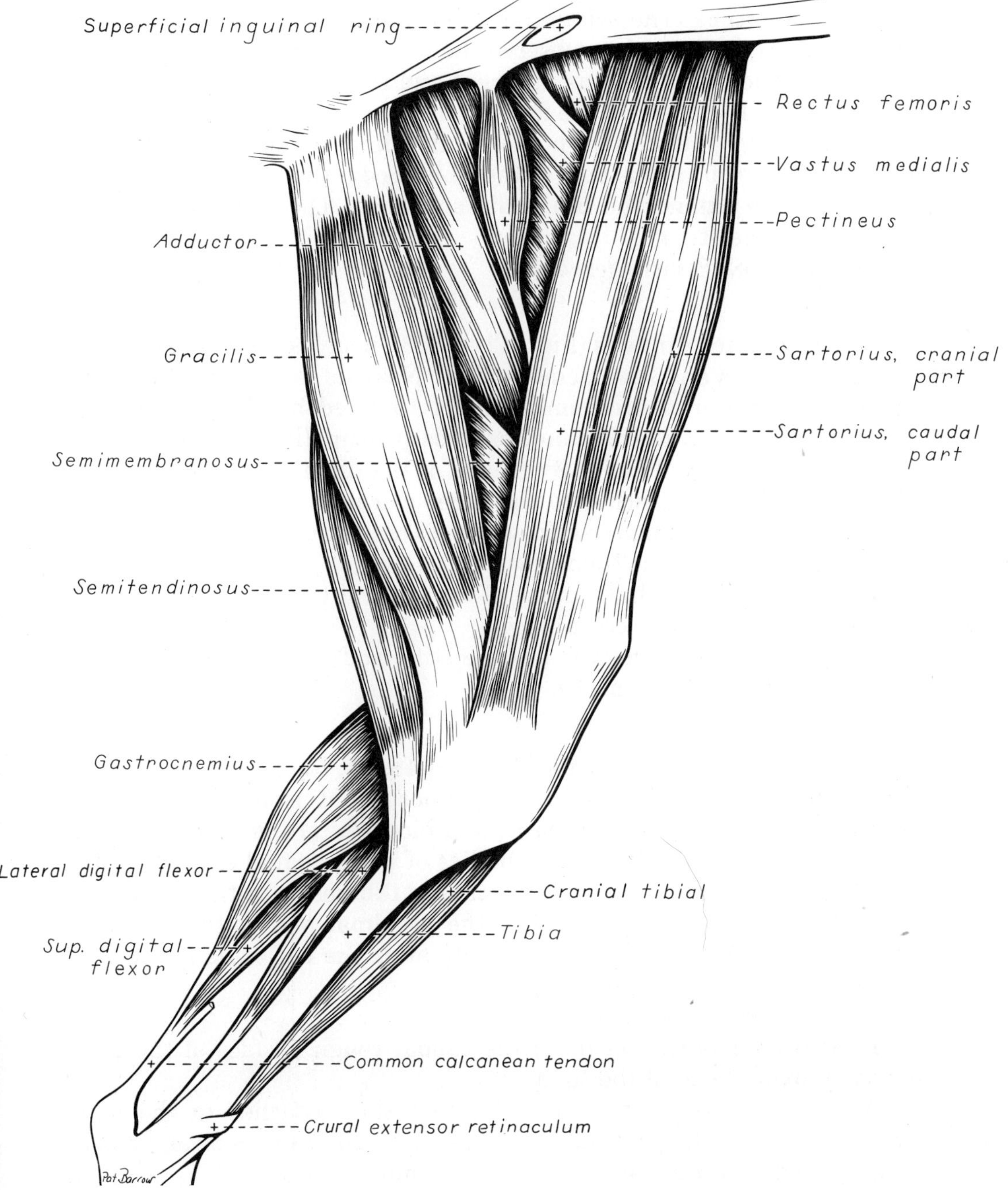

FIGURE 43. Superficial muscles of left pelvic limb, medial view.

1. The **biceps femoris** (Figs. 42, 45, 48, 53, 57) is the longest and widest of the muscles of the thigh. The bulk of its fibers runs craniodistally, although caudally there are fibers that run directly distally. Cranially it inserts by means of the fascia lata and crural fascia. Carefully clean the caudal border and the adjacent surface of the muscle. The lymph node lying in the fat at its caudal border directly caudal to the stifle is the **popliteal lymph node.** Do not cut the cranially lying fascia lata, which serves as part of its insertion. Caudally a strand of heavy fascia runs to the tuber calcanei and helps to form the **common calcanean tendon.** Transect the biceps. The sciatic nerve is interposed between the biceps and the underlying muscles. Observe the insertions of the biceps femoris.

ORIGIN: The sacrotuberous ligament and the ischiatic tuberosity.

INSERTION: By means of the fascia lata and crural fascia to the patella, patellar ligament and cranial border of the tibia; by means of the crural fascia to the subcutaneous part of the tibial body; the tuber calcanei.

ACTION: To extend the hip, stifle and hock. The caudal part of the muscle flexes the stifle.

INNERVATION: Sciatic nerve.

2. The **semitendinosus** (Figs. 42–48, 53), near its origin, lies between the biceps femoris and semimembranosus. Near its insertion it lies on the medial head of the gastrocnemius and is covered by the gracilis. It is nearly as wide as it is thick and extends principally from the ischiatic tuberosity to the tibial body. By means of the distal continuation of the crural fascia it also attaches to the tuber calcanei. Completely free this muscle from surrounding structures but do not transect it.

ORIGIN: The ischiatic tuberosity.

INSERTION: The medial surface of the body of the tibia and the tuber calcanei by means of the crural fascia.

ACTION: To extend the hip, flex the stifle and extend the hock.

INNERVATION: Sciatic nerve.

3. The **semimembranosus** (Figs. 42–48, 51, 53) is greater in transverse-sectional area than the semitendinosus but is not as long. It is wedged between the semitendinosus and biceps laterally and the gracilis and adductor medially. It has two bellies of nearly equal size. This short but thick muscle extends from the ischiatic tuberosity to the medial side of the distal end of the femur and the proximal end of the tibia. The insertions may be seen more adequately after the sartorius and gracilis muscles have been dissected.

ORIGIN: The ischiatic tuberosity.

INSERTION: The medial lip of the caudal rough surface of the femur and the proximal end of the tibia.

ACTION: To extend the hip. The part that attaches to the femur extends the stifle; the part that attaches to the tibia flexes or extends the stifle, depending upon the position of the limb.

INNERVATION: Sciatic nerve.

Text continued on page 75

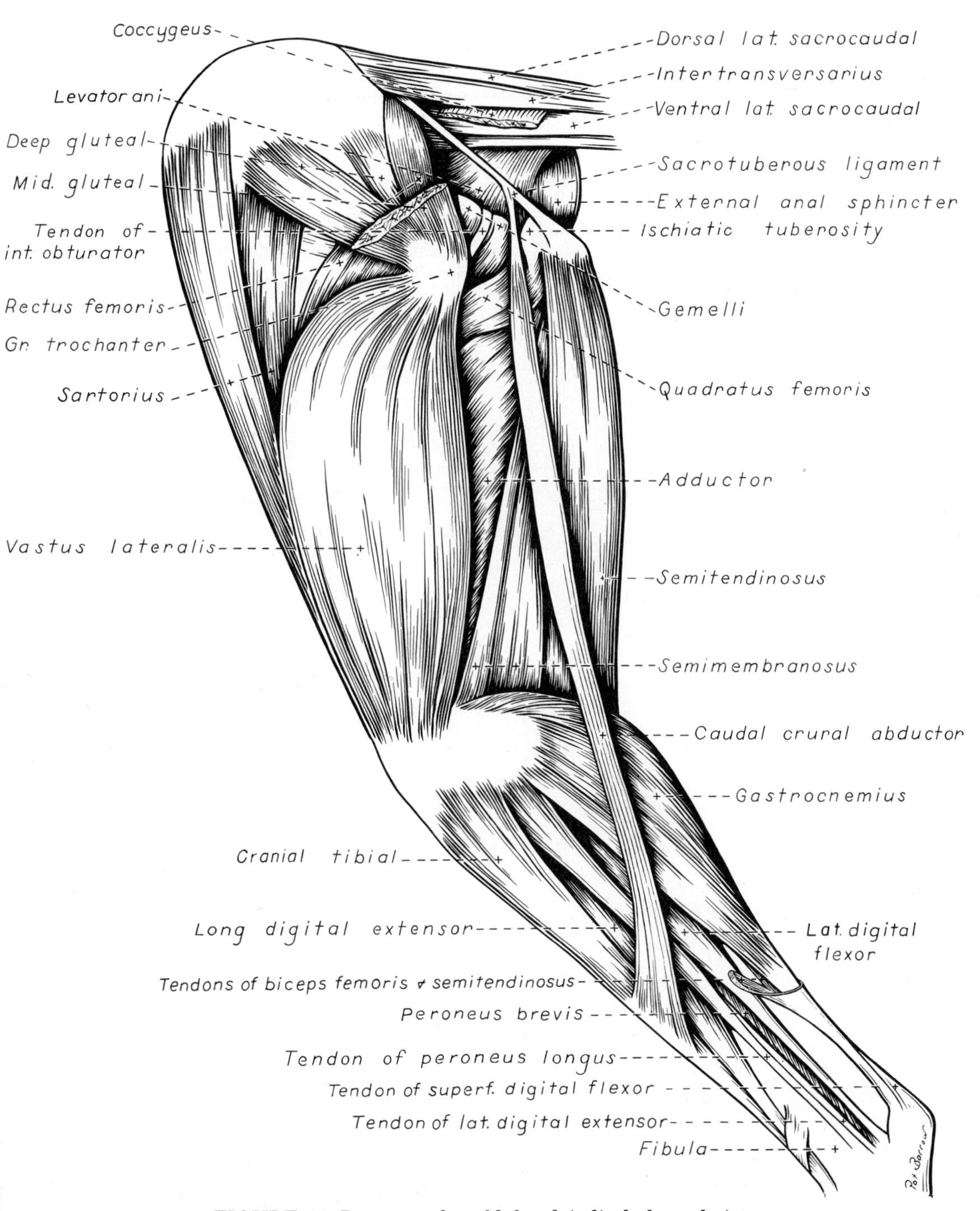

FIGURE 44. Deep muscles of left pelvic limb, lateral view.

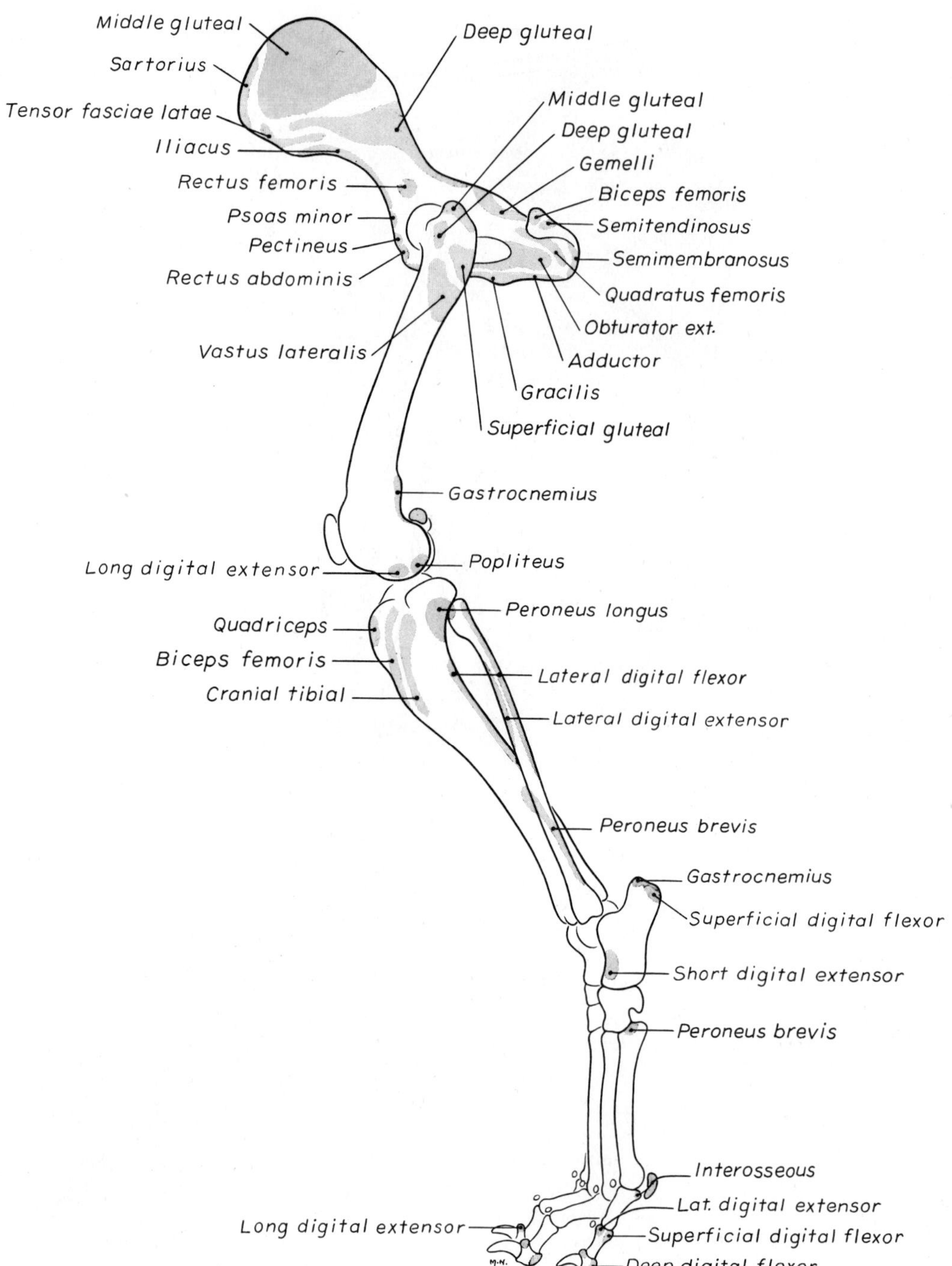

FIGURE 45. Muscle attachments on pelvis and left pelvic limb, lateral view.

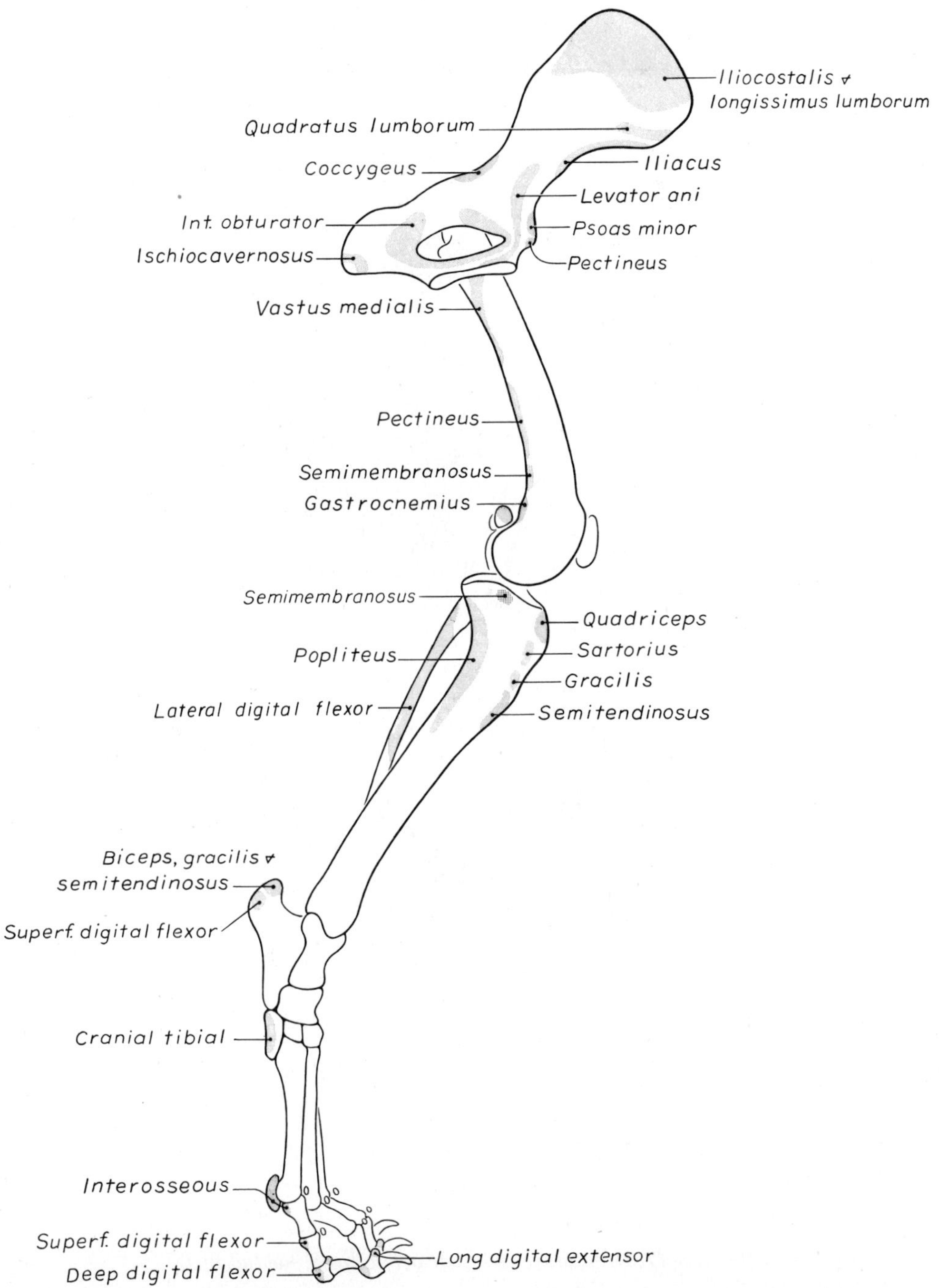

FIGURE 46. Muscle attachments on pelvis and left pelvic limb, medial view.

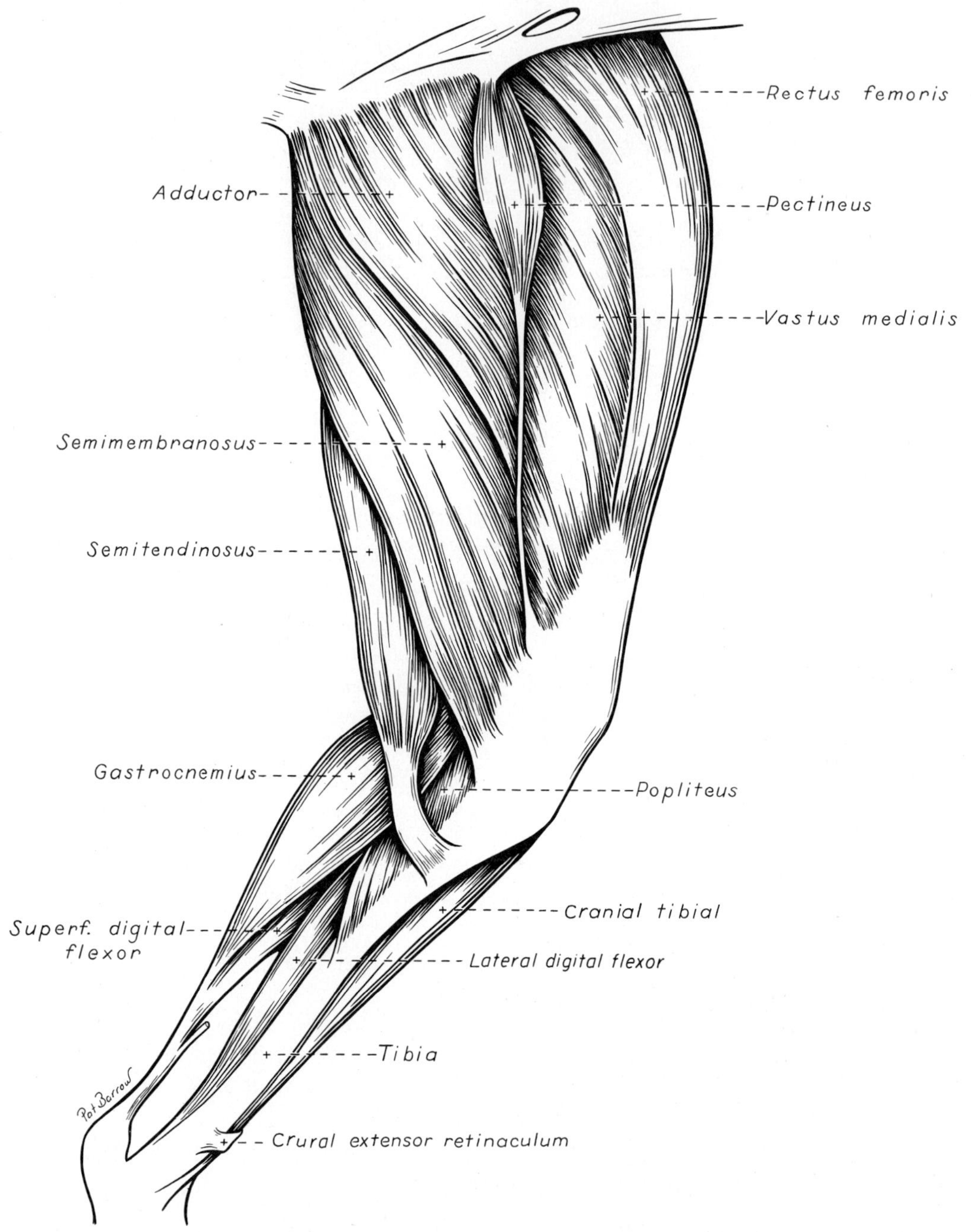

FIGURE 47. Deep muscles of left pelvic limb, medial view.

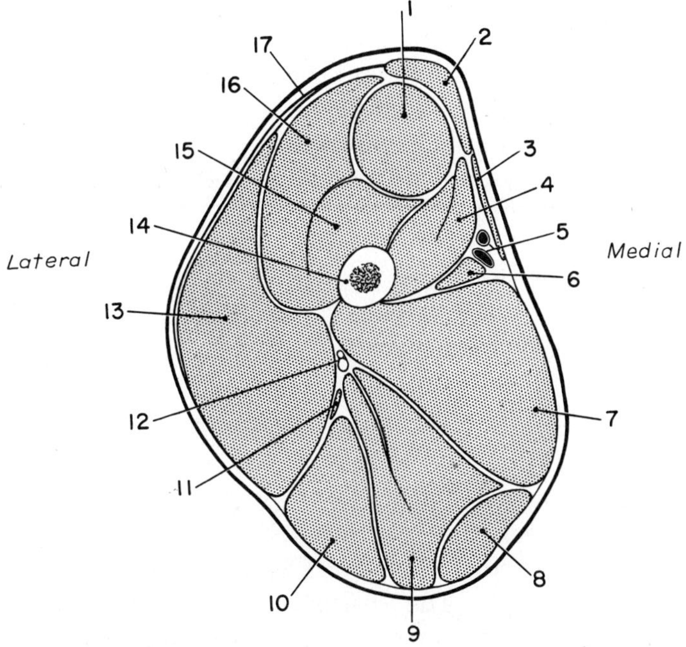

FIGURE 48. Transverse section of left thigh.

1. *Rectus femoris*
2. *Sartorius, cranial part*
3. *Sartorius, caudal part*
4. *Vastus medialis*
5. *Femoral a. and v.*
6. *Pectineus*
7. *Adductor*
8. *Gracilis*
9. *Semimembranous*
10. *Semitendinosus*
11. *Caudal crural abductor*
12. *Sciatic n.*
13. *Biceps femoris*
14. *Femur*
15. *Vastus intermedius*
16. *Vastus lateralis*
17. *Fascia lata*

Medial Muscles of the Thigh

1. The **sartorius** (Figs. 42–46, 48, 53) consists of two straplike parts that lie on the cranial and craniomedial surfaces of the thigh. These parts extend from the ilium to the tibia. The **cranial part** forms the cranial contour of the thigh and may be nearly a centimeter thick there. The **caudal part** is on the medial side of the thigh and is thinner, wider and longer than the cranial part. Both muscle parts lie predominantly on the medial side of the large quadriceps femoris. Transect both parts and reflect the distal parts to their insertions.

ORIGIN: Cranial part—the crest of the ilium and the thoracolumbar fascia; caudal part—the cranial ventral iliac spine and the adjacent ventral border of the ilium.

INSERTION: Cranial part—the patella, in common with the rectus femoris of the quadriceps; caudal part—the cranial border of the tibia, in common with the gracilis.

ACTION: To flex the hip. The cranial part extends the stifle; the caudal part flexes the stifle.

INNERVATION: Femoral nerve.

2. The **gracilis** (Figs. 43, 45, 46, 48, 53) arises from the **symphysial tendon,** a thick, flat tendon attached ventrally to the symphysis pelvis. The aponeurosis of the gracilis covers the adductor. Transect the gracilis through its aponeurotic origin. Reflect it distally and observe its insertion as well as that of the semimembranosus.

ORIGIN: The pelvic symphysis by means of the symphysial tendon.

INSERTION: The cranial border of the tibia and, with the semitendinosus, the tuber calcanei.

ACTION: To adduct the limb, flex the stifle and extend the hip and hock.

INNERVATION: Obturator nerve.

The **femoral triangle** is the shallow triangular space through which the femoral vessels run to and from the pelvic limb. It is located on the proximal medial surface of the thigh with its base at the abdominal wall. The triangle lies between the caudal belly of the sartorius cranially and the pectineus and adductor caudally. The iliopsoas forms the proximal lateral part of the triangle. The vastus medialis forms the distal lateral part. This triangle contains, among other structures, the femoral artery and vein. Remove the medial femoral fascia and adipose tissue that covers and fills in around the femoral artery and vein. Notice that the vein lies caudal to the artery. The pulse is usually taken from the femoral artery here.

3. The **pectineus** (Figs. 43, 45–48, 50, 51) is a small, spindle-shaped muscle that, with the adductor, belongs to the deep medial muscles of the thigh. It lies in large part between the adductor caudally and the vastus medialis cranially. It arises from the iliopubic eminence and the **cranial pubic ligament.** The latter is composed of transverse fibers that connect the pecten of the pubis on one side with that of the other side. The tendon of insertion of the pectineus lies interposed between the adductor and vastus medialis. By blunt dissection with the handle of the scalpel isolate the tendon of insertion. It inserts on the caudomedial surface of the distal end of the femur. Transect the pectineus in the middle of its belly.

ORIGIN: The cranial pubic ligament and the iliopubic eminence.

INSERTION: The distal end of the medial lip of the caudal rough face of the femur.

ACTION: To adduct the limb.

INNERVATION: Obturator nerve.

4. The **adductor** (Figs. 43–45, 47, 48, 50, 51) consists of two muscles (adductor magnus et brevis and adductor longus) that are often not clearly divisible. It is a large pyramidal muscle compressed between the semimembranosus and pectineus that extends from the pelvic symphysis to the caudal aspect of the femur. It is partly covered by the biceps femoris laterally and gracilis medially. Transect the adductor at its origin along the symphysis. Do not transect the external obturator which lies even deeper.

ORIGIN: The entire pelvic symphysis by means of the symphysial tendon, the adjacent part of the ischiatic arch and ventral surface of the pubis and ischium.

INSERTION: The entire lateral lip of the caudal rough face of the femur.

ACTION: To adduct the limb and extend the hip.

INNERVATION: Obturator nerve.

Lateral Muscles of the Rump

1. The **tensor fasciae latae** (Figs. 42, 45) is a triangular muscle that attaches proximally to the tuber coxae. It lies between the sartorius cranially, the middle gluteal caudodorsally and the quadriceps distomedially. Part of its caudodorsal surface is attached to the middle gluteal near its origin. The muscle can be divided into two portions. The cranial, more superficial portion is inserted on the lateral femoral fascia, which radiates over the quadriceps and blends with the fascial insertion of the biceps femoris. The deeper caudal portion is inserted on a layer of lateral femoral fascia that runs deep to the biceps toward the stifle on the lateral surface of the vastus lateralis. Transect the tensor fasciae latae across its middle and reflect the two halves toward their attachments.

ORIGIN: The tuber coxae and adjacent part of the ilium; the aponeurosis of the middle gluteal muscle.

INSERTION: The lateral femoral fascia.

ACTION: To tense the lateral femoral fascia, flex the hip and extend the stifle.

INNERVATION: Cranial gluteal nerve.

2. The **superficial gluteal** (Figs. 42, 45, 51) is small and lies caudal to the middle gluteal. Its fibers run distally, from the deep gluteal fascia that covers the middle gluteal and from the sacrum and the first caudal vertebra to the level of greater trochanter of the femur, where they converge before forming an aponeurosis that runs under the biceps to the third trochanter. Clean the superficial gluteal and transect it 1 centimeter from the beginning of its aponeurosis of insertion. Do not transect the caudally lying **sacrotuberous ligament.** This is a collagenous band that runs from the sacrum to the lateral angle of the ischiatic tuberosity. Notice that the superficial gluteal arises from the proximal half of this ligament. Sever the deep gluteal fascia 1 centimeter cranial to its junction with the muscle fibers of the superficial gluteal.

ORIGIN: The lateral border of the sacrum and the first caudal vertebra, partly by means of the sacrotuberous ligament; the cranial dorsal iliac spine by means of the deep gluteal fascia.

INSERTION: The third trochanter.

ACTION: To extend the hip and abduct the limb.

INNERVATION: Caudal gluteal nerve.

3. The **middle gluteal** (Figs. 42, 44, 45, 51) is a large, ovoid muscle that lies between the tensor fasciae latae and the superficial gluteal. Clean the surface of the muscle by reflecting the deep gluteal fascia

dorsocranial to the iliac crest. This reveals that the fibers of the muscle for the most part parallel its long axis. Carefully separate the cranioventral part of the middle gluteal from the underlying deep gluteal. The entire caudodorsal border of the middle gluteal is covered by the superficial gluteal. The deep caudal portion of the middle gluteal is readily separated from the main muscle mass. Starting at the middle of the cranioventral border of the middle gluteal, transect the entire muscle and reflect the distal half toward its insertion on the apex of the greater trochanter of the femur.

ORIGIN: The crest and gluteal surface of the ilium.

INSERTION: The greater trochanter.

ACTION: To extend and abduct the hip, and to rotate the pelvic limb medially.

INNERVATION: Cranial gluteal nerve.

4. The **deep gluteal** (Figs. 44, 45, 49, 51) is fan-shaped and completely covered by the middle gluteal. Superficially it is covered by an aponeurosis whose fibers converge to insert on the cranial face of the greater trochanter.

ORIGIN: The body of the ilium; the ischiatic spine.

INSERTION: The cranial aspect of the greater trochanter.

ACTION: To extend and abduct the hip, and to rotate the pelvic limb medially.

INNERVATION: Cranial gluteal nerve.

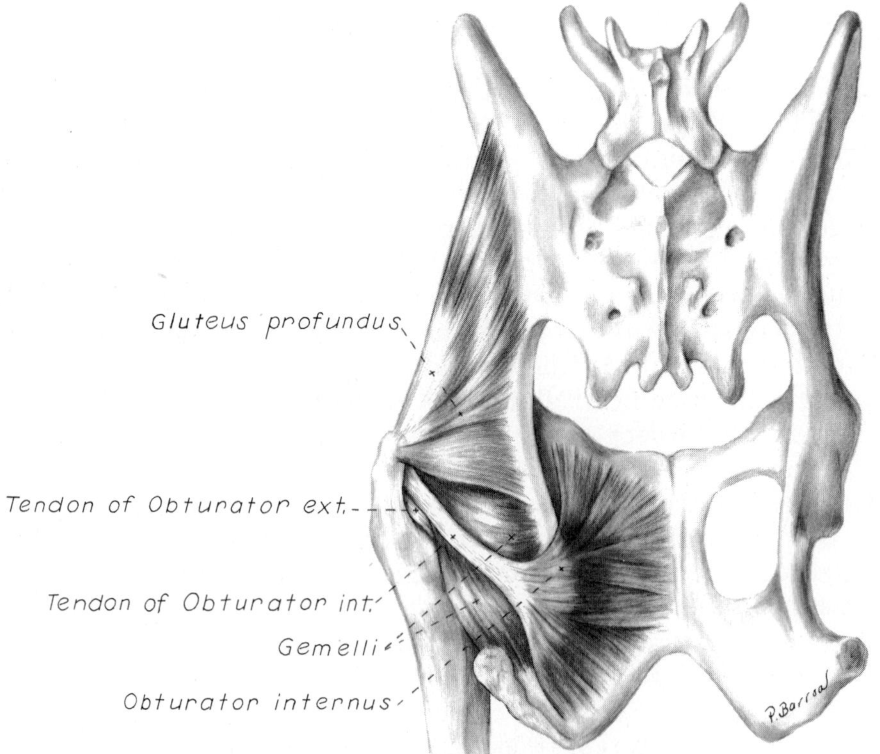

FIGURE 49. Muscles of left hip joint, dorsal aspect.

The **articularis coxae** is a small, spindle-shaped muscle lying on the craniolateral aspect of the hip joint capsule. It is covered by the deep gluteal muscle. It arises from the lateral surface of the ilium along with the rectus femoris and inserts on the neck of the femur. Transect the deep gluteal to observe it. Although of little functional consequence, it is used as a landmark for entering the joint capsule.

Caudal Hip Muscles

The four muscles of this group are important because of their proximity to the hip. They lie caudal to the hip and extend from the inner and outer surfaces of the ischium to the femur. All rotate the femur laterally.

1. The **internal obturator** (Figs. 46, 49, 51) is fan-shaped on the dorsal surface of the ischium and pubis. Its muscle fibers converge toward the lesser ischiatic notch. The body of the muscle may be exposed at its origin on the dorsal surface of the ischium by removing the loose fat and the fascia caudomedial to the sacrotuberous ligament. The most caudal fibers of the internal obturator run craniolaterally toward the lesser ischiatic notch, where the tendon of the muscle begins. The tendon of the internal obturator passes over the lesser ischiatic notch ventral to the sacrotuberous ligament. Transect the sacrotuberous ligament and reflect the adjacent soft tissues to expose the tendon of insertion of the internal obturator muscle running to the trochanteric fossa. Transect the tendon of the internal obturator as it crosses the gemelli and reflect it to observe the bursa that lies between the tendon and the lesser ischiatic notch.

ORIGIN: The symphysis pelvis and the dorsal surface of the ischium and pubis.

INSERTION: The trochanteric fossa of the femur.

ACTION: To rotate the pelvic limb laterally at the hip.

INNERVATION: Sciatic nerve.

2. The **gemelli** (Figs. 44, 45, 49), two muscles fused together, lie under the tendon of the internal obturator. They are interposed between the quadratus femoris and external obturator distally and the deep gluteal proximally. The gemelli are deeply grooved by the tendon of the internal obturator so that their edges overlap this tendon.

ORIGIN: The lateral surface of the ischium, caudal to the acetabulum and ventral to the lesser ischiatic notch.

INSERTION: The trochanteric fossa.

ACTION: To rotate the pelvic limb laterally at the hip.

INNERVATION: Sciatic nerve.

3. The **quadratus femoris** (Figs. 44, 45, 50) is short and thick. It lies under the biceps femoris, where it is interposed between the adductor distally and the external obturator and gemelli proximally. Its fibers are at right angles to the long axis of the thigh. It should be examined from

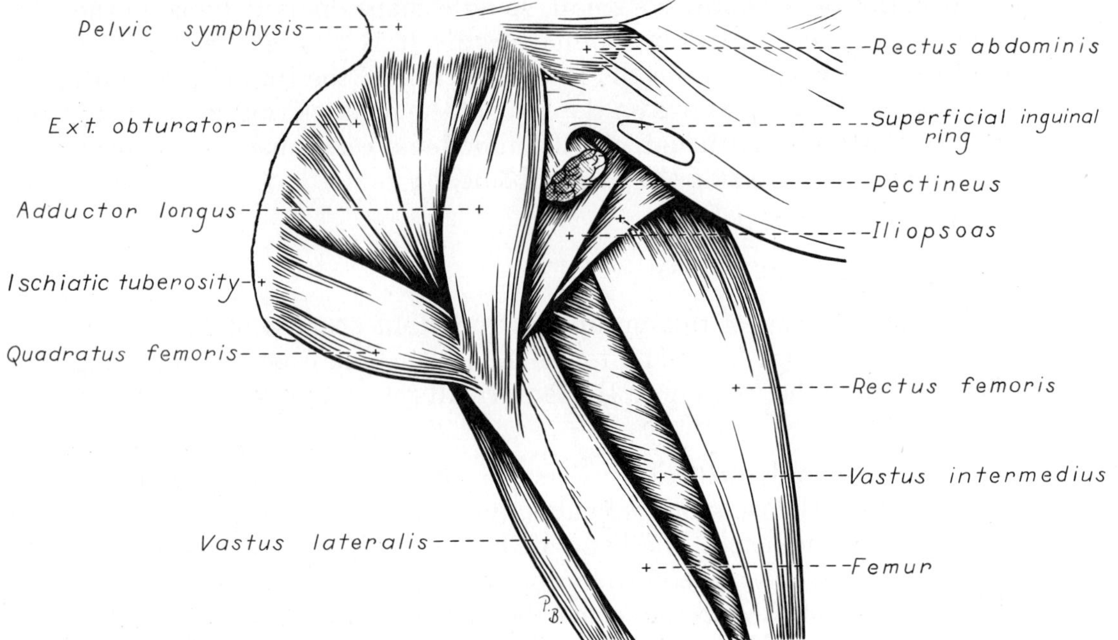

FIGURE 50. Deep muscles medial to left hip.

both the medial and lateral sides. The dorsal border of the quadratus femoris lies closely applied to the ventral border of the gemelli.

ORIGIN: The ventral surface of the caudal part of the ischium.

INSERTION: Just distal to the trochanteric fossa.

ACTION: To extend the hip and rotate the pelvic limb laterally.

INNERVATION: Sciatic nerve.

4. The **external obturator** (Figs. 45, 50, 51) is fan-shaped and arises on the ventral surface of the pubis and ischium. It covers the obturator foramen. Its caudal border is covered by the quadratus femoris, while its cranial border is hidden by the adductor. Follow the external obturator to its insertion.

ORIGIN: The ventral surface of the pubis and ischium.

INSERTION: The trochanteric fossa.

ACTION: To rotate the pelvic limb laterally.

INNERVATION: Obturator nerve.

Cranial Muscles of the Thigh

1. The **quadriceps femoris** (Figs. 43–48, 50, 51, 53) is divided into four heads of origin, which are fused distally. It arises from the femur and the ilium and is inserted on the tibial tuberosity. The patella lies in the tendon of insertion. This muscle is the most powerful extensor of the stifle.

The **rectus femoris** (Figs. 43–45, 47, 48, 50) is the most cranial component of the quadriceps femoris and the only one to arise from the

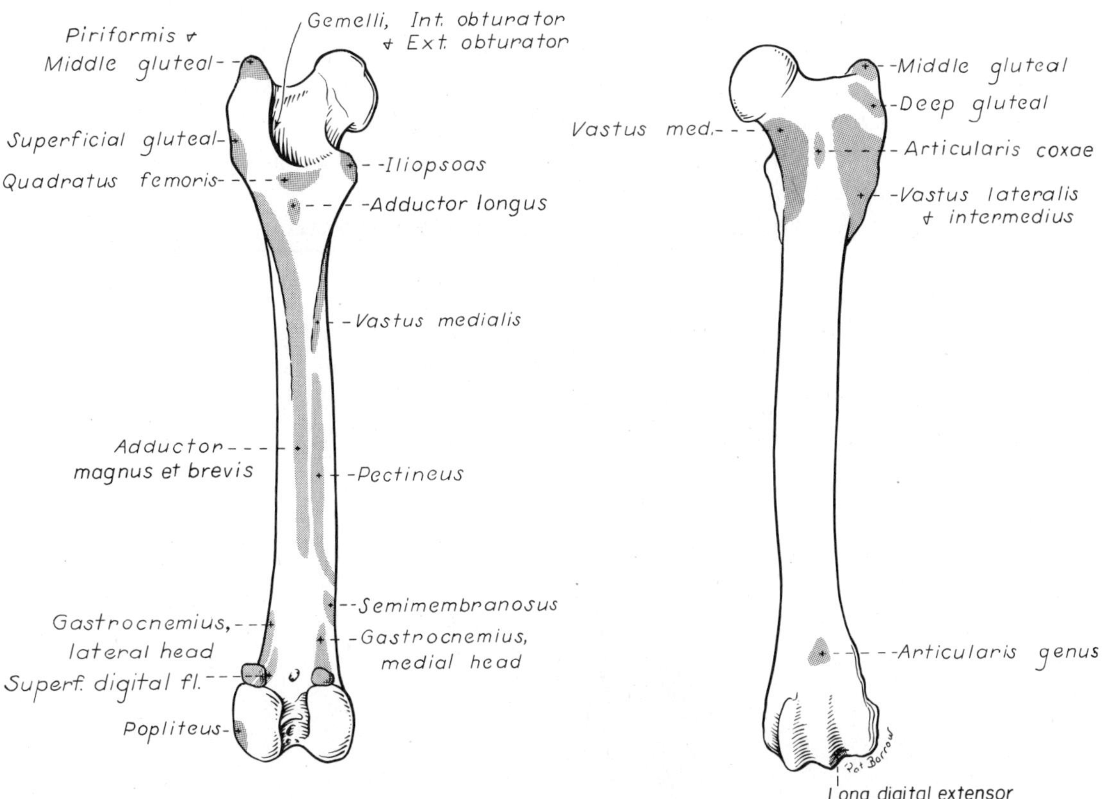

FIGURE 51a. Left femur with muscle attachments, caudal view.

FIGURE 51b. Left femur with muscle attachments, cranial view.

ilium. Proximally it is circular in transverse section and passes between the vastus medialis and the vastus lateralis. Uncover the rectus femoris near its origin. Transect and reflect the proximal part. The rectus arises from the ilium cranial to the acetabulum and inserts on the tibial tuberosity. It is a flexor of the hip as well as an extensor of the stifle.

The patella is intercalated in the strong tendon of insertion of the quadriceps. The **patellar ligament,** which extends from the patella to the tibial tuberosity, is but the distal end of the tendon of insertion of the quadriceps.

The **vastus lateralis** (Figs. 44, 45, 48, 50, 51) lies lateral and caudal to the rectus femoris, to which it is fused distally. The vastus lateralis is partly separated from the vastus intermedius by a scantly developed intermuscular septum. Notice that the vastus lateralis arises from the proximal part of the lateral lip of the caudal rough surface of the femur. It is inserted with the rectus femoris on the tibial tuberosity.

The **vastus intermedius** (Figs. 48, 50, 51) lies directly on the smooth cranial surface of the femur and is quite intimately fused with the other two vasti. It arises with the vastus lateralis, which covers it, from the lateral side of the proximal end of the femur. It inserts on the tibial tuberosity with the other members of the group.

The **vastus medialis** (Figs. 43, 46–48, 51) arises from the medial side of the proximal end of the cranial surface of the femur and the proximal end of the medial lip of the caudal rough surface. It is inserted with the other heads of the quadriceps on the tibial tuberosity.

ORIGIN: Rectus femoris—ilium; vasti muscles—proximal femur.

INSERTION: Tibial tuberosity.

ACTION: To extend the stifle and flex the hip (rectus).

INNERVATION: Femoral nerve.

2. The **iliopsoas** (Figs. 45, 46, 50–52), a sublumbar muscle, is now visible at its insertion on the lesser trochanter of the femur. This end of the muscle lies between the pectineus medially and the rectus femoris laterally. The iliopsoas represents a fusion of the psoas major and iliacus

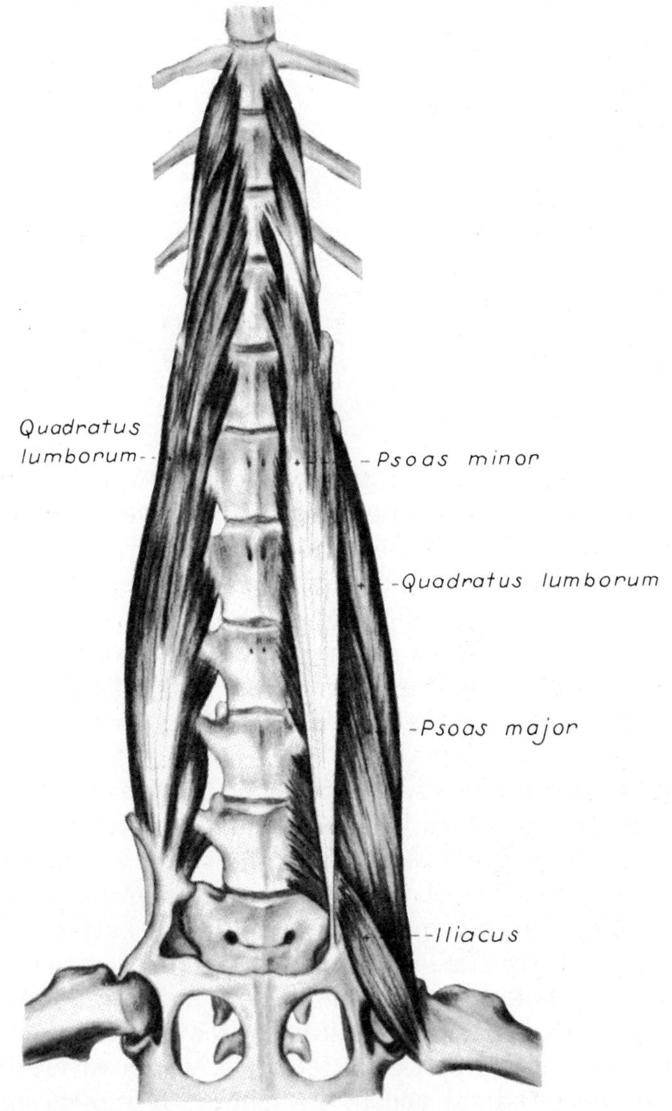

FIGURE 52. Sublumbar muscles, deep dissection, ventral aspect.

muscles. The **psoas major** arises from transverse processes and bodies of lumbar vertebrae. It passes caudally and ventrally under the cranioventral aspect of the ilium, where it joins the iliacus. The **iliacus** arises from the smooth ventral surface of the ilium between the arcuate line and the lateral border of the ilium. The two muscle masses continue caudoventrally as the iliopsoas to their conjoined insertion on the lesser trochanter. The action of the iliopsoas is to flex the hip and the lumbar vertebral column. It is the major flexor of the hip.

ORIGIN: Psoas major—lumbar vertebrae; iliacus—cranioventral ilium.
INSERTION: Lesser trochanter.
ACTION: To flex the hip.
INNERVATION: Ventral branches of lumbar nerves, and femoral nerve.

Muscles of the Leg (Crus)

The muscles of the leg, or crus, the region between the stifle and hock, are divided into craniolateral and caudal groups. Remove the skin that remains on the distal part of the pelvic limb to the level of the proximal interphalangeal joints.

The **superficial crural, tarsal, metatarsal** and **digital fasciae** are similar to the superficial fasciae of the corresponding regions of the forelimb. Cutaneous vessels and nerves course in the superficial fascia. One such vessel is the cranial branch of the lateral saphenous vein, which is used for venipuncture.

The medial and lateral femoral fasciae blend over the stifle and are continued distally in the leg as the **deep crural fascia.** The deep crural fascia covers the muscles of the leg and the free-lying surfaces of the crural skeleton. Septae from this fascia extend between the muscles to attach to the bone. Laterally the fibers of the caudal branch of the biceps femoris radiate into it. Caudally the crural fascia contributes to the common calcanean tendon.

Just proximal to the flexor surface of the tarsus the deep crural fascia is thickened to form an oblique band of about 0.5 centimeter, the **proximal extensor retinaculum.** As it stretches obliquely from the distal third of the fibula to the medial malleolus of the tibia, it binds down the tendons of the long digital extensor and the cranial tibial muscles.

The deep crural fascia decreases in thickness as it passes over the tarsus and becomes the deep tarsal fascia. The deep fascia extends into the metatarsal and digital pads and closely joins these pads with the overlying skeletal and ligamentous parts.

Make an incision through the cranial crural fascia and reflect it to the common calcanean tendon and the tibia.

CRANIOLATERAL MUSCLES OF THE LEG

1. The **cranial tibial** (Figs. 43–47, 53–55, 57) is the most cranial muscle of this group. Its medial margin is in contact with the tibia. It arises from the cranial border and the adjacent proximal articular margin

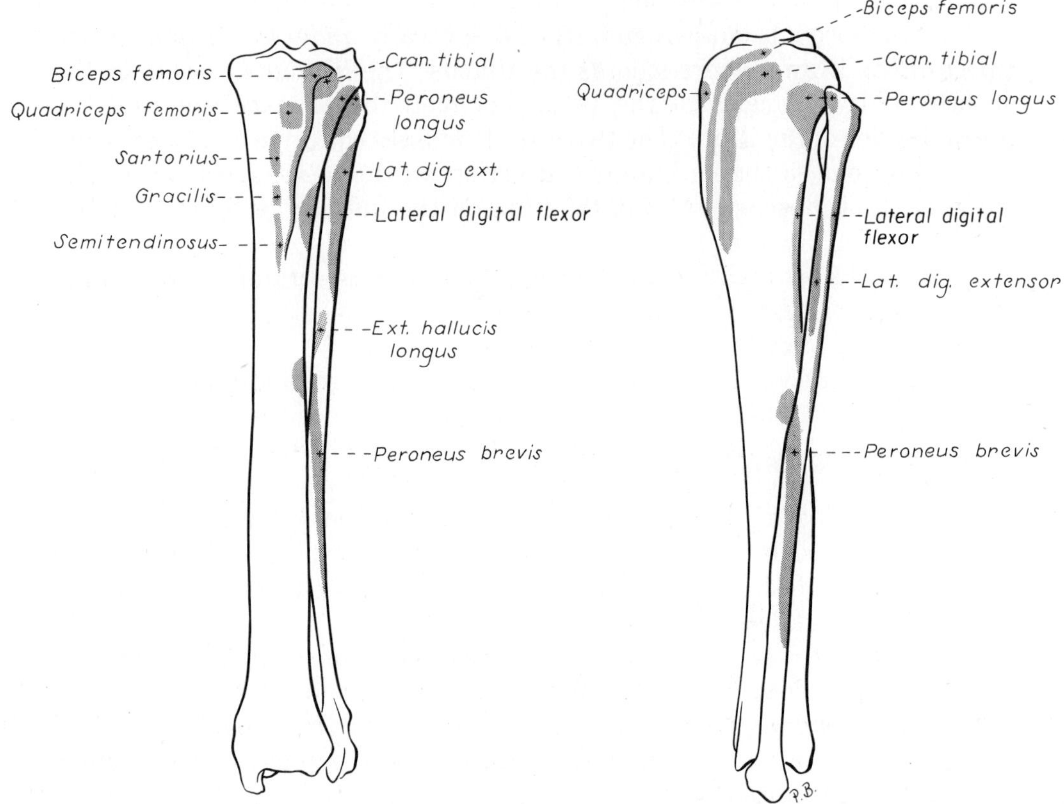

FIGURE 53a. Left tibia and fibula with muscle attachments, cranial view.

FIGURE 53b. Left tibia and fibula with muscle attachments, lateral view.

of the tibia. Its tendon inserts on the proximal plantar surface of the first and second metatarsals. The tendon of the cranial tibial runs under the proximal extensor retinaculum and is provided with a synovial sheath over most of the flexor surface of the tarsus.

ORIGIN: The extensor groove and the adjacent articular margin of the tibia; the lateral edge of the cranial tibial border.

INSERTION: The proximal plantar surface of metatarsals I and II.

ACTION: To flex the tarsus and to rotate the paw laterally.

INNERVATION: Peroneal nerve.

2. The **long digital extensor** (Figs. 44–46, 51, 54, 57) is a spindle-shaped muscle that is partly covered by the cranial tibial medially and the peroneus longus laterally. Expose the muscle and its tendon of origin from the extensor fossa of the femur. The tendon runs over the articular margin of the tibia in the extensor groove and is lubricated by an extension of the stifle joint capsule. Observe the four tendons of insertion. As the tendons pass over the tarsus, they are surrounded by a synovial sheath and are held in place by the extensor retinacula. In the metatarsus the four tendons diverge toward their respective digits.

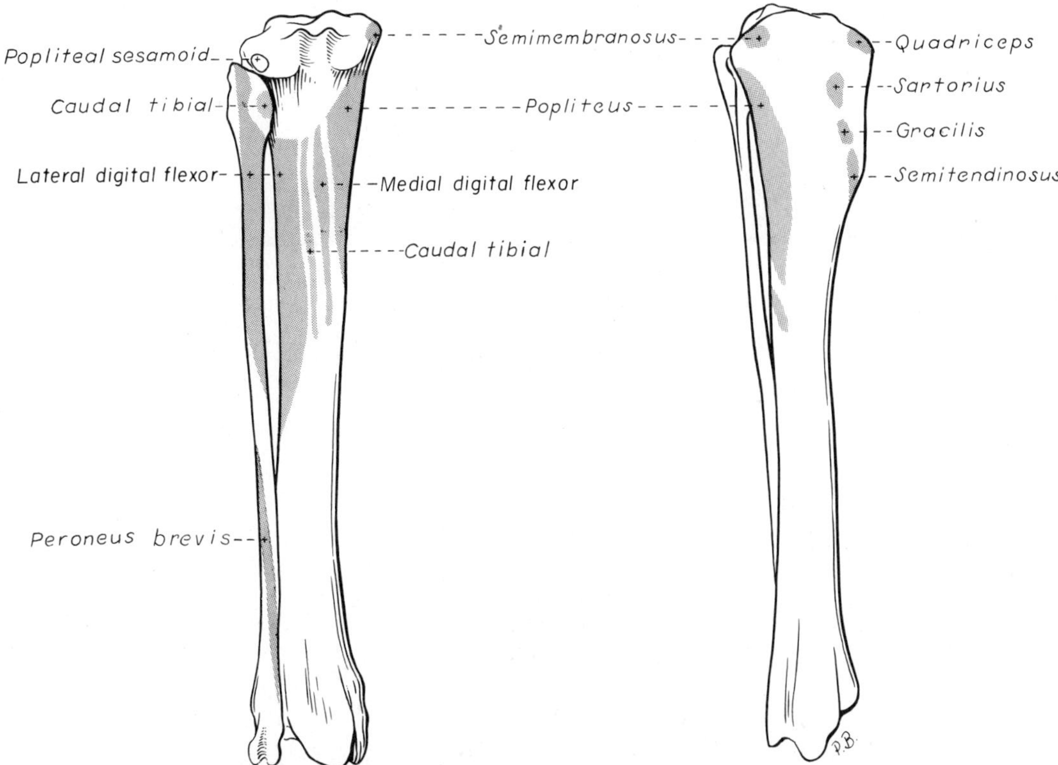

FIGURE 53c. Left tibia and fibula with muscle attachments, caudal view.

FIGURE 53d. Left tibia and fibula with muscle attachments, medial view.

ORIGIN: The extensor fossa of the femur.

INSERTION: The extensor processes of the distal phalanges of digits II, III, IV and V.

ACTION: To extend the digits and flex the tarsus.

INNERVATION: Peroneal nerve.

3. The **peroneus longus** (Figs. 44, 45, 53–57) lies just caudal to the long digital extensor, where a triangular portion of its short belly lies directly under the crural fascia. It is a short, thick, wedge-shaped muscle that lies in large part cranial to the fibula. It arises from the lateral collateral ligament of the stifle and the adjacent parts of the tibia and fibula. Its stout tendon courses distally on the lateral side of the fibula caudal to the proximal extensor retinaculum. It has a long synovial sheath that begins at a plane through the proximal extensor retinaculum and extends to its insertion on the fourth tarsal and the plantar surfaces of all the metatarsals. Open the synovial sheath at its proximal end. Preserving all ligaments and tendons that lie superficial to the tendon, trace it to its insertion on the fourth tarsal. Do not dissect the tendon beyond this attachment. Notice that it lies in a sulcus of the lateral malleolus of the fibula and that at the distal end of the tarsus it makes nearly a right angle as it turns medially around a groove in the fourth tarsal bone and courses to the plantar side.

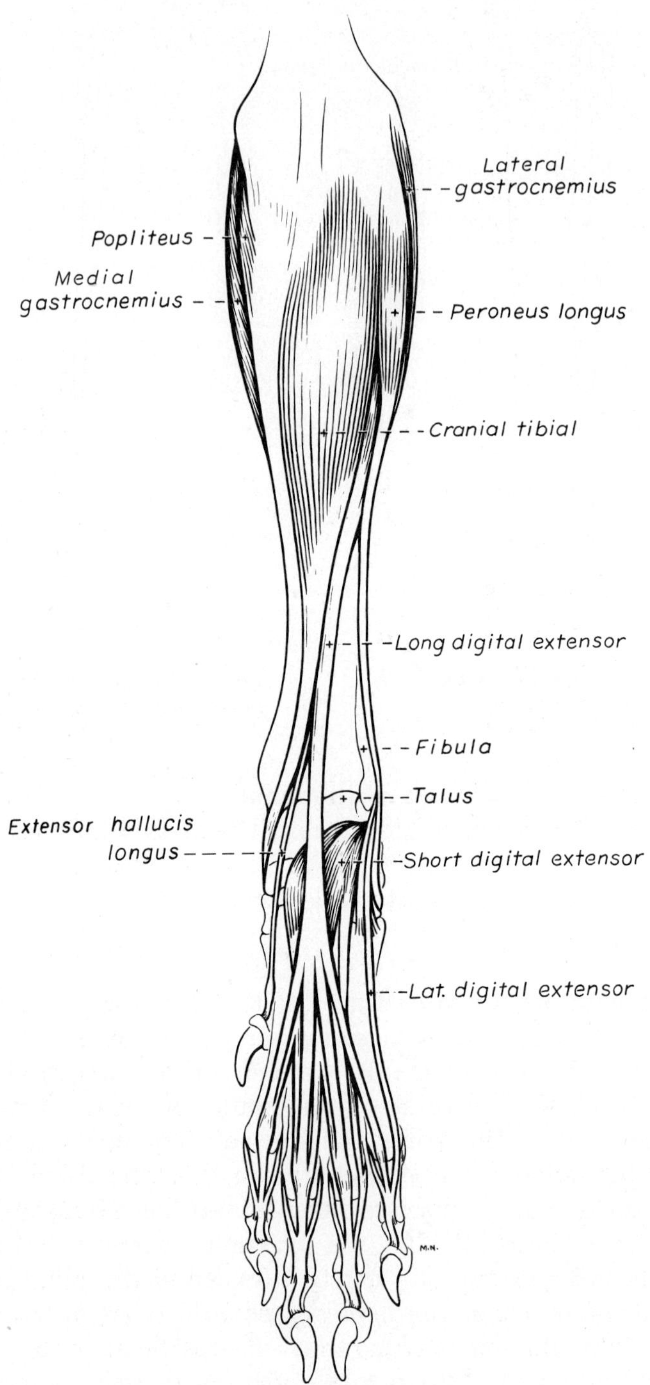

FIGURE 54. Muscles of left pelvic limb, cranial view.

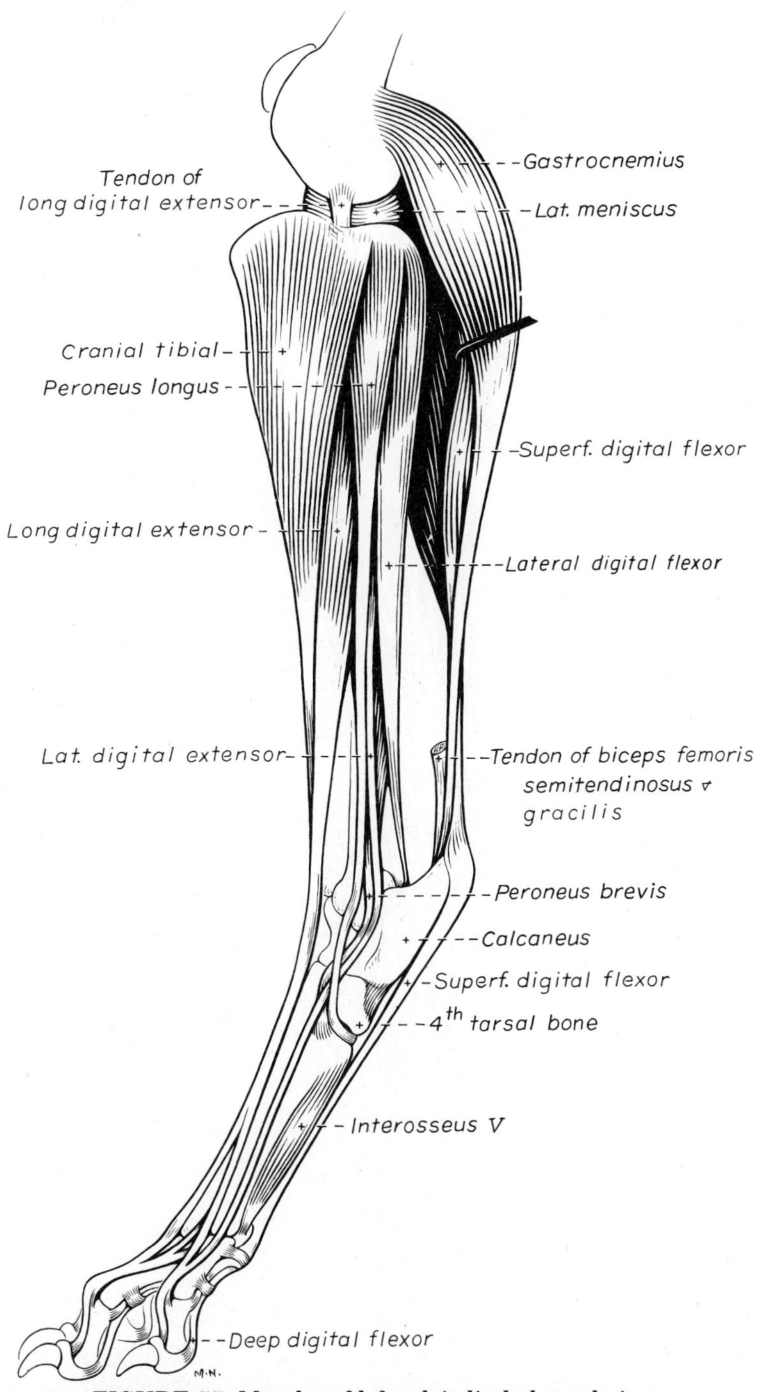

FIGURE 55. Muscles of left pelvic limb, lateral view.

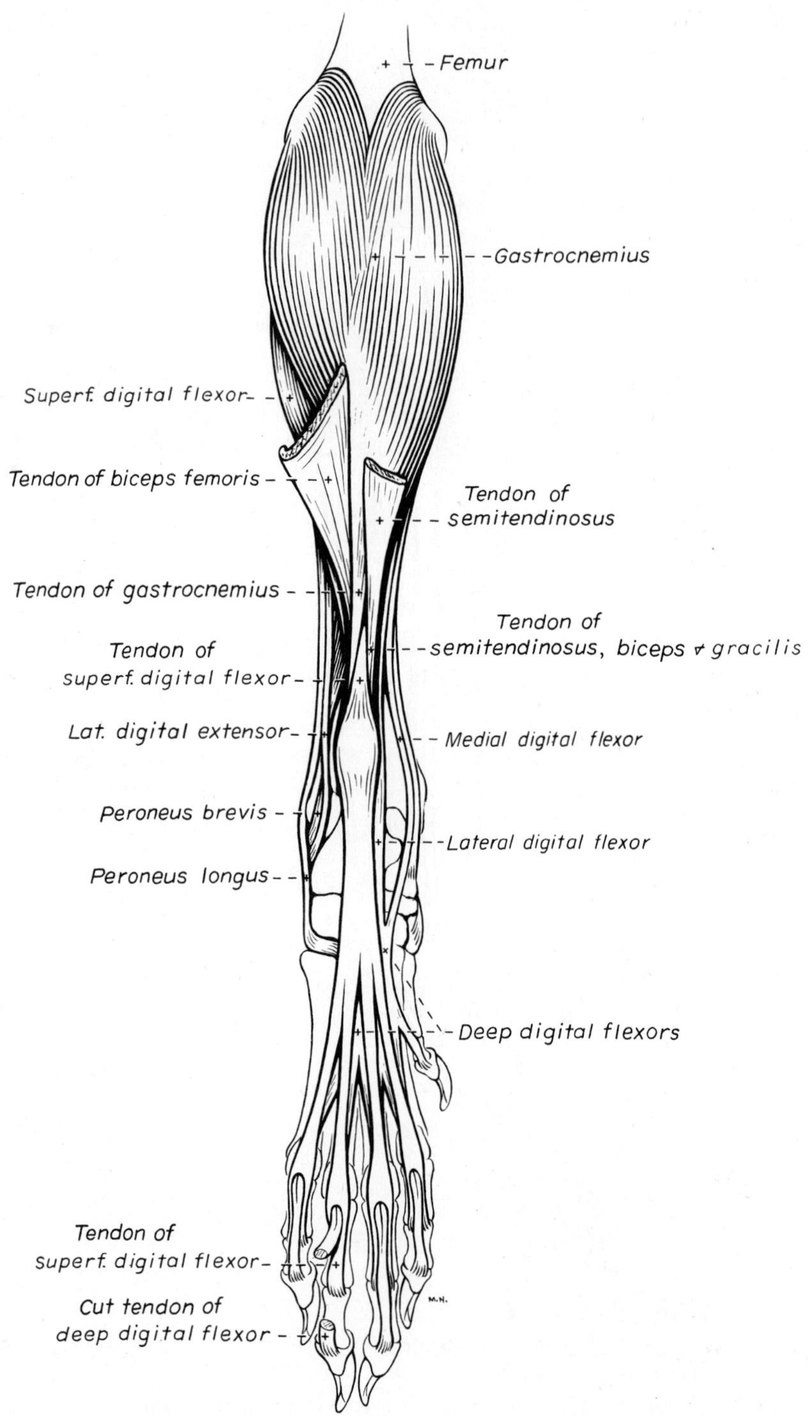

FIGURE 56. Muscles of left pelvic limb, caudal view.

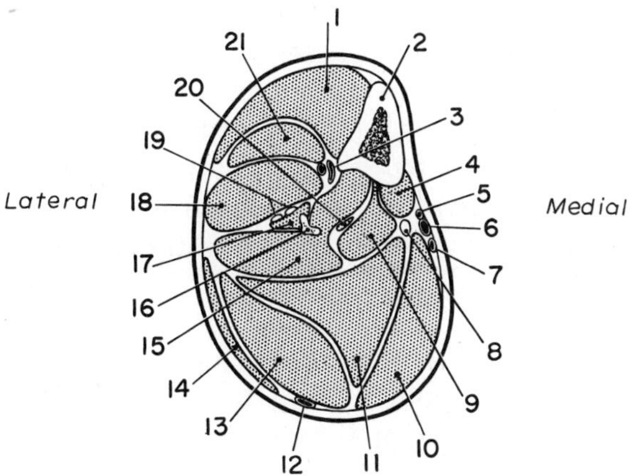

FIGURE 57. Transverse section of left leg.

1. *Cranial tibial*
2. *Tibia*
3. *Cranial tibial a. and v.*
4. *Popliteus*
5. *Saphenous a., cranial branch*
6. *Medial saphenous v.*
7. *Saphenous a., caudal branch*
8. *Tibial n.*
9. *Medial digital flexor*
10. *Gastrocnemius, medial head*
11. *Superficial digital flexor*
12. *Lateral saphenous v.*
13. *Gastrocnemius, lateral head*
14. *Biceps femoris*
15. *Lateral digital flexor*
16. *Fibula*
17. *Lateral digital extensor*
18. *Peroneus longus*
19. *Peroneal nerves*
20. *Caudal tibial*
21. *Long digital extensor*

ORIGIN: The lateral condyle of the tibia, the proximal end of the fibula and the lateral epicondyle of the femur by means of the lateral collateral ligament of the stifle.

INSERTION: The fourth tarsal bone; the plantar aspect of the proximal ends of the metatarsals.

ACTION: To flex the tarsus and rotate the paw medially so that the plantar surface faces laterally.

INNERVATION: Peroneal nerve.

The lateral digital extensor and the peroneus brevis (Figs. 44, 45, 53–57) are located beneath the peroneus longus on the lateral aspect of the leg and need not be dissected.

CAUDAL MUSCLES OF THE LEG

1. The **gastrocnemius muscle** (Figs. 43–47, 51, 54–57) consists of two heads which enclose the superficial digital flexor between them. These muscles form the caudal bulge of the leg (calf) and contribute the major component of the **calcanean tendon.**

The two heads of the gastrocnemius arise from the medial and lateral supracondylar tuberosities of the femur. In each tendon of origin there is a sesamoid bone (fabella) that articulates with the caudodorsal aspect of the femoral condyle.

Identify the lateral and medial heads of the gastrocnemius and follow them to their union as a common subcutaneous tendon which inserts on the tuber calcanei. The superficial digital flexor muscle, which is hidden proximally between the two heads of the gastrocnemius, emerges distally as a superficial tendon which passes over the medial surface of the gastrocnemial tendon, caps the tuber calcanei and continues over the plantar surface of the paw. With a probe separate the gastrocnemius from the superficial digital flexor by entering between their tendons. Distally, free one side of the superficial flexor tendon from the tuber calcanei to see the bursa between them. Proximally, bluntly separate the superficial digital flexor from the heads of the gastrocnemius.

The superficial digital flexor has a common origin with the lateral head of the gastrocnemius—the lateral supracondylar tuberosity of the femur—and is closely adherent to that head. Transect each head of the gastrocnemius and reflect them proximally to see their origin and the associated sesamoid bones.

ORIGIN: The medial and lateral supracondylar tuberosities of the femur.

INSERTION: The tuber calcanei.

ACTION: To extend the tarsus and flex the stifle.

INNERVATION: Tibial nerve.

2. The **superficial digital flexor** (Figs. 43–47, 51, 55–57) is a spindle-shaped muscle that arises from the lateral supracondylar tuberosity of the femur with the lateral head of the gastrocnemius. Its deep surface is in apposition to the deep digital flexor and the popliteus, while its other surfaces are largely covered by the gastrocnemius. Proximal to the calcaneal process its tendon twists across the medial surface of the gastrocnemius. Farther distally the tendon widens, caps the tuber calcanei, attaches on each side and continues distally. Make a sagittal incision through the superficial digital flexor tendon from the place where it gains the caudal side of the gastrocnemius to the tuber calcanei. Continue the incision distal to the tuber calcanei for an equal distance. Observe the large **calcaneal bursa** of the superficial digital flexor that lies under its tendon as it crosses the tuber calcanei. Notice the distinct medial and lateral attachments of the tendon on the tuber calcanei. Opposite the distal plantar surface of the tarsus the tendon bifurcates; each of these branches in turn bifurcates, thus forming four tendons of nearly equal size. Each tendon is disposed in its digit as the corresponding tendon in the forepaw. They need not be dissected.

ORIGIN: The lateral supracondylar tuberosity of the femur.

INSERTION: The tuber calcanei and the bases of the middle phalanges of digits II, III, IV and V.

ACTION: To flex the first two digital joints of the four principal digits; flex the stifle; extend the tarsus.

INNERVATION: Tibial nerve.

3. The **deep digital flexor** (Figs. 43–47, 53, 55–57) and the popliteus are the principal muscles yet to be dissected on the caudal surface of the leg. The separation between them runs distomedially. Transect the superficial digital flexor proximal to its tendon to expose these muscles. The

muscle division starts at the lateral tibial condyle and extends to the proximal third of the tibia on the medial side of the leg. The two muscles that compose the deep digital flexor are now exposed.

The **lateral digital flexor** (flexor hallucis longus) (Figs. 43–47, 53, 55–57) arises from the caudolateral border of the proximal two-thirds of the tibia, most of the proximal half of the fibula and the adjacent interosseous membrane. Its tendon begins as a wide expanse on the caudal side of the muscle but condenses distally. Medial to the tuber calcanei it is surrounded by the tarsal synovial sheath and bound in the groove over the sustentaculum tali of the calcaneus by the **flexor retinaculum.** At the level of the distal row of tarsal bones observe the tendon of the lateral digital flexor joining that of the medial digital flexor to form a common tendon. The courses, relations and attachments of the tendons distal to the tarsus are similar to those of the deep flexor tendon of the forelimb. Their dissection is not necessary.

The **medial digital flexor** (flexor digitorum longus) (Figs. 53, 56, 57) is smaller and lies between the lateral digital flexor and the popliteus. From the head of the fibula and the proximal end of the tibia it runs distomedially. Its tendon lies on the caudomedial side of the tibia. At the distal row of tarsal bones it unites with the tendon of the lateral digital flexor.

ORIGIN: The proximal two-thirds of the tibia, the proximal half of the fibula and the adjacent interosseous membrane.

INSERTION: The plantar surface of the base of each of the distal phalanges.

ACTION: To flex the digits and extend the tarsus.

INNERVATION: Tibial nerve.

4. The **popliteus** (Figs. 45–47, 51, 53, 54, 57) is covered by the gastrocnemius and the superficial digital flexor and lies on the stifle joint capsule and the proximal tibia. It arises from the lateral condyle of the femur by a long tendon that should be isolated just cranial to the lateral collateral ligament of the stifle. It courses caudally, medial to the lateral collateral ligament. At the junction of the tendon with the muscle there is a **sesamoid** that articulates with the lateral condyle of the tibia. Transect the popliteus at the tendomuscular junction and reflect it proximally to observe the sesamoid. The popliteus inserts on the proximal third of the tibia.

ORIGIN: The lateral condyle of the femur.

INSERTION: The proximal third of the caudal surface of the tibia.

ACTION: To flex the stifle and rotate the leg medially.

INNERVATION: Tibial nerve.

Live Dog

Palpate the middle gluteal on the lateral side of the wing of the ilium. Follow it to its termination on the greater trochanter. Palpate the quadriceps femoris cranially in the thigh and feel the straplike sartorius along its cranial edge. Follow the quadriceps distally to its patella and the patellar ligament with its termination on the tibial tuberosity. Proximally in the thigh feel the tensor fascia lata between the sartorius and the quadriceps. Palpate the muscle mass in the caudal thigh. This includes from lateral to medial the biceps femoris, semitendinosus, semimembranosus and gracilis. Feel the spindle-shaped pectineus proximally in the medial thigh on the caudal border of the femoral triangle. Abduct the limb and feel this muscle tighten. Palpate the pulse in the femoral artery just cranial to this muscle. The pulse rate and quality are usually determined here.

Grasp the tail and press down on the floor of the pelvis just cranial to the ischial arch. The muscle here is the internal obturator. The depression here between the tail and anus medially, superficial gluteal laterally and internal obturator ventrally is the ischiorectal fossa where perineal hernias occur.

Caudal to the stifle feel for the popliteal lymph node between the distal ends of the biceps femoris and semitendinosus. In the leg feel the craniolateral muscles that are the flexors of the tarsus and extensors of the digits. Proximally their muscle bellies fill the concave surface of the body of the tibia. Caudally feel the two gastrocnemius muscles proximally. Trace them from their origin on the distal femur to where their tendons join and form part of the common calcanean tendon. Follow this common tendon to the tuber calcanei. Feel the division between the gastrocnemius tendon and the tendon of the superficial digital flexor. Follow the latter on the plantar side of the tuber calcanei. Palpate the digital flexors distally in the caudal leg area and plantar aspect of the tarsus.

JOINTS OF THE PELVIC LIMB

The ischium and pubis of the right and left sides are joined on the median plane at the **symphysis pelvis.**

The **sacroiliac joint** (Figs. 59, 60) is an articulation of stability rather than mobility. The right and left wings of the ilia articulate with the broad right and left wings of the sacrum. In the adult most of the apposed articular surfaces are united by fibrocartilage. Around the periphery of the articular areas, bands of strong collagenous tissue, the **dorsal** and **ventral sacroiliac ligaments,** reinforce the fibrocartilage. Do not dissect this joint.

The **sacrotuberous ligament** (Fig. 59) runs from the transverse processes of the last sacral and first caudal vertebrae to the lateral angle of the ischiatic tuberosity. It serves as an origin for several muscles which have been dissected.

The **hip** (Figs. 59, 60) is a ball-and-socket joint whose main movements are flexion and extension. The joint capsule passes from the neck of the femur to a line peripheral to the acetabular lip. Transect the deep gluteal and iliopsoas muscles at their insertions. Cut the hip joint capsule to expose the joint and associated ligaments.

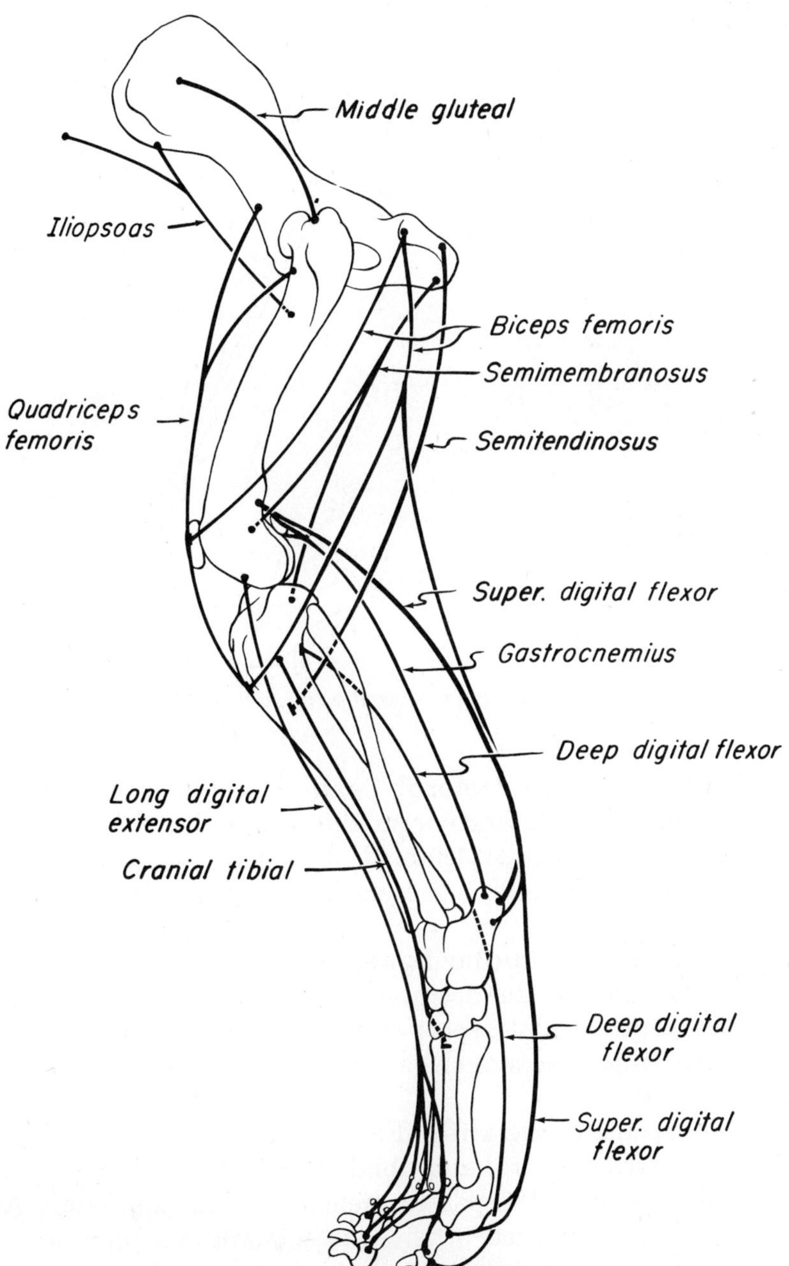

FIGURE 58. Major flexors and extensors of pelvic limb.

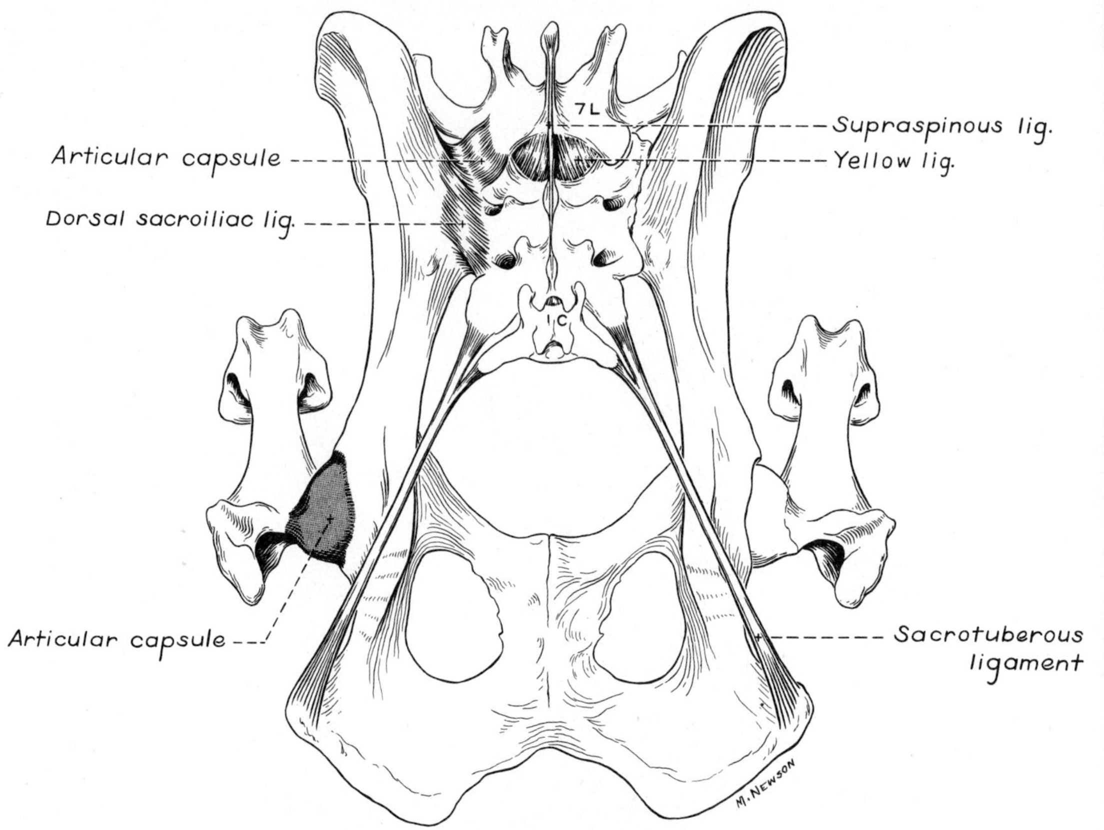

FIGURE 59. Ligaments of pelvis, dorsal view.

The **ligament of the femoral head** (Fig. 60) is a thick band of collagenous tissue that extends from the acetabular fossa to the fovea capitis. At its acetabular attachment it may blend slightly with the transverse acetabular ligament. A synovial membrane covers it. Transect this ligament.

The **transverse acetabular ligament** (Fig. 60) is a small band that extends from one side of the acetabular notch to the opposite side. It is located at the ventrocaudal aspect of the acetabulum and continues the **acetabular lip,** which deepens the acetabulum by forming a fibrocartilaginous border around it.

The joint capsule of the **stifle** (Figs. 61–65) forms three sacs. Two of these are between the femoral and tibial condyles (femorotibial joint sacs), and the third is beneath the patella (femoropatellar joint sac). All three sacs communicate with each other. The femorotibial joint sacs extend caudally and dorsally to incorporate the articulation of the gastrocnemial sesamoids. The lateral femorotibial sac continues distally through the extensor groove as the tendon sheath for the tendon of origin of the long digital extensor (Fig. 61c). It also surrounds the tendon of origin of the popliteus. Between each femoral condyle and the corresponding tibial condyle there is a **meniscus,** or **semilunar fibrocartilage** surrounded by the joint sac. These are C-shaped disks with thick peripheral margins and

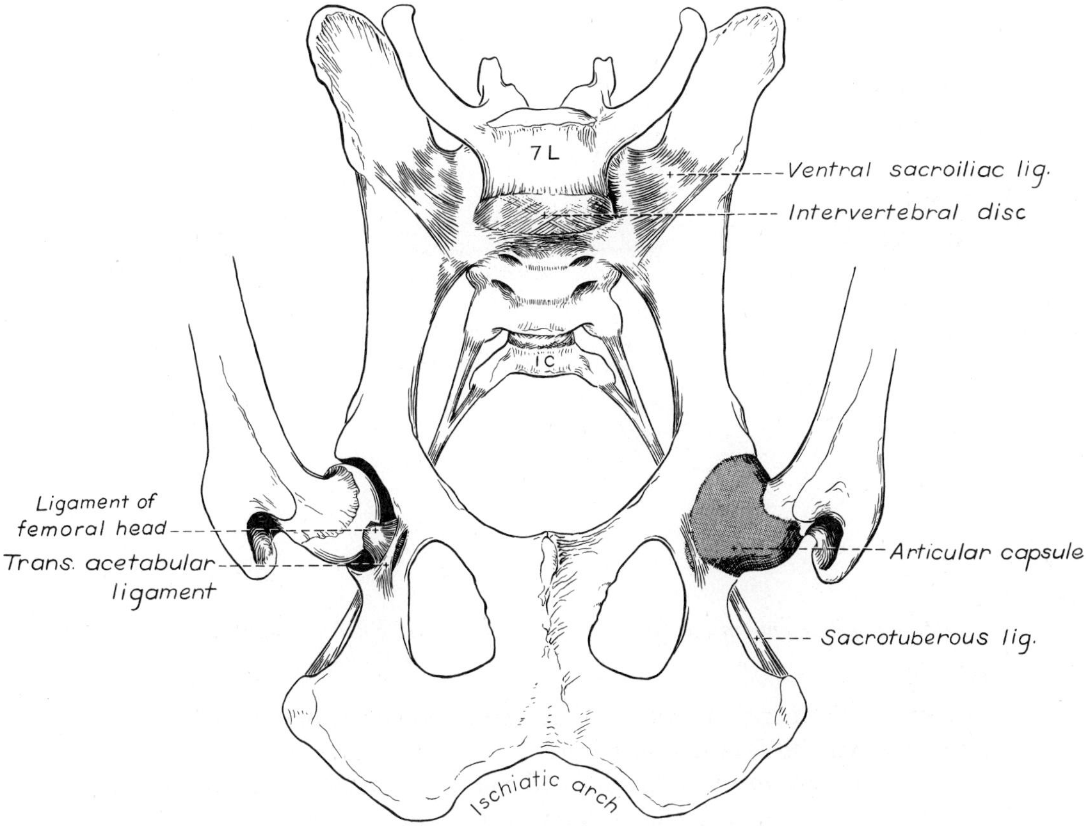

FIGURE 60. Ligaments of pelvis, ventral view.

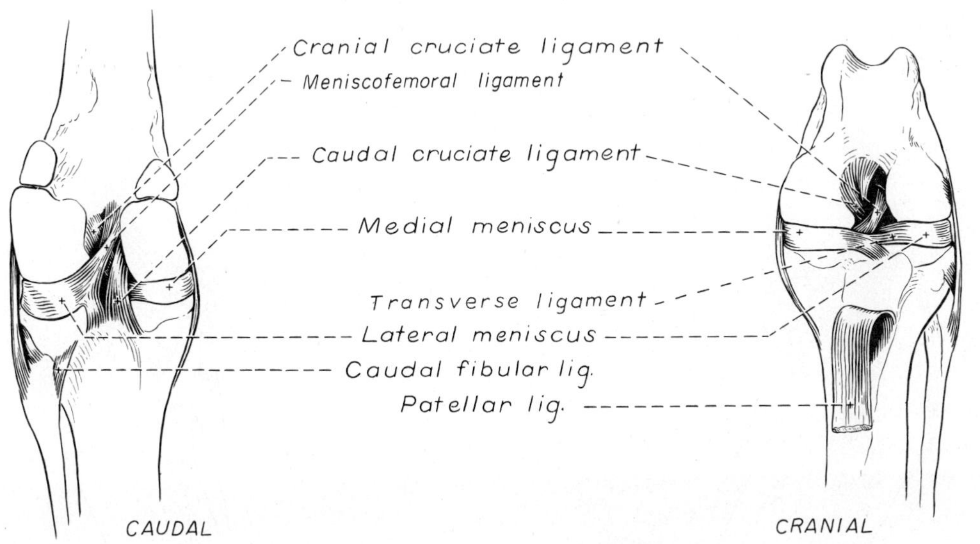

FIGURE 61a. Ligaments of left stifle, caudal view.

FIGURE 61b. Ligaments of left stifle, cranial view.

Illustration continued on following page

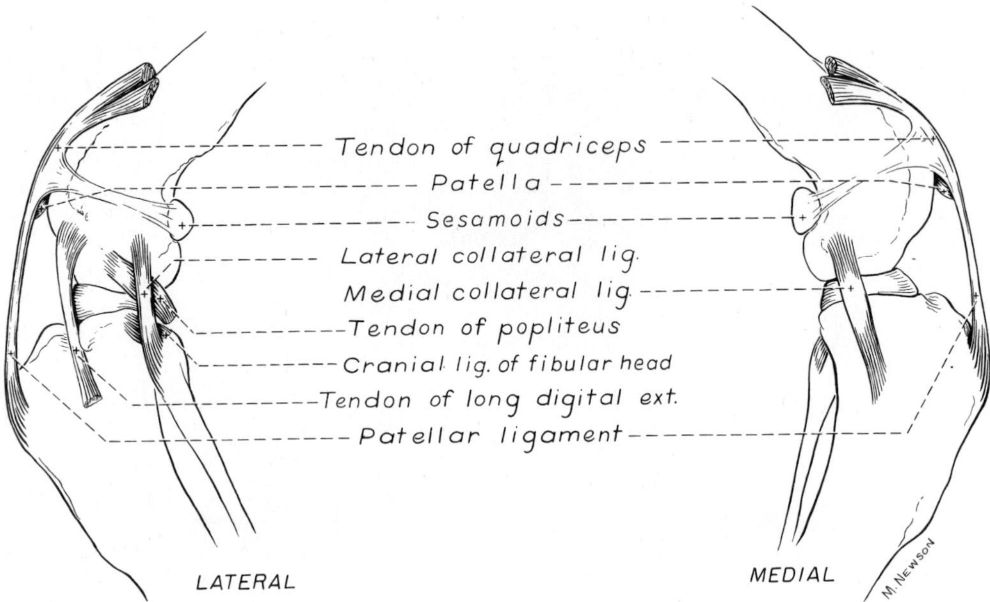

FIGURE 61c. Ligaments of left stifle, lateral view.

FIGURE 61d. Ligaments of left stifle, medial view.

thin, concave central areas that compensate for the incongruence that exists between the femur and tibia.

Clean the remaining fascia and related muscle attachments away from the stifle. Transect the patellar ligament and reflect it proximally. Note the large quantity of fat between the patellar ligament and the joint capsule. Medial and lateral parapatellar fibrocartilages extend from the patella (Fig. 65). Note the proximal extent of the femoropatellar joint sac. Remove the fat around the joint, open the joint capsule and remove as much joint capsule as necessary to observe the following ligaments.

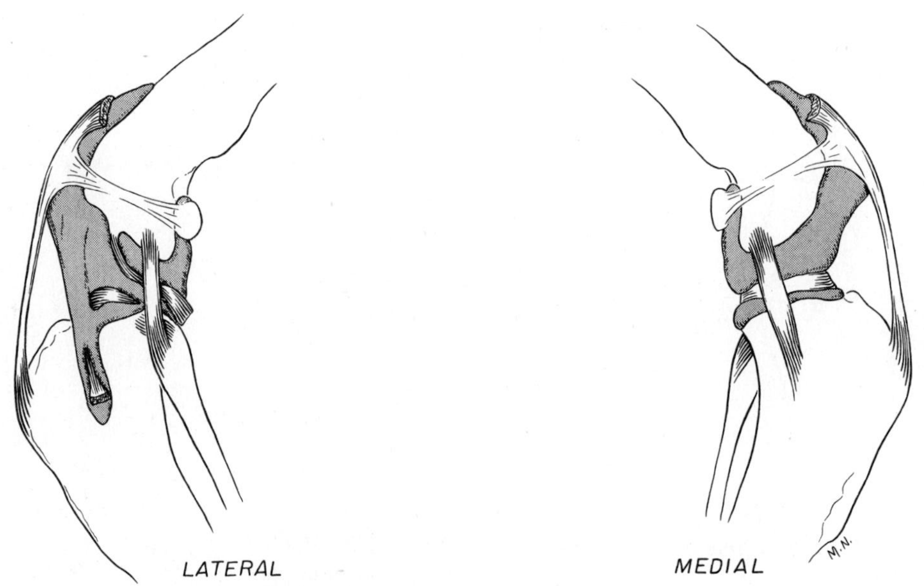

FIGURE 62. Capsule of left stifle joint.

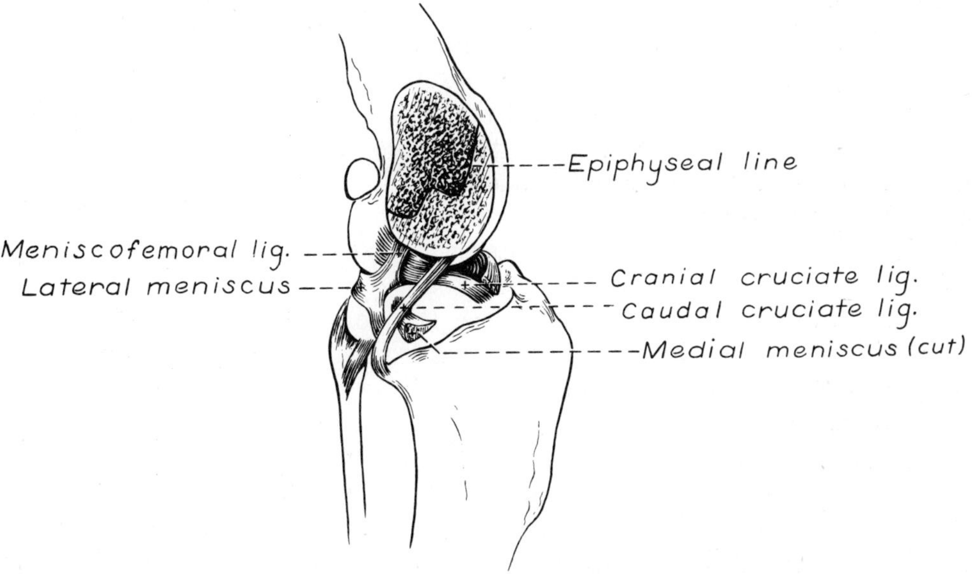

FIGURE 63. Cruciate and meniscal ligaments of left stifle, medial aspect.

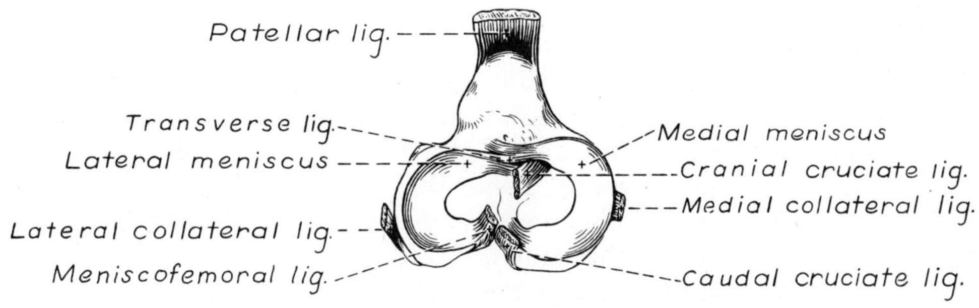

FIGURE 64. Menisci and ligaments, proximal end of left tibia.

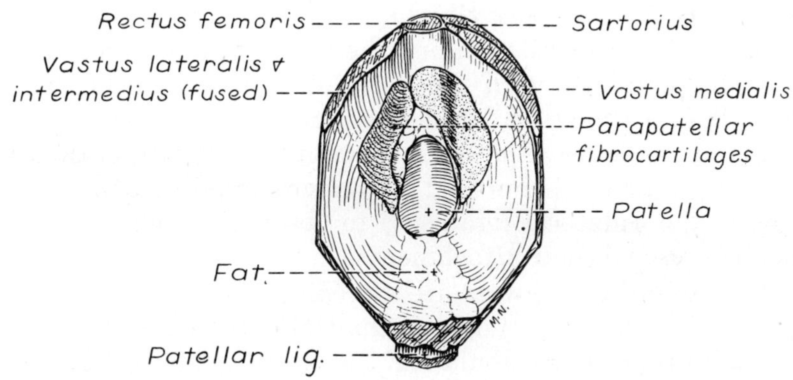

FIGURE 65. Left patella and fibrocartilages, articular surface.

Each meniscus attaches to the cranial and caudal intercondylar areas of the tibia (Fig. 64). A **transverse ligament** connects the cranial ends of the menisci. The caudal part of the lateral meniscus is attached to the intercondylar fossa of the femur by a **meniscofemoral ligament.** The medial meniscus is attached to the medial collateral ligament and moves only slightly when the stifle is flexed. The lateral meniscus is separated from the lateral collateral ligament by the tendon of origin of the popliteus. When the stifle is flexed the tibia rotates medially and the lateral meniscus moves caudally on the tibial condyle.

The **femorotibial ligaments** are the collateral and cruciate ligaments. The **medial collateral ligament** extends from the medial epicondyle of the femur to the medial side of the tibia distal to the medial condyle. It fuses with the joint capsule and the medial meniscus.

The **lateral collateral ligament** extends from the lateral epicondyle of the femur over the tendon of origin of the popliteus to the head of the fibula and adjacent lateral condyle of the tibia. Clean these ligaments to observe their attachments. With the stifle extended these ligaments prevent abduction, adduction and rotation of the stifle. When the stifle is flexed the lateral ligament is loosened.

The **cruciate ligaments** pass between the intercondylar areas of the tibia and femur and limit craniocaudal motion of these bones. The ligaments cross each other near their attachments in the intercondylar fossa of the femur.

The **cranial cruciate ligament** attaches within the intercondylar fossa of the femur to the caudomedial part of the lateral condyle. It extends distocranially to attach to the cranial intercondylar area of the tibia. This attachment is just behind the cranial attachment of the medial meniscus. The cranial cruciate ligament keeps the tibia from sliding cranially beneath the femur when the limb bears weight. It also limits medial rotation of the tibia when the stifle is flexed.

The **caudal cruciate ligament** attaches proximally within the intercondylar fossa of the femur to the medial condyle of the femur. Distally it attaches to the medial edge of the popliteal notch of the tibia behind the caudal attachment of the medial meniscus. This ligament prevents caudal movement of the tibia beneath the femur when the limb bears weight. Observe the attachments of these ligaments. They are named in accordance with the position of their tibial attachments.

Flex the stifle and while holding the femur firmly in one hand try to move the tibia cranially with the other. In the normal dog there is no movement. Transect the tibial attachment of the cranial cruciate ligament and repeat the procedure. The excessive cranial excursion of the tibia that results is diagnostic for a ruptured cranial cruciate ligament.

Transect the collateral ligaments to observe the effect on the joint. Note the menisci and their attachments.

The two **tibiofibular joints** are a proximal joint between the head of the fibula and the lateral condyle of the tibia and a distal joint between the lateral malleolus of the fibula and the lateral surface of the distal end of the tibia. Throughout the length of the interosseous space between the tibia and fibula is a sheet of fibrous tissue uniting the two bones, the **interosseous membrane** of the crus.

The **tarsal joint** is a composite of several articulations and joint sacs. The tarsocrural joint sac is the largest and incorporates the articulations of the distal tibia and fibula with that of the talus and calcaneus. Greatest movement in the tarsus occurs between the cochlea of the tibia and the trochlea of the talus. Puncture to obtain synovia is made into this joint sac. Other joint sacs include the proximal and distal intertarsal sacs and the tarsometatarsal sac. Medial and lateral collateral ligaments cross the tarsus from the tibia and fibula to the metatarsal bones. Numerous ligaments join the individual tarsal bones to each other. These will not be dissected.

Live Dog

Place the thumb of one hand behind the greater trochanter. With the other hand grasp the femur and flex and extend and rotate the hip. Try to lift the femoral head out of the acetabulum. This will be prevented by the ligament of the femoral head and joint capsule.

Stand behind the dog and palpate each stifle. Feel the patellar ligament and notice the normal slight depression caudal to it on the medial side. When the joint is swollen this is less evident. Remember there is fat between the patellar ligament and the stifle joint capsule.

With the dog recumbent, extend the stifle and palpate the patella. Move it laterally and medially. You should not be able to luxate the normal patella. Palpate the edges of the trochlea. Follow them distally to the femoral condyles. Palpate the tibial condyles and the collateral ligaments of the stifle. With the stifle extended rotate the tibia medially and laterally. Flex the stifle and repeat this rotation and note the normal increase in rotation, especially medially. Keep the stifle flexed and hold the femur with one hand while attempting to move the tibia cranially and caudally with the other hand to determine the integrity of the cruciate ligaments.

Palpate the tarsus. Flex and extend the tarsocrural joint and feel the ridges of the trochlea of the talus. Inflammation of this joint will cause this joint sac to enlarge.

BONES OF THE VERTEBRAL COLUMN

The **vertebral column** (see Fig. 2) consists of approximately 50 irregular bones. The vertebrae are arranged in five groups: **cervical, thoracic, lumbar, sacral** and **caudal.** The first letter or abbreviation of the word designating each group followed by the number of vertebrae in each group expresses a vertebral formula. That of the dog is C_7 T_{13} L_7 S_3 Cd_{20}. The number 20 for the caudal vertebrae is arbitrary; many dogs have fewer. The three sacral vertebrae fuse to form what is considered a single bone, the **sacrum.**

A typical vertebra consists of a body, a vertebral arch consisting of right and left pedicles and laminae, and transverse, spinous and articular processes (see Figs. 69, 70).

The **body** of a vertebra is constricted centrally. The cranial extremity is convex and the caudal extremity is concave. Adjacent vertebral bodies are connected by intervertebral disks, each of which is a fibrocartilaginous structure composed of a soft center, the **nucleus pulposus,** surrounded by concentric layers of dense fibrous tissue, the **anulus fibrosus.** These intervertebral disks will be dissected when the spinal cord is studied.

The **vertebral arch** is subdivided into basal parts, the **pedicles,** and the dorsal portion, formed by two **laminae.** Together with the body the vertebral arch forms a short tube, the **vertebral foramen.** All the vertebral foramina join to form the **vertebral canal.** The pedicles of each vertebra extend from the dorsolateral surface of the body. They present smooth-surfaced notches. The **cranial vertebral notches** are shallow; the **caudal vertebral notches** are deep. When the vertebral column is articulated the notches of adjacent vertebrae and the intervening fibrocartilage form the right and left **intervertebral foramina.** Through these pass the spinal nerves and blood vessels. The dorsal part of the vertebral arch is composed of right and left laminae, which unite to form a spinous process. Each typical vertebra has, in addition to the dorsally located **spinous process,** or spine, paired **transverse processes** that project laterally from the region where the arch joins the vertebral body. Farther dorsally on the arch, at the junction of pedicle and lamina, are located the **articular processes.** There are two of these on each side of the vertebra—a cranial pair whose articulating surfaces point dorsally or medially and a caudal pair whose surfaces are directed ventrally or laterally.

Cervical Vertebrae

There are seven cervical vertebrae in domestic mammals. The **atlas** (Fig. 66), or first cervical vertebra, is atypical in both structure and function. It articulates with the skull cranially. Its chief peculiarities are modified articular processes, a lack of a spinous process and a reduction of its body. The lateral parts are thick, forming the **lateral masses.** These are united by the **body** (ventral arch) ventrally and **arch** dorsally. The shelflike transverse processes, or wings, project from the lateral masses. The two **cranial articular foveae** articulate with the occipital condyles of the skull to form the atlanto-occipital joint, of which the main movement is flexion and extension. The **caudal articular foveae** consist of two shallow glenoid cavities that form a freely movable articulation with the second cervical vertebra. Rotatory movement occurs at this joint. Examine the caudal part of the dorsal surface of the body for the **fovea dentis.** This is concave from side to side and articulates with the dens of the second cervical vertebra. This articular area blends with those of the caudal surfaces of the lateral masses. Besides the large vertebral foramen through which the spinal cord passes, there are two pairs of foramina in the atlas. The **transverse foramina** are actually short canals located just

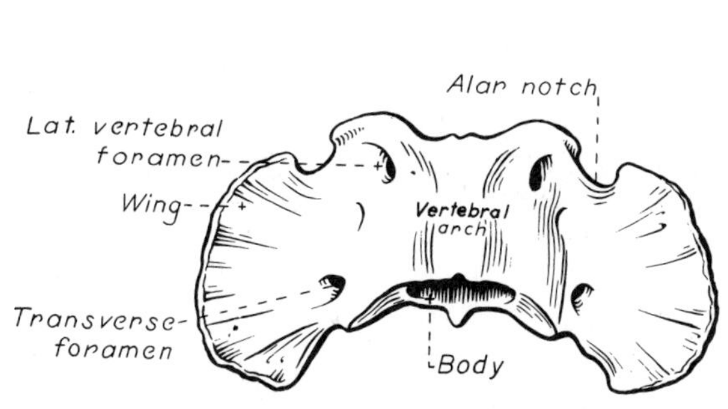

FIGURE 66. Atlas, dorsal view.

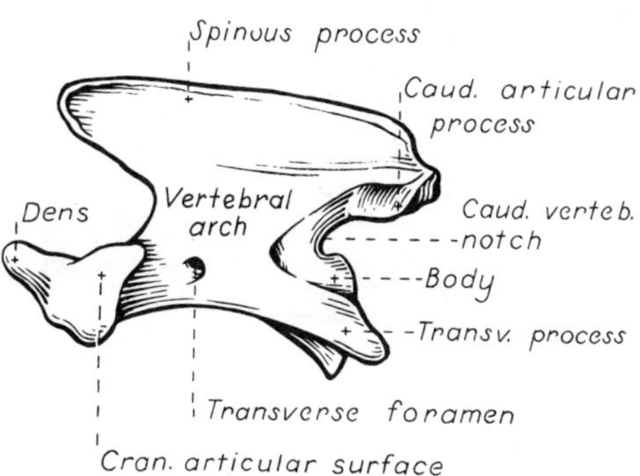

FIGURE 67. Axis, left lateral view.

lateral to the lateral masses. They pass obliquely through the transverse processes of the atlas. The **lateral vertebral foramina** perforate the cranial part of the vertebral arch. The first cervical spinal nerves pass through these foramina.

The **axis,** or second cervical vertebra (Fig. 67), presents an elongated, ridgelike spinous process as its most prominent characteristic. It is also peculiar in that cranially the body projects forward in a peglike eminence, the **dens.** The ventral surface of the dens is articular, while the tip and dorsal surface may be rough because of ligamentous attachment. The **cranial articular surface** is located on the body and is continuous with the articular area of the dens. The caudal part of the vertebral arch has two articular processes that face ventrolaterally. At the root of the transverse process is the small transverse foramen. The cranial vertebral notch concurs with that of the atlas to form the first intervertebral foramen for the transmission of the second cervical nerve (Fig. 68). The caudal notch concurs with that of the third cervical vertebra to form the second intervertebral foramen, through which the third cervical nerve passes. Other features of this bone are similar to those of a typical vertebra.

The middle three cervical vertebrae (Fig. 69) differ little from a typical vertebra. The spinous processes are low but gradually increase in

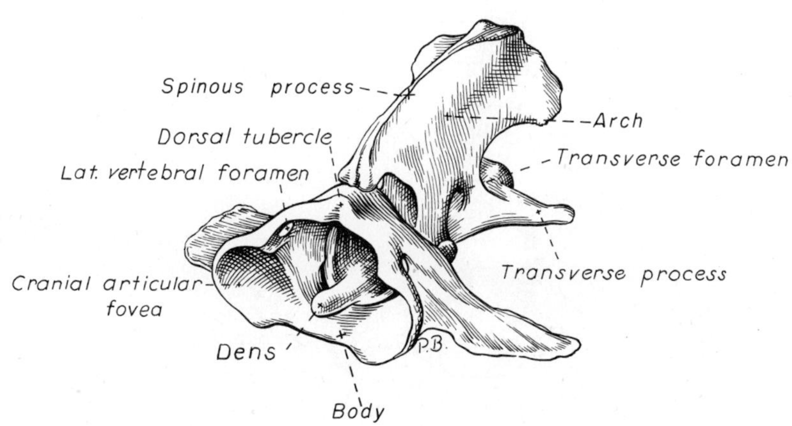

FIGURE 68. Atlas and axis articulated, craniolateral aspect.

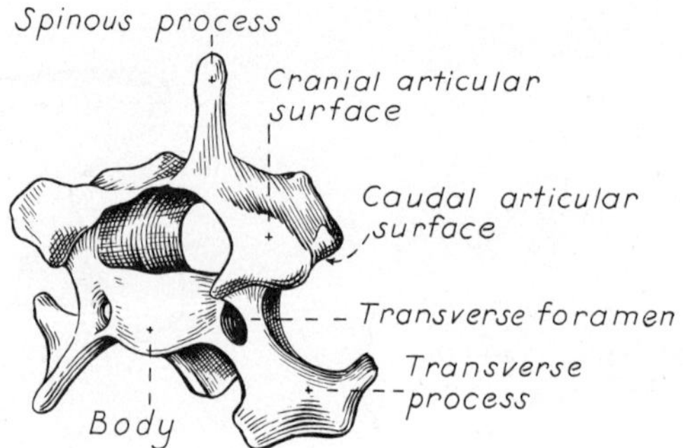

FIGURE 69. Fifth cervical vertebra, craniolateral aspect.

height from third to fifth. The transverse processes are two-pronged and are perforated at the base by a transverse foramen.

The sixth cervical vertebra has a high spine and an expanded **ventral lamina** of the transverse process. The seventh cervical vertebra (Fig. 70) lacks transverse foramina and has the highest cervical spine.

The **cranial articular processes** of cervical vertebrae 3 to 7 face dorsally and cranially in apposition to the **caudal articular processes** of adjacent vertebrae. There is a prominent **ventral crest** on the caudal midventral aspect of the vertebral body.

Thoracic Vertebrae

There are 13 thoracic vertebrae (Fig. 71); the first nine are similar. The bodies of the thoracic vertebrae are short. The first through the tenth have a **cranial** and a **caudal costal fovea** on each side for rib articulation. Since the foveae of two adjacent vertebrae form the articular surface for the head of one rib, they are sometimes called demifacets. The body of the tenth vertebra frequently lacks caudal foveae, while the eleventh through the thirteenth have only one complete fovea on each side. The head of the first rib articulates between the last cervical and first thoracic vertebrae. The tubercles of the ribs articulate with the transverse processes of the thoracic vertebrae of the same number in all instances.

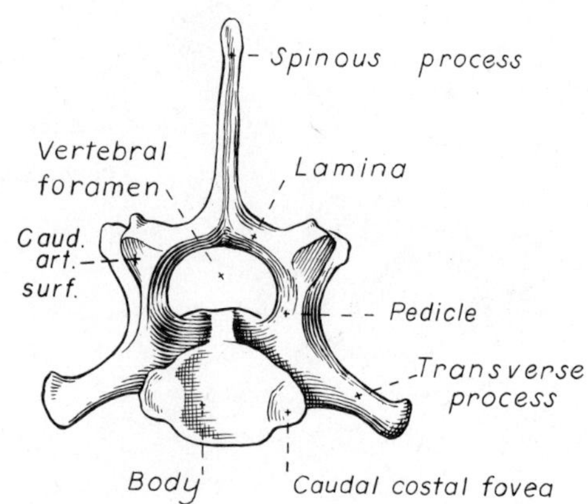

FIGURE 70. Seventh cervical vertebra, caudal aspect.

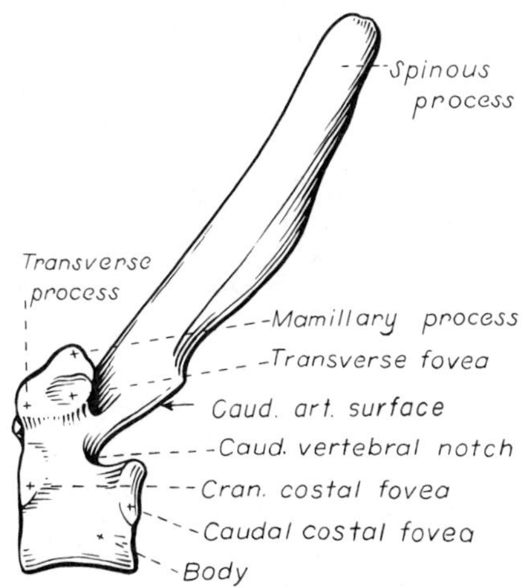

FIGURE 71a. Sixth thoracic vertebra, left lateral view.

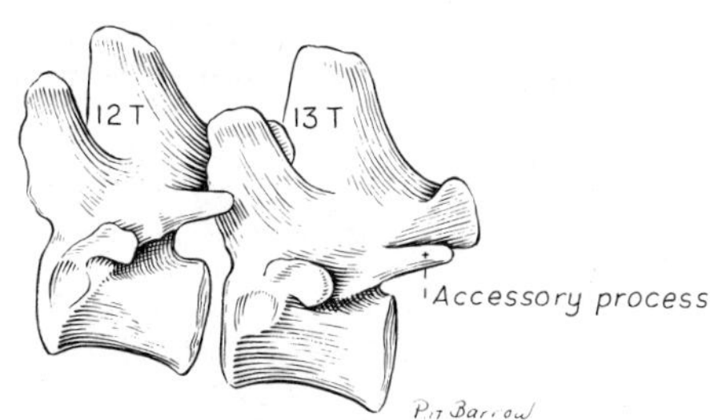

FIGURE 71b. The 12th and 13th thoracic vertebrae, left lateral view.

The **spine** is the most conspicuous feature of each of the first nine thoracic vertebrae. The massiveness of the spines gradually decreases with successive vertebrae, but there is little change in length and direction until the seventh or eighth is reached. The spines then become progressively shorter and incline caudally through the ninth and tenth. The spine of the eleventh thoracic vertebra is nearly perpendicular to the long axis of that bone. This vertebra is designated the **anticlinal vertebra.** All spines caudal to the eleventh point cranially; all spines cranial to the eleventh point caudally (Fig. 76).

The transverse processes are short, blunt and irregular. All contain costal foveae for articulation with the tubercles of the ribs.

The **articular processes** are located at the junctions of the pedicles and the laminae. The cranial pair are nearly confluent at the median plane on thoracic vertebrae 3 through 10. Similar to cervical vertebrae 3 through 7, the cranial articular surfaces of thoracic vertebrae 1 through 10 face dorsally and slightly cranially. Since the caudal articular processes articulate with the cranial ones of the vertebra behind, they are similar in shape but face in the opposite direction. The joints between thoracic vertebrae 10 to 13 are conspicuously modified because the articular surfaces of the caudal articular processes are located on the lateral side and the articular surfaces of the cranial articular processes of thoracic vertebrae 11 through 13 face medially. A similar arrangement is seen in all lumbar vertebrae. This type of interlocking articulation allows flexion and extension of the caudal-thoracic and lumbar region while limiting lateral movement.

The **accessory process** (Fig. 71b) projects caudally from the pedicle ventral to the caudal articular process and over the dorsal aspect of the intervertebral foramen. It is present from the midthoracic region to the fifth or sixth lumbar vertebra. The **mamillary process** is a knoblike dorsal projection of the transverse processes of the second through tenth thoracic vertebrae and cranial articular processes of the eleventh thoracic

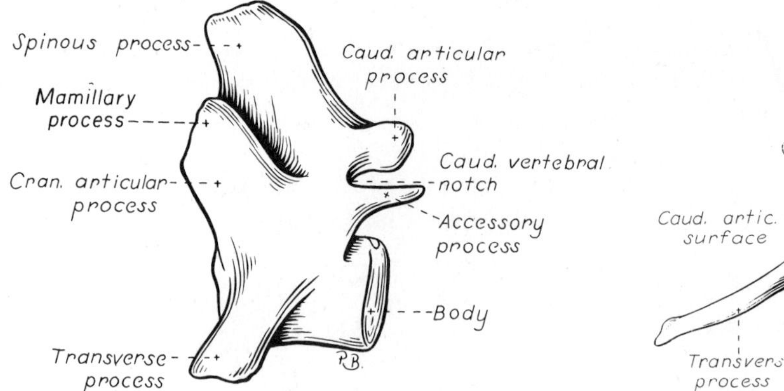

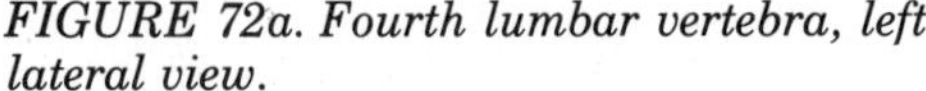
FIGURE 72a. Fourth lumbar vertebra, left lateral view.

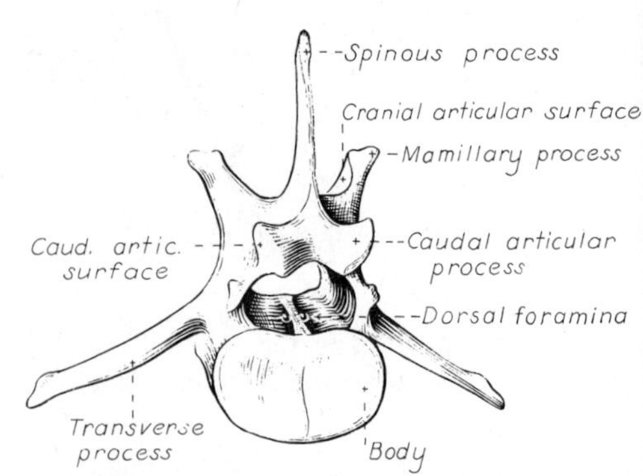

FIGURE 72b. Fifth lumbar vertebra, caudolateral view.

through the caudal vertebrae. Epaxial muscles of the transversospinalis system attach to these processes.

Lumbar Vertebrae

The lumbar vertebrae (Fig. 72) have longer bodies than the thoracic vertebrae. The transverse processes are directed cranially as well as ventrolaterally. The articular processes lie mainly in sagittal planes. The caudal processes protrude between the cranial ones of succeeding vertebrae. The prominent blunt spinous processes are largest in the midlumbar region and have a slight cranial direction.

Sacrum

The sacrum (Fig. 73) results from the fusion of the bodies and processes of three vertebrae. This bone lies between the ilia and firmly articulates with them. The body of the first segment is larger than the combined bodies of the other two segments. The three are united to form a concave ventral surface.

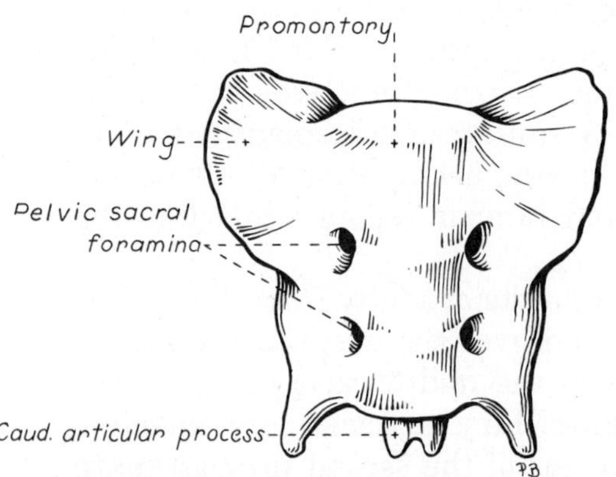

FIGURE 73a. Sacrum, ventral view.

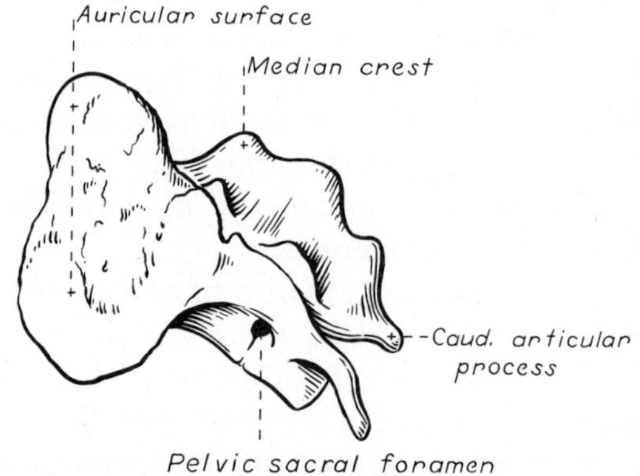

FIGURE 73b. Sacrum, left lateral view.

The dorsal surface presents several markings that result from the fusion of the three sacral vertebrae. The **median sacral crest** represents the fusion of the three spinous processes. The dorsal surface also bears two pairs of **dorsal sacral foramina,** which transmit the dorsal branches of the sacral spinal nerves.

The pelvic (ventral) surface has two pairs of **pelvic sacral foramina.** They transmit the ventral branches of the first two sacral spinal nerves. The wing of the sacrum is the enlarged lateral part that bears a large, rough surface, the **auricular face,** which articulates with the ilium.

The **base** of the sacrum faces cranially. The ventral part of the base has a transverse ridge, the **promontory.** This, with the ilia, forms the dorsal boundary of the pelvic inlet.

Caudal Vertebrae

The average number of caudal vertebrae (Fig. 74) in the dog is 20. These lose their distinctive features as one proceeds caudally.

Ribs

All 13 pairs of ribs (Figs. 75, 76) have dorsal bony and ventral cartilaginous parts which meet at a costochondral junction. This junction may be expanded in rapidly growing animals. The cartilaginous parts are called **costal cartilages.** The first nine pairs of ribs articulate directly with the sternum. The costal cartilages of the tenth, eleventh and twelfth unite with one another to form the **costal arch.** The thirteenth rib often ends freely in the flank.

The bony part of a typical rib presents a head, neck, tuberculum and body.

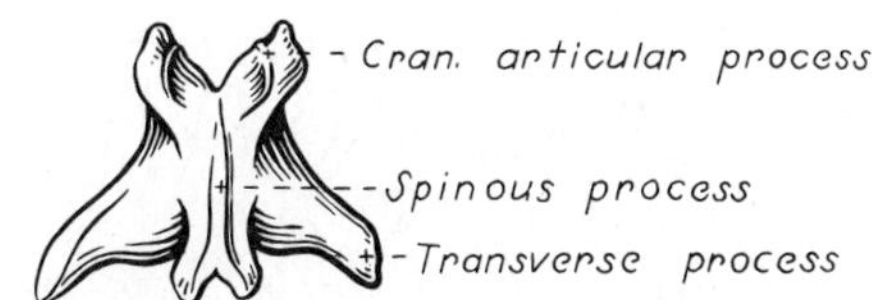

FIGURE 74a. Third caudal vertebra, dorsal view.

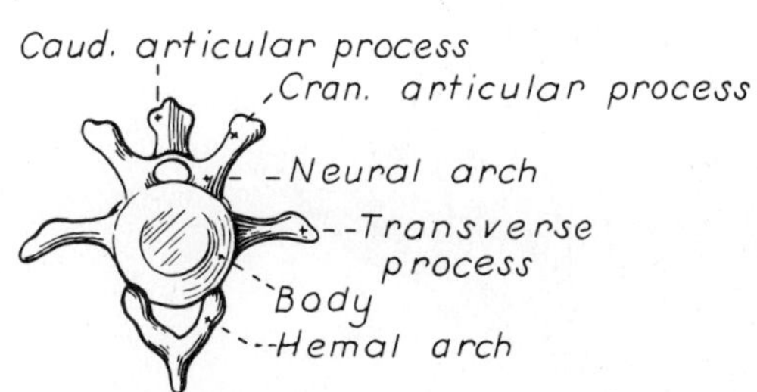

FIGURE 74b. Fourth caudal vertebra, cranial view.

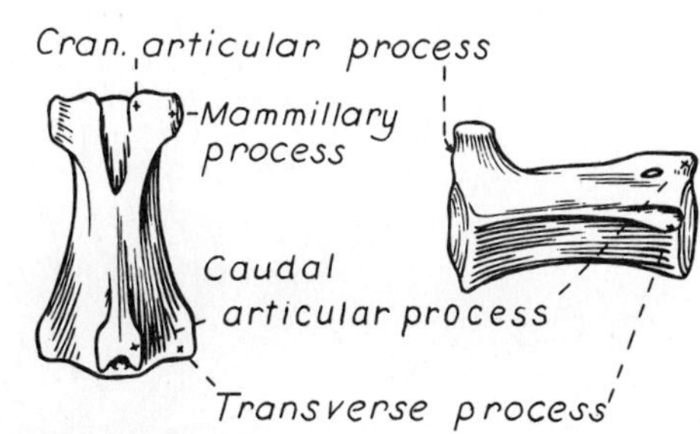

FIGURE 74c. Sixth caudal vertebra, dorsal and lateral views.

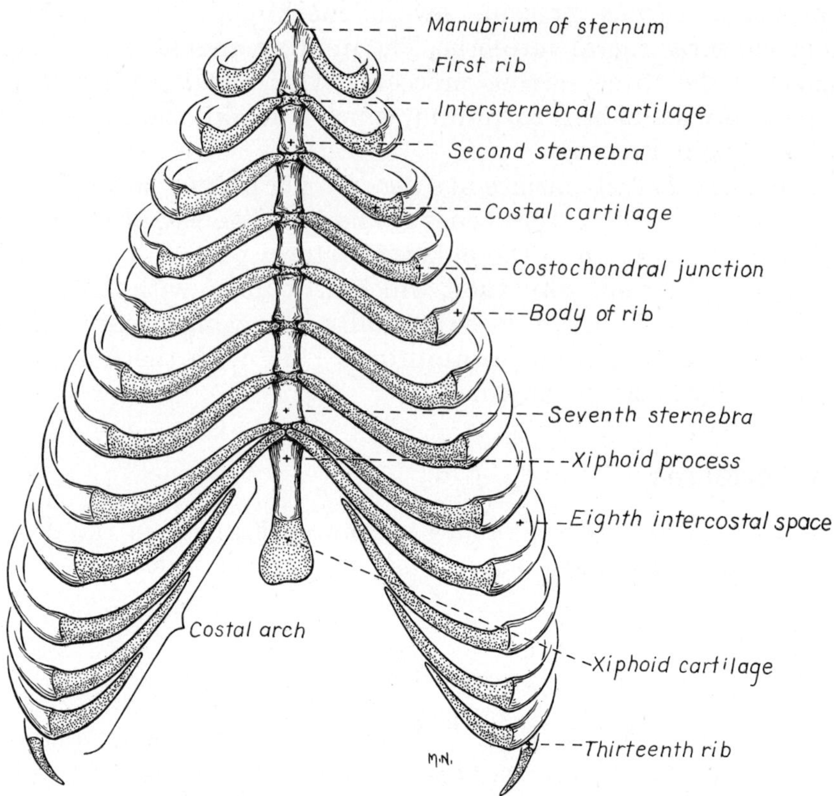

FIGURE 75. Rib cage and sternum, ventral view.

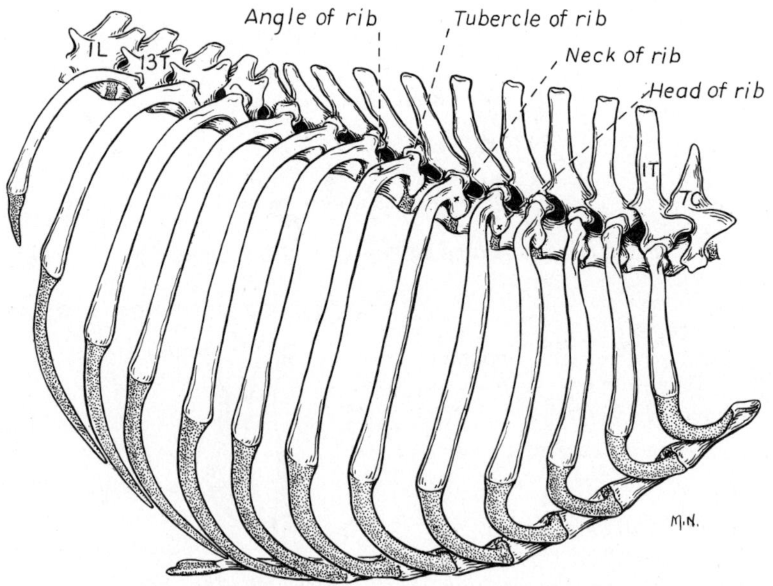

FIGURE 76. Rib cage and sternum, right lateral view.

The **head** of ribs 1 through 10 articulates with the costal foveae of two contiguous vertebrae and the intervening fibrocartilage. For ribs 11 through 13 the head articulates only with the costal fovea on the body of the vertebra of the same number. The **tuberculum** of the rib articulates with the costal fovea of the transverse process of the vertebra of the same number. Between the head and tuberculum of the rib is the **neck.**

Sternum

The sternum is composed of eight unpaired segments, the **sternebrae** (Fig. 75). Consecutive sternebrae are joined by the **intersternebral cartilages.** The first sternebra, also known as the **manubrium,** ends cranially in a clublike enlargement. The last sternebra is flattened dorsoventrally and is called the **xiphoid process.** The caudal end of this process is continued by a thin plate of cartilage.

Live Dog

Palpate the wings of the atlas. While holding these firmly, move the head up and down to extend and flex the atlanto-occipital joint. Where a transverse line drawn across the cranial edge of the wings of the atlas intersects the medial plane is the site for puncture to obtain cerebrospinal fluid. Palpate the spine of the axis. Move each wing of the atlas up and down to rotate the atlantoaxial joint. Palpate the transverse processes of cervical vertebrae 3 through 6. Note their relative position in the neck and the large volume of epaxial muscles dorsal to them. Feel the ventral edge of the broad transverse processes of the sixth cervical vertebra.

Palpate the spines of the thoracic and lumbar vertebrae. Note the prominence of the cranial thoracic and midlumbar spines. Palpate the transverse processes of the lumbar vertebrae. Locate the iliac crests and feel the spine of the seventh lumbar vertebra between them. This spine is shorter than that of the sixth vertebra and is used to locate the site for lumbosacral puncture for epidural anesthesia.

Palpate the ribs and compress the thorax noting the flexibility of the ribs. Palpate the sternum from manubrium to the xiphoid cartilage.

MUSCLES OF THE TRUNK AND NECK

The muscles of the trunk and neck, or axial muscles, are classified morphologically into hypaxial and epaxial groups. The epaxial muscles lie dorsal to the transverse processes of the vertebrae and function mainly as extensors of the vertebral column. The hypaxial group embraces all other trunk muscles not included in the epaxial division. These muscles are located ventral to the transverse processes and include those of the abdominal and thoracic walls.

The **superficial fascia of the trunk** covers the thorax and abdomen subcutaneously. It is continuous cranially and caudally with the superficial fasciae of the thoracic limb, the neck and the pelvic limb. Clean this fascia from the thorax and abdomen. It contains the cutaneous trunci muscle and numerous cutaneous vessels and nerves.

The **deep fascia of the trunk** (Fig. 41), the thoracolumbar fascia, is attached to the ends of the spinous and transverse processes of the thoracic and lumbar vertebrae. It passes over the epaxial musculature to the lateral thoracic and abdominal wall, where it serves as origin for several muscles. Detach the origin of the latissimus dorsi from the last few ribs and reflect it to the middorsal line where it arises from the lumbar vertebral spines via the superficial leaf of the thoracolumbar fascia.

Hypaxial Muscles

Muscles of the Neck

The **longus capitis** (see Fig. 83) lies on the lateral and ventral surfaces of cervical vertebrae, lateral to the longus colli. It arises from the transverse processes of the cervical vertebrae and inserts on the muscular tubercle on the ventral surface of the basioccipital bone of the skull.

The **longus colli** covers the ventral surfaces of the vertebral bodies from the sixth thoracic vertebra cranially to the atlas. It consists of many overlapping fascicles that attach to the vertebral bodies or the transverse processes. The most cranial cervical bundles attach to the atlas. Expose this muscle by reflecting the trachea and the esophagus and related soft tissues. Pass your finger into the thoracic inlet dorsally and feel the thoracic part of the longus colli muscle on the vertebral bodies. The longus colli must be reflected to expose the cervical intervertebral disks for surgical purposes.

Muscles of the Thoracic Wall

1. The **scalenus** (Figs. 77, 83) lies ventral to the origin of the serratus ventralis. It attaches to the first few ribs and the transverse processes of the cervical vertebrae and is divided into several slips. It is a muscle of inspiration.

2. The **serratus ventralis** (Fig. 77) is a large, fan-shaped muscle with an extensive origin on the neck and trunk. Its insertion on the serrated face of the scapula has been observed (see Fig. 15).

3. The **serratus dorsalis** arises by a broad aponeurosis from the tendinous raphe of the neck and from the thoracic and lumbar spines and inserts on the proximal portions of the ribs. It consists of two portions: the serratus dorsalis cranialis and the serratus dorsalis caudalis.

The **serratus dorsalis cranialis** (Fig. 77) lies on the dorsal surface

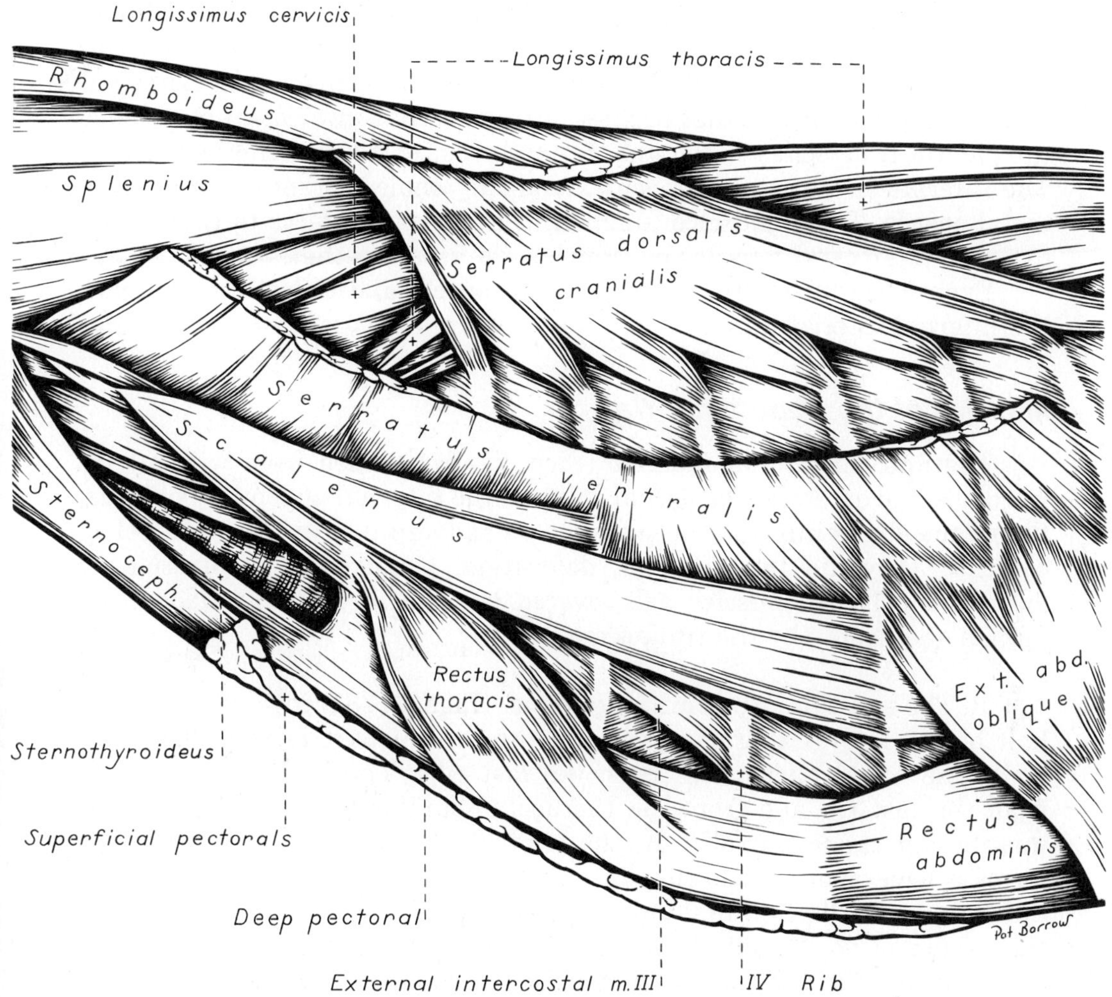

FIGURE 77. Muscles of neck and thorax, lateral view.

of the cranial thorax. It arises by a broad aponeurosis from the thoracolumbar fascia. It runs caudoventrally and inserts by distinct serrations on the craniolateral surfaces of the ribs. It lifts the ribs for inspiration. Transect this muscle at the beginning of its muscle fibers and reflect both portions.

The **serratus dorsalis caudalis** is smaller and is found on the dorsal surface of the caudal thorax. It consists of distinct muscle leaves that arise by an aponeurosis from the thoracolumbar fascia, course cranioventrally and insert on the caudal borders of the last three ribs. It functions in drawing the last three ribs caudally in expiration.

4. There are 12 **external intercostal muscles** (Fig. 77) on each side of the thoracic wall. Their fibers run caudoventrally from the caudal border of one rib to the cranial border of the rib behind. Their ventral border is near the costochondral junction. They function in respiration by drawing the ribs together, and their overall effect depends upon the fixation of the rib cage.

5. The **internal intercostal muscles** (see Fig. 92) are easily differentiated from the external intercostal muscles because their fibers run cranioventrally from the cranial border of one rib to the caudal border of the rib in front of it. Medial to most of the internal intercostal muscles is the pleura. It attaches to the muscles and ribs by the endothoracic fascia. The internal intercostal muscles extend the whole distance of the intercostal spaces. These muscles function in a manner similar to that of the external intercostal muscles by drawing the ribs together.

Expose the fifth external intercostal muscle and reflect it to observe the internal intercostal.

Muscles of the Abdominal Wall

The four abdominal muscles (Figs. 77–81) named from without inward, are the external abdominal oblique, the internal abdominal oblique, the rectus abdominis and the transversus abdominis (Fig. 78). When they contract, they aid in urination, defecation, parturition, respiration or locomotion. These muscles are covered superficially by the abdominal fasciae and deeply by the transverse fascia.

1. The **external abdominal oblique** (Figs. 77–81) covers the ventral half of the lateral thoracic wall and the lateral part of the abdominal wall. The costal part arises from the last ribs. The lumbar part arises from the last rib and from the thoracolumbar fascia. The fibers of this muscle run caudoventrally. In the ventral abdominal wall this muscle forms a wide aponeurosis that inserts on the linea alba and the cranial pubic ligament. The cranial pubic ligament extends transversely between the pecten of each pubic bone on the cranial surface of the pubis. The **linea alba** (Fig. 79) is the midventral aponeurosis of the abdominal muscles and extends from the xiphoid process to the symphysis pelvis. The aponeurosis of the external abdominal oblique, combined with that of the internal abdominal oblique, forms most of the external lamina of the sheath of the rectus abdominis.

Caudoventrally, just cranial to the iliopubic eminence and lateral to the midline, the aponeurosis of the external abdominal oblique separates into two parts, which then come together to form the **superficial inguinal ring** (Figs. 80, 81). This is the external opening of a very short, natural passageway through the abdominal wall, the **inguinal canal.** The internal opening and the boundaries of the canal will be seen when the abdominal cavity is opened. A blind extension of peritoneum protrudes through the inguinal canal to a subcutaneous position outside the body wall. This is the **vaginal tunic** (Figs. 81, 128) in the male and the **vaginal process** in the female (see Fig. 125). In the male it is accompanied by the testis and spermatic cord, which it envelops. In the female the vaginal process envelops the round ligament of the uterus and a varying amount of fat and ends blindly a short distance from the vulva. In both sexes the external pudendal artery and vein and the genitofemoral nerve also pass through the inguinal canal (Figs. 80, 168).

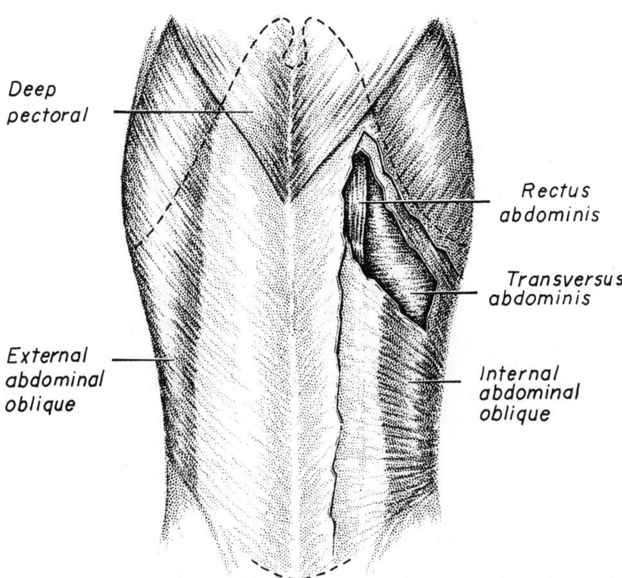

FIGURE 78. *Muscles of abdominal wall, ventral view.*

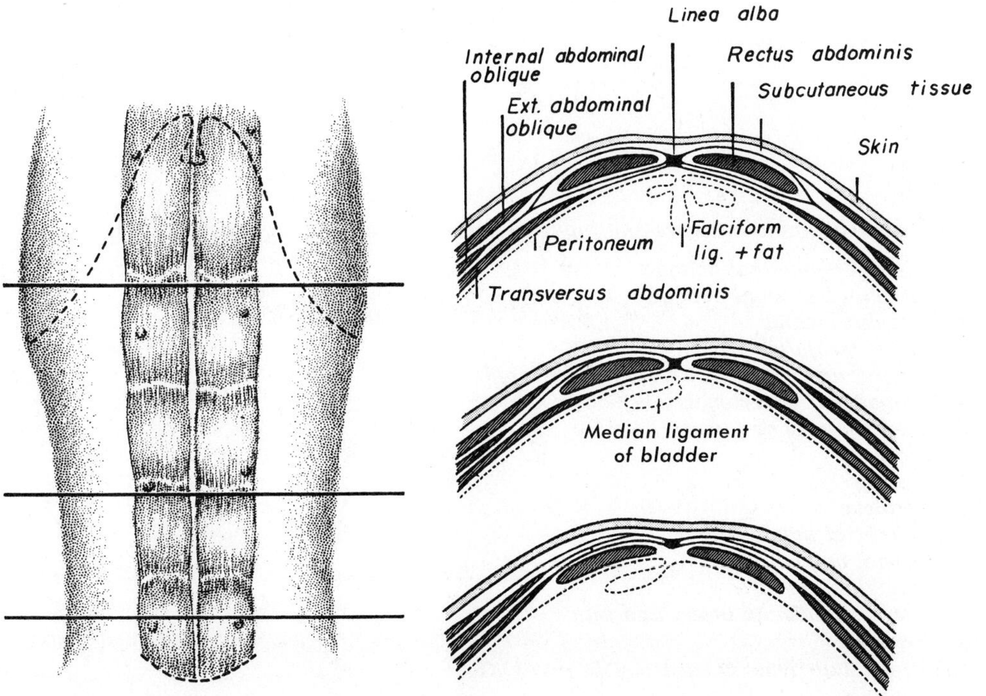

FIGURE 79. *Abdominal wall in ventral view with transections at three levels.*

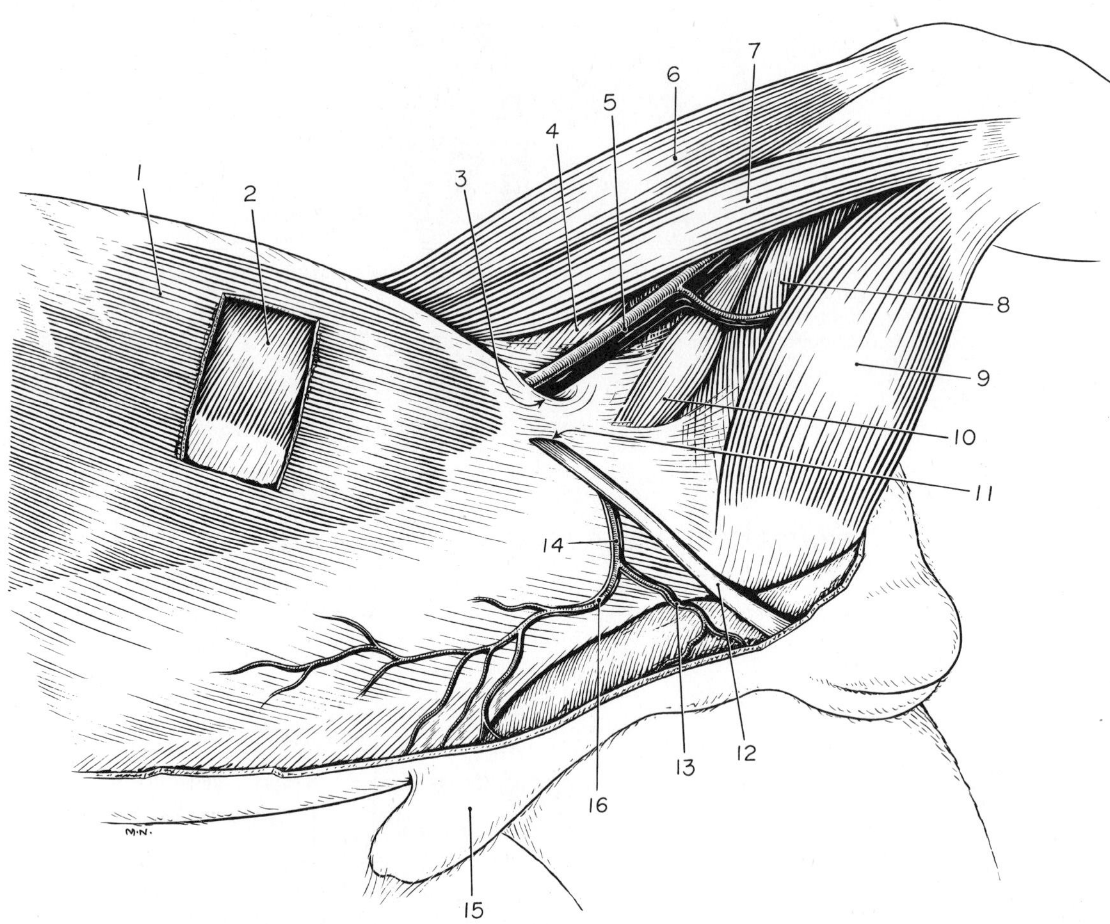

FIGURE 80. Abdominal muscles and inguinal region of the male, superficial dissection, left side.

1. *External abdominal oblique*
2. *Internal abdominal oblique*
3. *Vascular lacuna*
4. *Vastus medialis*
5. *Femoral artery and vein in femoral triangle*
6. *Cranial part of sartorius*
7. *Caudal part of sartorius*
8. *Adductor*
9. *Gracilis*
10. *Pectineus*
11. *Superficial inguinal ring*
12. *Parietal vaginal tunic*
13. *Cranial scrotal artery and vein*
14. *External pudendal artery and vein*
15. *Prepuce*
16. *Caudal superficial epigastric artery and vein*

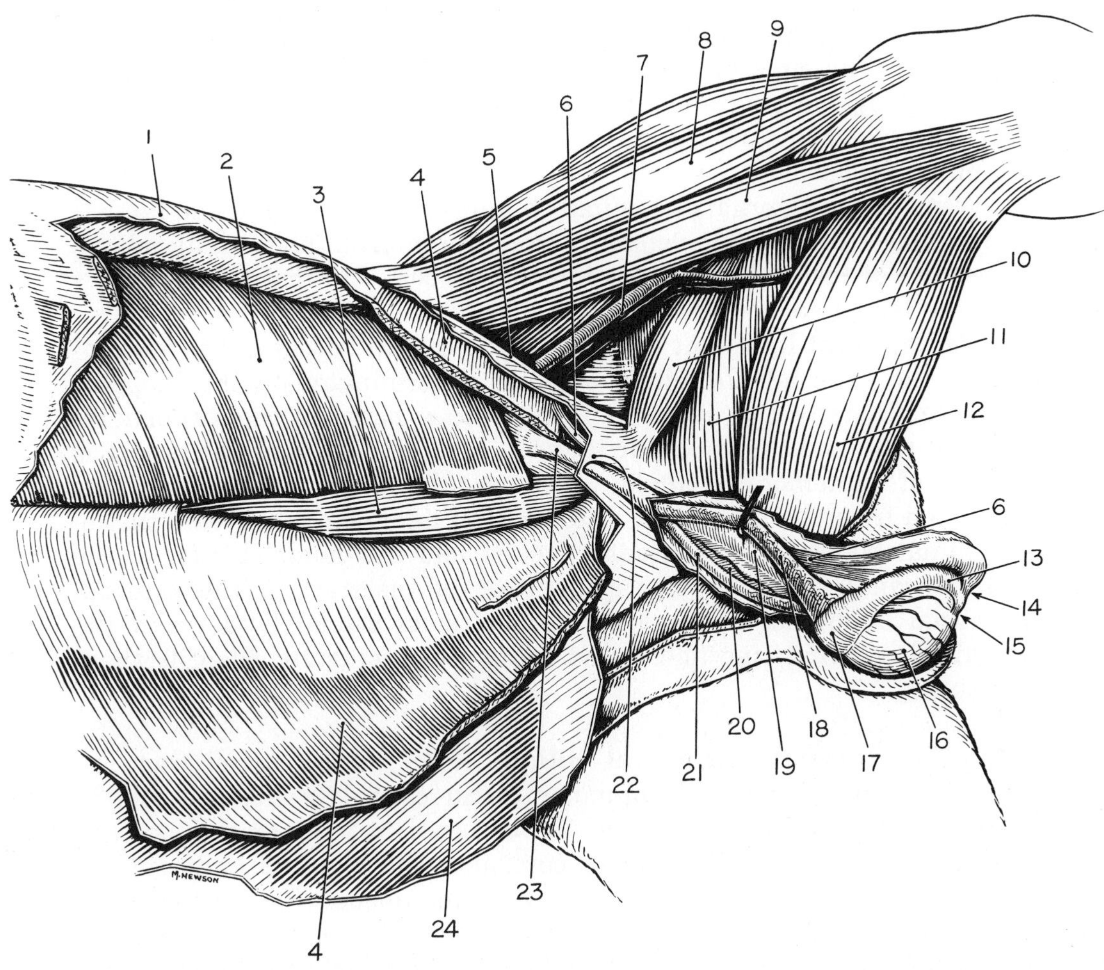

FIGURE 81. Abdominal muscles and inguinal region of the male, deep dissection, left side.

1. *Thoracolumbar fascia*
2. *Transversus abdominis*
3. *Rectus abdominis*
4. *Internal abdominal oblique (transected and reflected)*
5. *Inguinal ligament*
6. *Cremaster muscle*
7. *Femoral artery and vein*
8. *Cranial part of sartorius*
9. *Caudal part of sartorius*
10. *Pectineus*
11. *Adductor*
12. *Gracilis*
13. *Tail of epididymis*
14. *Ligament of tail of epididymis*
15. *Proper ligament of testis*
16. *Testis in visceral vaginal tunic*
17. *Head of epididymis*
18. *Testicular artery and vein in visceral vaginal tunic (mesorchium)*
19. *Mesorchium*
20. *Mesoductus deferens*
21. *Ductus deferens in visceral vaginal tunic*
22. *Superficial inguinal ring, lateral crus*
23. *Parietal vaginal tunic*
24. *External abdominal oblique (reflected)*

Clean the surface of the aponeurosis of the external abdominal oblique and identify the superficial inguinal ring and vaginal tunic. Be careful not to destroy the ring and tunic in cleaning this muscle. The vaginal tunic is covered by the spermatic fascia, which is continuous with the abdominal fascia.

Transect the external abdominal oblique close to its costal and lumbar origin. Reflect this muscle ventrally from the internal abdominal oblique to the line of fusion between their aponeuroses. The fused aponeuroses cover the rectus abdominis.

The **inguinal ligament** (Fig. 81) is the caudal border of the aponeurosis of the external abdominal oblique. It terminates on the iliopubic eminence. Distally the ligament is interposed between the **superficial inguinal ring** and the **vascular lacuna** (Fig. 80). The vascular lacuna is the base of the femoral triangle, which contains the femoral vessels that run to and from the hind limb. The inguinal ligament thus forms part of the cranial border of the vascular lacuna and the caudal border of the inguinal canal. Transect the aponeurosis of insertion of the external abdominal oblique 2 centimeters cranial to its caudal border and the superficial inguinal ring and reflect the muscle ventrally. Identify these structures.

2. The **internal abdominal oblique** (Figs. 80, 81) arises from the superficial leaf of the thoracolumbar fascia caudal to the last rib, in common with the lumbar part of the external abdominal oblique, and from the tuber coxae and adjacent portion of the inguinal ligament. Its fibers run cranioventrally. It inserts by a wide aponeurosis on the costal arch, on the rectus abdominis and on the linea alba, in common with the aponeurosis of the external abdominal oblique to which it is fused to form the external sheath of the rectus abdominis.

Transect the muscle 2 centimeters from its origin (Fig. 81) and reflect it ventrally to the rectus abdominis. In doing so, detach its insertions on the ribs. Separate it from all underlying structures except the rectus abdominis. Study the aponeurosis of the internal abdominal oblique, which contributes to the external lamina of the sheath of the rectus abdominis. Note how the caudal border of the muscle forms the cranial border of the inguinal canal. In the male note that fibers from the caudal border of the internal oblique form the cremaster muscle, which accompanies the vaginal tunic (Fig. 81).

3. The **transversus abdominis** (Figs. 41, 81) is medial to the internal abdominal oblique and the rectus abdominis. Its fibers run transversely. The muscle arises dorsally from the medial surfaces of the last four or five ribs and from the transverse processes of all the lumbar vertebrae by means of the deep leaf of the thoracolumbar fascia. Its aponeurosis attaches to the linea alba after crossing the internal surface of the rectus abdominis. Except for its most caudal part, the aponeurosis of the transversus abdominis forms the internal sheath of the rectus abdominis.

4. The **rectus abdominis** (Figs. 77–79, 81) extends from the pubis, where it forms the prepubic tendon, to the sternum. It flexes the thoracolumbar part of the vertebral column. Observe its cranial aponeurosis from the first few ribs and sternum. The rectus abdominis has distinct transverse tendinous intersections. In the umbilical region the external lamina of the sheath of the rectus abdominis is formed by the fused aponeuroses of the oblique muscles. The internal lamina is formed by the aponeurosis of the transversus abdominis.

The Inguinal Canal

The inguinal canal is a slit between the abdominal muscles. It extends from the deep to the superficial inguinal ring (see Fig. 128). The **superficial inguinal ring** in the aponeurosis of the external abdominal oblique has already been dissected. The **deep inguinal ring** is formed on the inside of the abdominal wall by the annular reflection of transversalis fascia onto the vaginal process. This fascia lies between the transversus abdominis and peritoneum. The ring is a boundary and not a distinct anatomical structure. The inguinal canal is bounded laterally by the aponeurosis of the external abdominal oblique, cranially by the internal abdominal oblique, caudally by the caudal border of the external abdominal oblique and medially by the lateral border of the rectus abdominis and by the transversalis fascia and peritoneum. The vaginal tunic and the **spermatic cord** pass obliquely caudoventrally through the inguinal canal.

Epaxial Muscles

The dorsal trunk musculature associated with the vertebral column and ribs may be divided into three parallel longitudinal muscle masses on each side. Each is composed of many overlapping fascicles. These three columns include the lateral **iliocostalis** system, the intermediate **longissimus** system and the medial **transversospinalis** system (Fig. 82). Various fusions occur between these columns, giving rise to different muscle patterns which are difficult to separate. These muscles act as extensors of the vertebral column and also produce lateral movements of the trunk when contracting on only one side.

Iliocostalis System

1. The **iliocostalis lumborum** (Fig. 82) arises from the wing of the ilium in common with the longissimus lumborum and inserts on the transverse processes of the lumbar vertebrae and the last four or five ribs. In the lumbar region this muscle is fused with the longissimus lumborum. The thoracolumbar fascia covers these muscles. Remove this fascia and any underlying fat to expose the glistening aponeurosis of these fused muscles, which attaches to the crest of the ilium and the spines of the lumbar and last four or five thoracic vertebrae. The cranial end of this

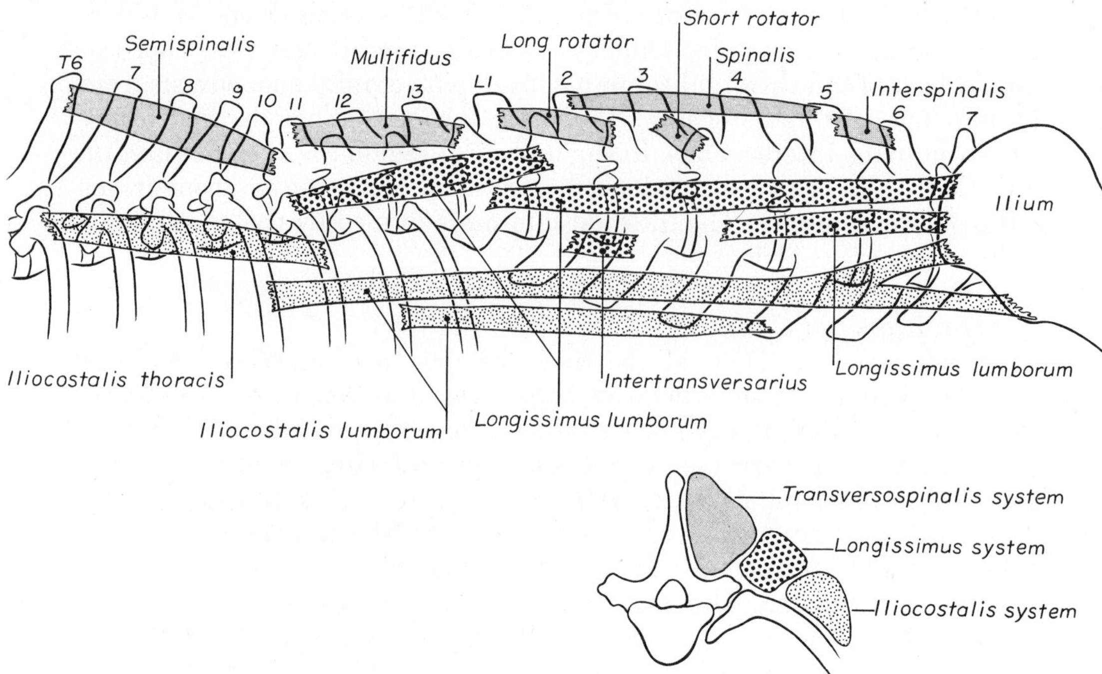

FIGURE 82. Schema of epaxial muscles.

muscle is distinctly separated from the longissimus lumborum as it inserts on the ribs. Expose this insertion.

2. The **iliocostalis thoracis** (Figs. 82, 83) is a long, narrow muscle mass extending from the twelfth rib to the transverse process of the seventh cervical vertebra. Individual components of the muscle extend between and overlap the ribs. Identify the boundaries of this muscle mass.

Longissimus System

The longissimus is the intermediate portion of the epaxial musculature. Lying medial to the iliocostalis, its overlapping fascicles extend from the ilium to the head. It consists of three major regional divisions: **thoracolumbar, cervical** and **capital.**

1. The **longissimus thoracis et lumborum** (Figs. 82, 83) arises from the crest and medial surface of the wing of the ilium and, by means of an aponeurosis, from the supraspinous ligament and the spines of the lumbar and thoracic vertebrae. Its fibers course craniolaterally. Superficially, only a shallow furrow is seen to separate the longissimus and iliocostalis muscles in the lumbar region. This division of the longissimus inserts on various processes of the lumbar and thoracic vertebrae. The thoracic portion may be seen inserting on the ribs, just medial to the iliocostalis thoracis.

2. The **longissimus cervicis** (Fig. 83), the cranial continuation of the longissimus muscle into the neck, consists of four fascicles so arranged

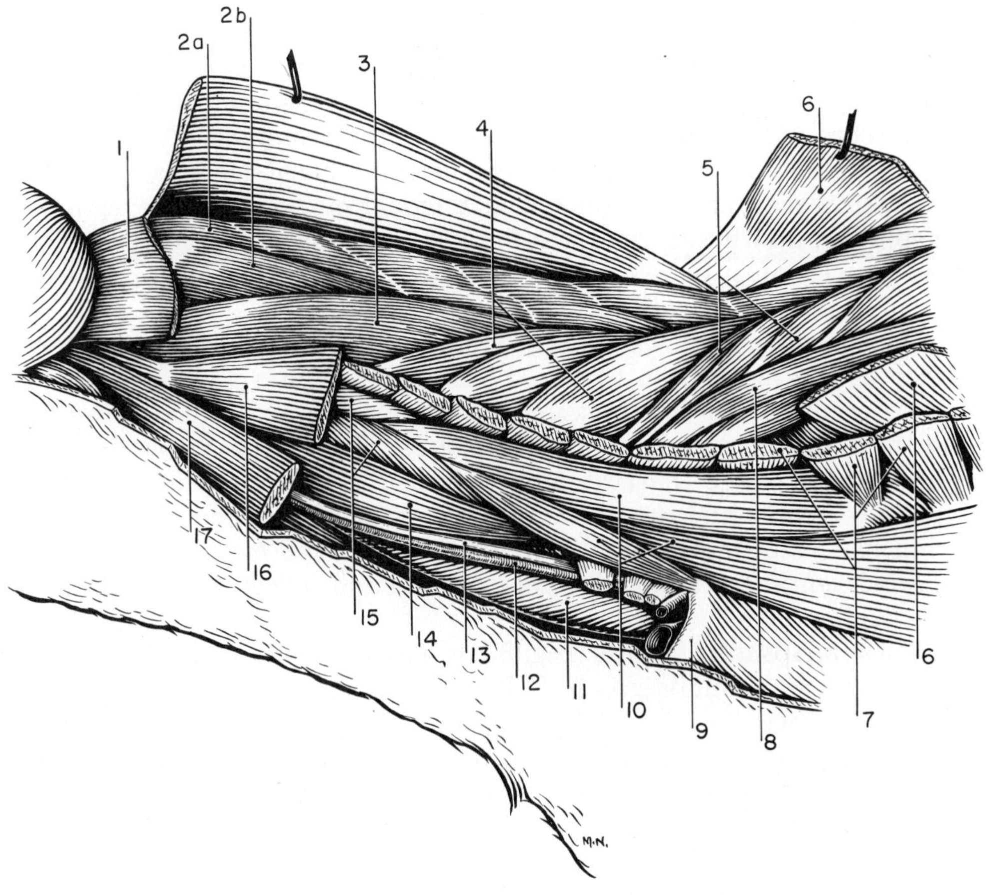

FIGURE 83. Deep muscles of neck, left side.

1. *Splenius*
2. *Semispinalis capitis:*
   a. *Biventer cervicis*
   b. *Complexus*
3. *Longissimus capitis*
4. *Longissimus cervicis*
5. *Longissimus thoracis*
6. *Serratus dorsalis cranialis*
7. *Serratus ventralis*
8. *Iliocostalis thoracis*
9. *First rib*
10. *Scalenus*
11. *Esophagus*
12. *Common carotid a.*
13. *Vagosympathetic trunk*
14. *Longus capitis*
15. *Intertransversarius*
16. *Omotransversarius*
17. *Cleidomastoideus*

that the caudal fascicles partially cover those that lie directly cranioventral to them. They lie in the angle between the cervical and thoracic vertebrae and insert on the transverse processes of the last few cervical vertebrae.

3. The **longissimus capitis** (Fig. 83) is a distinct muscle medial to the longissimus cervicis and splenius muscles. It extends from the first three thoracic vertebrae to the mastoid part of the temporal bone. It is firmly united with the splenius as it passes over the wing of the atlas to its insertion.

Transversospinalis System

The transversospinalis system, the most medial and deep epaxial muscle mass, consists of a number of different groups of muscles that join one vertebra with another or span one or more vertebrae. This complex system extends from the sacrum to the head. Included are muscles whose names depict their attachments or the functions of their fascicles: spinalis, semispinalis, multifidus, rotatores, interspinalis and intertransversarius. These muscles must be reflected to perform a surgical laminectomy. Only a few of these muscles will be dissected in the neck.

The **splenius** (Figs. 77, 83) is a rather large muscle on the dorsolateral surface of the neck, deep to the rhomboideus capitis and the serratus dorsalis cranialis. Its fibers extend in a slightly cranioventral direction from the third thoracic vertebra to the skull. The muscle arises from the cranial border of the thoracolumbar fascia, the spines of the first three thoracic vertebrae and the entire median raphe of the neck. It inserts on the nuchal crest and mastoid part of the temporal bone. Transect the splenius 2 centimeters caudal to its insertion and reflect the muscle mass to the median plane.

The **semispinalis capitis** (Fig. 83) is a member of the cervical portion of the transversospinalis group. It lies deep to the splenius and extends from the thoracic vertebrae to the head. It consists of two muscles, the dorsal biventer cervicis and the ventral complexus.

The **biventer cervicis** is dorsal to the complexus and has tendinous intersections. It arises from thoracic vertebrae and inserts on the caudal surface of the skull. Transect the muscle and reflect it.

The **complexus** is ventral to the biventer and arises from cervical vertebrae. It inserts on the nuchal crest. Transect this muscle and reflect it.

The **nuchal ligament** may be seen extending from the tip of the spinous process of the first thoracic vertebra to the broad caudal end of the spine of the axis. It is a laterally compressed, paired, yellow elastic band lying between the medial surfaces of the two semispinalis capitis muscles.

The **supraspinous ligament** continues the nuchal ligament caudally and extends from the spinous process of the first thoracic vertebra to the caudal vertebrae. It passes from one spinous process to another.

Live Dog

Palpate the cervical epaxial muscles and note the volume of these muscles dorsal to the cervical vertebrae. Continue caudally and feel the symmetry of the epaxial muscles on either side of the vertebral spines. In racing Greyhounds these thoracolumbar epaxial muscles are extremely well-developed. Palpate the abdominal wall and visualize the individual layers, the direction of their muscle fibers and the extent of the aponeuroses of these muscles. Occasionally the peritoneal cavity is opened by a surgical approach that separates the abdominal muscles in the plane of their fibers—the grid technique. Palpate the rectus abdominis and visualize the aponeuroses that sheath it and the direction of the rectus fibers. Feel the linea alba where most abdominal incisions are made.

With the dog in lateral recumbency, abduct one pelvic limb. Follow the pectineus muscle proximally to its origin from the prominent iliopubic eminence. The medial crus of the superficial inguinal ring extends cranially from this eminence. This crus has a slightly firm feeling and the spermatic cord in the male and external pudendal vessels in both sexes may be felt passing over this medial crus as the structures emerge from the inguinal canal.

JOINTS OF THE AXIAL SKELETON

Some of these joints will be seen now and others will be observed when the vertebral column and spinal cord are exposed.

Vertebral Joints

The **atlanto-occipital joint** (Fig. 84) is continuous with the **atlantoaxial joint** through the articulation of the dens with the body of the axis. The dens is held against the fovea dentis by the **transverse ligament of the atlas,** which passes dorsal to it and is attached to the body on both sides. Apical and alar ligaments pass from the cranial end of the dens to the basioccipital bone between the occipital condyles. The spine of the axis is joined to the arch of the atlas by a thick **dorsal atlantoaxial ligament** (Fig. 85).

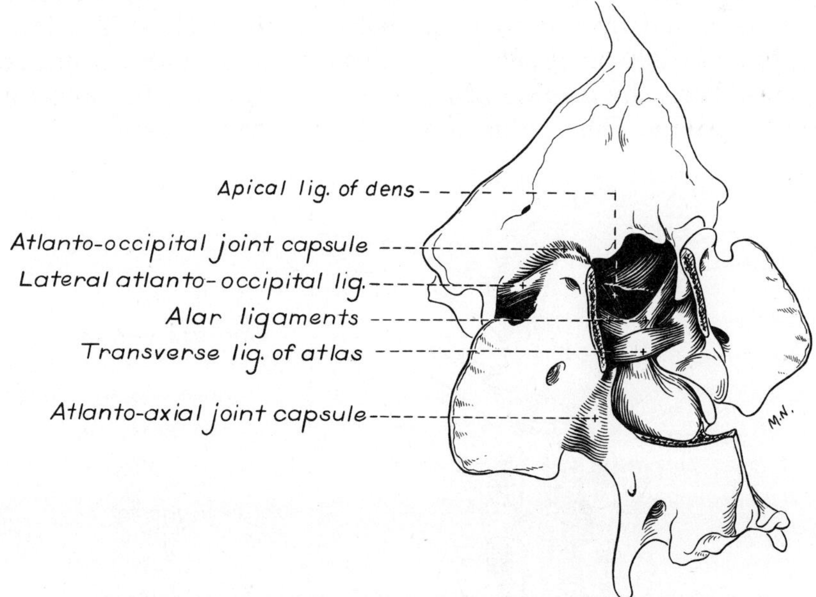

FIGURE 84. Ligaments of atlas and axis, dorsolateral view.

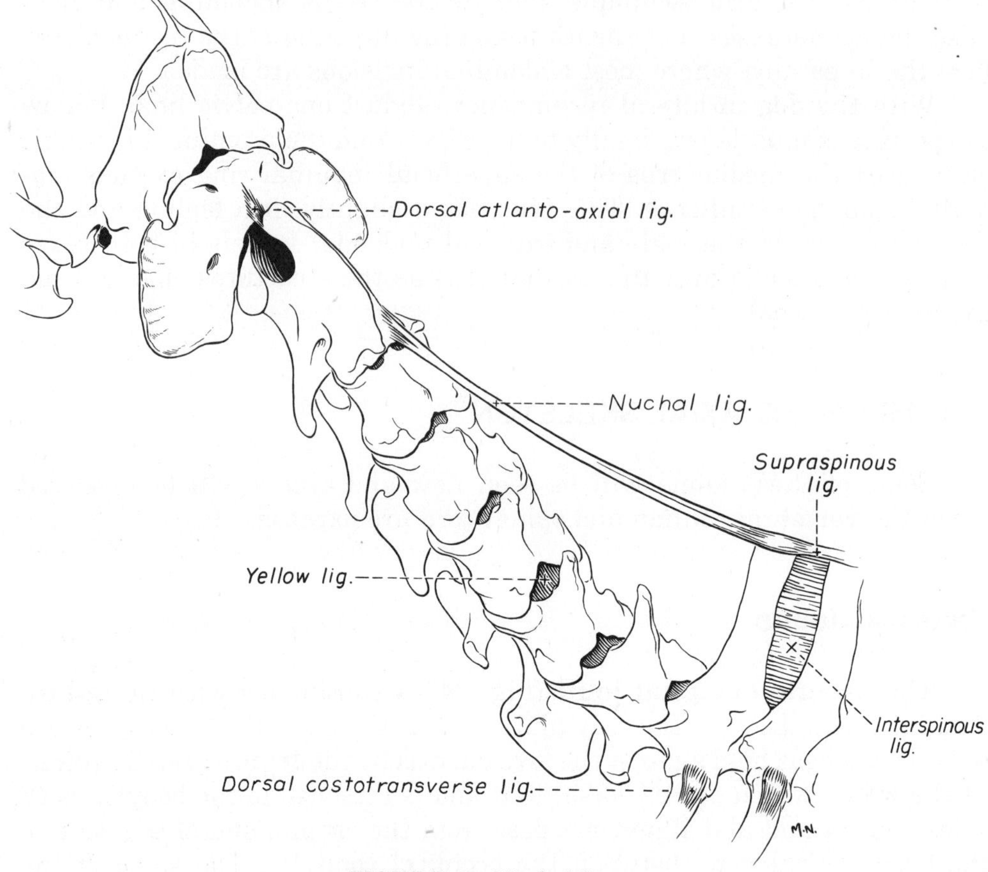

FIGURE 85. Nuchal ligament.

The remaining vertebrae articulate by synovial joints between articular processes, and by fibrous joints between the bodies. The latter are **intervertebral disks** (Figs. 86–88), which consist of outer circumferential collagenous fibers, the **anulus fibrosus,** and an inner gelatinous core, the **nucleus pulposus.** The anulus is usually thicker ventrally.

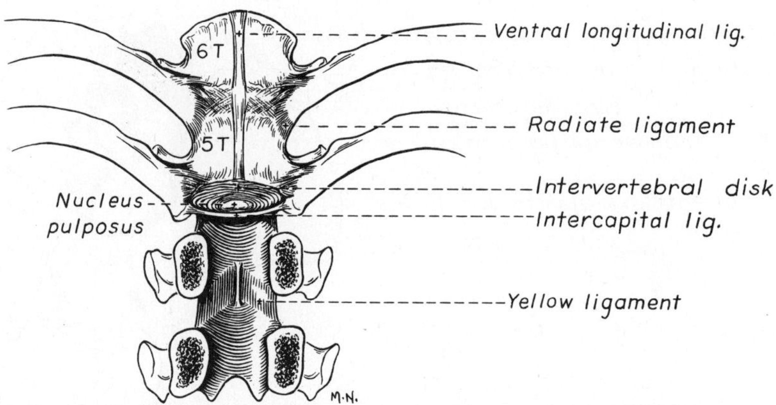

FIGURE 86. Ligaments of vertebral column and ribs, ventral view.

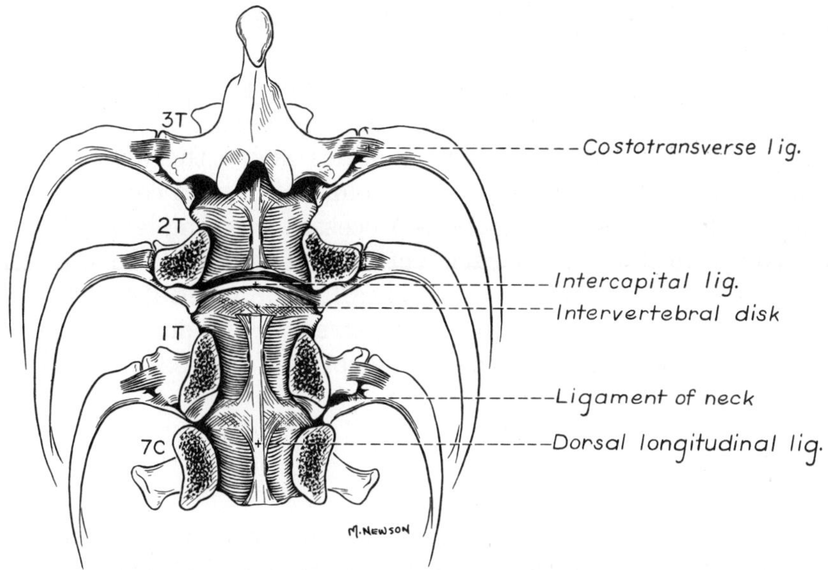

FIGURE 87. Ligaments of vertebral column and ribs, dorsal view.

Narrow longitudinal ligaments, one within the vertebral canal and the other beneath the vertebrae, extend across all of the vertebral bodies. The **ventral longitudinal ligament** (Fig. 86) is on the ventral surface of the vertebral bodies and extends from the sacrum to the axis. It is best developed in the caudal thoracic and lumbar regions. The thicker **dorsal longitudinal ligament** (Fig. 87) is on the midline of the floor of the vertebral canal ventral to the spinal cord. It widens where it passes over

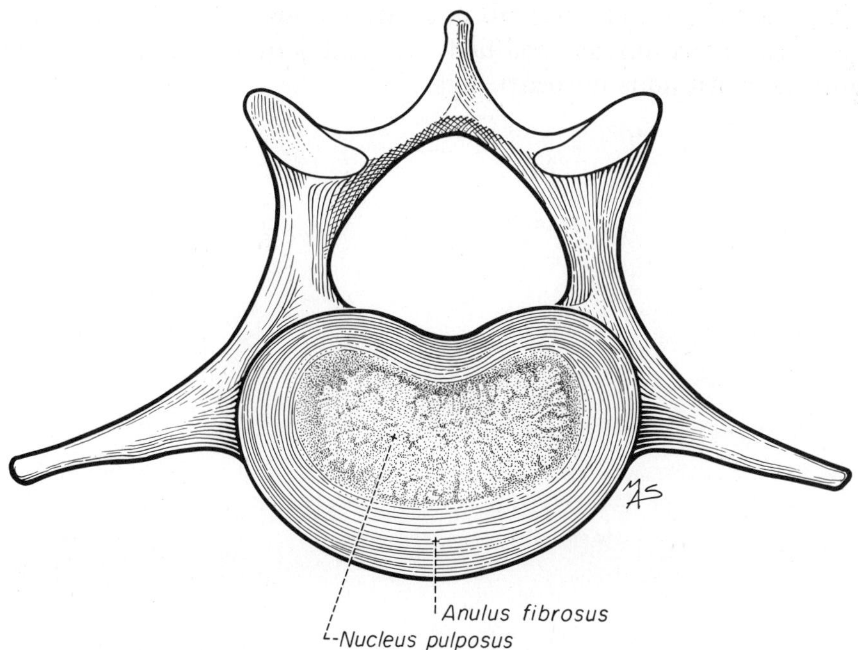

FIGURE 88. Intervertebral disk in lumbar region of ten-week-old pup, cranial view.

and attaches to the anulus fibrosis of the intervertebral discs. It extends cranially to the axis.

Yellow ligaments extend between vertebral arches to cover the epidural interarcuate space between the articular processes. **Interspinous ligaments** (Fig. 85) connect adjacent spines above the arches. The **supraspinous ligament** is a longitudinal band of fibrous connective tissue that connects the apices of all spinous processes from the level of the third caudal vertebra to the first thoracic vertebra. The cranial continuation of this ligament into the cervical region is called the **nuchal ligament** (Fig. 85). The nuchal ligament is elastic and paired. In the dog it extends from the apex of the first thoracic spine to the spine of the axis.

Ribs

The head of each rib articulates in a synovial joint with the craniodorsal aspect of the vertebra of the same number. For the first 10 ribs this articulation includes the caudodorsal portion of the next cranial vertebral body. These adjacent vertebral articular surfaces are demifacets. The tuberculum of each rib except the last few has a fibrous articulation with the transverse process.

From the second through the tenth ribs, where the rib heads articulate between vertebral bodies, an **intercapital ligament** (Figs. 86, 87) connecting right and left rib heads extends across the dorsal aspect of the anulus fibrosis ventral to the dorsal longitudinal ligament. This ligament holds the head of the rib in its socket and may provide additional containment for the intervertebral disk, which has a lower incidence of extrusion in this region.

The sternal part of each rib is cartilaginous. The second to seventh ribs join the sternum at modified synovial joints. Other ribs join the sternum as continuous fibrocartilages.

THE NECK, THORAX AND THORACIC LIMB

Make a skin incision from the midventral line at the level of the thoracic limb to the medial side of the right elbow joint. Make a circular skin incision at the elbow joint. Make transverse skin incisions from midventral to middorsal lines at the level of the umbilicus and at the cranial part of the neck. Reflect the skin flap to the dorsal median plane, leaving the cutaneous muscles and superficial fascia on the dog.

In the following dissection of vessels and nerves, the arteries can be recognized by the red latex that was injected into the arterial system. The veins will sometimes contain dark-colored clotted blood.

VESSELS AND NERVES OF THE NECK

There are eight pairs of cervical spinal nerves in the dog. The first cervical nerve passes through the lateral vertebral foramen of the atlas. The remaining nerves pass through succeeding intervertebral foramina. The eighth cervical nerve emerges from the intervertebral foramen between the seventh cervical and first thoracic vertebrae. Immediately upon leaving the foramina the nerves divide into large ventral and small dorsal branches. The dorsal branches supply structures dorsal to the vertebrae (see Fig. 93) and will not be dissected. When dissecting nerves it is helpful to bluntly separate the tissue by inserting a scissors and opening it to spread the tissue. Fascial strands of connective tissue will break but nerves usually stretch. Another method for exposing nerves is to scratch along the most likely path over the surface of a muscle or between muscles with a fine pointed forceps. As soon as any nerve becomes visible, follow it in both directions.

Palpate the wing of the atlas and dissect the fascia near its caudoventral border to uncover the ventral branch of the **second cervical nerve** (Fig. 89). This lies along or deep to the middle of the caudoventral border of the platysma, which is dorsal to the external jugular vein.

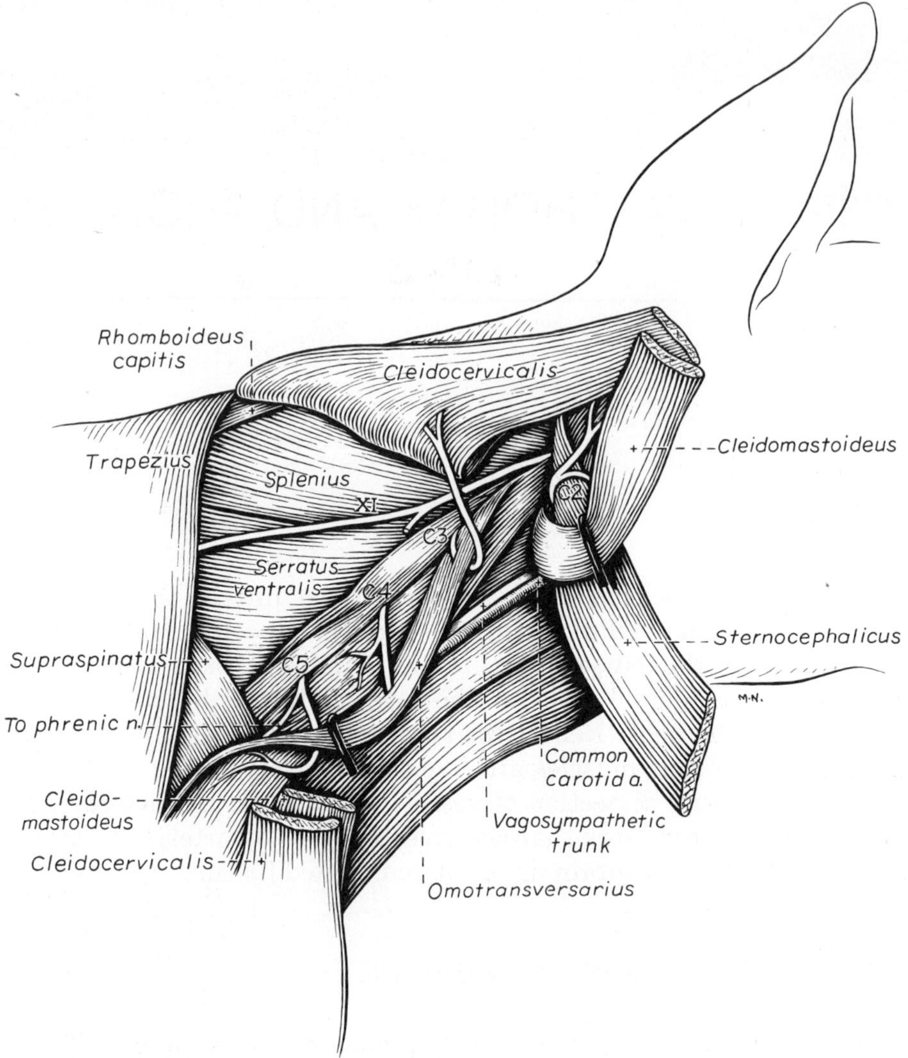

FIGURE 89. Ventral branches of cervical spinal nerves emerging through the lateral musculature.

Separate the overlying fascia until the nerve is found. It emerges between the cleidomastoideus and the omotransversarius. The ventral branch of the second cervical nerve divides into two cutaneous branches: (1) The **great auricular nerve** extends toward the ear. It branches and supplies the skin of the neck, the ear and the back of the head with sensory branches. Trace the nerve as far as present muscle and skin reflections will allow. (2) The **transverse cervical nerve** branches to the skin of the cranioventral part of the neck and need not be dissected.

The **external jugular vein** (Fig. 90), on the side of the neck, is formed by the **linguofacial** and **maxillary** veins. The ovoid body that lies in the fork formed by these veins is the **mandibular salivary gland** (see Fig. 13). The **mandibular lymph nodes** (see Fig. 13) lie on both sides of the linguofacial vein, ventral to the mandibular salivary gland.

Ligate and transect the external jugular vein at its approximate

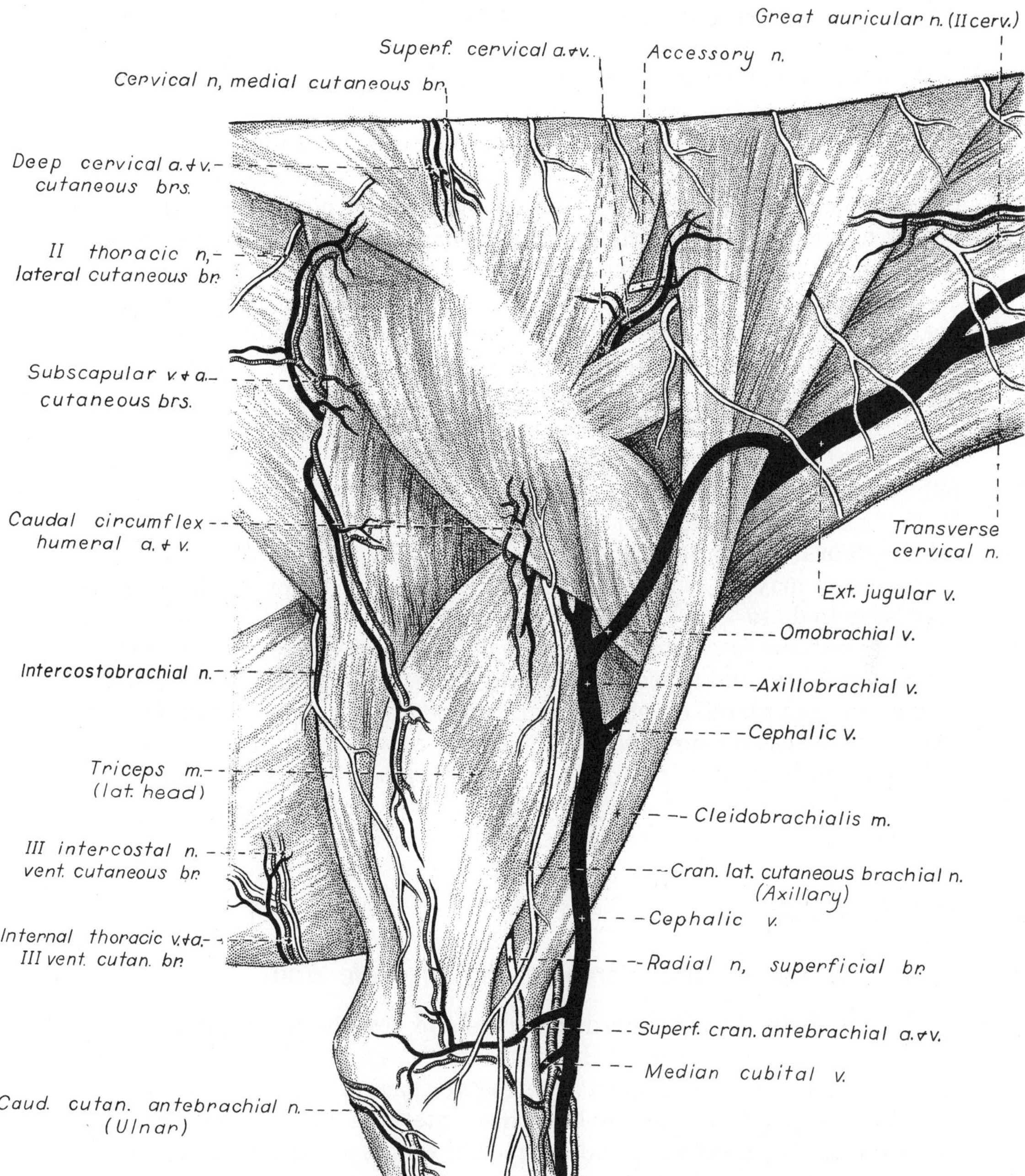

FIGURE 90. Superficial structures of scapula and arm, lateral view.

middle and reflect each end. In some specimens the omobrachial and cephalic veins (Fig. 90) may be observed entering the jugular vein after crossing the shoulder. These may be transected and reflected. Free the sternocephalicus and transect it 3 centimeters from its origin. Reflect it craniodorsally to a point cranial to the place where the second cervical nerve crosses the muscle. Transect the brachiocephalicus 1 centimeter cranial to the clavicular intersection. Reflect it toward its cervical and mastoid attachments.

The **superficial cervical lymph nodes** lie in the areolar tissue cranial to the shoulder. They lie deep to the cervical parts of the brachiocephalic and omotransverse muscles and receive lymph drainage from the cutaneous area of the head, neck and thoracic limb.

The **accessory** or **eleventh cranial nerve** (Fig. 90) is a large nerve found deep to the cranial part of the sternocephalicus. As it emerges from the cleidomastoideus, it crosses the second cervical nerve, runs along the dorsal border of the omotransversarius and terminates in the thoracic part of the trapezius. Dissect between the trapezius and cleidocervicalis and identify this nerve coursing caudally to the trapezius. The accessory nerve is the only motor nerve to the trapezius. In addition it supplies in part the omotransversarius, the mastoid and cervical portion of the brachiocephalicus and the sternocephalicus.

Free the ventral border of the omotransversarius and lift it. Look for the ventral branches of the **third, fourth** and **fifth cervical nerves** (Fig. 89) which are distributed in a segmental manner to the muscles and skin of the neck. The third and fourth nerves, after emerging from the intervertebral foramina, pass through the deep fascia and the omotransversarius. It may be difficult to identify each cervical nerve, and it is not necessary to do so.

Transect the fused sternohyoideus and sternothyroideus 2 centimeters from their origin and reflect them to their insertions. Parts of the trachea, larynx, thyroid gland, esophagus and carotid sheath are exposed. Identify these structures on your specimen. Note the common carotid artery dorsal to the sternothyroideus. Bound to its medial side is the **vagosympathetic nerve trunk.** The **medial retropharyngeal lymph node** (see Fig. 13) lies opposite the larynx, ventrolateral to the carotid sheath.

THE THORAX

Superficial Vessels and Nerves of the Thoracic Wall

Before dissecting the thoracic nerves and vessels, study Figures 91 to 94, which show the pattern of distribution of these structures. Notice that the artery and nerve of each intercostal space divide into dorsal and ventral branches. The dorsal branches enter the epaxial muscles. The ventral branches descend in the intercostal spaces along the caudal border of each rib. The dorsal and ventral arterial branches are derived from the dorsal intercostal arteries. The first three dorsal intercostal arteries come from a branch of the costocervical trunk, the remaining nine come from the aorta. The dorsal intercostal arteries and veins have lateral cutaneous branches that perforate the intercostal and adjacent muscles to supply cutaneous structures, including the thoracic mammary glands. The dorsal intercostal artery and vein pass ventrally where they anastomose with ventral intercostal branches from the internal thoracic artery and vein. At the ventral aspect of each intercostal space, perforating branches of the internal thoracic vessels emerge and supply cutaneous structures and the thoracic mammary glands.

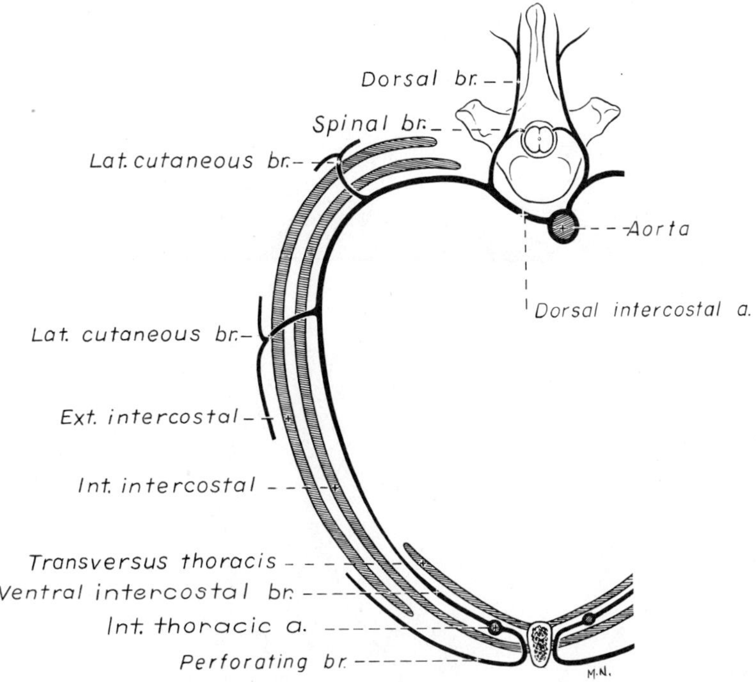

FIGURE 91. Schematic transection of thoracic wall to show distribution of an intercostal artery.

The dorsal and ventral nerve branches are derived from the spinal nerve as it emerges from the intervertebral foramen. The ventral branches of the first 12 thoracic nerves are intercostal nerves and have lateral and ventral cutaneous branches and branches medial to these that go largely to muscles.

Dorsal and lateral rows of lateral cutaneous branches of intercostal nerves, arteries and veins emerge at regular intervals between the ribs and supply the cutaneous muscle, subcutaneous tissue and skin. The nerves of the dorsal row arise from the dorsal branches of the thoracic nerves. A row of ventral cutaneous branches emerge through the origin of the deep pectoral muscle after having penetrated the ventral ends of the intercostal spaces. These vessels are perforating branches of the internal thoracic artery and vein. The nerves are terminal branches of the intercostal nerves. Although these emerge at regular intervals, not all will be seen in the dissection.

The **cranial thoracic mamma** is supplied by the fourth, fifth and sixth ventral and lateral cutaneous vessels and nerves, and by branches of the lateral thoracic vessels. The latter are from the axillary vessels, which will be dissected later.

The **caudal thoracic mamma** is supplied in a similar manner from the sixth and seventh cutaneous nerves and vessels. In addition, mammary branches of the cranial superficial epigastric vessels supply this mamma.

The **axilla** is the space between the thoracic limb and the thoracic wall. It is bounded ventrally by the pectoral muscles and dorsally by the

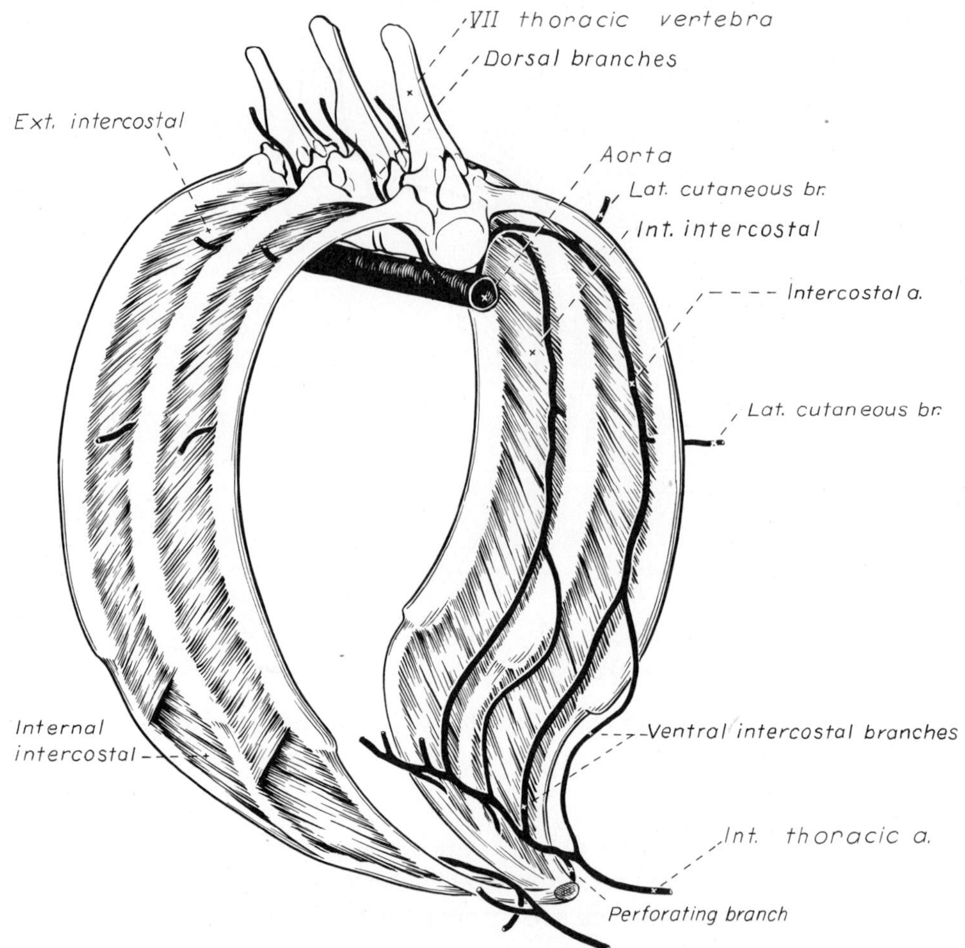

FIGURE 92. Intercostal arteries as seen within rib cage.

attachment of the serratus ventralis to the scapula. Cranially it extends under the muscles that extend from the arm to the neck. Caudally, a similar extension is found under the latissimus dorsi and cutaneus trunci.

The **lateral thoracic artery, vein** and **nerve** emerge from the axilla between the latissimus dorsi and deep pectoral muscles. The nerve is motor to the cutaneus trunci and may be found on its ventral border. It consists of fibers from the ventral branches of the eighth cervical and first thoracic spinal nerves. The vessels supply the muscle, skin and subcutaneous tissues, including the cranial thoracic mamma. If these vessels and this nerve are not readily identified in your dissection, you may find them later when the axillary vessels and the brachial plexus are dissected.

Transect the pectoral muscles close to the sternum. Reflect them toward the forelimb to expose the axilla.

The **axillary lymph node** lies dorsal to the deep pectoral muscle and caudal to the large vein coming from the arm. Most of the afferent lymph vessels of the thoracic wall and deep structures of the limb drain into this node.

Medial br.
Spinal cord
Thoracic n.
Dorsal br.
Lateral br.
Ventral br.
Lat. cutaneous br.
Ramus communicans to sympathetic trunk
Intercostal n.
Lat. cutaneous br.
Ext. intercostal
Int. intercostal
Transversus thoracis
Ventral cutaneous br.

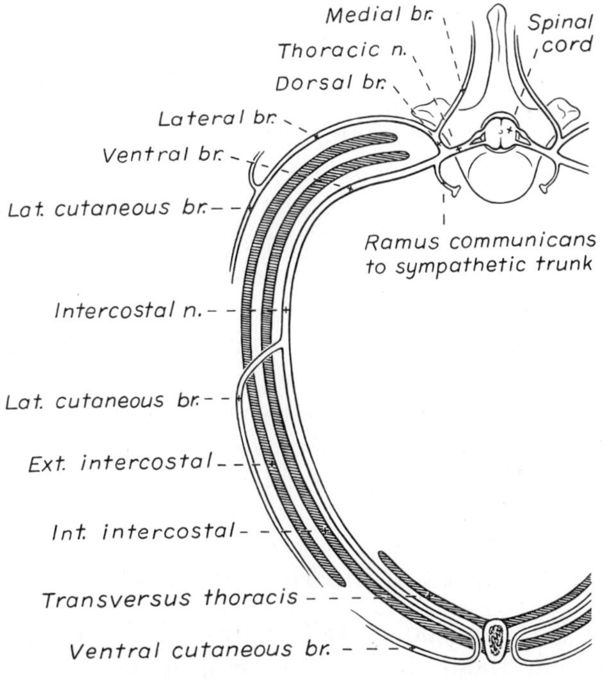

FIGURE 93. Schema of a thoracic spinal nerve.

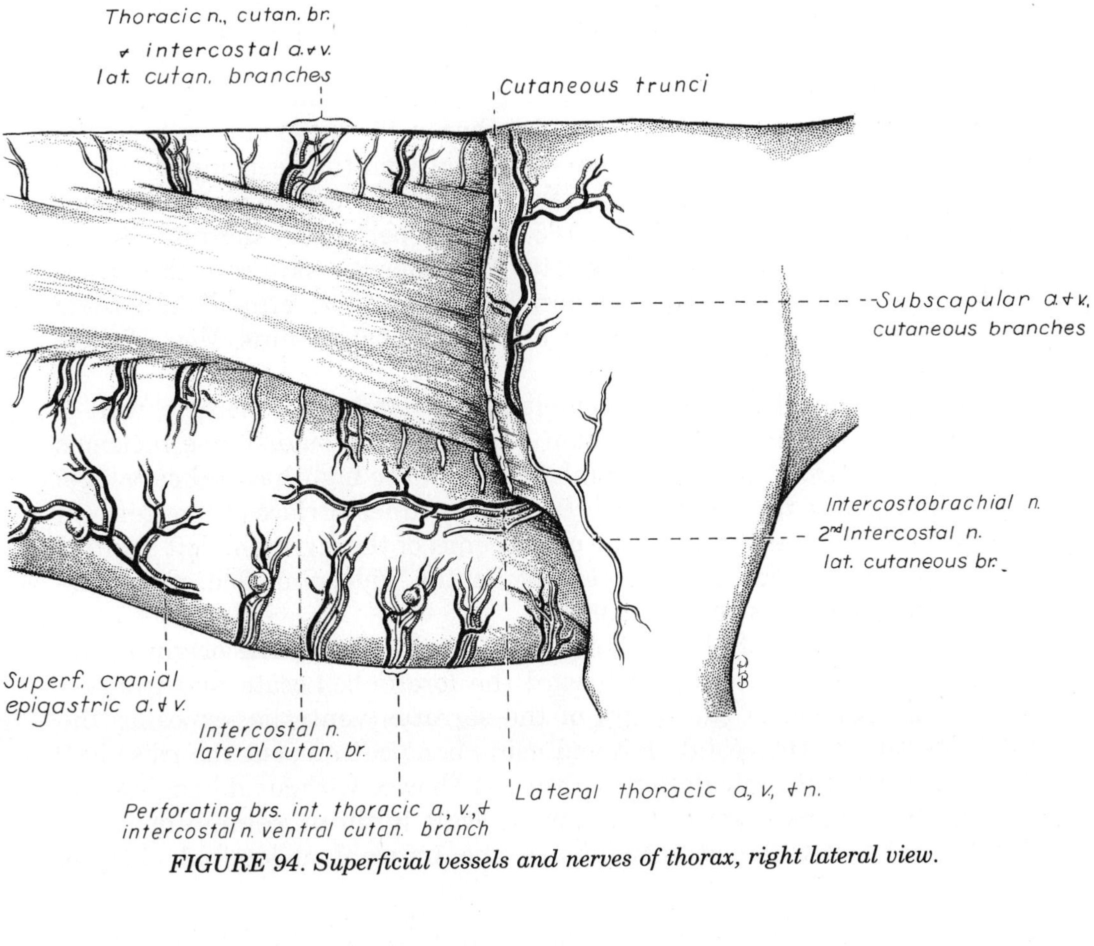

Thoracic n., cutan. br.
& intercostal a.&v.
lat. cutan. branches
Cutaneous trunci
Subscapular a.&v.
cutaneous branches
Intercostobrachial n.
2nd Intercostal n.
lat. cutaneous br.
Superf. cranial
epigastric a.&v.
Intercostal n.
lateral cutan. br.
Lateral thoracic a, v, & n.
Perforating brs. int. thoracic a., v.,&
intercostal n. ventral cutan. branch

FIGURE 94. Superficial vessels and nerves of thorax, right lateral view.

Live Dog

Palpate the structures in the neck ventral to the cervical vertebrae. The larynx and trachea are readily palpable. The esophagus is usually too soft to feel but should be on the left of the trachea in the middle and caudal cervical region. Try to palpate a pulse in the carotid artery. It usually courses along the dorsolateral side of the trachea but is too deep to allow a pulse to be felt regularly. Cranially feel the firm oval mandibular salivary gland and the smaller, looser mandibular lymph nodes. The latter can be felt subcutaneously at the angle of the mandible. Caudally in the neck, feel the superficial cervical lymph nodes cranial to the shoulder and deep to the omotransversarius or brachiocephalicus muscle. Extend the neck and compress the vessels that enter the thorax at the thoracic inlet to try to distend the external jugular vein so that it is visible. This is more difficult to observe in long-haired breeds without removing the hair.

Deep Vessels and Nerves of the Thoracic Wall

Expose the lumbar and costal origins of the external abdominal oblique and detach them. Reflect the muscle ventrally to the rectus abdominis. Reflect the mammae if necessary. Free the thoracic attachment of the rectus abdominis close to the sternum and first costal cartilage. Reflect the rectus abdominis caudally, noting and cutting any nerves or vessels that enter the deep face of the muscle from any of the intercostal spaces.

The **cranial epigastric artery** is a terminal branch of the internal thoracic artery that emerges from the thorax in the angle between the costal arch and the sternum. It passes caudally on the deep surface of the rectus abdominis. The cranial epigastric artery gives rise to the **cranial superficial epigastric** (see Fig. 125), which perforates the muscle and runs caudally on its surface. This artery supplies the skin over the rectus abdominis and the cranial abdominal mamma. The cranial epigastric vessels continue on the deep surface of the rectus abdominis. Most of their branches terminate in this muscle.

Make a sagittal incision completely through the thoracic wall 1 centimeter from the ventral median plane on each side. These incisions should extend from the thoracic inlet through the eighth costal cartilage. The **transversus thoracis** muscle lies on the inner surface of the sternum and costal cartilages. Connect the caudal ends of the right and left sagittal incisions and free the sternum, except for the wide, thin fold of mediastinum that is now its only attachment.

On the right half of the thorax clean and transect the origin of the latissimus dorsi and reflect it toward the forelimb. Locate and transect the caudal portion of the origin of the serratus ventralis, exposing the ribs. Starting at the costal arch and using bone cutters, cut the ribs close to their vertebral articulation within the thorax without damaging the sympathetic trunk. Reflect the thoracic wall **without removing it.** As this is done, cut the attachments of the internal abdominal oblique, transversus abdominis and diaphragm from the ribs along the costal arch. Reflect the left thoracic wall in a similar manner.

On the internal surface of the thoracic wall, notice the intercostal vessels and nerves coursing along the caudal border of the ribs. Ventrally the vessels bifurcate and anastomose with the intercostal branches of the internal thoracic artery and vein. The intercostal nerves supply the intercostal musculature. Their sensory branches were seen as lateral and ventral cutaneous branches.

The **pleurae** (Figs. 95, 96) are serous membranes that cover the lungs and line the walls of the thorax. These form right and left sacs that enclose the pleural cavities.

The **pulmonary pleura** closely attaches to the surfaces of the lungs, following all their small irregularities as well as the fissures that divide the two lobes.

The **parietal pleura** is attached to the thoracic wall by the endothoracic fascia. This pleura may be divided into costal, diaphragmatic and mediastinal parts. Each of these is named after the region or surface it covers and all are continuous, one with another. The **costal pleura** covers the inner surfaces of the ribs and intercostal muscles. The **mediastinal pleurae** are the layers that cover the sides of the partition between the two pleural cavities. The **mediastinum** includes the two mediastinal pleurae and the space between them. Enclosed in the mediastinum are the thymus, the lymph nodes, the heart, the aorta, the trachea, the

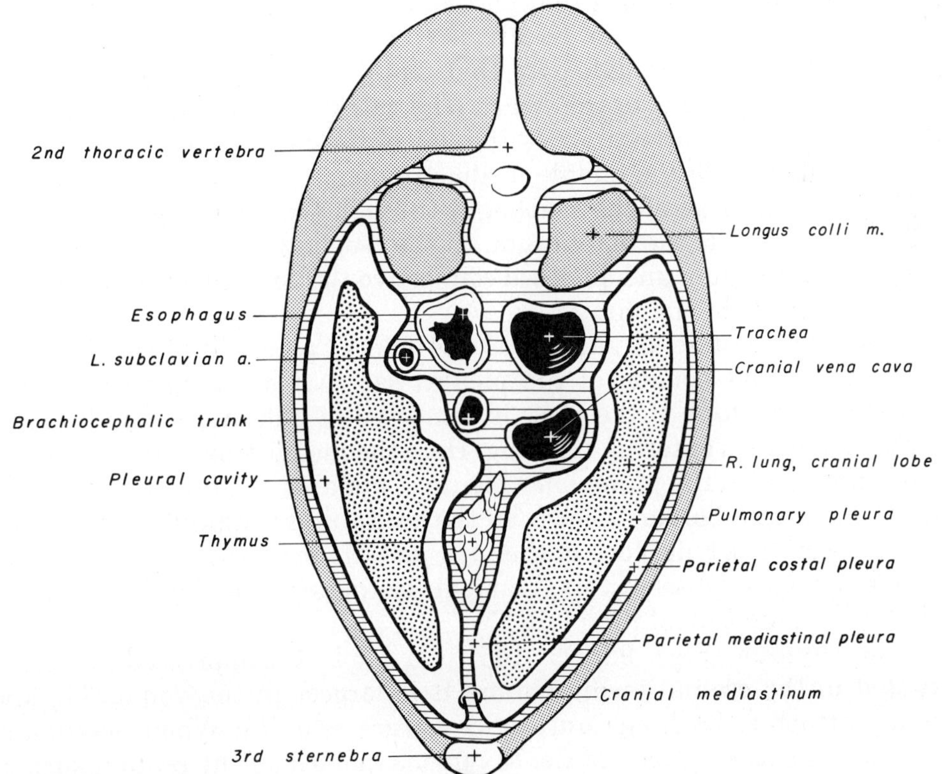

FIGURE 95. Schematic transverse section of thorax through cranial mediastinum, caudal view.

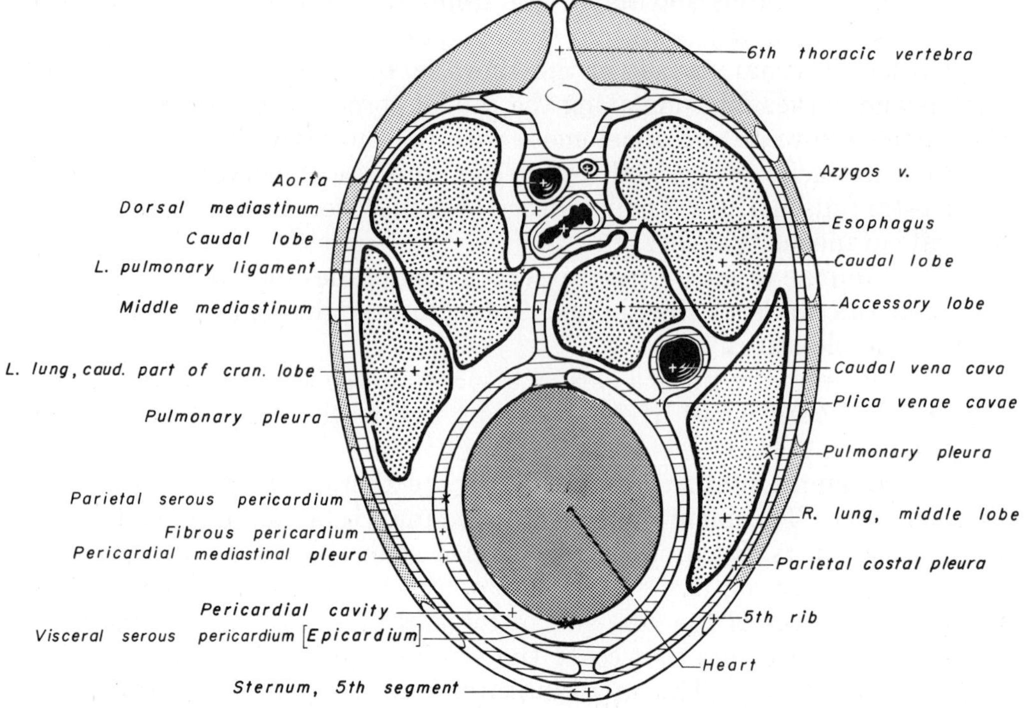

FIGURE 96. Schematic transverse section of thorax through heart, caudal view.

esophagus, the vagus nerves and other nerves and vessels. The **pericardial mediastinal pleura** is that portion covering the heart.

The **mediastinum** can be divided into a cranial part, that lying cranial to the heart; a middle part, that containing the heart; and a caudal part, that lying caudal to the heart. The caudal mediastinum is thin. It attaches to the diaphragm far to the left of the median plane. Cranially it is continuous with the middle mediastinum.

Note the passage of the esophagus through the mediastinum and the esophageal hiatus of the diaphragm. At the esophageal hiatus a thin layer of pleura, peritoneum and enclosed connective tissue attach the esophagus to the muscle of the diaphragm.

The **plica venae cavae** is a loose fold of pleura derived from the right mediastinal portion of the pleural sac that surrounds the caudal vena cava. The **root** of the lung is composed of pleura and the bronchi, vessels and nerves entering the lung. Here the mediastinal parietal pleura is continuous with the pulmonary pleura. Caudal to the hilus this connection forms a free border, known as the **pulmonary ligament** (Figs. 96, 97), between the caudal lobe of the lung and the mediastinum at the level of the esophagus. Observe this ligament. In thoracic surgery this must be cut to reflect the caudal lung lobe.

The **thymus** (Figs. 98, 99, 105) is a bilobed, compressed structure situated in the cranial mediastinum. It is largest in the young dog and usually atrophies with age until only a trace remains. When maximally developed, the caudal part of the thymus is molded on the cranial surface of the pericardium.

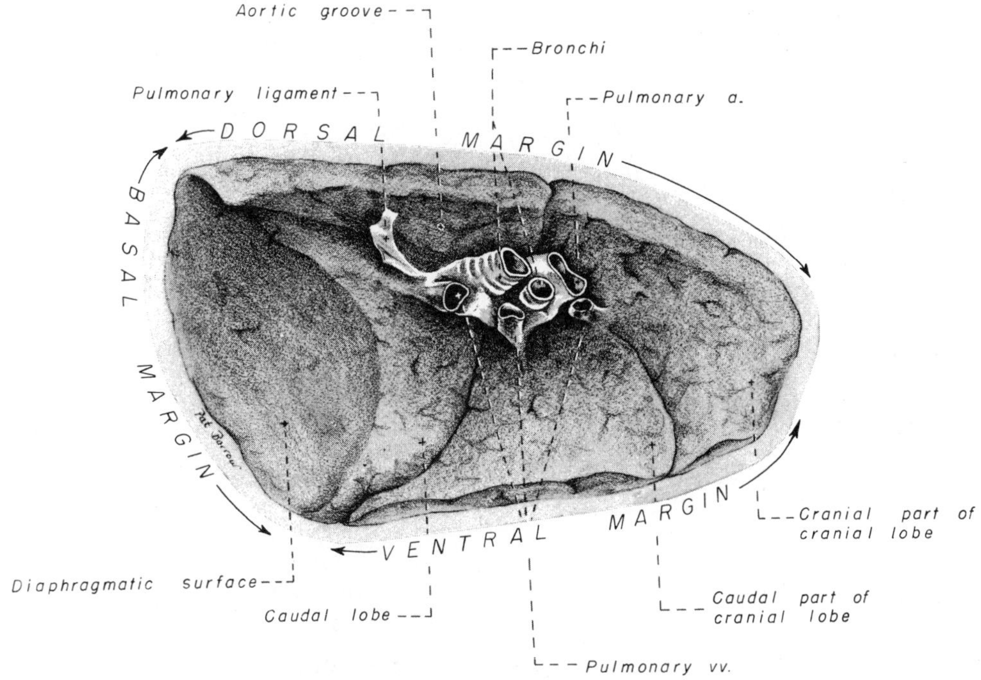

FIGURE 97. Left lung, medial view.

The **internal thoracic artery** (Figs. 100–105) leaves the subclavian artery, courses ventrocaudally in the cranial mediastinum and disappears deep to the cranial border of the transversus thoracis muscle. It supplies many branches to surrounding structures—the phrenic nerve, the thymus, the mediastinal pleurae and the intercostal spaces. The perforating branches to the superficial structures of the ventral third of the thorax

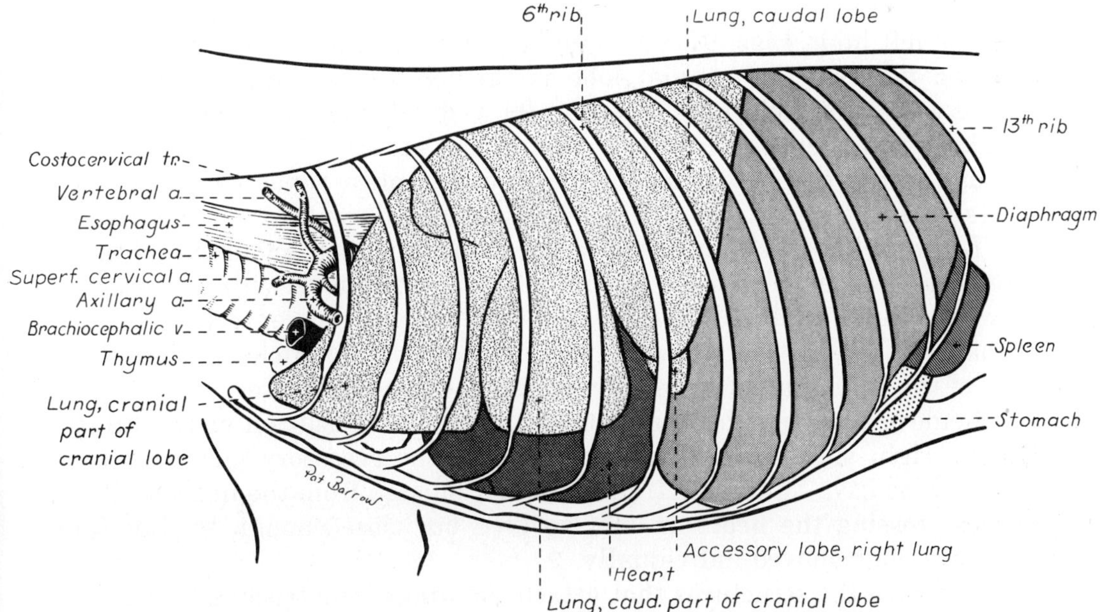

FIGURE 98. Thoracic viscera within the rib cage, left lateral view.

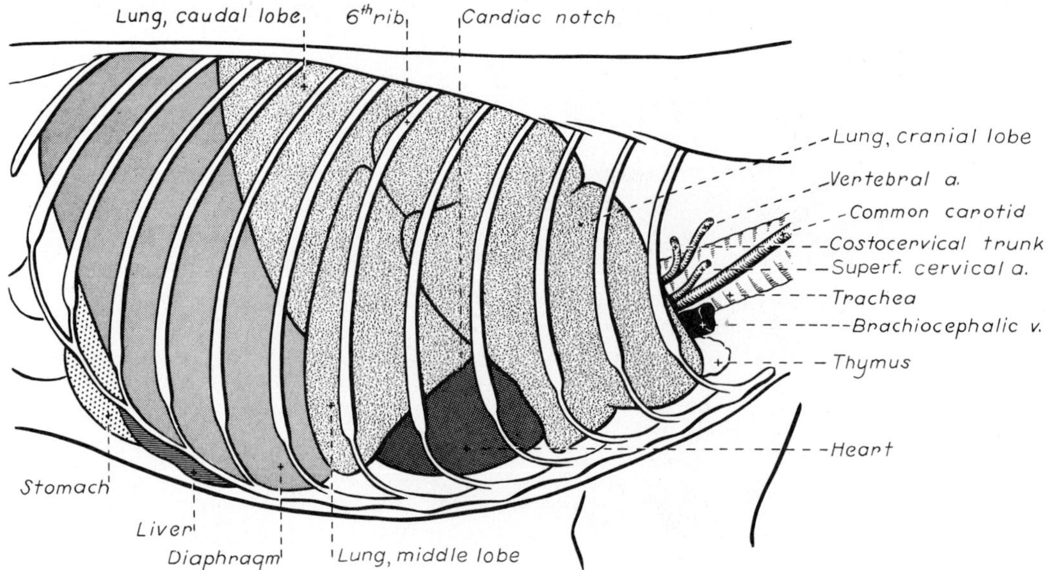

FIGURE 99. Thoracic viscera within the rib cage, right lateral view.

have been seen. The anastomoses with the intercostal arteries on the medial side of the thoracic wall have been seen. Near the attachment of the costal arch with the sternum, the internal thoracic artery terminates in the musculophrenic artery and the larger cranial epigastric artery. The latter has been dissected along with its cranial superficial epigastric branch. The **musculophrenic artery** runs cardodorsolaterally in the angle formed by the diaphragm and lateral thoracic wall. Dissect its origin. Cut the mediastinum and reflect the sternum cranially.

The Lungs

The left lung (Figs. 97–99) is divided into **cranial** and **caudal lobes** by deep fissures. The cranial lobe is further divided into cranial and caudal parts. The right lung (Figs. 98, 99) is divided into **cranial, middle, caudal** and **accessory lobes.** A part of the accessory lobe can be seen through either the caudal mediastinum or the plica venae cavae, where it lies in the space between these two structures.

Examine the **cardiac notch** of the right lung at the fourth and fifth intercostal spaces. The apex of the notch is continuous with the fissure between cranial and middle lobes. A larger area of the ventral convexity of the heart is exposed on the right side. The right ventricle occupies this area of the heart and is accessible for cardiac puncture here.

Remove the lungs by transecting all structures that enter the hilus. On the right side this will involve slipping the accessory lobe over the caudal vena cava. Make the transection far enough from the heart so that nerves crossing the heart are not severed but close enough so that the lobes are not removed individually.

Examine the structures that attach the lungs. The trachea bifurcates into left and right principal bronchi. The **carina** is the partition between

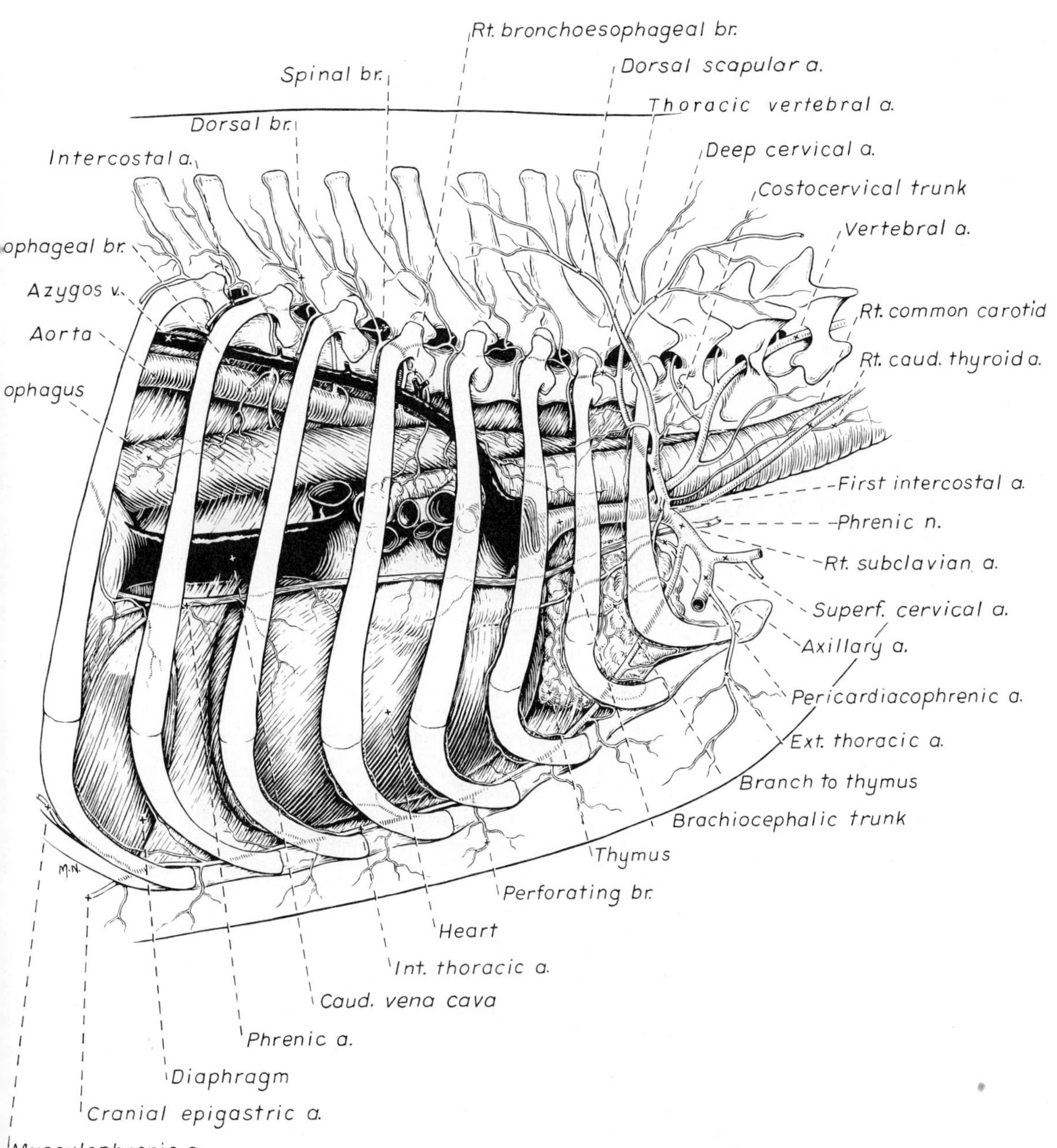

FIGURE 100. Arteries of thorax, right lateral view.

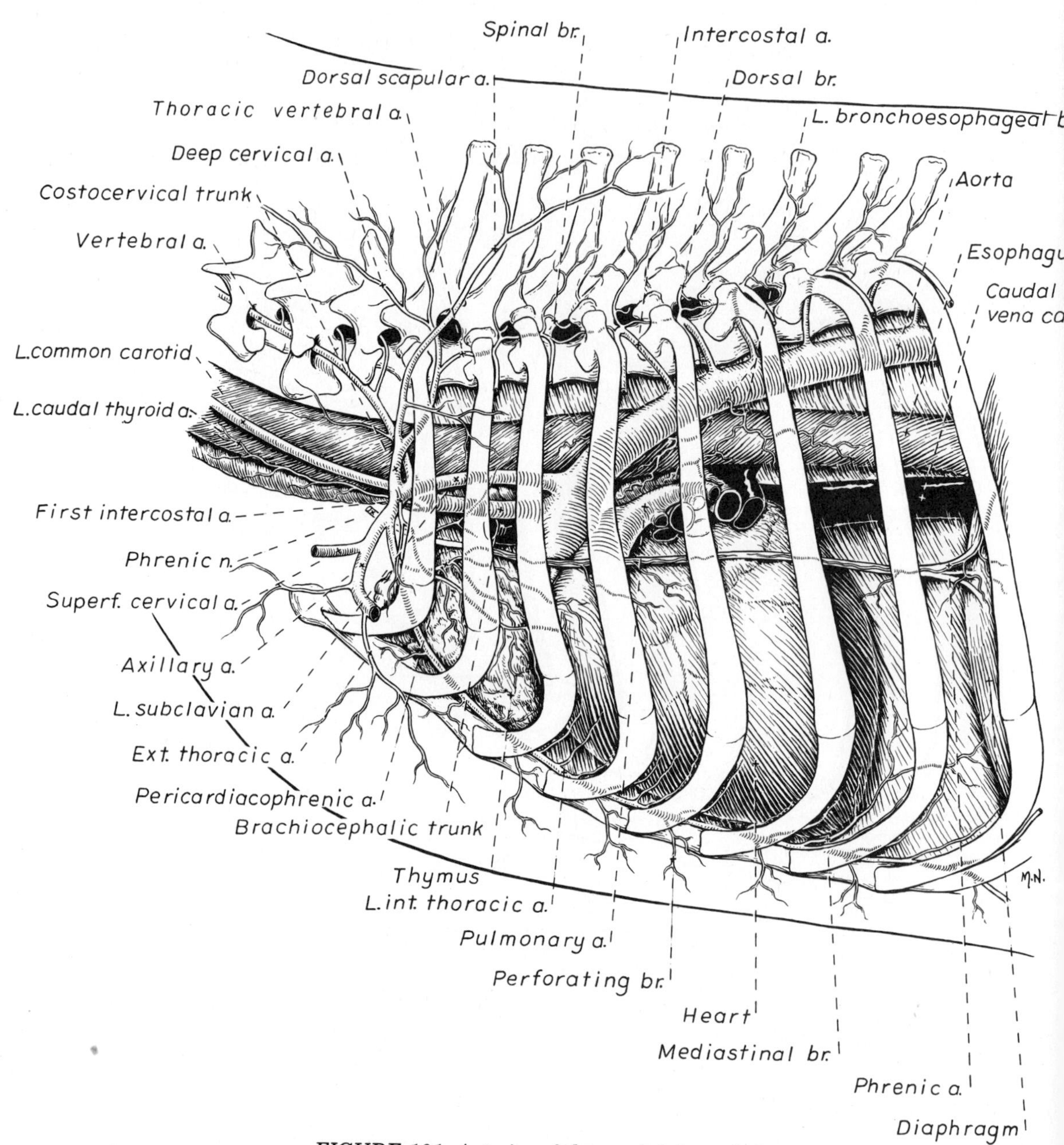

FIGURE 101. Arteries of thorax, left lateral view.

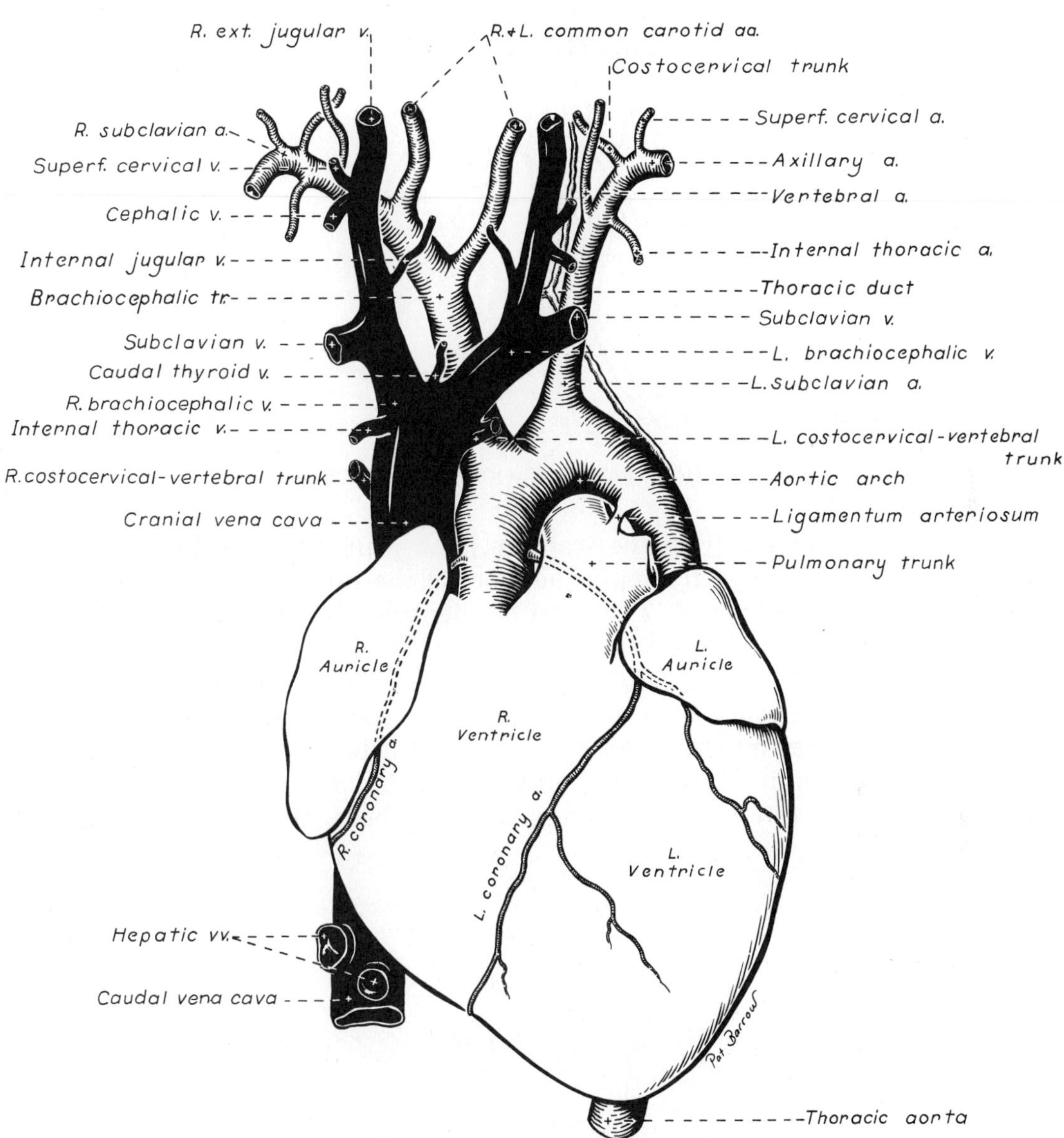

FIGURE 102. Heart and great vessels, ventral view.

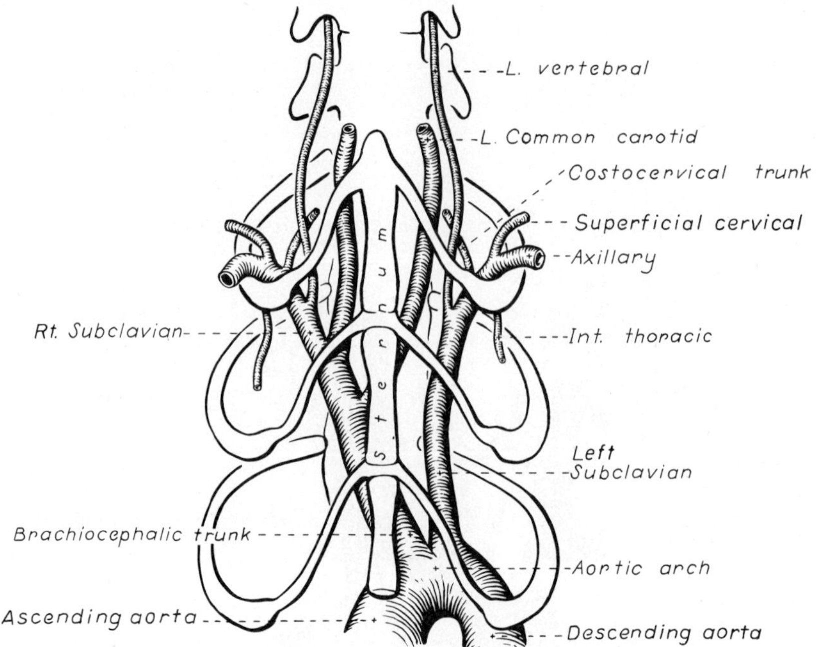

FIGURE 103. Branches of aortic arch, ventral view.

them at their origin from the trachea. Each principal bronchus divides into lobar bronchi which supply the lobes of the lung. Find these on the lungs that were removed. They can be identified by the cartilage rings within their walls. Notice that there is usually a single pulmonary vein from each lobe that drains directly into the left atrium of the heart. (The pulmonary veins contain red latex because the specimen was prepared by injecting the latex into the carotid artery. Moving in a retrograde direction, the latex in turn filled the aorta, left ventricle, left atrium and pulmonary veins. Because latex does not cross capillary beds, there is usually no

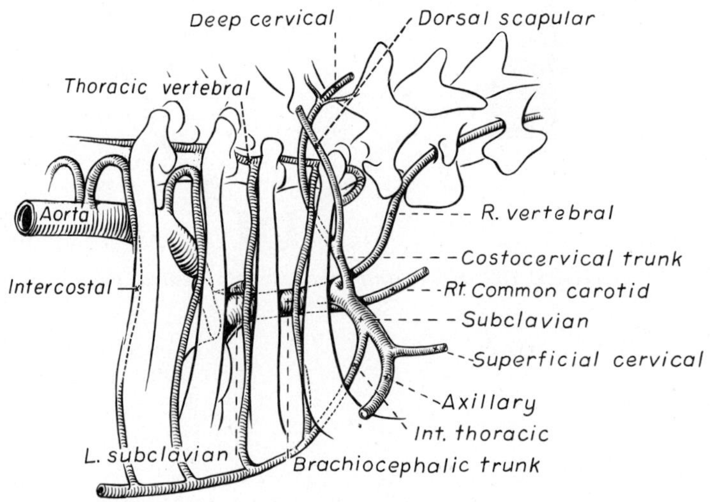

FIGURE 104. Branches of brachiocephalic trunk, right lateral view.

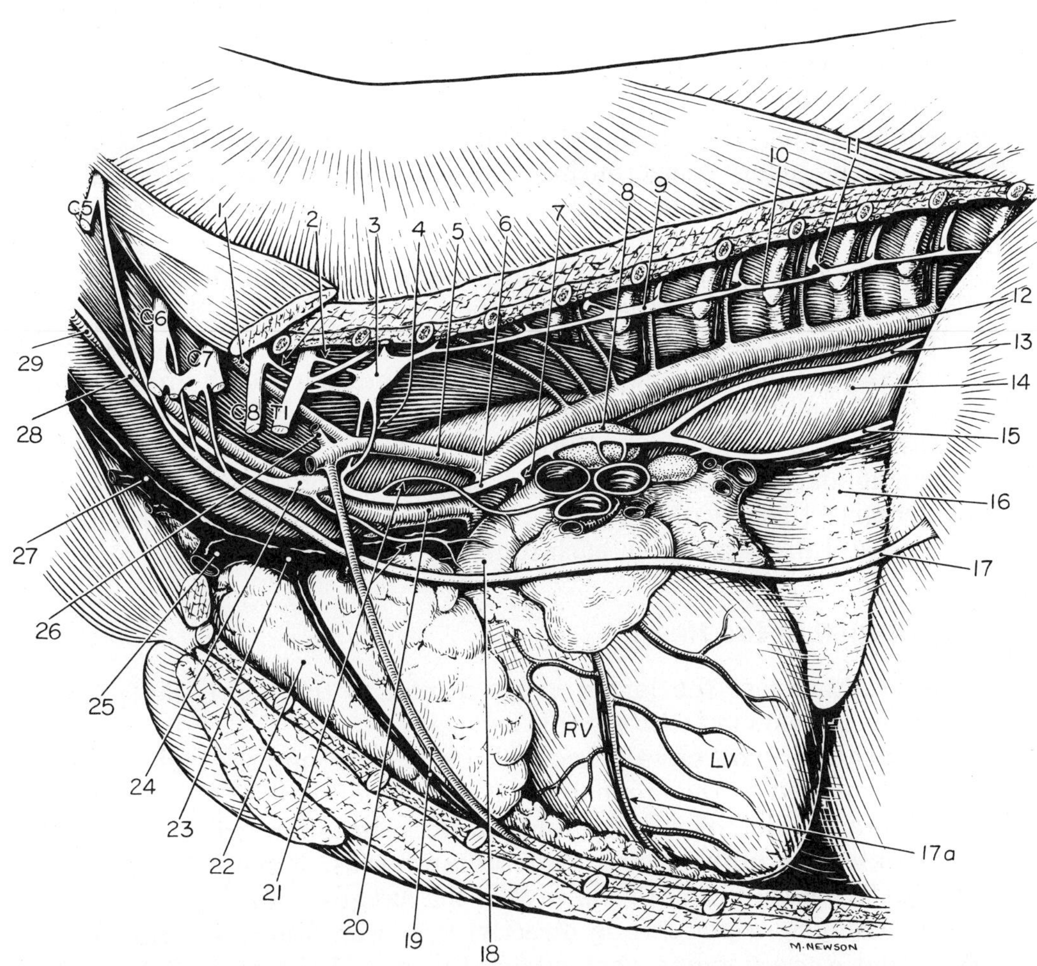

FIGURE 105. Thoracic autonomic nerves, left lateral view, lung removed.

1. *Vertebral artery and nerve*
2. *Communicating rami from cervicothoracic ganglion to ventral branches of cervical and thoracic nerves*
3. *Left cervicothoracic ganglion*
4. *Ansa subclavia*
5. *Left subclavian artery*
6. *Left vagus nerve*
7. *Left recurrent laryngeal nerve*
8. *Left tracheobronchial lymph node*
9. *Sympathetic trunk ganglion*
10. *Sympathetic trunk*
11. *Ramus communicans*
12. *Aorta*
13. *Dorsal branch of vagus nerve*
14. *Esophagus*
15. *Ventral trunk of vagus nerve*
16. *Accessory lobe of lung (through caudal mediastinum)*
17. *Phrenic nerve to diaphragm*

17a. *Paraconal inter. groove*

18. *Pulmonary trunk*
19. *Internal thoracic artery and vein*
20. *Brachiocephalic trunk*
21. *Cardiac autonomic nerves*
22. *Thymus*
23. *Cranial vena cava*
24. *Middle cervical ganglion*
25. *Left subclavian vein*
26. *Costocervical trunk*
27. *External jugular vein*
28. *Vagosympathetic trunk*
29. *Common carotid artery*

RV—Right ventricle
LV—Left ventricle

latex in the pulmonary arteries. Occasionally the pressure of injection ruptures the interatrial or interventricular septum in the heart, flooding the right chambers with the latex and thus filling the pulmonary arteries as well as the veins.)

The pulmonary trunk supplies each lung with a pulmonary artery. At the root of the lung, the left pulmonary artery usually lies cranial to the left bronchus. The right pulmonary artery is ventral to the right bronchus. The artery and bronchus are at a more dorsal level than the veins. Using a scissors or scalpel, open a few of the major bronchi to observe the lumen.

Note the **tracheobronchial lymph nodes** located at the bifurcation of the trachea and also farther out on the bronchi.

Determine which structures form the various grooves and impressions by replacing the lungs in the thorax. Observe the long **aortic impression** of the left lung. The most marked impressions on the right lung are on the accessory lobe. This lobe is interposed between the caudal vena cava on one side and the esophagus on the other, and both leave impressions on it. Observe the vascular impressions on the cranial lobes of the lungs and the costal impressions on each lung.

Veins Cranial to the Heart

Carefully expose the larger veins cranial to the heart. Reflect the sternum to one side to facilitate this exposure.

The **cranial vena cava** (Figs. 100–102) drains into the right atrium after its formation by the union of the right and left brachiocephalic veins at the thoracic inlet. The **brachiocephalic vein** is formed on each side by the **external jugular** and **subclavian veins.** Sometimes the last branch entering the cranial vena cava is the **azygos vein** (Fig. 100). This usually enters the right atrium directly. It is seen from the right in the mediastinal space winding ventrocranially around the root of the right lung. It originates dorsally in the abdomen and collects all of the dorsal intercostal veins on each side as far cranially as the third or fourth intercostal space.

The **thoracic duct** is the chief channel for the return of lymph from lymphatic capillaries and ducts to the venous system. It begins in the sublumbar region between the crura of the diaphragm as a cranial continuation of the **cisterna chyli.** The latter is a dilated structure that receives the lymph drainage from the viscera and pelvic limbs. The thoracic duct runs cranially on the right dorsal border of the thoracic aorta and the ventral border of the azygos vein to the level of the sixth thoracic vertebra. (It may not be visible.) Here it crosses the ventral surface of the fifth thoracic vertebra and courses on the left side of the middle mediastinal pleura. It continues cranioventrally through the cranial mediastinum to the left brachiocephalic vein, where it usually terminates (Fig. 102). The thoracic duct also receives the lymph drainage from the left thoracic limb and the **left tracheal duct** (from the left side of the head and neck). The lymph drainage from the right thoracic limb and the **right tracheal duct** (from the right side of the head and neck) enters the venous system in the vicinity of the right brachiocephalic vein. There are often multiple terminations of a complicated nature, which may include swellings or anastomoses. All lymphatic channels will be difficult to see unless they are congested with lymph or refluxed blood. They are frequently double.

Look for the thoracic duct. It is not always visible but it may be identified by the reddish-brown or straw color of its contents and the numerous random constrictions in its wall. The tracheal ducts may be found in each carotid sheath or parallel to the sheath and its contents.

Arteries Cranial to the Heart

The **aorta** (Figs. 100–105) is the large, unpaired vessel that emerges from the left ventricle medial to the pulmonary trunk. As the **ascending aorta** it extends cranially, covered by the pericardium; it makes a sharp bend dorsally and to the left as the **aortic arch;** it runs caudally as the **descending aorta** located ventral to the vertebrae. The part cranial to the diaphragm is the thoracic aorta and the caudal part is the abdominal aorta. Cranial to the heart are several branches of the aorta; reflect the veins that were dissected cranial to the heart to observe these arteries.

The right and left **coronary arteries** are branches of the ascending aorta that supply the heart muscle. They will be studied with the heart.

The **brachiocephalic trunk** (Figs. 100–105), the first branch from the aortic arch, passes obliquely to the right across the ventral surface of the trachea. It gives rise to the **left common carotid artery** and terminates as the **right common carotid artery** and the **right subclavian artery.**

The **left subclavian artery** (Figs. 101–105) originates from the aortic arch beyond the level of the brachiocephalic trunk and passes obliquely to the left across the ventral surface of the esophagus.

The branches of the right and left subclavian arteries are similar; only the right subclavian artery will be described. For each artery described there is a comparable vein with a similar area of distribution. The terminations of the veins are variable and they will not be dissected. Remove them when necessary to expose the arteries. The right subclavian artery has four branches that arise medial to the first rib or intercostal space. They are the vertebral artery, the costocervical trunk, the superficial cervical artery and the internal thoracic artery. Do not sever the nerves or arteries.

The **vertebral artery** (Figs. 100–105) crosses the medial surface of the first rib and disappears dorsally between the longus colli and the scalenus muscles. It passes through the transverse foramina of the first six cervical vertebrae. It supplies both muscular branches to the cervical muscles and also spinal branches at each intervertebral foramen to the spinal cord and its coverings. At the level of the atlas it terminates by entering the vertebral canal through the lateral vertebral foramen and contributes to the ventral spinal and basilar arteries. These will be seen later in the dissection of the nervous system.

The **costocervical trunk** (Figs. 100–105) arises distal to the vertebral artery, crosses its lateral side and extends dorsally as far as the vertebral end of the first rib. By its various branches it supplies the structures of the first, second and third intercostal spaces, the muscles at the base of the neck and the muscles dorsal to the first few thoracic vertebrae. These need not be dissected.

The **superficial cervical artery** (Figs. 100–104) arises from the subclavian opposite the origin of the internal thoracic artery, medial to the first rib. It emerges from the thoracic inlet to supply the base of the neck and the adjacent scapular region.

The **internal thoracic artery** has been studied.

Branches of the Thoracic Aorta

The **esophageal** and **bronchial arteries** vary in number and origin. Usually the small **bronchoesophageal artery** leaves the right fifth intercostal artery close to its origin and crosses the left face of the esophagus, which it supplies. It terminates shortly afterwards in the **bronchial arteries,** which supply the lung.

There are nine pairs of dorsal **intercostal arteries** that leave the aorta (Figs. 91, 94). These start with either the fourth or the fifth intercostal artery and continue caudally, there being an artery in each of the remaining intercostal spaces. Each lies close to the caudal border of the rib. The costocervical trunk supplies the first three or four intercostal spaces. The dorsal costoabdominal artery courses ventrally, caudal to the last rib.

The **phrenic nerves** (Figs. 100, 101, 105) supply the diaphragm. Find each nerve as it passes through the thoracic inlet. The nerve arises from the ventral branches of the fifth, the sixth and usually the seventh cervical nerves. Follow the phrenic nerves to the diaphragm. Each is both motor and sensory to the corresponding half of the diaphragm except at its periphery. This part of the muscle receives sensory fibers from the caudal intercostal nerves.

INTRODUCTION TO THE AUTONOMIC NERVOUS SYSTEM

The **nervous system** is highly organized both anatomically and functionally. It is composed of a **central nervous system** and a **peripheral nervous system.** The central nervous system includes the **brain** and the **spinal cord.** The peripheral nervous system comprises the **cranial nerves,** which connect the base of the brain (brain stem) with structures of the head and body, and the **spinal nerves,** which connect the spinal cord to structures of the neck, trunk, tail and extremities.

The peripheral nervous system can be further classified on the basis of anatomy and function. The peripheral nerves contain axons that conduct impulses *to* the central nervous system—**sensory, afferent axons**—and axons that conduct impulses *from* the central nervous system to muscles and glands of the body—**motor, efferent axons.** Most peripheral nerves have both sensory and motor axons. When one speaks of a motor nerve, it is an indication of the primary function of the majority of the neurons, but it is understood that sensory neurons are also present. Likewise, so-called sensory nerves also contain motor neurons. This duality is the basis of feedback regulation. All the nerves you dissect are bundles of neuronal processes belonging to both sensory and motor neurons.

The **motor portion** of the peripheral nervous system is classified according to the type of tissue being innervated. Motor neurons supplying voluntary, striated, skeletal muscle are **somatic efferent** neurons. Somatic refers to the body, body wall or neck, trunk and limbs where these striated skeletal muscles are located. Those supplying involuntary, smooth muscle of viscera, blood vessels, cardiac muscle and glands are **visceral efferent** neurons.

A **neuron** is composed of a cell body and its processes. Ordinarily a motor neuron of the peripheral nervous system has its cell body located in the gray matter of the spinal cord (or brain stem), and its process, or axon, courses through the peripheral spinal (or cranial) nerve to end in the muscle innervated. Thus there is only one neuron spanning the distance from the central nervous system to the innervated structure.

The visceral efferent system is the peripheral motor part of the **autonomic nervous system.** It differs anatomically from the other peripheral motor systems in having a second neuron interposed between the central nervous system and the innervated structures. One neuron has its cell body located in the gray matter of the central nervous system. Its axon courses in the peripheral nerves only part way toward the structure to be innervated. Along the course of the peripheral nerve is a gross enlargement called a **ganglion.** This is a collection of neuronal cell bodies located outside the central nervous system. Some ganglia have a motor function, others a sensory function. Groups of neuronal cell bodies within the central nervous system are called **nuclei.** Autonomic ganglia contain the cell bodies of the second motor neurons in the pathway of the visceral efferent system. Their axons complete the pathway to the structure being innervated. Because of its relationship to the cell bodies in the autonomic ganglia, the first neuron with its cell body in the central nervous system is called the **preganglionic neuron.** The cell body of the second neuron is in an autonomic ganglion. Its axon is **postganglionic.** A synapse occurs between these two neurons where the preganglionic axon meets the cell body of the postganglionic axon.

The visceral efferent system is divided into two subdivisions on the basis of anatomical, pharmacological and functional characteristics. They are the sympathetic and parasympathetic divisions. (It should be kept in mind that the autonomic nerves we observe also have visceral afferent or sensory axons within them).

In the **sympathetic division** the preganglionic cell bodies are limited to the segments of the spinal cord from approximately the first thoracic to the fifth lumbar segments—the **thoracolumbar** portion. The cell bodies of postganglionic axons are located in ganglia that are usually only a short distance from the spinal cord. At most of the postganglionic nerve endings of this portion of the autonomic nervous system a humoral transmitter substance, norepinephrine, is released, which causes a response in the structures innervated. The overall effect of this system is to help the body withstand unfavorable environmental conditions or conditions of stress.

In the **parasympathetic division** the preganglionic cell bodies are located in specific nuclei in the brain stem associated with cranial nerves III, VII, IX, X and XI and in the three sacral segments of the spinal cord—the **craniosacral** portion. The cell bodies of the postganglionic axons are often located in terminal ganglia on or in the wall of the structure being innervated. Others are found in specifically named ganglia near the innervated structure. At the postganglionic nerve endings a humoral transmitter substance, acetylcholine, is released, which causes a response in the structures innervated. This system is associated with the normal homeostatic activity of the visceral body functions, the conservation and restoration of body resources and reserves.

The anatomy of the sympathetic division of the peripheral autonomic nervous system requires further description before it is dissected. The preganglionic cell bodies are located in the gray matter of the thoracic and first five lumbar spinal cord segments. Their axons leave the spinal cord along with those of other motor neurons in the ventral rootlets of each of these spinal cord segments. Each **ventral root** unites with the corresponding sensory **dorsal root** at the level of the intervertebral foramen to form the **spinal nerve.** The dorsal branch of the spinal nerve branches off immediately. Just beyond this point a nerve leaves the ventral branch of the spinal nerve, the **ramus communicans.** It courses a short distance ventrally to join the **sympathetic trunk,** which runs in a craniocaudal direction just lateral to the vertebral column. A ganglion is usually located in the trunk at the point where each ramus joins it. This is the **sympathetic trunk ganglion** and contains the cell bodies of postganglionic axons.

Leaving the caudal thoracic and lumbar portions of the sympathetic trunk are nerves that course into the abdominal cavity, the **splanchnic nerves.** These form plexuses around the main blood vessels of the abdominal organs. Additional sympathetic ganglia are located in association with these plexuses and blood vessels. The cell bodies of sympathetic postganglionic axons are located here. These axons follow the terminal branching of the blood vessels of the abdominal organs to reach the organ innervated. These abdominal plexuses and ganglia are named according to the major artery with which they are associated. They will be dissected later.

Each preganglionic sympathetic axon must pass through the ramus communicans of its spinal nerve to reach the sympathetic trunk. Its fate from here is variable and mostly dependent on the structure to be innervated. A few examples will illustrate this.

Smooth muscle of blood vessels, piloerector muscles and sweat glands are innervated by postganglionic sympathetic axons in spinal nerves. The preganglionic axon enters the sympathetic trunk via the ramus communicans. It may synapse in the ganglion where it entered or it may pass up or down the sympathetic trunk a few segments and synapse in the ganglion of that segment. The postganglionic axons then return to the segmental spinal nerve via the ramus communicans, usually of the segment in which the synapse occurred. The postganglionic axon then courses with the distribution of the spinal nerve to the smooth muscle and sweat glands. Thus the rami communicantes of spinal cord segments T1 to L5 contain both preganglionic and postganglionic axons.

For smooth muscle and glands of the head, the preganglionic axons enter the sympathetic trunk in the cranial thoracic region. Some may synapse in ganglia as they enter the sympathetic trunk. Many others continue as preganglionic axons up the sympathetic trunk in the neck, where it courses in the same fascial sheath as the common carotid artery and vagus nerve. At the cranial end of this trunk, just ventral to the base of the skull, a ganglion is located—the **cranial cervical ganglion.** All remaining preganglionic axons to the head will synapse here. The postganglionic axons are then distributed with the blood vessels to the structures of the head innervated by this sympathetic system.

The sympathetic trunk is located along the full length of the vertebral column on both sides. Throughout the thoracic, lumbar and sacral levels it is joined to each segmental spinal nerve by a ramus communicans. Only those from spinal cord segments T1 to L5 contain preganglionic axons.

For smooth muscle and glands in the abdominal and pelvic cavities, the preganglionic sympathetic axon reaches the sympathetic trunk via the ramus communicans. It may synapse with a postganglionic neuron in a trunk ganglion, but more often it continues through the ganglion without synapsing and enters a splanchnic nerve. The preganglionic (or occasionally postganglionic) axon courses through the appropriate splanchnic nerve to the abdominal plexuses and their ganglia. Preganglionic axons synapse in one of these ganglia with a cell body of a postganglionic axon. The postganglionic axons follow the terminal branches of abdominal blood vessels to the organs innervated.

Dissection

Selected portions of the autonomic nervous system will be dissected as they are exposed in the regions being studied.

Examine the dorsal aspect of the interior of the pleural cavities (Fig. 105). Notice the sympathetic trunks coursing longitudinally across the ventral surface of the necks of the ribs. The small enlargements in these trunks at each intercostal space are sympathetic trunk ganglia. Dissect a portion of the trunk and a few ganglia on either side. Notice the fine filaments that run dorsally between the vertebrae to join the spinal nerve of that space. These are the rami communicantes of the sympathetic trunk.

Follow the thoracic portion of the sympathetic trunk cranially. Notice the irregular enlargement of the trunk medial to the dorsal end of the first intercostal space on the lateral side of the longus colli. This is the **cervicothoracic ganglion** (stellate ganglion). It is formed by a collection of cell bodies from a fusion of caudal cervical and the first two or three thoracic ganglia. Locate this ganglion on both sides.

Many branches leave the cervicothoracic ganglion. Rami communicantes connect to the ventral branches of the first and second thoracic spinal nerves and to the ventral branches of the seventh and eighth cervical spinal nerves. These spinal nerves contribute to the formation of the **brachial plexus,** which provides a pathway for postganglionic axons to reach the thoracic limb. A branch or plexus from the cervicothoracic ganglion follows the vertebral artery through the transverse foramina—the **vertebral nerve.** This is a source of postganglionic axons for the remaining cervical spinal nerves via branches at each intervertebral space from the vertebral nerve to the cervical spinal nerve. Postganglionic axons may leave the cervicothoracic ganglion and course directly to the heart.

Cranial to the cervicothoracic ganglion a branch of the sympathetic trunk, the **ansa subclavia,** passes ventral to the subclavian artery. The sympathetic trunk and the ansa reunite and join at the **middle cervical ganglion.** This ganglion lies at the junction of the ansa and the vagosympathetic trunk and appears as a swelling of the combined structures. Locate these structures on both sides. Numerous branches, **cardiac nerves,** leave the ansa and middle cervical ganglion and course to the heart.

The **vagosympathetic trunk** in the neck lies in the carotid sheath. Its sympathetic portion carries preganglionic and postganglionic sympathetic axons cranially to structures in the head. The **cranial cervical ganglion** is located at its most cranial end. This is at the level of the base of the ear, just caudomedial to the tympanic bulla. It will be dissected later. The tenth cranial or **vagus nerve** contains parasympathetic preganglionic axons that course caudally down the neck to thoracic and abdominal organs.

At the level of the middle cervical ganglion, notice the vagus nerve as it leaves the vagosympathetic trunk to continue its course caudally. Cardiac nerves leave the vagus to innervate the heart. Study the caudal course of the vagus nerve on each side.

At the middle cervical ganglion, or slightly caudal to it, the **right recurrent laryngeal nerve** leaves the vagus, curves dorsocranially around the right subclavian artery, reaches the dorsolateral surface of the trachea and courses cranially to the larynx. It may be found in the angle between the trachea and the longus colli. On the left side the **left recurrent laryngeal nerve** leaves the vagus caudal to the middle cervical ganglion, curves medially around the arch of the aorta and becomes related to the ventrolateral aspect of the trachea and the ventromedial edge of the esophagus. In this position it courses the neck to reach the larynx. As it ascends it reaches the dorsolateral aspect of the trachea. Each recurrent nerve sends branches to the heart, trachea and esophagus before terminating in the laryngeal muscles as the **caudal laryngeal nerve.** The laryngeal nerves will be dissected later.

Follow each **vagus nerve** as it courses over the base of the heart and supplies cardiac nerves to it. Branches are supplied to the bronchi as the vagus passes over the roots of the lungs. Between the azygos vein and the right bronchus on the right and the area just caudal to the base of the heart on the left, each vagus divides into dorsal and ventral branches. The right and left ventral branches soon unite with each other to form the **ventral vagal trunk** on the esophagus. The dorsal branches of each vagus do not unite until further caudally near the diaphragm where they form the **dorsal vagal trunk,** which lies dorsal to the esophagus. The termination of these trunks in the abdomen will be studied later.

HEART AND PERICARDIUM

The **pericardium** (Fig. 96) is the fibroserous covering of the heart. It is a thin layer consisting of three inseparable components. It is located in the middle part of the mediastinum from the level of the third to the sixth rib. The **fibrous pericardium** is the connective tissue between the parietal serous pericardium and the adjacent pericardial mediastinal pleura. The continuation of the fibrous pericardium to the sternum and diaphragm forms the **phrenicopericardial ligament.** The **serous pericardium** is a closed sac that envelops most of the heart. The **parietal layer** adheres to the fibrous pericardium. At the base of the heart it is continuous with the **visceral layer,** or **epicardium,** which tightly adheres to the heart. Between the parietal and visceral serous pericardium is the pericardial cavity, which contains a small amount of pericardial fluid. Incise the combined pericardial mediastinal pleura, fibrous pericardium and parietal serous pericardium to expose the heart.

The **heart** (Figs. 105–107) consists of a dorsal base where the great vessels are attached and an apex that faces ventrally, caudally and usually to the left depending on the shape of the thorax. The side of the heart facing the left thoracic wall is called the **auricular surface** because the tips of the two auricles project on this side. The opposite side facing the right thoracic wall is the **atrial surface.** The thin-walled right ventricle winds across the cranial surface from the atrial surface of the heart.

Trace the **coronary groove** around the heart. It lies between the atria and ventricles and contains coronary vessels and fat. The **interventricular grooves** are the superficial separations of the right and left ventricles. The **subsinuosal interventricular groove** is a short furrow that lies caudodorsally on the atrial surface and marks the approximate position of the interventricular septum. It contains the terminal part of the left coronary artery, the subsinuosal interventricular branch. Obliquely transversing the auricular surface of the heart is the **paraconal interventricular groove.** Longer and more distinct than the subsinuosal, it begins at the base of the pulmonary trunk, where it is covered by the left auricle. It contains the paraconal interventricular branch of the left coronary artery.

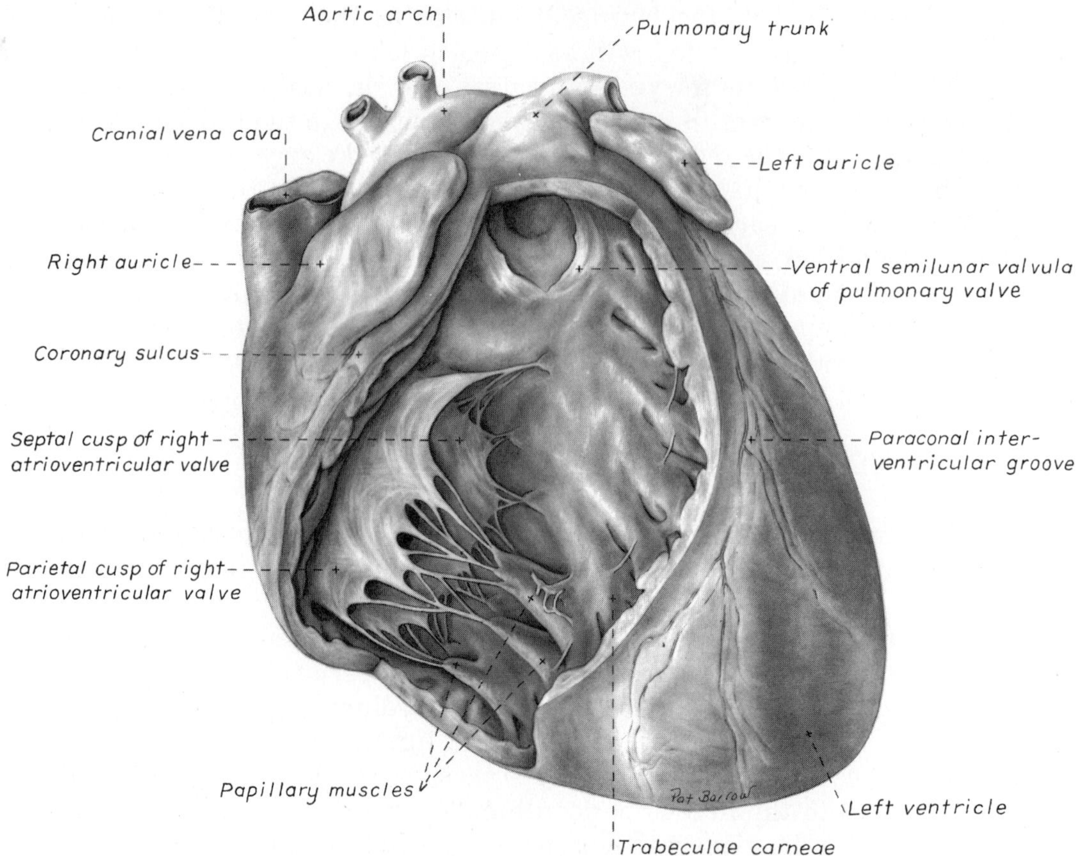

FIGURE 106. Interior of right ventricle, ventral aspect.

The **right atrium** receives the blood from the systemic veins and most of the blood from the heart itself. It lies dorsocranial to the right ventricle. It is divided into a main part, the **sinus venarum,** and a blind cranial part, the **right auricle.**

Open the right atrium by a longitudinal incision through its lateral wall from the cranial vena cava to the caudal vena cava. Extend a cut from the middle of the first incision to the tip of the auricle.

There are four openings into the sinus venarum of the right atrium. The **caudal vena cava** enters the atrium caudally. Ventral to this opening is the **coronary sinus,** the enlarged venous return for most of the blood from the heart. The subsinuosal interventricular groove is ventral to this sinus on the atrial surface of the heart. The **cranial vena cava** enters the atrium dorsally and cranially. Ventral and cranial to the coronary sinus is the large opening from the right atrium to the right ventricle, the **right atrioventricular orifice.** The valve will be described with the right ventricle.

Examine the dorsomedial wall of the sinus venarum, the **interatrial septum.** Between the two caval openings is a transverse ridge of tissue, the **intervenous tubercle.** It diverts the inflowing blood from the two caval veins toward the right atrioventricular orifice. Caudal to the inter-

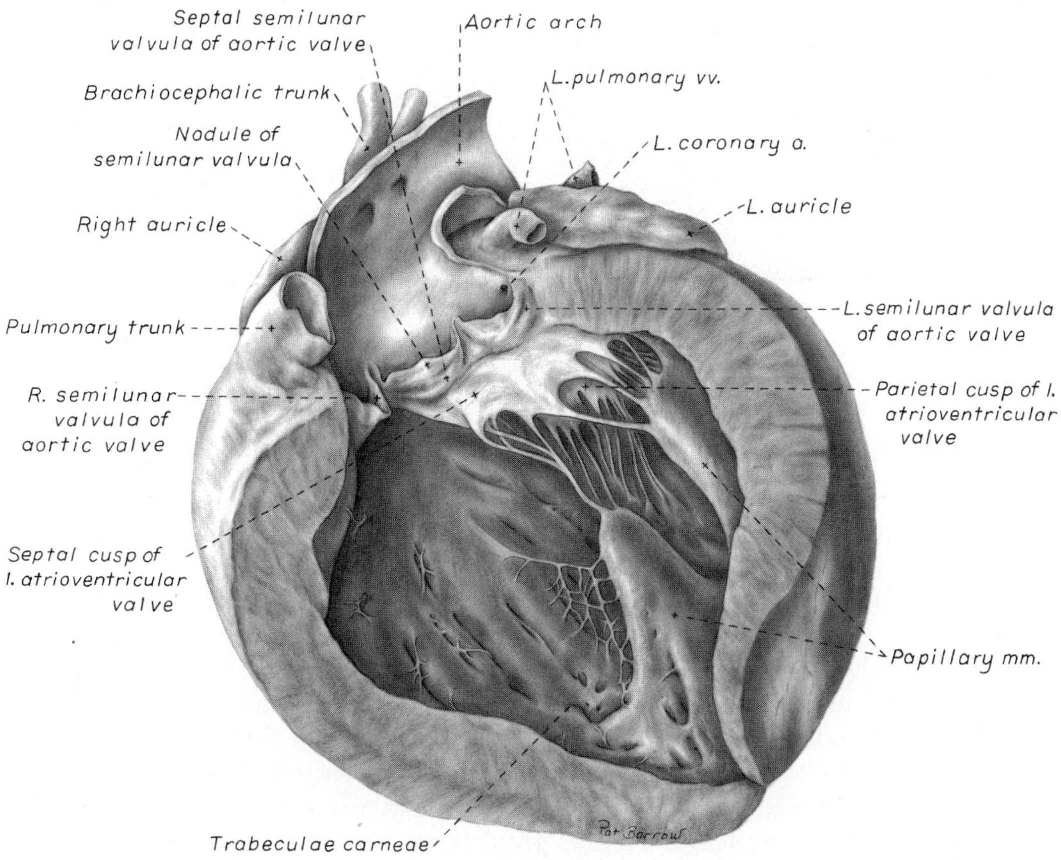

FIGURE 107. Interior of left ventricle, left lateral aspect.

venous tubercle is a slitlike depression, the **fossa ovalis.** In the fetus there is an opening at the site of the fossa, the foramen ovale, which allows blood to pass from the right to the left atrium.

The **right auricle** is the blind, ear-shaped pouch of the right atrium that faces cranially. The internal surface of the wall of the right auricle is strengthened by interlacing muscular bands, the **pectinate muscles.** These are also found on the lateral wall of the atrium proper. The internal surface of the heart is lined everywhere with a thin, glistening membrane, the **endocardium,** which is continued in the blood vessels as the tunica intima. The **crista terminalis** is the smooth-surfaced, thick portion of heart muscle shaped like a semilunar crest at the entrance into the auricle. Pectinate muscle bands radiate from this crest into the auricle.

Locate the **pulmonary trunk** leaving the right ventricle at the left craniodorsal angle of the heart. Begin at the cut end of the left pulmonary artery and extend an incision through the wall of this artery, the pulmonary trunk and the wall of the right ventricle along the paraconal interventricular groove. Continue this cut around the right ventricle following the interventricular septum to the origin of the subsinuosal interventricular groove. Cut through the caudal angle of the right atrioventricular valve and reflect the ventricular wall.

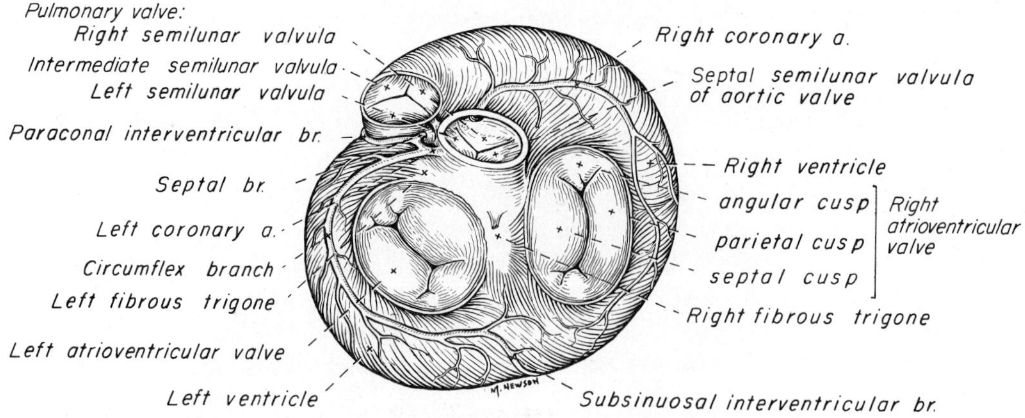

FIGURE 108. Atrioventricular, aortic and pulmonary valves, dorsoventral view.

The greater part of the base of the right ventricle communicates with the right atrium through the atrioventricular orifice. This opening contains the **right atrioventricular valve** (Figs. 106, 108). There are two main parts to the valve in the dog: a wide but short flap that arises from the parietal margin of the orifice, the **parietal cusp,** and a flap from the septal margin, the **septal cusp,** which is nearly as wide as it is long. Subsidiary leaflets are found at each end of the septal flap. The points of the flaps of the valve are continued to the septal wall of the ventricle by the **chordae tendineae.** The chordae tendineae are attached to the septal wall by means of conical muscular projections, the **papillary muscles,** of which there are usually three to four. The **trabeculae carneae** are the muscular irregularities of the interior of the ventricular walls. The **trabecula septomarginalis** is a muscular strand that extends across the lumen of the ventricle from the septal to the parietal wall. The septal attachment is often to a papillary muscle. The right ventricle passes across the cranial surface of the heart and terminates as the funnel-shaped **conus arteriosus,** which gives rise to the pulmonary trunk. This is at the left craniodorsal aspect of the heart. The paraconal interventricular groove is adjacent to the caudal border of the conus arteriosus on the auricular surface of the heart.

At the junction between the right ventricle and the pulmonary trunk is the pulmonary valve consisting of three **semilunar cusps.** A small fibrous **nodule** is located at the middle of the free edge of each cusp. The pulmonary trunk bifurcates into right and left pulmonary arteries, each going to its respective lung.

Open the left side of the heart with one longitudinal cut through the lateral wall of the left atrium, left atrioventricular valve and left ventricle midway between the subsinuosal and paraconal interventricular grooves. Extend the incision into the left auricle.

The **left atrium** is situated on the left dorsocaudal part of the base of the heart dorsal to the left ventricle. Five or six openings mark the entrance of the pulmonary veins into the atrium. The inner surface of the atrium is smooth except for pectinate muscles confined to the **left auricle.** A thin flap of tissue is present on the cranial part of the interatrial septal wall. This is the **valve of the foramen ovale,** a remnant of the passageway for blood from the right atrium to the left atrium in the fetus.

Notice the thickness of the left ventricular wall as compared with the right. The **left atrioventricular valve** is composed of two major cusps, the **septal** and **parietal,** but the division is indistinct (Fig. 108). Secondary cusps are present at the ends of the two major ones. Notice the two large papillary muscles and their chordae tendineae attached to the cusps. The trabeculae carneae are not as numerous in the left ventricle as in the right.

Remove the fat, pleura and pericardium from the aorta. In doing this, isolate the **ligamentum arteriosum,** a fibrous connection between the pulmonary trunk and the aorta. In the fetus it was patent **(ductus arteriosus)** and served to shunt the blood destined for the nonfunctional lungs to the aorta. Observe the left recurrent laryngeal nerve as it turns around the caudal surface of the ligamentum arteriosum. Isolate the origins of the pulmonary trunk and aorta.

From the left ventricle insert scissors into the aortic valve, located beneath the septal cusp of the left atrioventricular valve, and cut the aortic valve, aortic wall and left atrium. This exposes the aortic valve and the first centimeter of the ascending aorta. The **aortic valve,** like the pulmonary, consists of three semilunar cusps. Notice the nodules of the semilunar cusps in the middle of their free borders. Behind each cusp, the aorta is slightly expanded to form the **sinus** of the aorta.

The **right coronary artery** (Figs. 102, 108) leaves the right sinus of the aorta. It encircles the right side of the heart in the coronary groove and often extends to the subsinuosal interventricular groove. It sends many small and one or two large descending branches over the surface of the right ventricle. Remove the epicardium and fat from its surface and follow the artery to its termination.

The **left coronary artery** (Figs. 102, 108) is about twice as large as the right. It is a short trunk that leaves the left sinus and immediately terminates in (1) a **circumflex branch,** which extends caudally in the left part of the coronary groove, and (2) a **paraconal interventricular branch,** which obliquely crosses the auricular surface of the heart in the paraconal interventricular groove. Both of these branches send large rami over the surface of the left ventricle. Expose the artery and its large branches by removing the epicardium and fat. A **septal branch** courses into the interventricular septum, which it supplies.

The **coronary sinus** is the dilated terminal end of the great cardiac vein. The **great cardiac vein,** which begins in the paraconal interventricular groove, returns blood supplied to the heart by the left coronary artery. Clean the surface of the great cardiac vein and open the coronary sinus. Usually one or two poorly developed valves are present in the coronary sinus.

Live Dog

Observe the thorax and watch it expand and contract with each inspiration and expiration respectively. Place the middle finger of one hand over the dorsal aspect of the ninth or tenth intercostal space. Tap the distal end of this finger just behind the nail with the middle finger of the other hand. Listen for the sound produced by this method of percussion. Compare this with an area over the epaxial muscles or abdomen. A resonating sound will result where normal lung is beneath the region percussed. This method of physical examination can be used to define the extent of normal lung tissue in the thorax.

Place your hands over the ventral thorax and feel for the heart beat. It should be more evident on the left where the apex of the heart is directed.

It is important to know the relationship of the cardiac chambers and valve areas for auscultation. A simple hand rule may be helpful. Make a fist with your left hand and extend your thumb at the proximal interphalangeal joint. Your fist represents the left ventricle and your thumb is the aorta arising from it. The metacarpophalangeal joint of your thumb is at the position of the left atrioventricular valve. Hold your right hand with the fingers extended. Place the palm of your right hand against the closed fingers of your left fist. Wrap the fingers of your right hand around the front of your left fist and curve the second digit of your right hand around your left thumb. Your right hand is in the position of the right ventricle (right and cranial sides of the heart). Your right second digit represents the pulmonary valve and trunk on the left craniodorsal aspect of the heart to the left of the aortic valve. Your right thumb is in the position of the right atrioventricular valve. The paraconal interventricular groove is between the ends of your right fingers and the metacarpophalangeal joints of your left fist. The subsinuosal interventricular groove is between the base of the palms of your two hands.

VESSELS AND NERVES OF THE THORACIC LIMB

Primary Vessels of the Thoracic Limb

- Subclavian
  - Vertebral
  - Costocervical
  - Internal thoracic
  - Superficial cervical
- Axillary
  - External thoracic
  - Lateral thoracic
  - Subscapular
    - Thoracodorsal
    - Caudal circumflex humeral
  - Cranial circumflex humeral
- Brachial
  - Collateral ulnar
  - Superficial brachial
  - Common interosseous
    - Ulnar
    - Cranial interosseous
    - Caudal interosseous
  - Deep antebrachial
- Median
  - Radial
  - Superficial palmar arch

The main artery to the thoracic limb arises within the thorax as a terminal branch of the brachiocephalic on the right side and directly from the aorta on the left side. It is divided into three parts. That which extends from its origin to the first rib is the **subclavian artery** (Fig. 105). From the first rib to the conjoined tendons of the teres major and latissimus dorsi the vessel is the **axillary artery.** The vessel from there to the terminal **median** artery distal to the elbow is called the **brachial artery.**

The **superficial cervical artery** (Figs. 98–104, 109) arises from the subclavian just inside the thoracic inlet. It runs dorsocranially between the scapula and the neck. It supplies the superficial muscles of the base of the neck, the superficial cervical lymph nodes, the muscles of the scapula and the shoulder.

There are usually two **superficial cervical lymph nodes,** which lie cranial to the supraspinatus, covered by the omotransversarius and the brachiocephalicus. These nodes receive the afferent lymph vessels from the superficial part of the lateral surface of the neck, the caudal surface of the head, including the ear and pharynx, and the lateral surface of the thoracic limb.

Brachial Plexus

The brachial plexus (Fig. 110) is formed by the ventral branches of the sixth, seventh and eighth cervical and the first and second thoracic

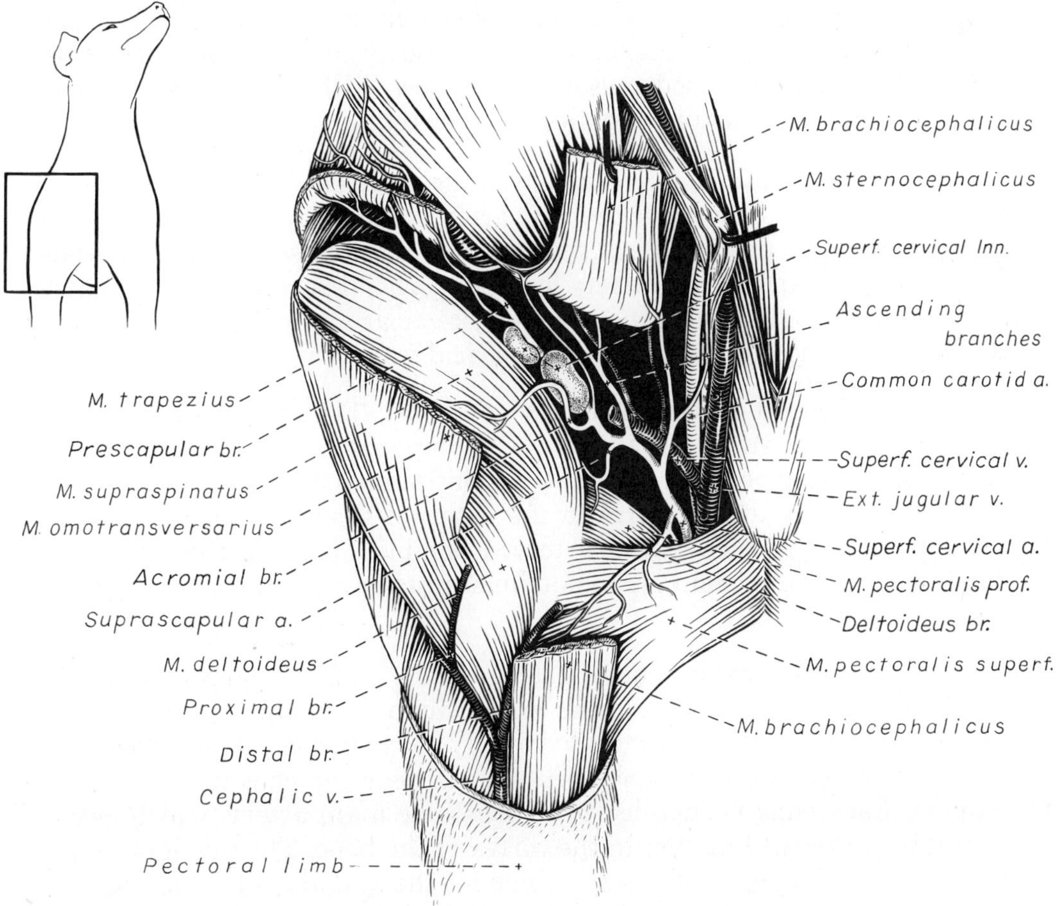

FIGURE 109. Branches of superficial cervical artery.

spinal nerves. In some dogs there is a small contribution from the ventral branch of the fifth cervical spinal nerve. These branches pass between vertebrae, emerge along the ventral border of the scalenus and extend across the axillary space to the thoracic limb. In the axilla, numerous branches of these nerves communicate to form the brachial plexus. From the plexus arise nerves of mixed origin that supply the structures of the thoracic limb and adjacent muscles and skin (Fig. 110).

The pattern of interchange in the brachial plexus is variable, but the specific spinal nerve composition of the named nerves that continue into the thoracic limb is consistent. These nerves include the suprascapular, subscapular, axillary, musculocutaneous, radial, median, ulnar, thoracodorsal, lateral thoracic and pectoral nerves. Expose the brachial plexus in the axilla.

Reflect the superficial and deep pectoral muscles toward their insertions to expose the vessels and nerves on the medial aspect of the arm.

Axillary Artery

The axillary artery (Figs. 110–112) is the continuation of the subclavian and extends from the first rib to the conjoined tendons of the teres major and latissimus dorsi. It has four branches: the external thoracic, the lateral thoracic, the subscapular and cranial circumflex humeral.

In the following dissection some variability may be encountered in the origin of specific blood vessels and nerves. Although the origin may vary, the area or structure supplied is consistent.

1. The **external thoracic artery** (Figs. 111, 112) leaves the axillary near its origin. The external thoracic artery curves around the craniomedial border of the deep pectoral with the nerve to the superficial pectorals and is distributed almost entirely to the superficial pectorals. It may arise from a common trunk with the lateral thoracic artery or it may arise from the deltoid branch of the superficial cervical artery.

2. The **lateral thoracic artery** (Figs. 94, 111, 112) runs caudally across the lateral surface of the axillary lymph node and along the dorsal border of the deep pectoral ventral to the latissimus dorsi. It usually arises from the axillary artery distal to the external thoracic. The vessel may arise distal to the subscapular artery. It supplies parts of the latissimus dorsi, deep pectoral and cutaneus trunci muscles and the thoracic mammae.

3. The **subscapular artery** (Figs. 111, 112) is larger than the continuation of the axillary in the arm. Only a short part of the subscapular is now visible. It passes caudodorsally between the subscapularis and the teres major and becomes subcutaneous near the caudal angle of the scapula. Each bone is supplied by at least one main artery, which enters through a nutrient foramen in the cortex of the bone. The nutrient artery is a branch of an adjacent artery, which for the scapula is the subscapular artery. Dissect the following branches of the subscapular artery:

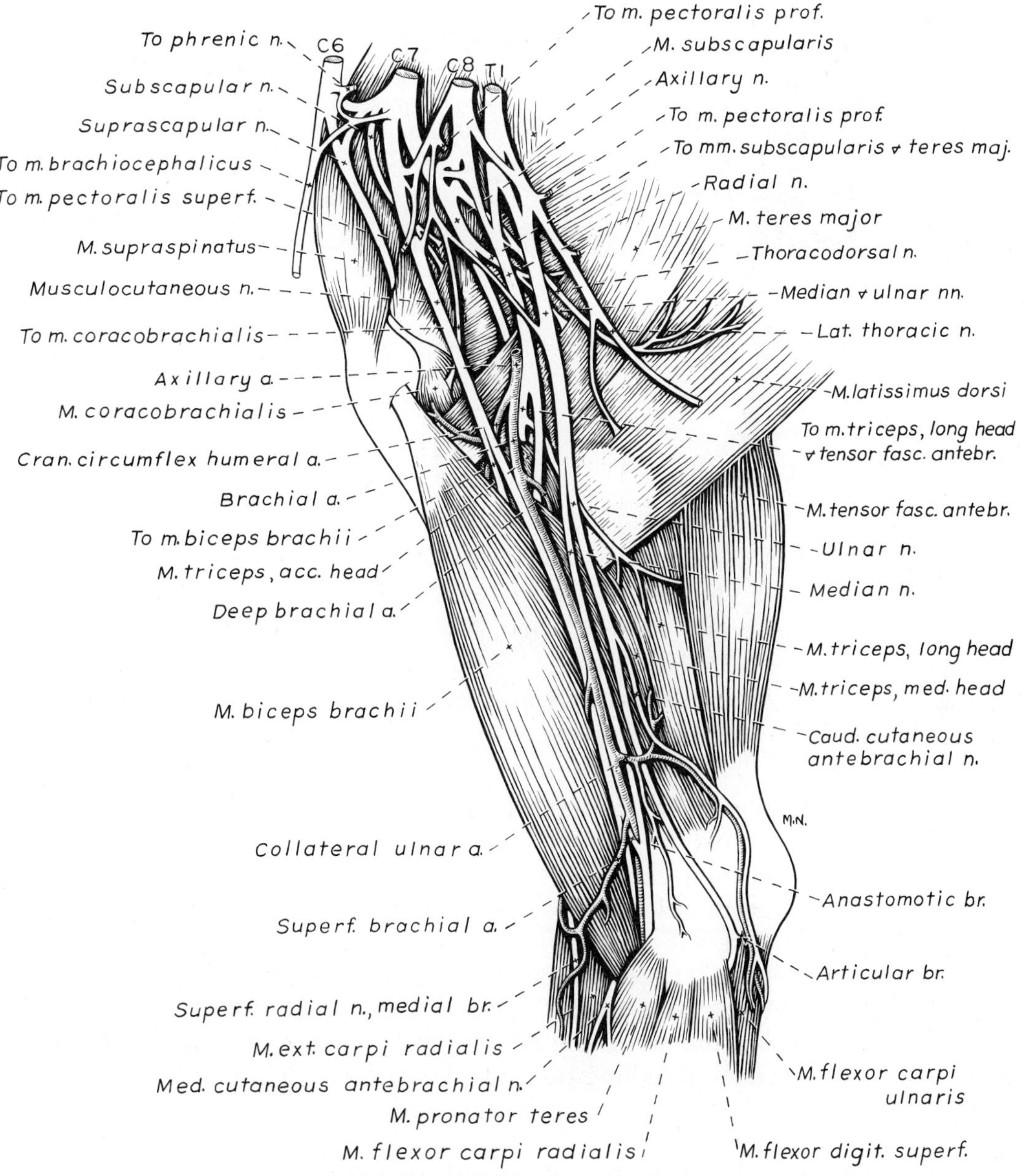

FIGURE 110. Brachial plexus, right thoracic limb, medial aspect.

The **thoracodorsal artery** (Figs. 111, 112) leaves the dorsal surface of the subscapular near its origin. It supplies a part of the teres major and latissimus dorsi and ends in the skin. It is readily seen on the deep surface of the latissimus dorsi. Transect the teres major and reflect both ends to expose the subscapular artery.

The **caudal circumflex humeral artery** (Figs. 111, 112) leaves the subscapular opposite the thoracodorsal and courses laterally between the head of the humerus and the teres major. Pull the subscapular artery medially to see this branch coursing laterally. Expose the caudal circumflex humeral artery from the lateral side where branches become superfi-

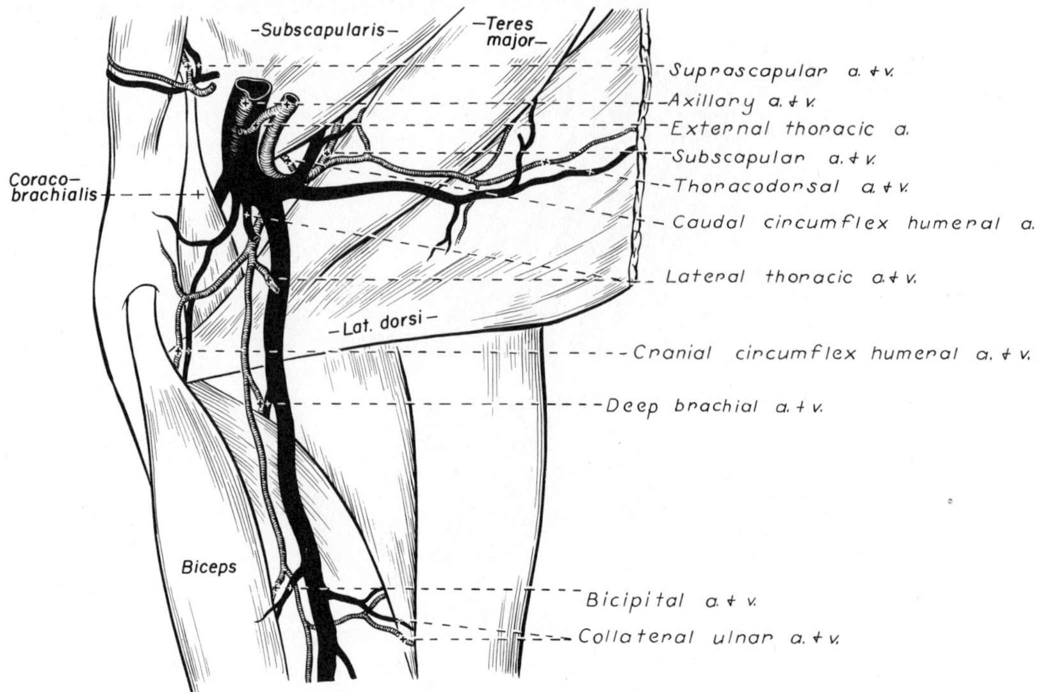

FIGURE 111. Vessels of right axillary region, medial view.

cial at the dorsal part of the lateral head of the triceps. Transect the insertion of the deltoideus. Reflect the deltoideus proximally and observe the axillary nerve and caudal circumflex humeral artery entering the deep surface of the muscle. Notice the large branch of the axillobrachial vein (Fig. 90) that travels with the artery and nerve. The caudal circumflex humeral supplies the triceps, deltoideus, coracobrachialis and infraspinatus muscles and the shoulder joint capsule. Transect the long head of the triceps at its origin. Reflect it and follow the continuation of the subscapular artery caudodorsally along the caudal surface of the scapula. Numerous branches supply the adjacent musculature and bone.

4. The **cranial circumflex humeral artery** (Figs. 111, 112) arises from the axillary artery proximal or distal to the subscapular artery. It courses cranially to supply the biceps brachii and the joint capsule of the shoulder.

Brachial Artery

The brachial artery (Figs. 111–113) is a continuation of the axillary from the conjoined tendons of the teres major and latissimus dorsi. It crosses the distal half of the humerus to reach the craniomedial surface of the elbow, passes under the pronator teres muscle, gives off its largest branch, the **common interosseous,** the smaller **deep antebrachial,** and becomes the **median artery** in the proximal half of the forearm. The deep

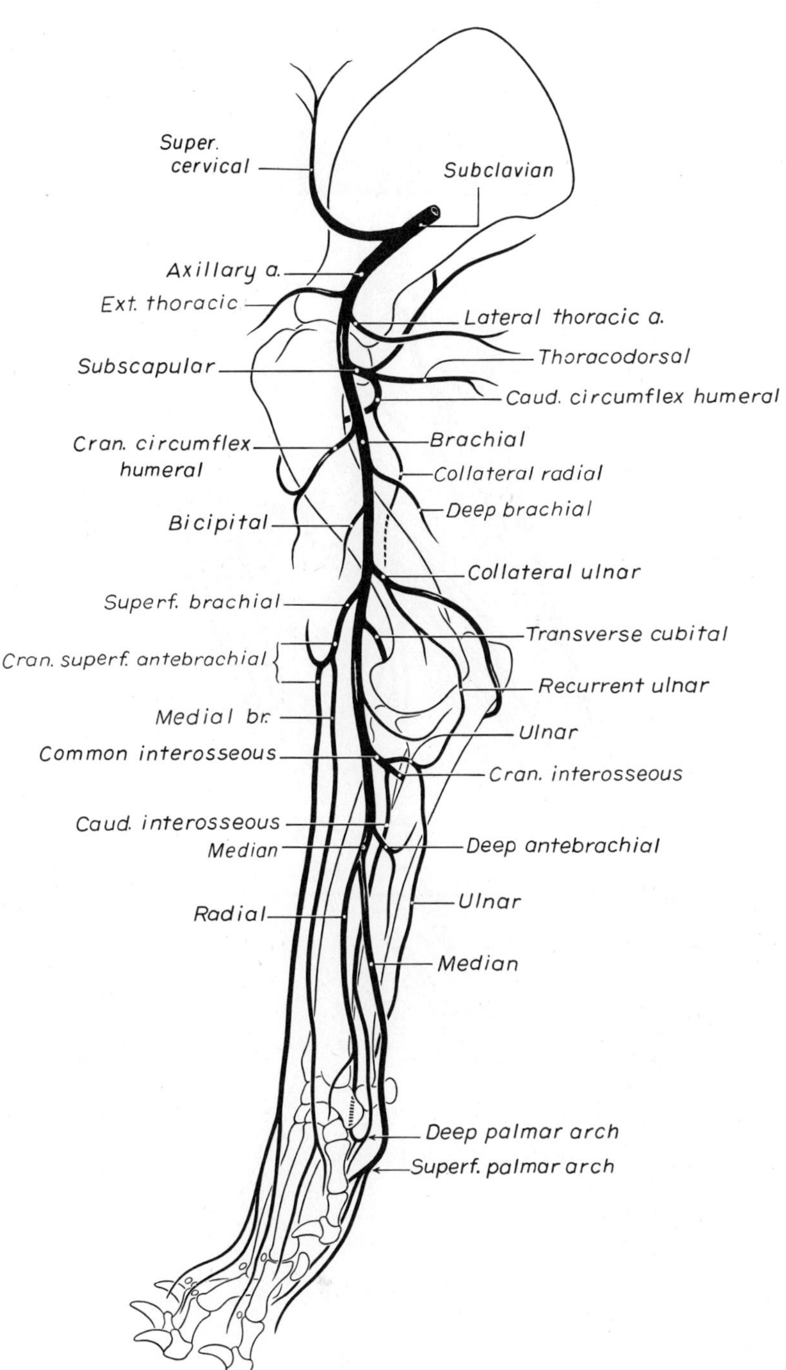

FIGURE 112. Arteries of right thoracic limb, schematic medial view.

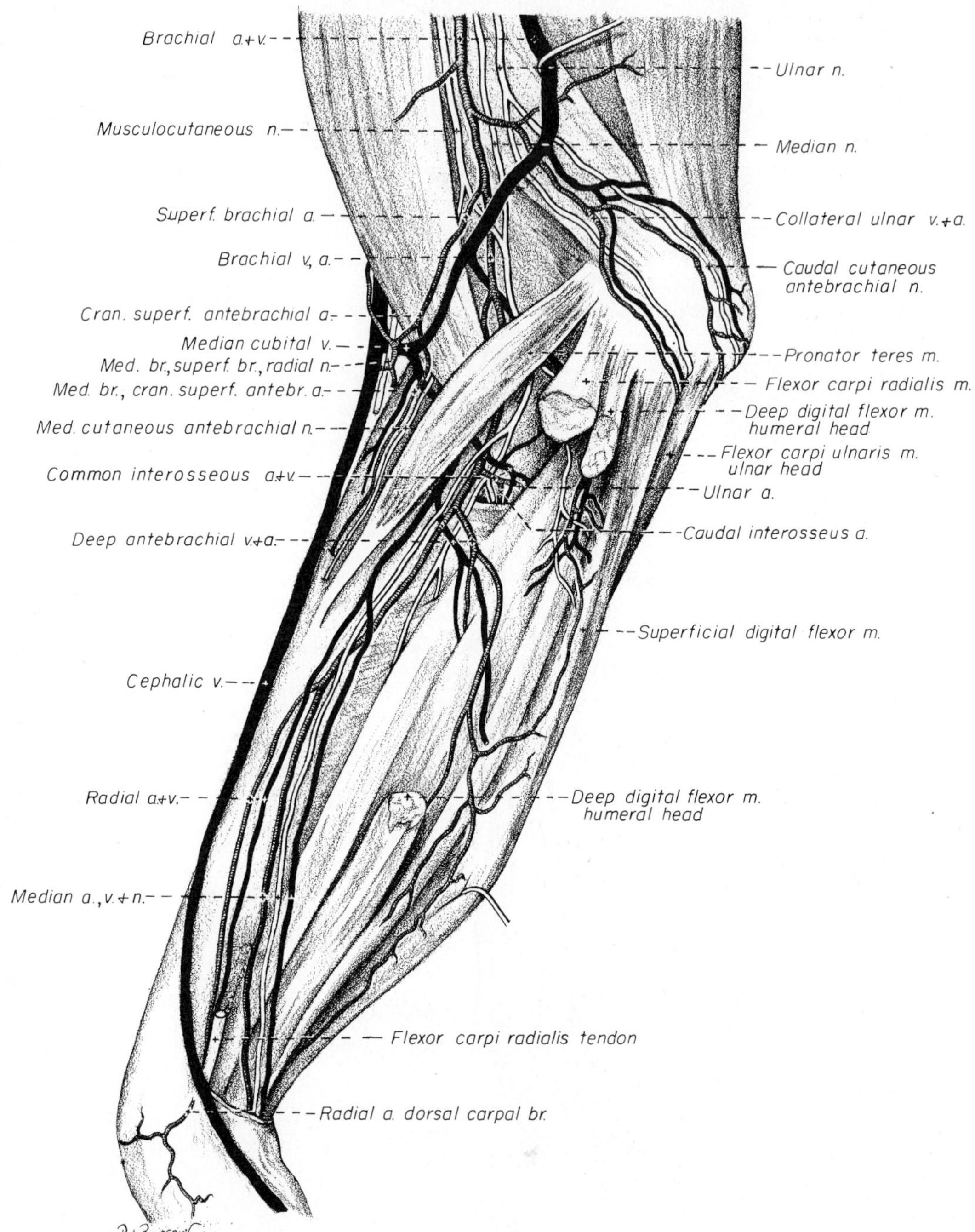

FIGURE 113. Deep structures, right forearm and elbow, medial view.

brachial and bicipital arteries are muscular branches of the brachial in the arm that are variable and need not be dissected. Dissect the following branches of the brachial artery:

1. The **collateral ulnar artery** (Figs. 112–114) is a caudal branch of the brachial in the distal third of the arm. It supplies the triceps, the ulnar nerve and the elbow.

2. The **superficial brachial artery** (Figs. 112, 113) loops around the cranial surface of the distal end of the biceps brachii, deep to the cephalic vein. It continues in the forearm as the **cranial superficial antebrachial**

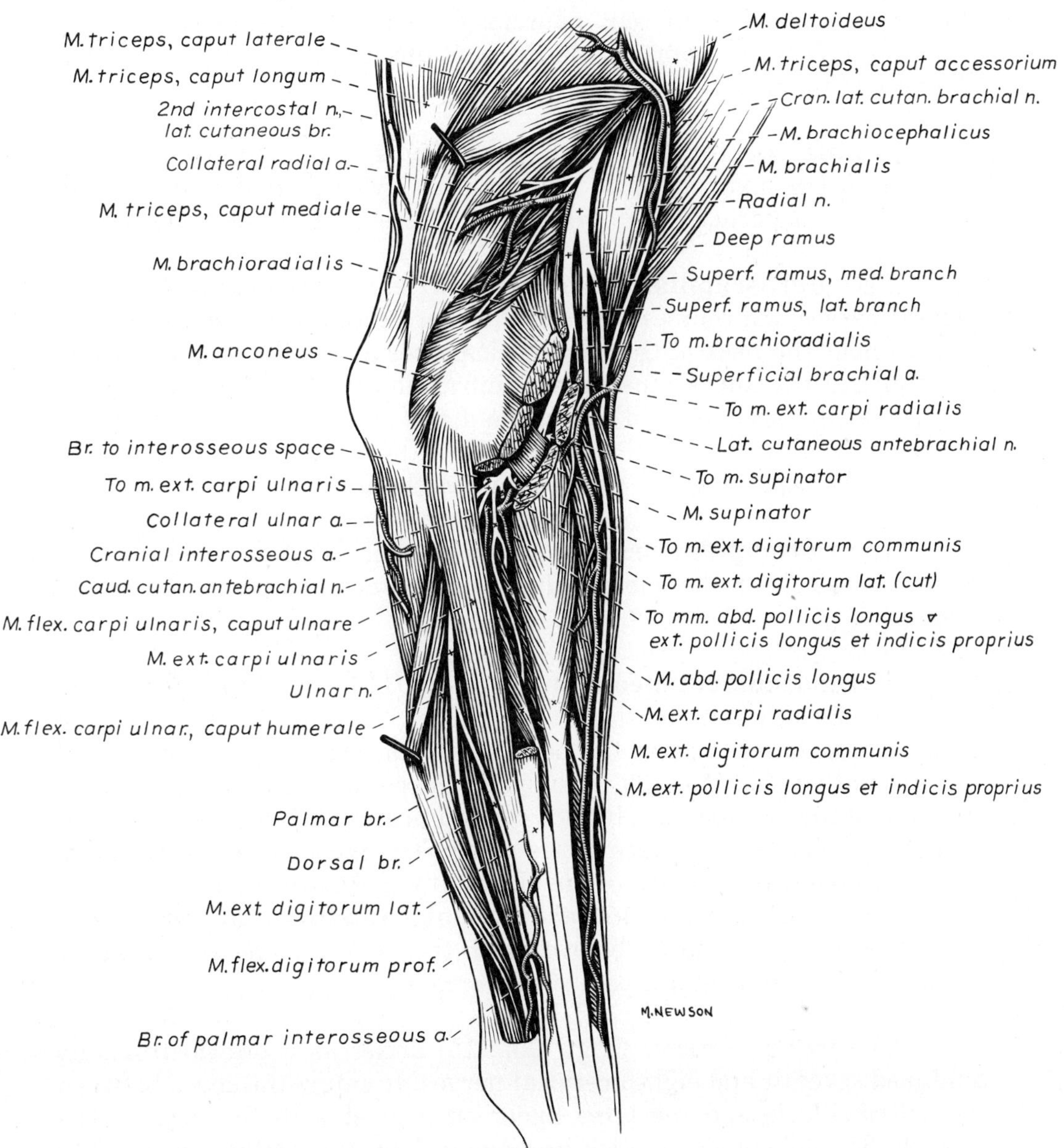

FIGURE 114. Deep structures, right forearm and elbow, lateral view.

artery. A medial ramus arises from the latter, and both course distally on either side of the cephalic vein accompanied by the medial and lateral branches of the superficial radial nerve. These vessels supply blood to the dorsum of the forepaw (see Fig. 120).

The transverse cubital artery supplies the elbow and adjacent muscles and need not be dissected. The brachial artery courses deep to the flexor carpi radialis and pronator teres. Its branches in the forearm will be dissected later.

Nerves of the Scapular Region and Arm

All of the following nine nerves contain somatic efferent neurons to striated muscles and afferent neurons from these muscles. Cutaneous somatic afferents are found only in the musculocutaneous, axillary, radial, median and ulnar nerves.

1. **Cranial pectoral nerves** (Fig. 110) are derived from ventral branches of the sixth, seventh and eighth cervical spinal nerves. They innervate the superficial pectoral muscle. These need not be dissected.

2. The **suprascapular nerve** (Fig. 110) leaves the sixth and seventh cervical nerves and courses between the supraspinatus and subscapularis muscles near the neck of the scapula. It passes across the scapular notch and supplies the supraspinatus and infraspinatus. Transect the supraspinatus at its insertion and reflect the distal end. Trace the branches of the suprascapular nerve to this muscle. Note the continuation of the nerve distal to the scapular spine, where it enters the infraspinatus muscle.

3. The **subscapular nerve** (Fig. 110) is a branch from the sixth and seventh cervicals to the subscapularis. Sometimes two nerves enter the muscle.

4. The **musculocutaneous nerve** (Figs. 110, 113, 115) arises from the sixth, seventh and eighth cervical nerves. Throughout the brachium the musculocutaneous nerve lies between the biceps brachii cranially and the brachial vessels caudally. It supplies the coracobrachialis, the biceps brachii and the brachialis. A branch communicates with the median nerve proximal to the flexor surface of the elbow. The musculocutaneous nerve courses deep to the insertion of the biceps. It supplies the distal end of the brachialis and gives off the **medial cutaneous antebrachial nerve** (Figs. 110, 113), which is usually removed with the skin. This nerve is sensory to the skin on the medial aspect of the forearm.

5. The **axillary nerve** (Figs. 90, 110) arises as a branch from the combined seventh and eighth cervical nerves. It enters the space between the subscapularis and the teres major on a level with the neck of the scapula. The following muscles are supplied by the axillary nerve: the teres major, the teres minor, the deltoideus and part of the subscapularis.

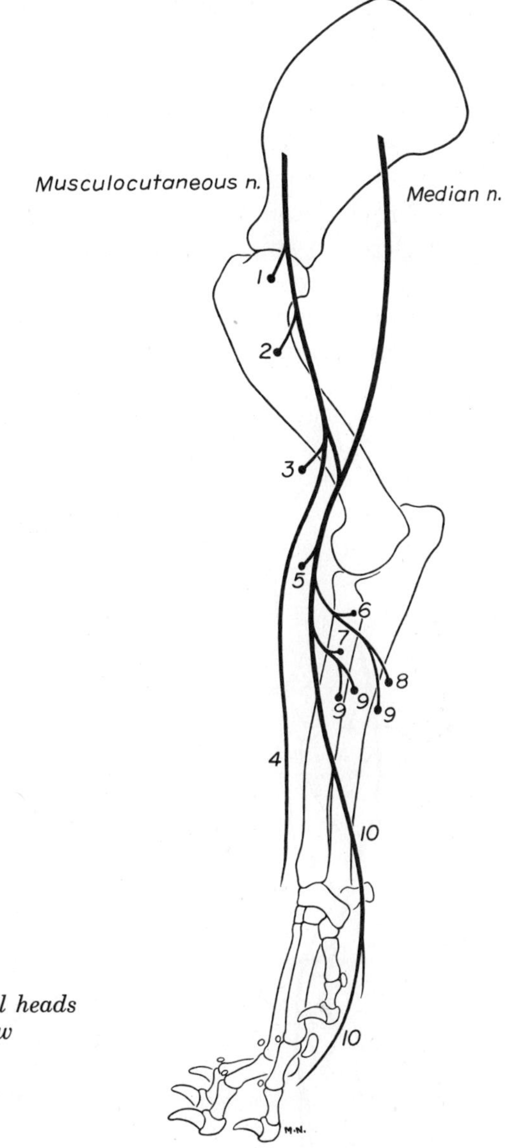

Musculocutaneous nerve
1. *Coracobrachialis*
2. *Biceps brachii*
3. *Brachialis*
4. *Skin of medial antebrachium*

Median nerve

5. *Pronator teres*
6. *Flexor carpi radialis*
7. *Pronator quadratus*
8. *Superficial digital flexor*
9. *Deep digital flexor, humeral, ulnar and radial heads*
10. *Skin of caudal antebrachium and palmar paw*

FIGURE 115. Distribution of musculocutaneous and median nerves, right thoracic limb, schematic medial view.

The **cranial lateral cutaneous brachial nerve** (Figs. 90, 114) appears subcutaneously on the lateral surface of the brachium. It supplies the skin on the lateral surface of the brachium and caudal scapular region. There are cranial cutaneous antebrachial branches that supply the skin on the cranial surface of the forearm. The latter overlap with cutaneous antebrachial branches of the superficial branch of the radial nerve laterally and the musculocutaneous nerve medially.

6. The **thoracodorsal nerve** (Fig. 110) arises primarily from the eighth cervical nerve. It supplies the latissimus dorsi muscle. It courses with the thoracodorsal vessels on the medial surface of the muscle.

7. The **radial nerve** (Figs. 90, 110, 113, 114, 116) arises from the last two cervical and first two thoracic nerves, runs a short distance distally with the trunk of the median and ulnar nerves and enters the triceps distal to the teres major. The radial nerve is motor to all the extensor muscles of the elbow, carpal and phalangeal joints. The muscles of the arm supplied by the radial nerve are the triceps, the tensor fasciae antebrachii and the anconeus. Observe the branches to the triceps. The radial nerve spirals around the humerus on the caudal and then on the lateral surface of the brachialis muscle. On the lateral side at the distal third of the arm the radial nerve terminates as a **deep** and a **superficial branch.** Transect the lateral head of the triceps at its origin and reflect

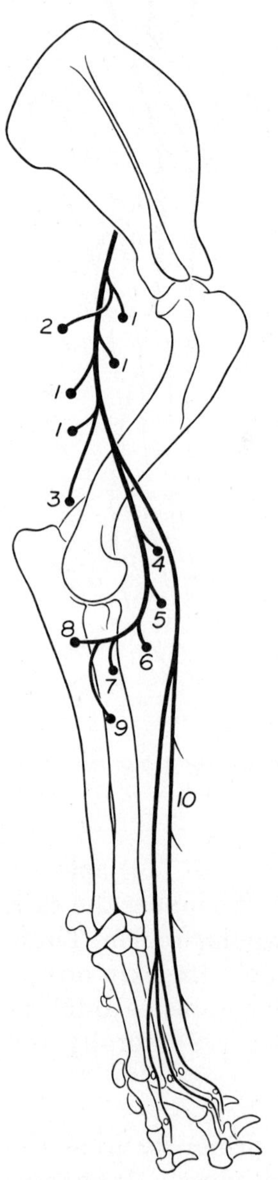

1. *Triceps brachii*
2. *Tensor fasciae antebrachii*
3. *Anconeus*
4. *Extensor carpi radialis*
5. *Supinator*
6. *Common digital extensor*
7. *Lateral digital extensor*
8. *Ulnaris lateralis*
9. *Abductor pollicis longus*
10. *Skin of cranial and lateral antebrachium and dorsal paw*

FIGURE 116. Distribution of radial nerve, right thoracic limb, schematic lateral view.

it to expose these terminal branches. The distribution in the antebrachium will be dissected later.

8. The **median** and **ulnar nerves** (Figs. 90, 110, 113–115, 117) arise by a common trunk from the eighth cervical and the first and second thoracic nerves. The common trunk lies on the medial head of the triceps between the brachial vein caudally and the brachial artery cranially. The **median nerve,** the cranial division of the common trunk, runs to the antebrachium in contact with the caudal surface of the brachial artery. It receives a branch from the musculocutaneous nerve at the level of the elbow. Its branches to several of the muscles of the forearm and to the skin of the palmar side of the paw will be dissected later.

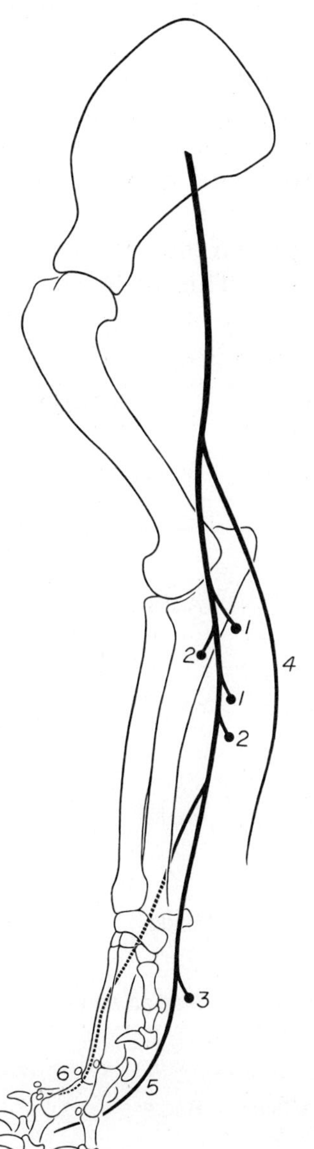

1. *Flexor carpi ulnaris, ulnar and humeral heads*
2. *Deep digital flexor, ulnar and humeral heads*
3. *Interossei*
4. *Skin of caudal antebrachium*
5. *Skin of palmar paw*
6. *Skin of fifth metacarpal, lateral surface of digit*

FIGURE 117. Distribution of ulnar nerve, right thoracic limb, schematic medial view.

The **ulnar nerve,** the caudal division of the common trunk, separates from the median nerve in the distal arm and crosses the elbow caudal to the medial epicondyle of the humerus. Trace it with the collateral ulnar artery to the cut edge of the skin. The **caudal cutaneous antebrachial nerve** (Figs. 90, 110, 114) leaves the ulnar near the middle of the arm and runs caudodistally across the medial surface of the triceps and olecranon. This supplies the skin of the distal medial aspect of the brachium and the caudal aspect of the antebrachium.

9. **Caudal pectoral nerves** are derived from the ventral branches of the eighth cervical and first and second thoracic nerves. They innervate the deep pectoral muscle and are often combined with the lateral thoracic nerve at their origin. They need not be dissected. The lateral thoracic nerve was dissected previously at its termination in the cutaneous trunci muscle.

Vessels and nerves of the thoracic limb are listed in Table 1.

Incise the skin from the olecranon to the palmar surface of the third digit. (Pass through the carpal, metacarpal and third digital pads.) Remove the skin from the forearm and paw, leaving the vessels and nerves on the specimen wherever possible.

The **cephalic vein** (Figs. 90, 113, 118–121) begins on the palmar side of the paw from the superficial palmar venous arch. This need not be

***Table 1.** VESSELS AND NERVES OF THE THORACIC LIMB*

Area	Arterial Supply	Nerve Supply
Lateral Muscles of Scapula and Shoulder		
Stabilizers, flexors and extensors of shoulder: Supraspinatus, infraspinatus	Superficial cervical	Suprascapular
Caudal Muscles of Scapula and Shoulder		
Flexors of shoulder: Deltoideus, Teres major, Teres minor	Subscapular	Axillary
Cranial Muscles of Arm		
Flexors of elbow, extensor of shoulder: Biceps brachii (Brachialis)	Superficial cervical, Axillary, Brachial	Musculocutaneous
Caudal Muscles of Arm		
Extensor of elbow: Triceps brachii	Axillary, Brachial	Radial
Cranial Muscles of Forearm		
Carpal extensors Digital extensors	Brachial: cranial interosseous	Radial
Caudal Muscles of Forearm Carpal flexors Digital flexors	Brachial: common interosseous, Deep antebrachial	Median and Ulnar
Dorsal Surface of Paw	Superficial brachial, Dorsal carpal rete	Radial
Palmar Surface of Paw	Median, Caudal interosseous	Median and Ulnar

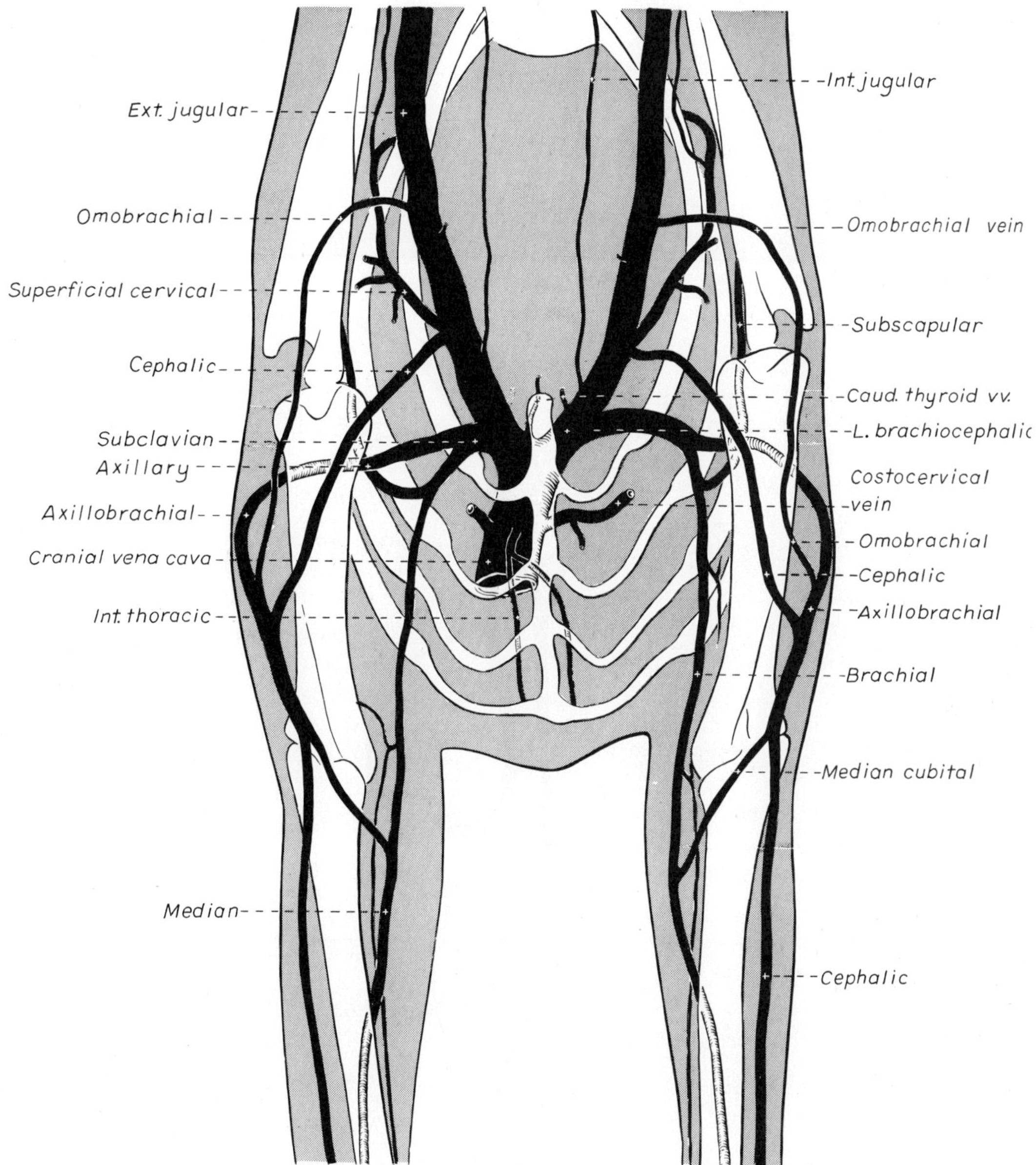

FIGURE 118. Veins of neck and shoulders, schematic cranial aspect.

dissected. The **accessory cephalic vein** joins the cephalic on the cranial surface of the distal third of the forearm. It arises from small veins on the dorsum of the paw. At the flexor surface of the elbow the **median cubital vein** forms a connection between the cephalic and brachial veins. From this connection the cephalic continues proximally on the craniolateral surface of the arm in a furrow between the brachiocephalicus cranially and the origin of the lateral head of the triceps caudally. In the middle of the arm the **cephalic vein** runs deep to the brachiocephalicus and enters the external jugular near the thoracic inlet. The **axillobrachial vein** continues proximally and passes deep to the caudal border of the deltoideus

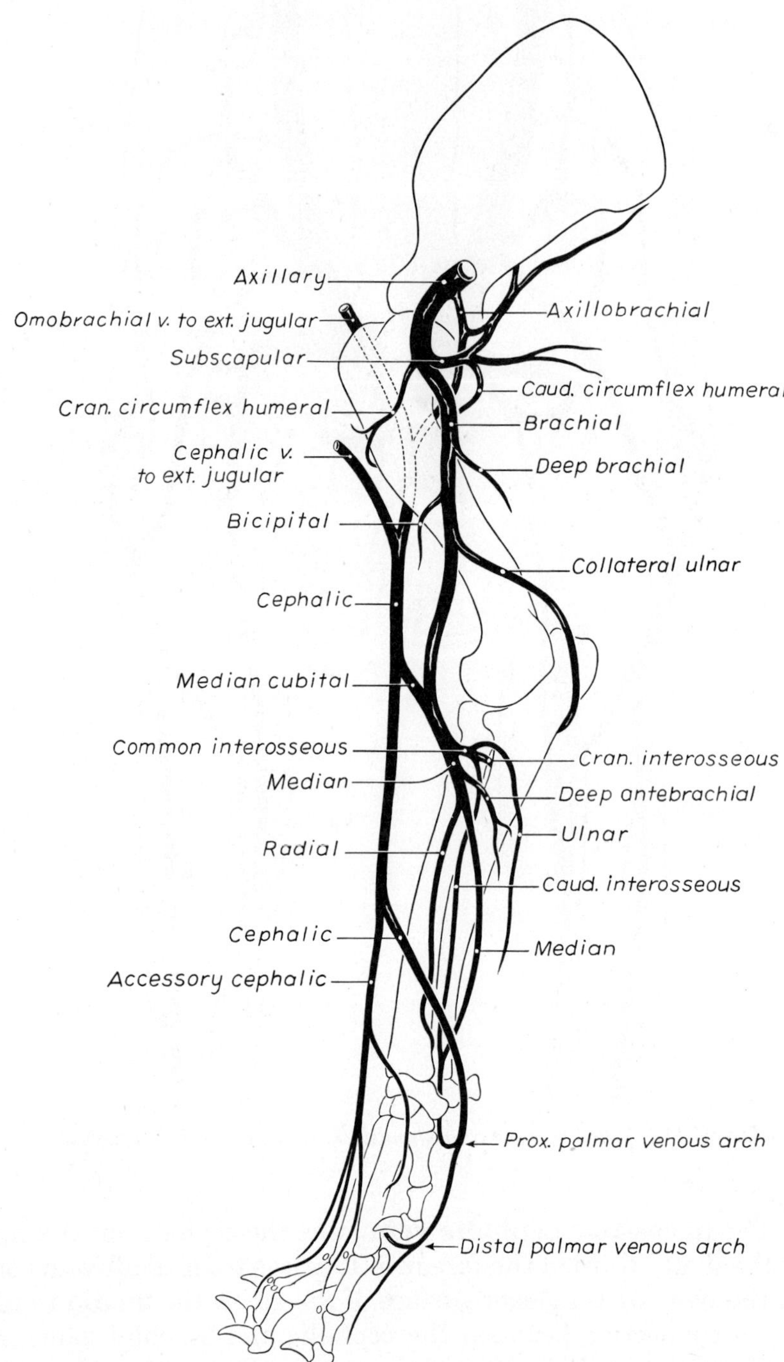

FIGURE 119. Veins of right thoracic limb, schematic medial view.

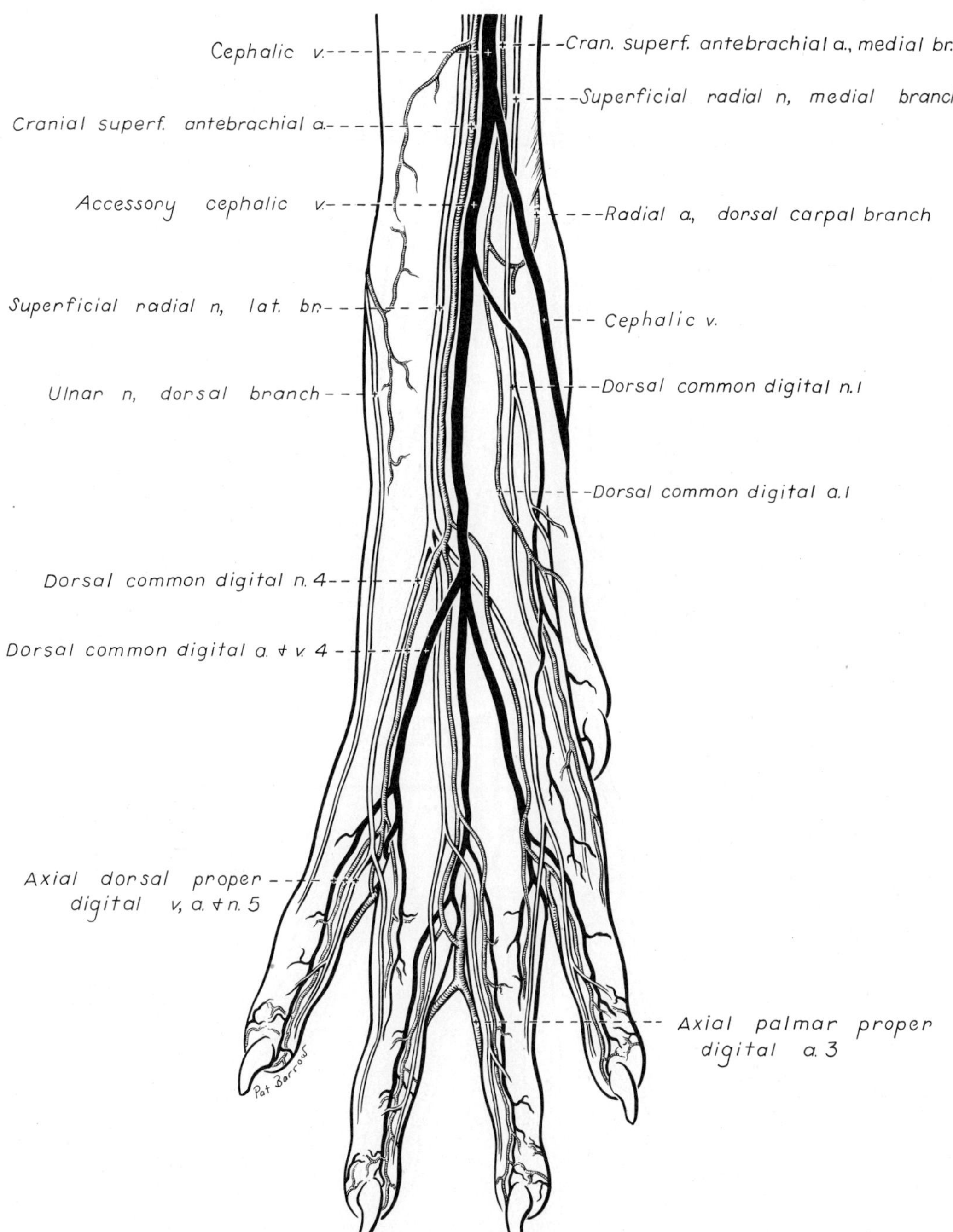

FIGURE 120. Superficial arteries, veins and nerves of right forepaw, dorsal view.

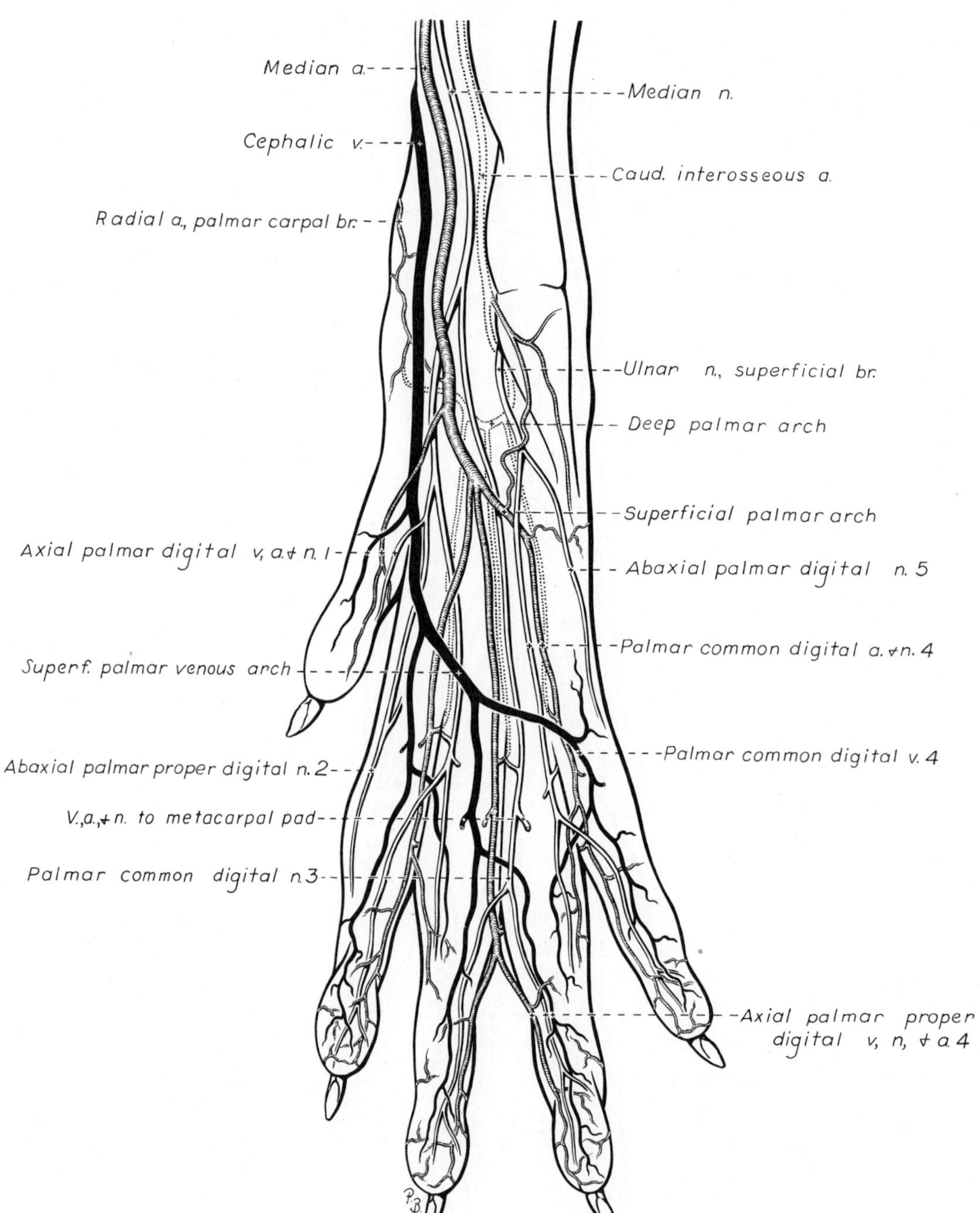

FIGURE 121. Superficial arteries, veins and nerves of right forepaw, palmar view.

to join the axillary vein. The **omobrachial vein** arises from the axillobrachial vein and continues subcutaneously across the cranial surface of the arm and shoulder before entering the external jugular cranial to the cephalic vein.

Arteries of the Forearm and Paw

Make a longitudinal incision through the medial part of the antebrachial fascia midway between the cranial and caudal borders. Extend this incision to the carpus. Remove the fascia from the forearm. Transect the pronator teres and flexor carpi radialis close to their origins and reflect them to uncover the brachial artery.

The **brachial artery** (Figs. 112, 113) in the forearm gives rise to the common interosseous and deep antebrachial arteries and continues as the median artery. The median extends to the superficial palmar arch in the paw.

1. The **common interosseous artery** (Figs. 112, 113) is short and passes to the proximal part of the interosseous space between the radius and ulna. Pull the brachial artery medially to see the branches of the common interosseous. The common interosseous and ulnar arteries may arise together from the brachial.

The **ulnar artery** courses caudally. Separate the muscles on the caudomedial side of the forearm to expose its course. A recurrent branch extends proximally between the humeral and ulnar heads of the deep digital flexor. The ulnar artery continues distally with the ulnar nerve between the humeral head of the deep digital flexor and the flexor carpi ulnaris. It supplies the ulnar and humeral heads of the deep digital flexor and the flexor carpi ulnaris.

The **caudal interosseous artery** lies between the apposed surfaces of the radius and ulna. On the medial side of the forearm expose the pronator quadratus muscle between the radius and ulna. Cut the attachments of this muscle between the two bones and scrape it with the scalpel handle to remove it from the interosseous space. The caudal interosseous artery lies deep in this space. In its course down the forearm the artery supplies many small branches to adjacent structures. It passes through the lateral side of the carpal canal and in the carpometacarpal region it joins with branches of the radial and median arteries to form arches that supply the palmar surface of the forepaw (Figs. 112, 121). These will not be dissected.

The **cranial interosseous artery** passes through the proximal part of the interosseous space cranially to supply the muscles lying laterally and cranially in the forearm. This artery will not be dissected.

2. The **deep antebrachial artery** (Figs. 112, 113) is a caudal terminal branch of the brachial. Follow it deep to the flexor carpi radialis and transect the humeral head of the deep digital flexor, which covers the

vessel. The deep antebrachial artery supplies the flexor carpi radialis, the deep digital flexor, the flexor carpi ulnaris and the superficial digital flexor.

The **median artery** (Figs. 112, 113, 121) is the continuation of the brachial artery beyond the origin of the deep antebrachial. It gives off the radial artery in the forearm and continues into the paw. It is the principal source of blood to the paw. It is accompanied by the median nerve along the humeral head of the deep digital flexor. The median artery passes through the carpal canal between the superficial and deep digital flexor tendons. Transect the flexor retinaculum and superficial digital flexor tendon. Reflect these and follow this artery through the carpal canal to the proximal end of the metacarpus, where it forms the superficial palmar arch with a branch of the caudal interosseous artery. This arch gives rise to the palmar common digital arteries, which course to the palmar surface of the forepaw (these need not be dissected).

The **radial artery** (Figs. 112, 113, 120, 121) arises from the median artery in the middle of the forearm. It follows the medial border of the radius. At the carpus it divides into palmar and dorsal carpal branches that supply the deep vessels of the forepaw. These need not be dissected.

Nerves of the Forearm and Paw

The skin of the antebrachium is innervated by four nerves: the cranial surface by the axillary and radial nerves; the lateral surface by the radial nerve; the caudal surface by the ulnar nerve; and the medial surface by the musculocutaneous nerve. There is considerable overlap in their cutaneous areas of distribution.

1. The **radial nerve** (Fig. 116) supplies the extensors of the elbow, carpus and digital joints and one flexor of the carpus. Reflect the lateral head of the triceps. Observe the radial nerve near the elbow, where it divides into superficial and deep branches. Transect the extensor carpi radialis and reflect the proximal end. The **deep branch** of the radial nerve crosses the medial surface of the extensor carpi radialis in its course into the forearm with the brachialis muscle. Transect the supinator, which lies over the radial nerve. The radial nerve innervates the extensor carpi radialis, the common digital extensor, the supinator, the lateral digital extensor, the abductor pollicis longus and the ulnaris lateralis.

The **superficial branch** divides into a **lateral cutaneous antebrachial nerve** and medial and lateral branches. The small **medial branch** follows the medial ramus of the cranial superficial antebrachial artery and continues distally in the forearm on the medial side of the cephalic vein. The **lateral branch** becomes associated with the lateral side of the cephalic vein and enters the forearm with the cranial superficial antebrachial artery. These branches continue to the paw on either side of the accessory cephalic vein. These medial and lateral branches are sensory to the skin on the cranial and lateral surface of the forearm and the dorsal surface of the carpus, metacarpus and digits (Fig. 122). They terminate in dorsal common digital nerves in the paw (Fig. 123). Do not dissect these nerves.

2. The **median nerve** (Figs. 113, 115, 121) runs distally into the antebrachium with the brachial artery. It innervates the pronator teres, the pronator quadratus, the flexor carpi radialis, the superficial digital flexor and the radial head and parts of the humeral and ulnar heads of the deep digital flexor.

Reflect the flexor carpi radialis and the humeral head of the deep digital flexor to observe the course of the nerve. The median nerve passes through the carpal canal with the median artery and branches to supply sensory innervation to the palmar surface of the forepaw. Note the course

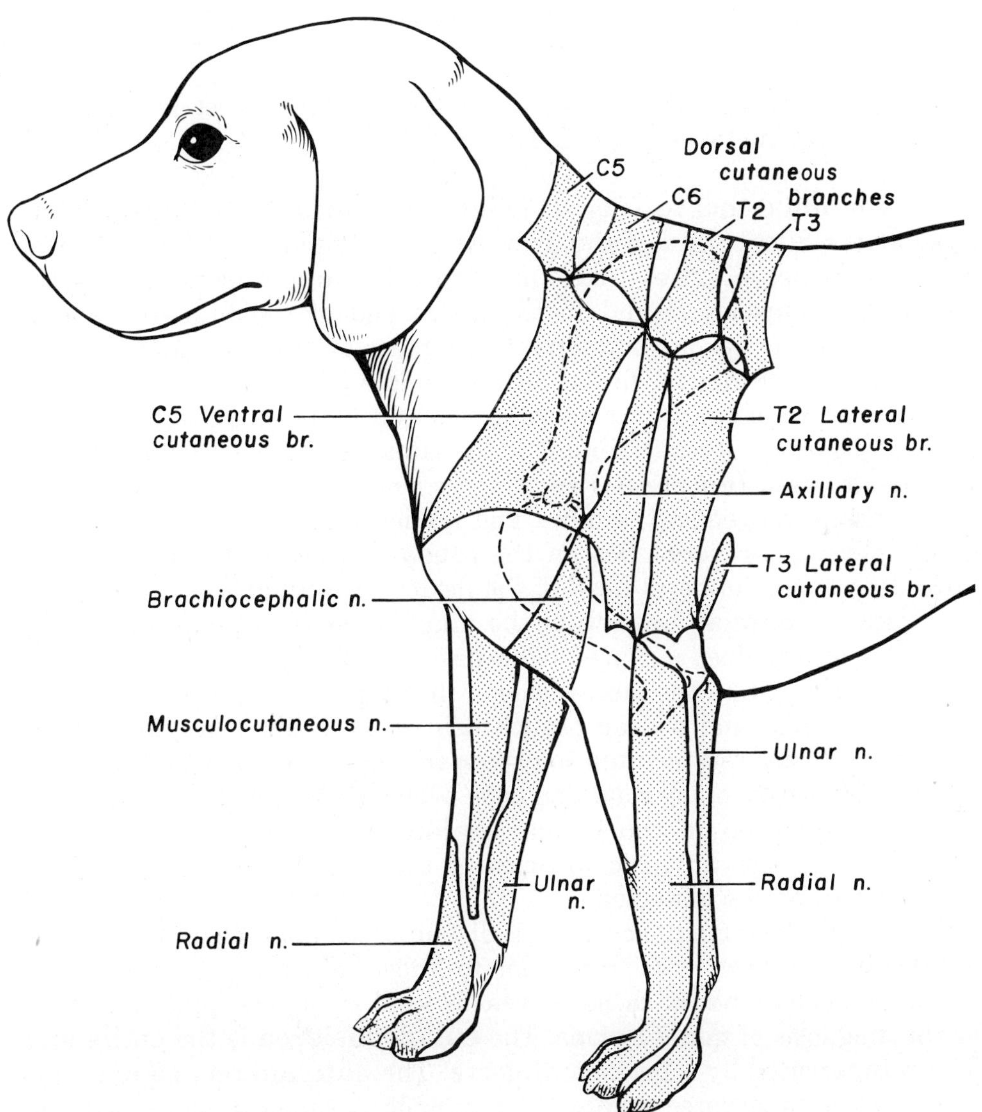

FIGURE 122. Autonomous zones of cutaneous innervation of thoracic limb.

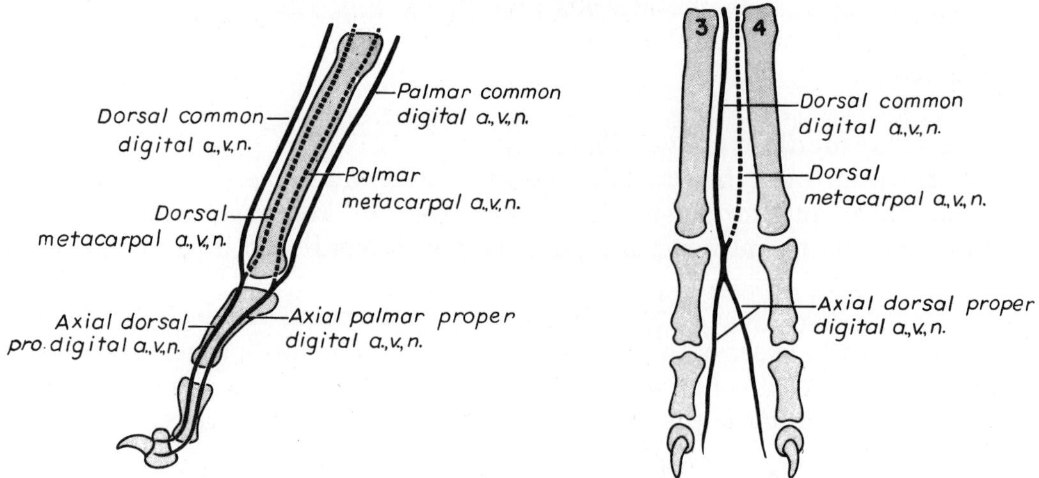

FIGURE 123. Schema of blood supply and innervation of digits.

of the nerve through the carpal canal but do not dissect its terminal branches in the metacarpus (Fig. 121).

3. The **ulnar nerve** (Fig. 117) diverges caudally from the median nerve at the distal third of the arm. At this point the caudal cutaneous antebrachial nerve arises from the ulnar. The ulnar nerve enters the antebrachial muscles caudal to the distal end of the humerus and is distributed to the flexor carpi ulnaris and parts of the ulnar and humeral heads of the deep digital flexor. In the middle of the antebrachium the **dorsal branch** of the ulnar nerve arises and becomes subcutaneous on the lateral surface. It is distributed to the lateral surface of the metacarpus and the fifth digit (Fig. 120).

Transect and reflect the ulnar and humeral heads of the flexor carpi ulnaris. The ulnar nerve lies on the caudal surface of the deep digital flexor deep to the humeral head of the flexor carpi ulnaris.

Trace the **palmar branch** of the ulnar nerve into the forepaw. The nerve lies on the deep surface of the flexor carpi ulnaris, just above the carpus. It then enters the lateral side of the carpal canal, where it divides into a superficial and a deep branch (Fig. 121). Reflect the flexor carpi ulnaris and flexor retinaculum to expose the nerve in the carpal canal. In the metacarpus the superficial and deep branches further divide to supply sensory innervation to the palmar surface of the forepaw and motor innervation to the intrinsic muscles of the forepaw (Fig. 122). These branches will not be dissected.

Although the terminal cutaneous distribution of nerves in the forelimb is difficult to dissect, it is important to know both the cutaneous areas and autonomous zones of these nerves for local anesthetic procedures and for the diagnosis of nerve lesions. The **cutaneous area** is the entire area of skin innervated by a peripheral nerve. The **autonomous zone** is that area of skin innervated solely by a specific peripheral nerve with no overlap from adjacent nerves. Figure 122 shows the autonomous zones for the nerves of the forelimb and adjacent neck and trunk regions.

The distribution of vessels and nerves to the digits will not be dissected, but you should understand the pertinent terminology (Fig. 123). In the metacarpus there are superficial and deep branches on both the dorsal and palmar sides, but there are no deep nerves dorsally. Superficial branches are called **dorsal** or **palmar common digital** vessels or nerves and deep branches are called **dorsal** or **palmar metacarpal** vessels and nerves. These four branches are oriented in a sagittal plane between the metacarpal bones. At the metacarpophalangeal joint the common digital and metacarpal branches unite to form a common trunk on the dorsal and palmar surfaces. Each common trunk then divides into medial and lateral branches to the digits. These digital branches are called axial or abaxial dorsal or palmar proper digital vessels or nerves. Axial or abaxial refers to whether they are on the surface of the digit that faces towards the axis (axial) or away from the axis (abaxial) of the paw. The axis of the paw passes between digits 3 and 4. As a rule the largest digital vessels are the palmar axial proper digitals. (See Table 2 for the arrangement of these vessels and nerves in the forepaw.)

Table 2. *BLOOD SUPPLY AND INNERVATION OF THE DIGITS OF THE THORACIC LIMB*

	Dorsal Surface		
Arteries			
(Superficial)	Superficial brachial a.: Cranial superficial antebrachial Medial br. of cranial superficial antebrachial	Dorsal common digital a.	Axial or abaxial dorsal digital a.
(Deep)	Dorsal carpal rete: Radial a., dorsal carpal br. Caudal interosseous a.	Dorsal metacarpal a.	
Nerves			
(Superficial only)	Superficial br. radial n., medial and lateral br.; Ulnar n., dorsal br.	Dorsal common digital n.	Axial or abaxial dorsal digital n.
	Palmar Surface		
Arteries			
(Superficial)	Superficial palmar arch, Median a.; Br. caudal interosseous a.	Palmar common digital a.	Axial or abaxial palmar proper digital a.
(Deep)	Deep palmar arch, Caudal interosseous a.; Radial a., palmar carpal br.	Palmar metacarpal a.	
Nerves			
(Superficial)	Median n.; Ulnar n., superficial br.	Palmar common digital nn.	Axial or abaxial palmar proper digital n.
(Deep)	Ulnar n., deep br.	Palmar metacarpal nn. I–IV	

Live Dog

On the medial side of the arm feel the vessels and nerves coursing distally between the biceps brachii cranially and the medial head of the triceps brachii caudally. Palpate a pulse in the brachial artery. Arterial injections can be made into this artery.

Palpate the medial epicondyle of the humerus. The brachial vessels and median nerve course just cranial to this and pass beneath the pronator teres. These cannot be felt here. The ulnar nerve courses caudal to the medial epicondyle and can be palpated by stroking the skin in a caudal to cranial direction towards the epicondyle. The ulnar nerve will be felt when it slips beneath your fingers.

On the lateral side of the distal arm, the radial nerve can be felt where it emerges from under the distal border of the lateral head of the triceps on the brachialis muscle and divides into superficial and deep branches. Press firmly on the skin over the brachialis muscle here and stroke from proximal to distal. The superficial branch will be felt as it slips out from under your fingers.

Place your fingers across the flexor surface of the elbow and compress this area. This will distend the cephalic vein in the forearm, which is commonly used for venipuncture. Remember that a small artery and sensory nerve accompany this vein on both sides. Repeated needle punctures may injure them and contribute to subcutaneous hemorrhage, as well as being a painful experience for the patient. In some short-haired breeds the cephalic, axillobrachial and omobrachial veins may be visible in the arm and shoulder region.

Review the location of the autonomous zones of the peripheral nerves in the forelimb of the dog.

THE ABDOMEN, PELVIS AND PELVIC LIMB

The surface anatomy of the abdomen is divided into cranial, middle and caudal regions (Fig. 124). Reflect the skin from the right abdominal wall, leaving the mammary papillae in the female, the prepuce in the male and the cutaneus trunci. Extend a perpendicular incision from the ventral midline to the middle of the medial surface of the right thigh and thence to its cranial border. Continue this incision dorsally along the cranial edge of the thigh past the crest of the ilium to the middorsal line. Starting on the medial surface of the thigh, reflect or remove the skin of the right side of the abdomen.

VESSELS AND NERVES OF THE VENTRAL AND LATERAL PARTS OF THE ABDOMINAL WALL

The arteries that supply the superficial part of the ventral abdominal wall are branches of the superficial epigastric arteries (Fig. 125). The origin of the cranial superficial epigastric artery is from the cranial epigastric (Fig. 146).

The subcutaneous tissue of the ventral abdominal wall contains the abdominal and inguinal mammae and the vessels and nerves that supply them. In the female the cranial superficial epigastric vessels are seen subcutaneously near the cranial abdominal papilla. By blunt dissection separate the right row of mammae from the fascia and turn them laterally.

Dissect the **external pudendal artery** (Figs. 80, 125, 146), which emerges from the superficial inguinal ring. Its origin from the pudendo-epigastric trunk, a branch from the deep femoral artery, will be seen later. The external pudendal artery courses caudoventrally to the cranial border of the gracilis. The **caudal superficial epigastric artery** is large and appears as a direct continuation of the external pudendal dorsal to the superficial inguinal lymph node. The caudal superficial epigastric artery (Fig. 125) runs cranially to the deep surface of the inguinal mamma

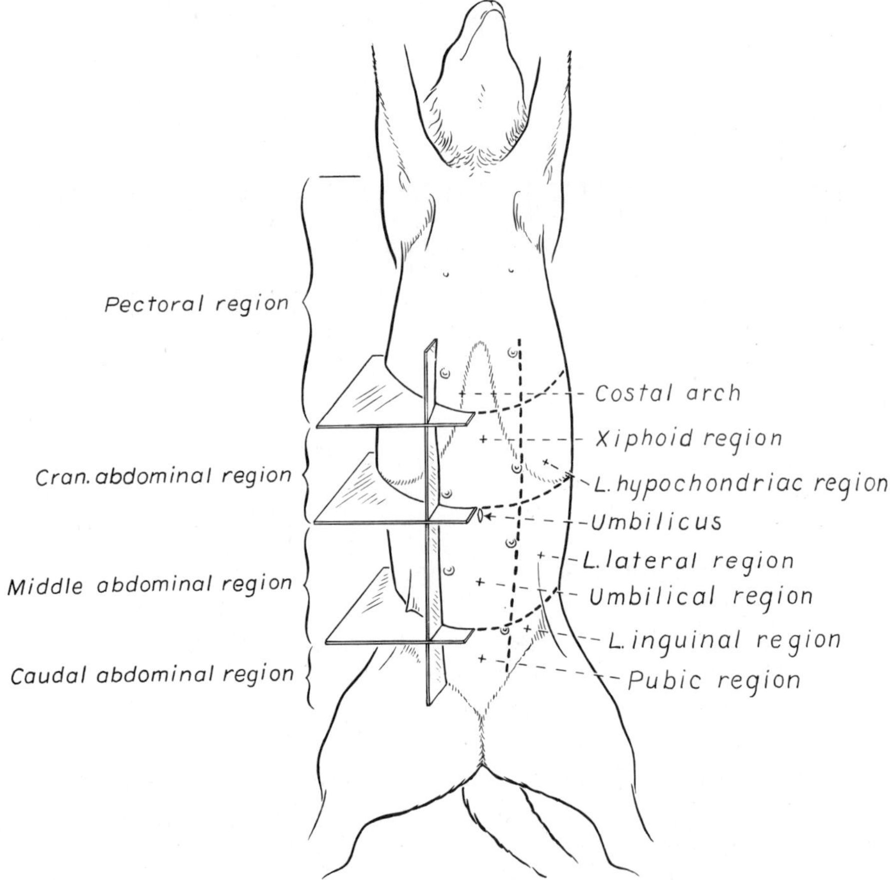

FIGURE 124. Topographic regions of thorax and abdomen.

and supplies the mammary branches. The artery continues, to supply the caudal abdominal mamma and anastomose with branches of the cranial superficial epigastric artery. In the male it supplies the prepuce. A small branch of the external pudendal courses caudally to supply the labia in the female and scrotum in the male.

Expose the **superficial inguinal lymph nodes** (see Fig. 153), which lie adjacent to the caudal superficial epigastric vessels and cranial to their origin from the external pudendal vessels. The afferent lymphatics of these nodes drain the mammae, the prepuce, the scrotum and the ventral abdominal wall as far cranially as the umbilicus.

The abdominal wall receives its vascular supply primarily from four vessels (see Fig. 146): the cranial abdominal artery (craniodorsal), cranial epigastric artery (cranioventral), caudal epigastric artery (caudoventral) and deep circumflex iliac artery (caudodorsal).

Reflect the superficial fascia from the lateral abdominal wall. Emerging from the dorsolateral abdominal wall, caudal to the last rib, are superficial branches of the **cranial abdominal artery** (see Fig. 146). The latter arises from a common origin with the caudal phrenic artery of the aorta, and perforates the abdominal musculature to reach the skin.

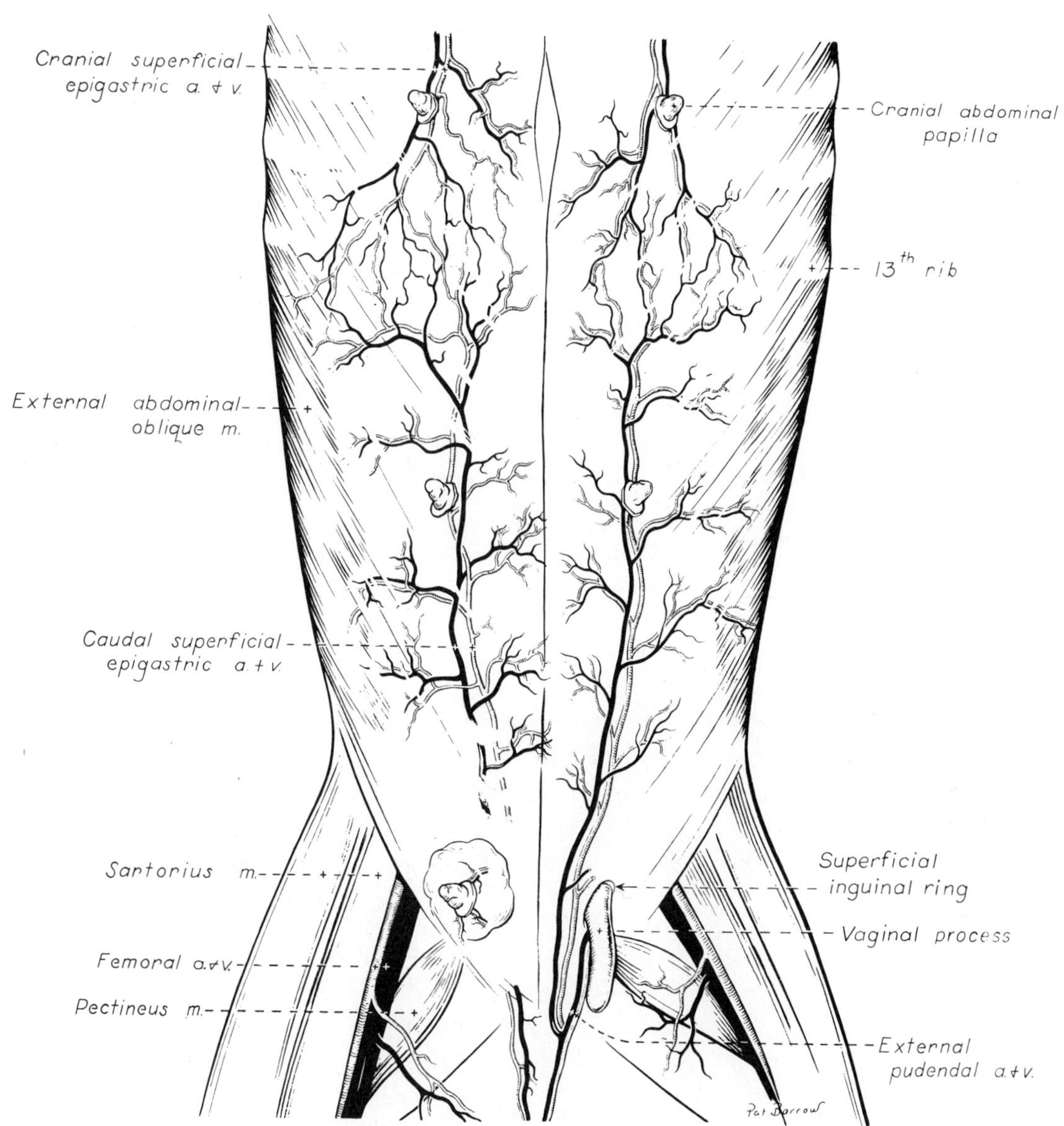

FIGURE 125. Superficial veins and arteries of abdomen. Left vaginal process exposed.

The **cutaneous nerves** of the abdomen differ somewhat from those of the thorax. The lateral cutaneous branches from the last five thoracic nerves do not follow the convexity of the costal arch but run in a caudoventral direction and supply most of the ventral and ventrolateral parts of the abdominal wall (Fig. 126). The cutaneous branches of the first three lumbar nerves perforate the lateral part of the abdominal wall and, as small nerves, run caudoventrally. They supply the skin of the caudolateral abdominal wall and the thigh in the region of the stifle. Do not dissect these cutaneous nerves. Cranial to the cranioventral iliac spine the **lateral cutaneous femoral nerve** (Fig. 126) and the **deep circumflex iliac artery** and **vein** (see Fig. 146) perforate the internal abdominal oblique and appear superficially. The nerve arises from the fourth lumbar nerve and is cutaneous to the cranial and lateral surfaces of the thigh.

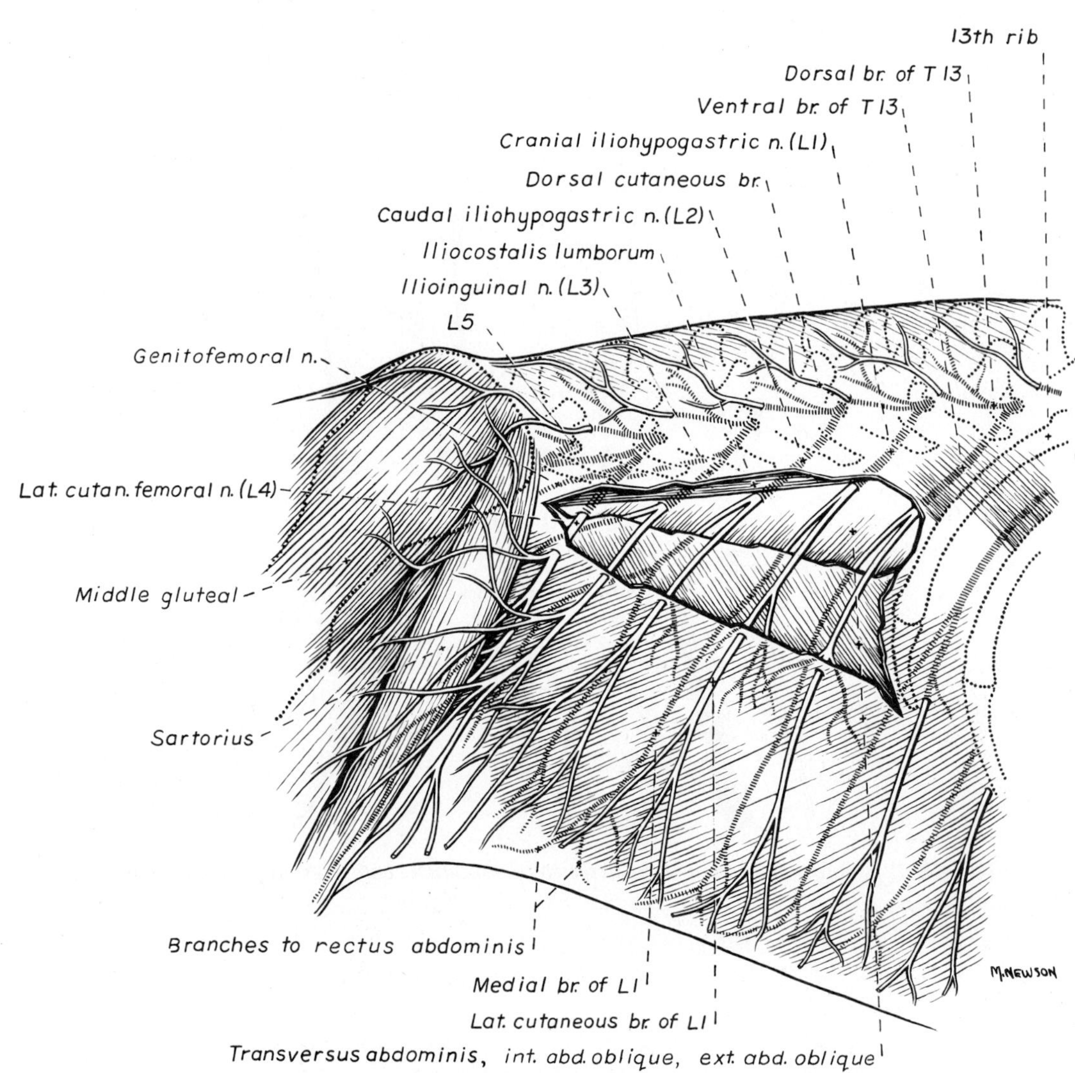

FIGURE 126. Lateral view of the first four lumbar nerves.

The artery arises from the aorta and supplies the caudodorsal abdominal wall. Dissect these vessels and trace the nerve as far as the present skin reflection will allow.

Transect the lumbar origin of the external abdominal oblique and reflect it ventrally.

Transect the internal abdominal oblique at the origin of the muscle fibers from the thoracolumbar fascia. Extend the transection caudally to the level of the deep circumflex iliac vessels and lateral cutaneous femoral nerve. Separate the internal abdominal oblique muscle from the underlying transverse muscle and reflect it ventrally to expose the ventral branches of the last few thoracic nerves and the first four lumbar nerves. These are parallel to each other and supply the ventral and lateral parts of the thoracic and abdominal wall. The first four lumbar nerves form the **cranial iliohypogastric, caudal iliohypogastric, ilioinguinal** and **lateral cutaneous femoral** nerves respectively. It may be difficult to differentiate between the ventral branches of T13 and L1 without tracing them to the intervertebral foramina, which is not necessary. Usually the ventral branch of T13 courses along the caudal aspect of the thirteenth rib.

The cranial and caudal iliohypogastric and ilioinguinal nerves (Fig. 126) pass through the aponeurosis of origin of the transversus abdominis. Each has a medial branch that descends between the transversus abdominis and the internal abdominal oblique to the rectus abdominis. The medial branches supply these muscles, the underlying peritoneum and the skin of the ventral abdominal wall. The lateral branches of these nerves perforate the internal abdominal oblique and descend between the oblique muscles. They may be seen on the deep surface of the external abdominal oblique. Each lateral branch supplies these muscles, perforates the external abdominal oblique and terminates subcutaneously as the lateral cutaneous branch to the abdominal wall in that region.

Inguinal Structures

Dissect the structures in the male that pass through the inguinal canal and the superficial inguinal ring (Fig. 80). Review pages 110–115, which concern this area.

Male

The **external pudendal artery** and **vein** leave the superficial inguinal ring caudal to the structures that supply the testis. Their branches have been dissected.

The **genitofemoral nerve** (Figs. 126, 166, 168, 172) arises from the ventral branches of the third and fourth lumbar nerves. It is bound by fascia to the external pudendal vein medial to the spermatic cord. It innervates the skin covering the inguinal region and proximal medial thigh of both sexes and part of the prepuce in the male.

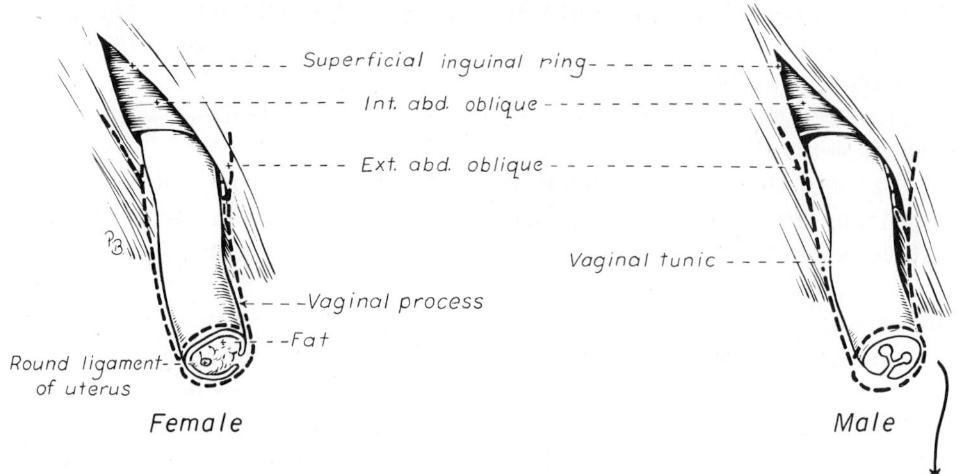

FIGURE 127. Diagram of transected vaginal process in male and female. (Dotted lines indicate spermatic fascia. In the male the contents of the vaginal tunic are not shown. See Figure 128.)

The **spermatic fascia,** a continuation of abdominal and transversalis fascia, surrounds the structures emerging from the superficial inguinal ring. This includes the spermatic cord and cremaster muscle (Figs. 127, 128).

The **cremaster muscle** is surrounded by this spermatic fascia as it courses along the caudal part of the vaginal tunic. The cremaster muscle arises from the caudal free border of the internal abdominal oblique and attaches to the vaginal tunic near the testis (Fig. 80). Reflect the spermatic fascia to expose the vaginal tunic, which can be seen extending from its emergence through the superficial inguinal ring to the testis.

The **vaginal process** (Figs. 127, 128) is a diverticulum of the peritoneum present in both sexes. In the male it envelops the testis and structures of the spermatic cord and is referred to as the vaginal tunic. It consists of the parietal vaginal tunic and visceral vaginal tunic.

The **parietal vaginal tunic,** the outer layer of the vaginal process, extends from the deep inguinal ring to the bottom of the scrotum. Incise this parietal tunic along the most ventral part of the testis and along the cranial border of the vaginal tunic to the superficial inguinal ring to expose the visceral vaginal tunic. The cavity entered is a continuation of the peritoneal cavity.

The **visceral vaginal tunic** is closely fused to the testis and epididymis and surrounds the ductus deferens and vessels and nerves that arise or terminate in the testis and epididymis. The **mesorchium** is part of the visceral vaginal tunic and contains the vessels and nerves of the testis. The **mesoductus deferens** is the part of the visceral vaginal tunic that attaches to the ductus deferens.

The **spermatic cord** (Figs. 127, 128) is carried through the inguinal canal by the descent of the testis and is composed of two distinct parts: the ductus deferens and the testicular artery and vein.

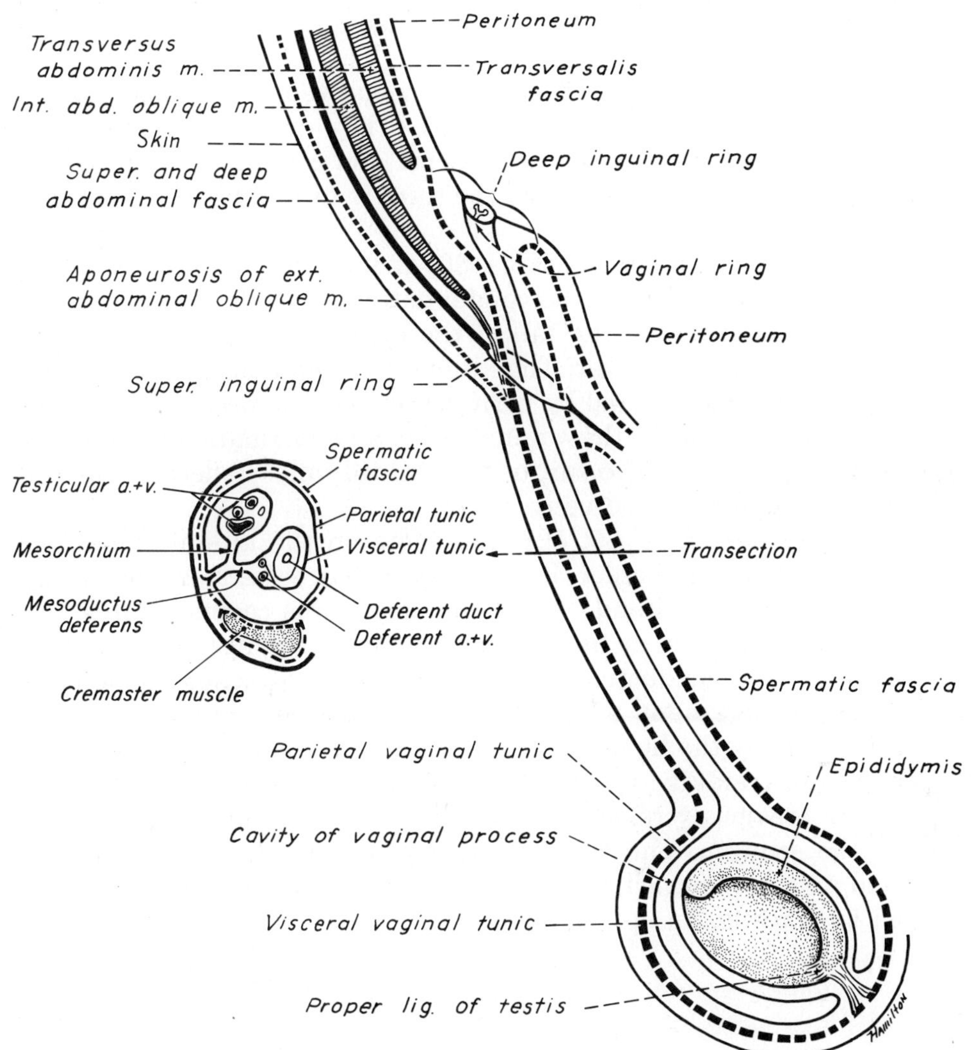

FIGURE 128. Schema of the vaginal tunic in the male.

The **ductus deferens** carries the spermatozoa from the epididymis to the urethra. It arises from the tail of the epididymis at the caudal end of the testis and is attached to the mesorchium by the **mesoductus deferens.** The small **deferent artery** and **vein** accompany the deferent duct.

The **testicular artery** and **vein,** as well as the testicular lymph vessels and the testicular plexus of autonomic nerves, are closely associated with each other. These vessels and nerves are covered by a fold of visceral vaginal tunic, the **mesorchium,** which is continuous with the parietal vaginal tunic. The artery is tortuous, and woven around it are the nerve plexus and the venous plexus. The venous plexus is the **pampiniform plexus.** The testicular artery and vein are branches of the aorta and caudal vena cava, respectively. They enter the testis at its cranial end. The nerve plexus is autonomic and contains postganglionic sympathetic axons, which arise from the third to fifth lumbar sympathetic ganglia.

The **testis** and **epididymis** (Figs. 80, 128) are intimately covered by the visceral vaginal tunic. At the caudal extremity of the epididymis the visceral peritoneum leaves the tail of the epididymis at an acute angle and becomes the parietal layer. Thus there is a small circumscribed area on the epididymis not covered by peritoneum. The connective tissue that attaches the epididymis to the vaginal tunic and spermatic fascia at this point is the **ligament of the tail of the epididymis.** Reflect the skin of the scrotum caudally to observe it.

The epididymis (Figs. 80, 128) lies more on the lateral side of the testis than on its dorsal border. For descriptive purposes it is divided into a cranial extremity, or **head,** where the epididymis communicates with the testis; a middle part, or **body;** and a caudal extremity, or **tail,** which is continuous with the ductus deferens. The tail is attached to the testis by the **proper ligament of the testis** and to the vaginal tunic and spermatic fascia by the ligament of the tail of the epididymis. The ductus deferens passes cranially over the testis medial to the epididymis.

Lay the previously reflected skin back over the inguinal region and examine the scrotum. The **scrotum** is a pouch divided by an external raphe and an internal median septum into two cavities, each of which is occupied by a testis, an epididymis and the distal part of the spermatic cord.

Female

In the female, locate the external pudendal blood vessels and the genitofemoral nerve emerging from the superficial inguinal ring. The **vaginal process** (Fig. 125) is the peritoneal diverticulum that is accompanied by the **round ligament of the uterus.** (The origin of this ligament from the mesometrium within the abdomen will be seen later.) These two structures, enclosed in fascia and surrounded by fat, may extend as far as the vulva.

The Inguinal Canal

The **inguinal canal** (Fig. 128) is a short fissure filled with connective tissue between the abdominal muscles. It extends between the deep and superficial inguinal rings. It is bounded laterally by the aponeurosis of the external abdominal oblique, cranially by the caudal border of the internal abdominal oblique, caudally by the caudal border of the aponeurosis of the external abdominal oblique (inguinal ligament) and medially, in part, by the superficial surface of the rectus abdominis. The vaginal tunic and spermatic cord in the male and the vaginal process and round ligament of the female pass obliquely caudoventrally through the canal. In both sexes the external pudendal vessels and genitofemoral nerve traverse the canal. Notice as many of these boundaries as possible before opening the abdomen.

Abdominal and Peritoneal Cavities

The abdominal cavity is formed by the muscles of the abdominal wall, the ribs and the diaphragm. It is lined by peritoneum, which encloses the peritoneal cavity.

The **peritoneal cavity,** like the pleural and pericardial cavities, is a closed space. It is lined by a serous membrane. Serous membranes are thin layers of loose connective tissue covered by a layer of mesothelium. The peritoneum is derived from the somatic and splanchnic mesodermal layers lining the embryonic celom.

The **parietal peritoneum** is the layer that lines the body wall and is incised to open the peritoneal cavity. It is continuous dorsally with the **visceral peritoneum,** which suspends and surrounds the organs of the abdominal cavity (Fig. 129). There are no organs in the peritoneal cavity because all organs are covered by visceral peritoneum.

The **transversalis fascia** reinforces the peritoneum and attaches it to the abdominal muscles and diaphragm. Make a sagittal incision through the abdominal wall on each side dorsal to the rectus abdominis from the costal arch to the level of the inguinal canal. Connect the cranial ends of these incisions and reflect the ventral abdominal wall. Observe the following structures:

The **falciform ligament** is a fold of peritoneum that passes from the umbilicus to the diaphragm. It is also attached to the liver between the left medial and quadrate lobes. In obese specimens a large accumulation of fat is found in this remnant of the embryonic ventral mesentery. In young animals the **round ligament of the liver** may still be visible in the free border of the falciform ligament. Caudal to the umbilicus the fold of peritoneum is the **median ligament of the bladder.** In the fetus the umbilical vein courses cranially in the free border of the falciform ligament to enter the liver, whereas the urachus and umbilical arteries are in the free border of the median ligament of the bladder.

Examine the caudoventral aspect of the inside of the peritoneal cavity at the level of the inguinal canal and observe the vaginal ring.

The **vaginal ring** (Fig. 128) is formed by the parietal peritoneum as it leaves the abdomen and enters the inguinal canal to form the vaginal process or tunic. It marks the position of the **deep inguinal ring,** which is formed by the reflection of the transversalis fascia outside the vaginal ring. A deposit of fat is usually present in the transversalis fascia around the vaginal ring.

In the male the **ductus deferens** is attached to the abdominal and pelvic walls by a fold of peritoneum, the mesoductus deferens. At the vaginal ring this fold joins the mesorchium, which contains the testicular artery and vein and the testicular nerve plexuses. The ductus deferens courses caudally from the vaginal ring to the urethra just beyond the neck of the bladder. In the female a fold of peritoneum from the mesometrium, which suspends the uterus, passes into the vaginal ring. This contains the round ligament of the uterus.

The **caudal epigastric artery** and **vein** course cranially on the deep face of the caudal part of the rectus abdominis. The origin of the artery from the pudendo-epigastric trunk of the deep femoral artery will be dissected later (see Fig. 146).

ABDOMINAL VISCERA

The **greater omentum** (Figs. 129–132) is the first structure seen after reflecting the abdominal wall. It is a caudoventral extension of the two layers of visceral peritoneum that pass from the dorsal body wall to the greater curvature of the stomach, the dorsal mesogastrium. As the stomach forms and rotates to its definitive position in the embryo, this mesogastrium grows extensively and forms a double layered sac that extends caudoventrally beneath many of the abdominal organs. The space contained within the folded mesogastrium is the omental bursa. The fold adjacent to the ventral body wall is the superficial leaf. The deep leaf is adjacent to the abdominal organs. The greater omentum is lacelike in appearance, with depositions of fat along the vessels. The greater omentum covers the jejunum and ileum, leaving the descending colon exposed on the left, the bladder exposed caudally and the descending duodenum exposed on the right. Reflect the omentum and, using your fingers on opposite sides, separate its superficial and deep walls to expose its cavity, the **omental bursa.** Follow the greater omentum from its ventral attachment on the greater curvature of the stomach to its dorsal attachment to the dorsal body wall. The spleen is enclosed in the superficial portion of the greater omentum on the left side, and the left lobe of the pancreas is enclosed in the deep leaf.

Three organs in the abdomen that are capable of considerable variation in size are the stomach, the urinary bladder and the uterus. If one or more of these are distended, the relations of the organs will be altered.

The **urinary bladder** (Figs. 130, 132, 133), when empty, is contracted and lies on the floor of the pelvic inlet. When distended, it lies on the floor

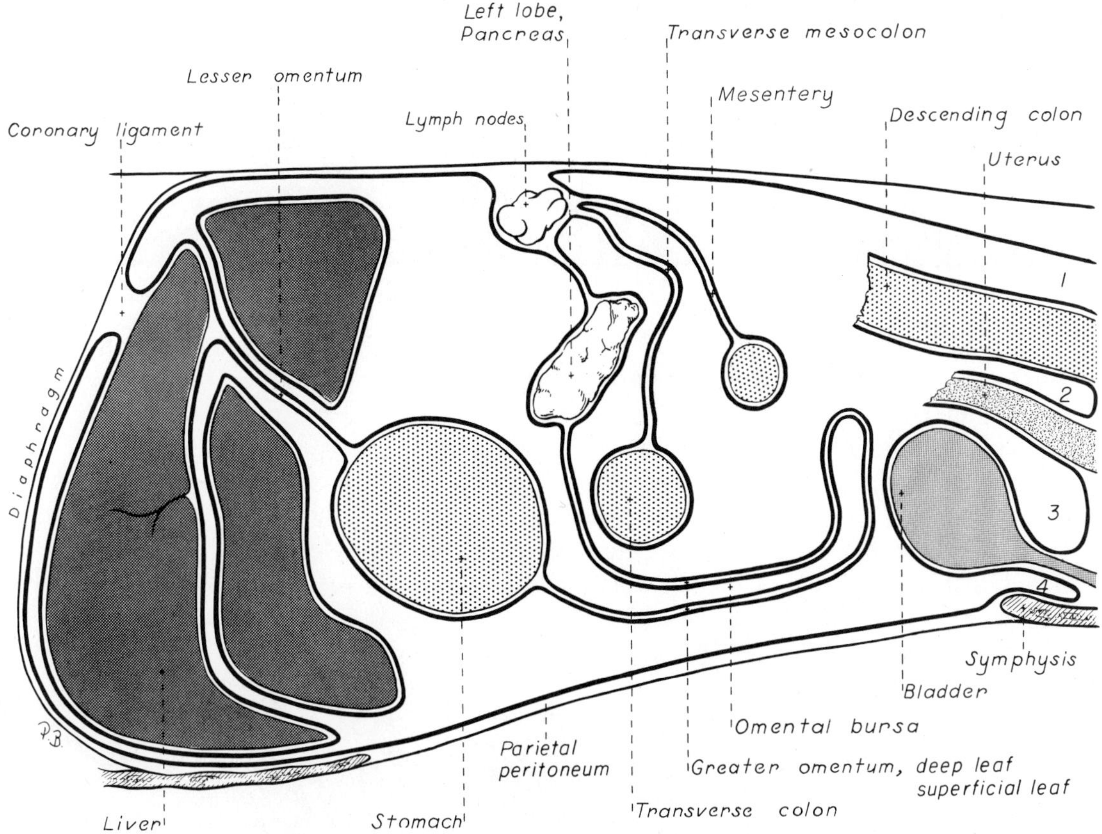

FIGURE 129. Diagram of peritoneal reflections, sagittal section.
1. Pararectal fossa
2. Rectogenital pouch
3. Vesicogenital pouch
4. Pubovesical pouch

of the abdomen and conforms in shape to the caudal part of the abdominal cavity, since it displaces all freely movable viscera. It frequently reaches a transverse plane through the umbilicus.

The nonpregnant **uterus** (Figs. 130, 133) is remarkably small even in a bitch that has had several litters. The uterus consists of a **cervix,** a **body** and two **horns.** The gravid uterus lies on the floor of the abdomen during the second month or last half of pregnancy. As the uterus enlarges, the middle parts of the horns gravitate cranially and ventrally and come to lie medial to the costal arches; thus the uterus bends on itself as the ovarian and vaginal ends move very little during enlargement.

The **spleen** (Figs. 130, 134) lies in the superficial leaf of the greater omentum to the left of the median plane along the greater curvature of the stomach. Its position, shape and degree of distension are variable. Its lateral surface lies against the parietal peritoneum of the left lateral abdominal wall and the liver. Its caudal part may reach to a transverse plane through the midlumbar region. Its cranial limit is usually marked by a plane passing between the twelfth and thirteenth thoracic vertebrae. It may reach the floor of the abdomen. The part of the greater omentum that attaches the spleen to the stomach is the **gastrosplenic ligament.**

Text continued on page 190

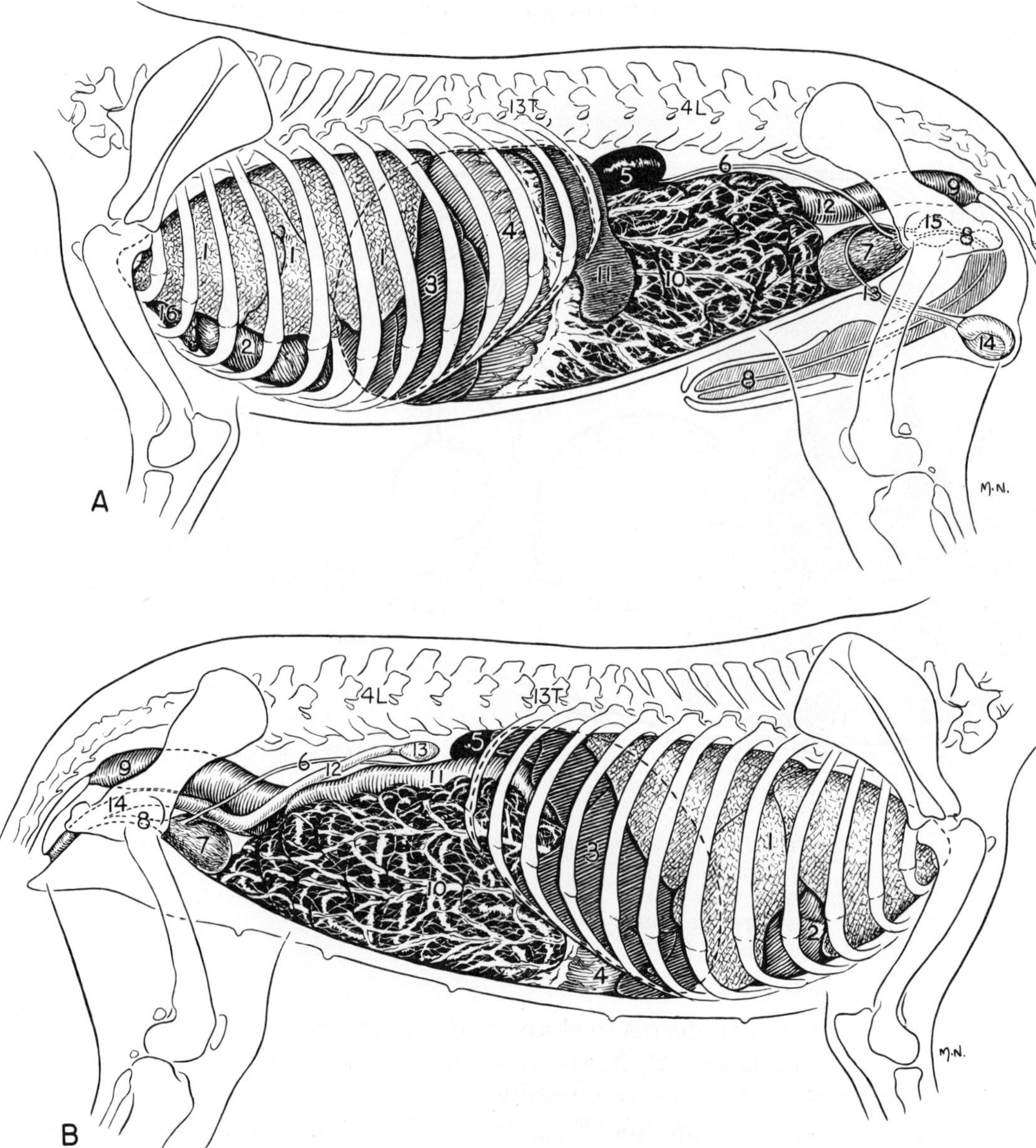

FIGURE 130. Viscera of the dog

A) Viscera of male dog, left lateral view.

1. *Left lung*
2. *Heart*
3. *Liver*
4. *Stomach*
5. *Left kidney*
6. *Ureter*
7. *Bladder*
8. *Urethra*
9. *Rectum*
10. *Greater omentum covering small intestine*
11. *Spleen*
12. *Descending colon*
13. *Ductus deferens*
14. *Left testis*
15. *Prostate*
16. *Thymus*

B) Viscera of female dog, right lateral view.

1. *Right lung*
2. *Heart*
3. *Liver*
4. *Stomach*
5. *Right kidney*
6. *Ureter*
7. *Bladder*
8. *Urethra*
9. *Rectum*
10. *Greater omentum covering small intestine*
11. *Descending duodenum*
12. *Right uterine horn*
13. *Right ovary*
14. *Vagina*

Ascending colon
Omental bursa
Pancreas
Jejunum
Descend. duod.
Ascend. duod.
Parietal peritoneum
Left
Spleen
Desc. colon
R. kidney
L. kidney
Renal v.
2L
Caudal vena cava
Root of mesentery

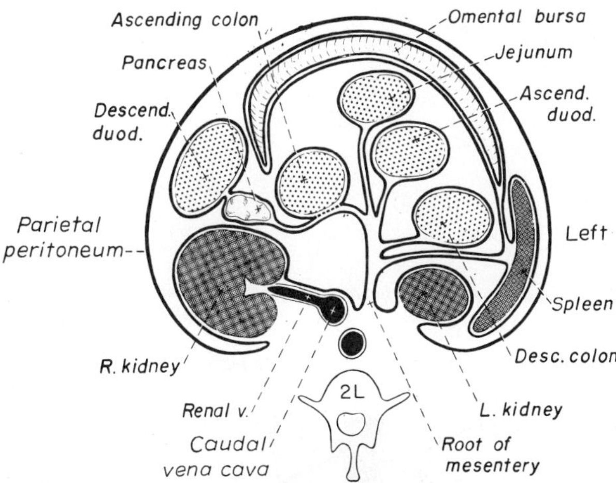

FIGURE 131. Abdominal mesenteries. Schematic transverse section at the level of the spleen.

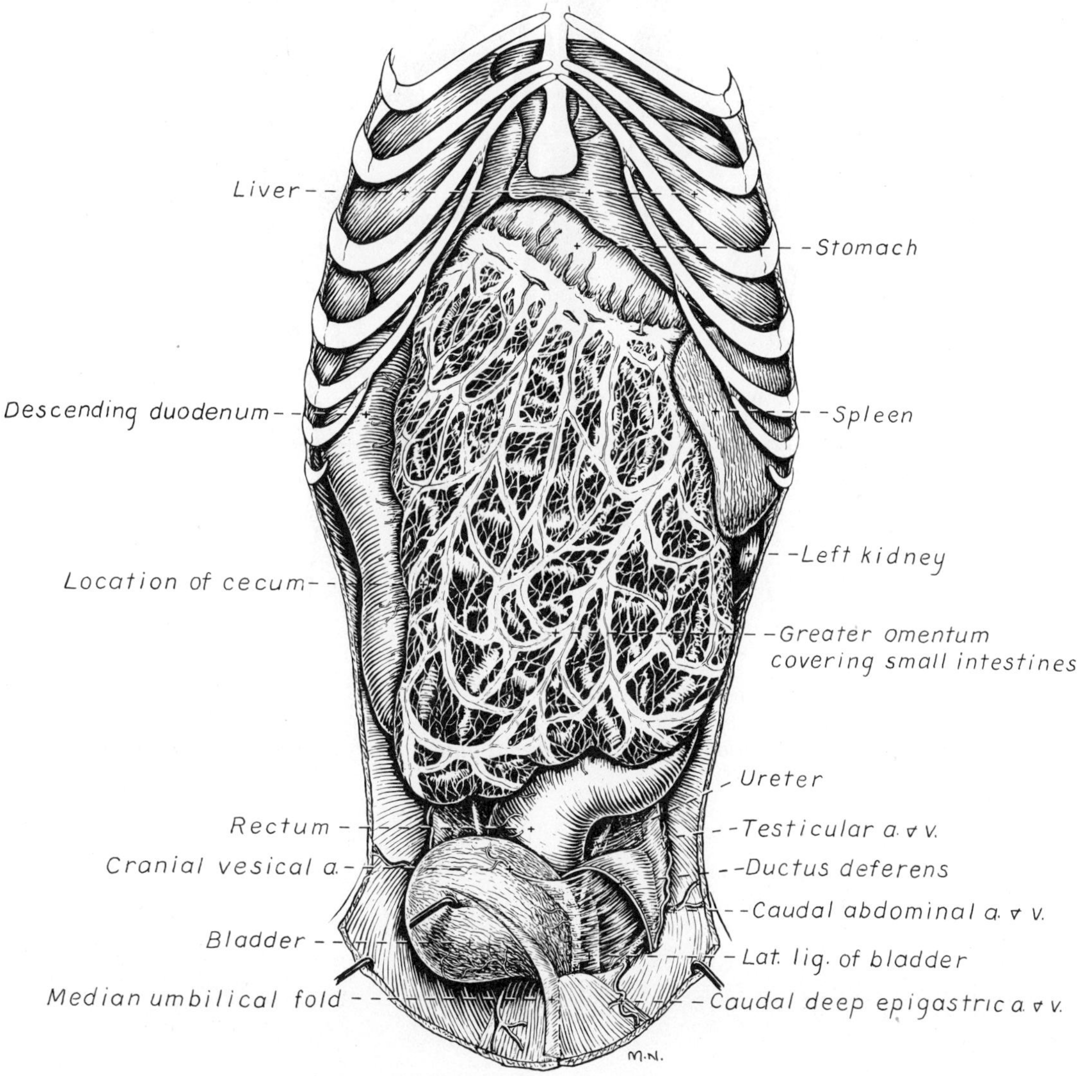

Liver
Stomach
Descending duodenum
Spleen
Left kidney
Location of cecum
Greater omentum covering small intestines
Ureter
Rectum
Testicular a. & v.
Cranial vesical a.
Ductus deferens
Caudal abdominal a. & v.
Bladder
Lat. lig. of bladder
Median umbilical fold
Caudal deep epigastric a. & v.
M.N.

FIGURE 132. Abdominal viscera of the male dog, ventral aspect.

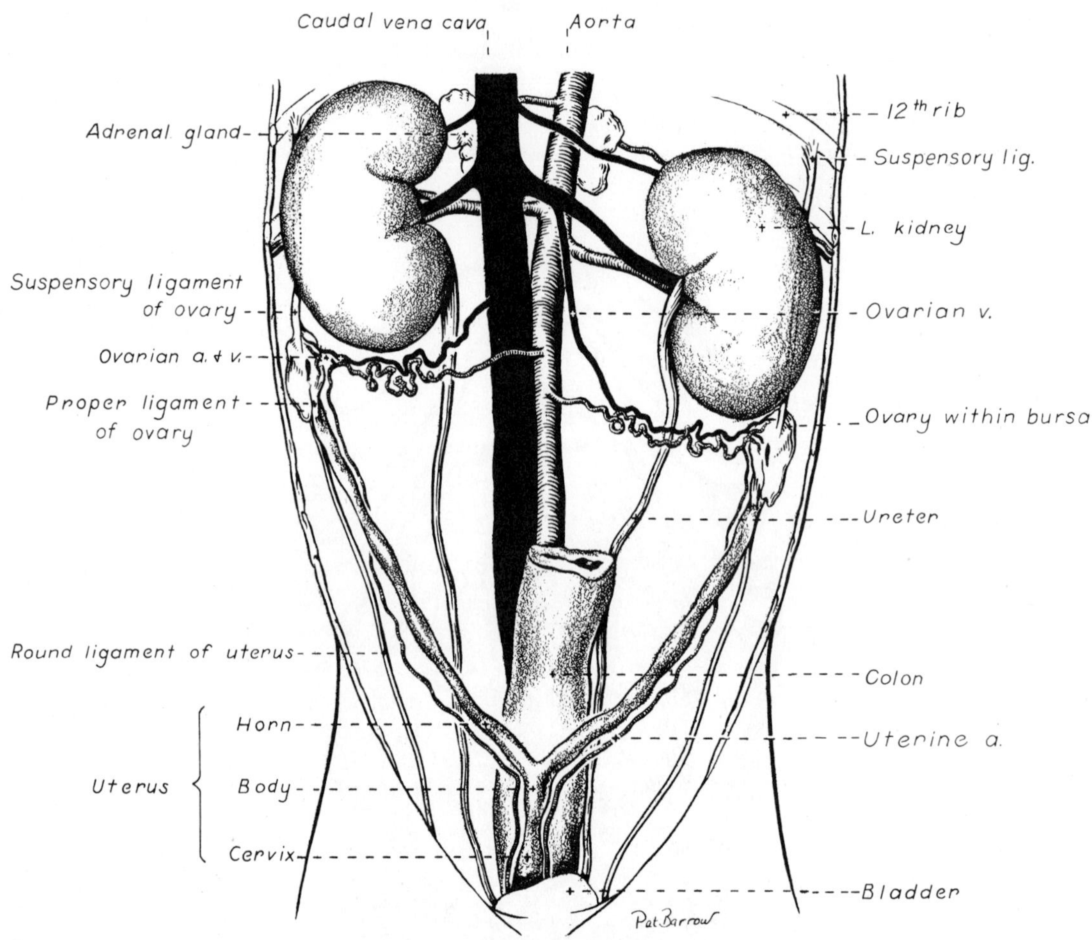

FIGURE 133. Topographic relations of kidney, adrenal gland, ovary and nonpregnant uterus, ventral view.

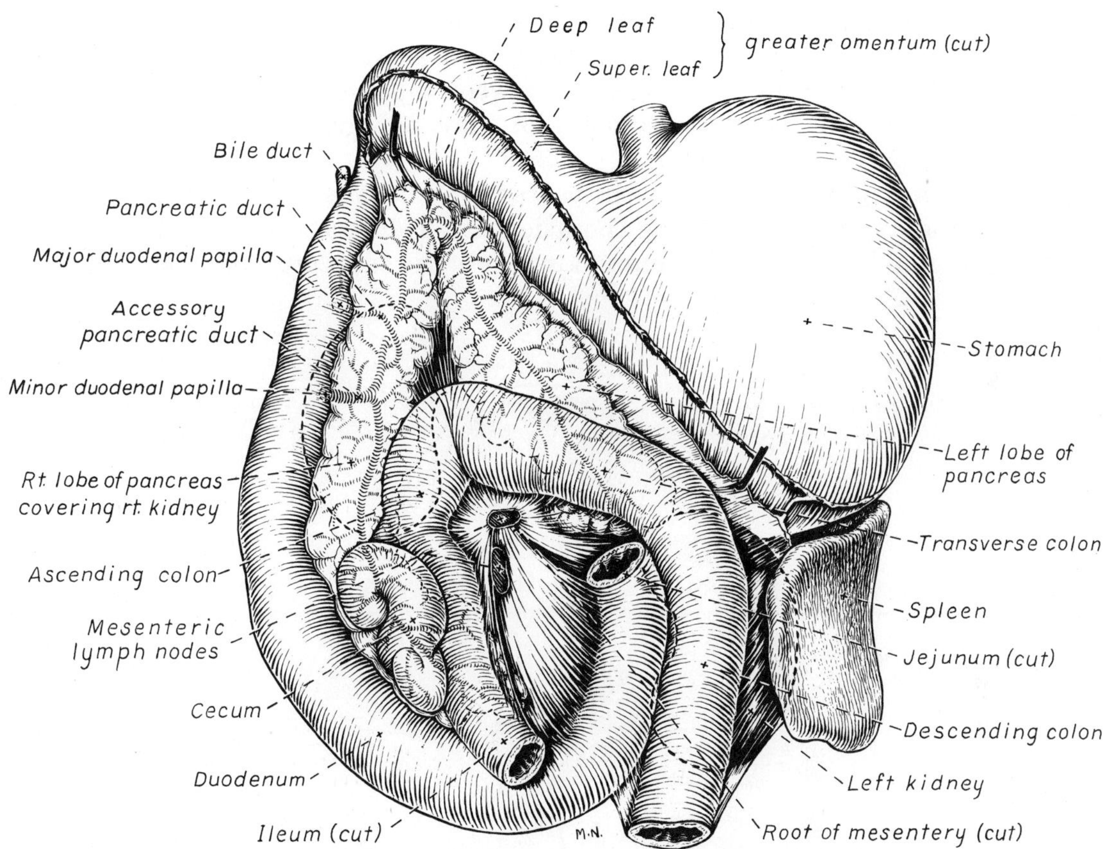

FIGURE 134. Duodenum and transverse colon in relation to the root of the mesentery. Pancreas in situ and position of kidneys indicated by dotted line.

If your specimen was anesthetized with a barbiturate, the spleen may be abnormally enlarged.

The **diaphragm** (Fig. 135), the muscular partition between the thoracic and the abdominal cavities, is a muscle of inspiration. It has an extensive muscular periphery and a small, V-shaped **tendinous center.** The muscular part of the diaphragm may be divided into three parts according to its attachments: lumbar, costal and sternal. The **lumbar part** forms the left and right crura that attach to the bodies of the third and fourth lumbar vertebrae by strong tendons. The right crus is larger than the left. The **costal part** of the diaphragm arises from the medial surfaces of the eighth to thirteenth ribs. It interdigitates with the transversus abdominis muscle. The **sternal part** arises from the dorsal surface of the sternum cranial to the xiphoid cartilage. The extensions of the V-shaped tendinous center run dorsally between the lumbar and costal parts of each side. The caudal mediastinum may be severed to expose the tendinous part of the muscle.

The **aortic hiatus** is a dorsal passageway between the crura for the aorta, the azygos vein and the thoracic duct. The more centrally located **esophageal hiatus** is in the muscular part of the right crus and transmits

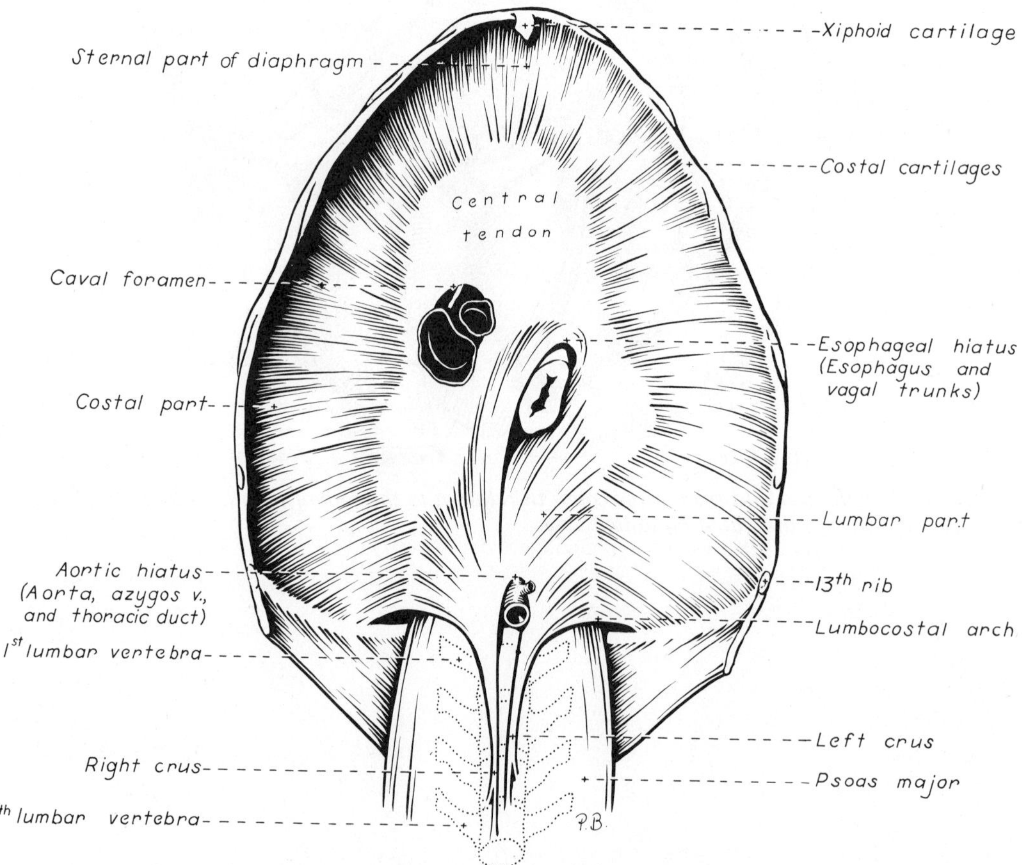

FIGURE 135. Diaphragm, abdominal view.

the esophagus, vagal nerve trunks and esophageal vessels. The **caval foramen** is located at the junction of the tendinous and muscular parts of the right side of the diaphragm. The caudal vena cava passes through it.

The **liver** (Figs. 130, 132, 136) has six lobes, and its parietal surface conforms to the abdominal surface of the diaphragm. The visceral surface of the liver is related on the left to the stomach and sometimes to the spleen; on the right to the pancreas, right kidney and duodenum; and ventrally to the greater omentum and through this to the small intestine. Its most caudal part covers the cranial extremity of the right kidney and reaches a transverse plane through the thirteenth thoracic vertebra. The liver rarely projects caudal to the costal arch. It undergoes slight longitudinal movement with each respiration.

The **right medial lobe** of the liver contains a fossa for the gall bladder. The **right lateral lobe,** which is smaller, is located next to the caudate lobe, which embraces the cranial end of the right kidney. The **quadrate lobe** is narrow and is located between the right and left medial lobes. It forms the left boundary of the fossa of the gall bladder. The **left medial lobe** is separated by a fissure from the right medial and quadrate lobes. The umbilical vein enters the liver through this fissure. The **left lateral lobe** is separated by a fissure from the left medial lobe. The free margin of the left lateral lobe is frequently notched. The visceral surface of the left lateral lobe is concave where it contacts the stomach. The **caudate lobe** is indistinctly separated from the central mass of the liver, which is cranial to it. It lies transversely, but is mainly to the right of and dorsal to the main bulk of the organ. It is constricted in its middle where the portal vein enters the liver ventral to it and the caudal vena cava crosses dorsal to it. Its extremities are in the form of two processes. The **caudate process** caps the cranial end of the right kidney and thus contains the deep **renal impression.** The **papillary process** can be seen through the lesser omentum if the liver is tipped forward. It lies in the lesser curvature of the stomach.

Biliary Passages

Much of the biliary duct system within the liver is microscopic. The bile, which is secreted by the liver cells, is collected into the canaliculi, which drain into interlobular ducts. The interlobular ducts of each lobe unite to form **hepatic ducts** (Fig. 137), which emerge from each lobe. The arrangement of the hepatic ducts is variable.

The **gall bladder** (Fig. 137) is located in a fossa between the quadrate and right medial lobes of the liver. The full gall bladder extends through the liver and contacts the diaphragm (which is often stained green in preserved specimens). The neck of the gall bladder is continued as the **cystic duct.**

The main duct formed by the union of the hepatic ducts and the cystic duct from the gall bladder is the **bile duct (ductus choledochus).** It courses through the wall of the descending duodenum and terminates on

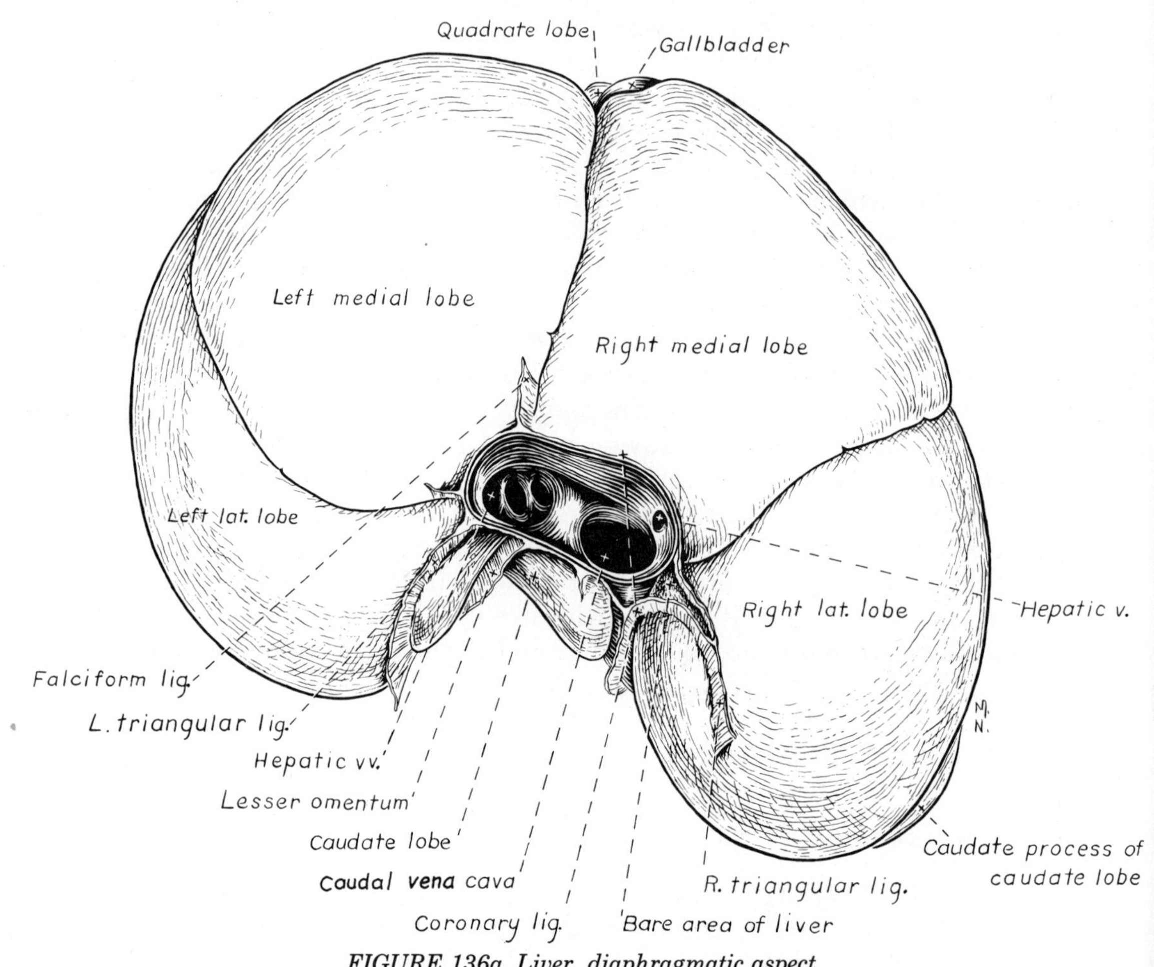

FIGURE 136a. Liver, diaphragmatic aspect.

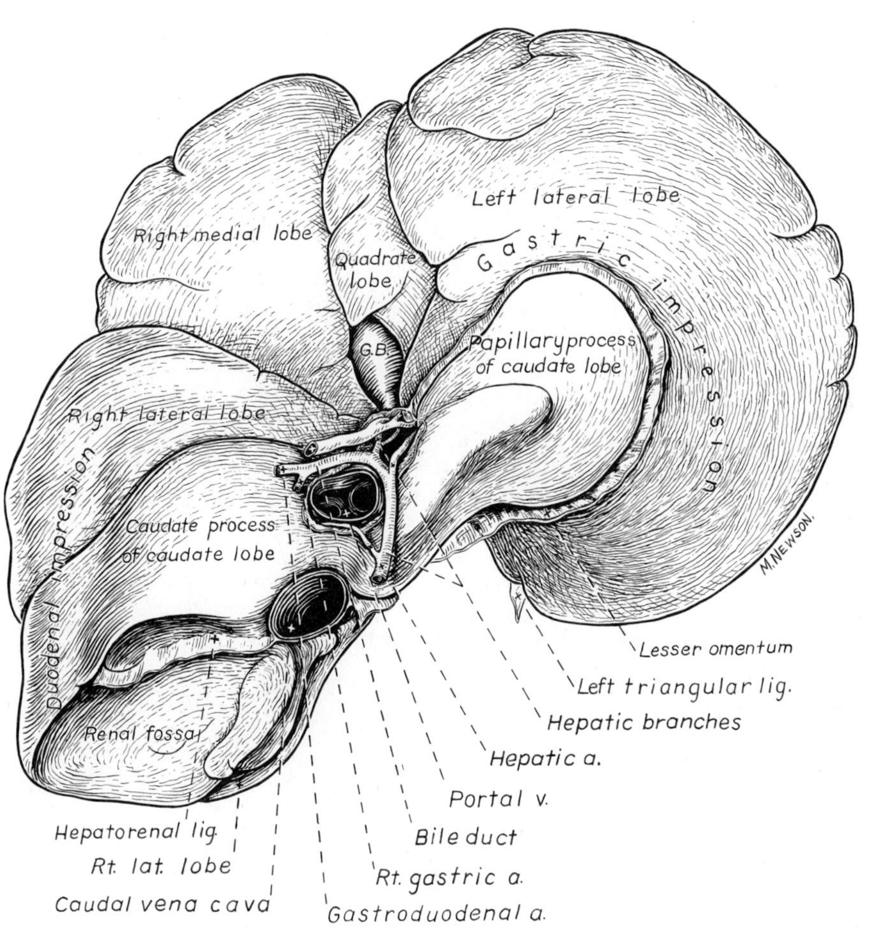

FIGURE 136b. Liver, visceral aspect.

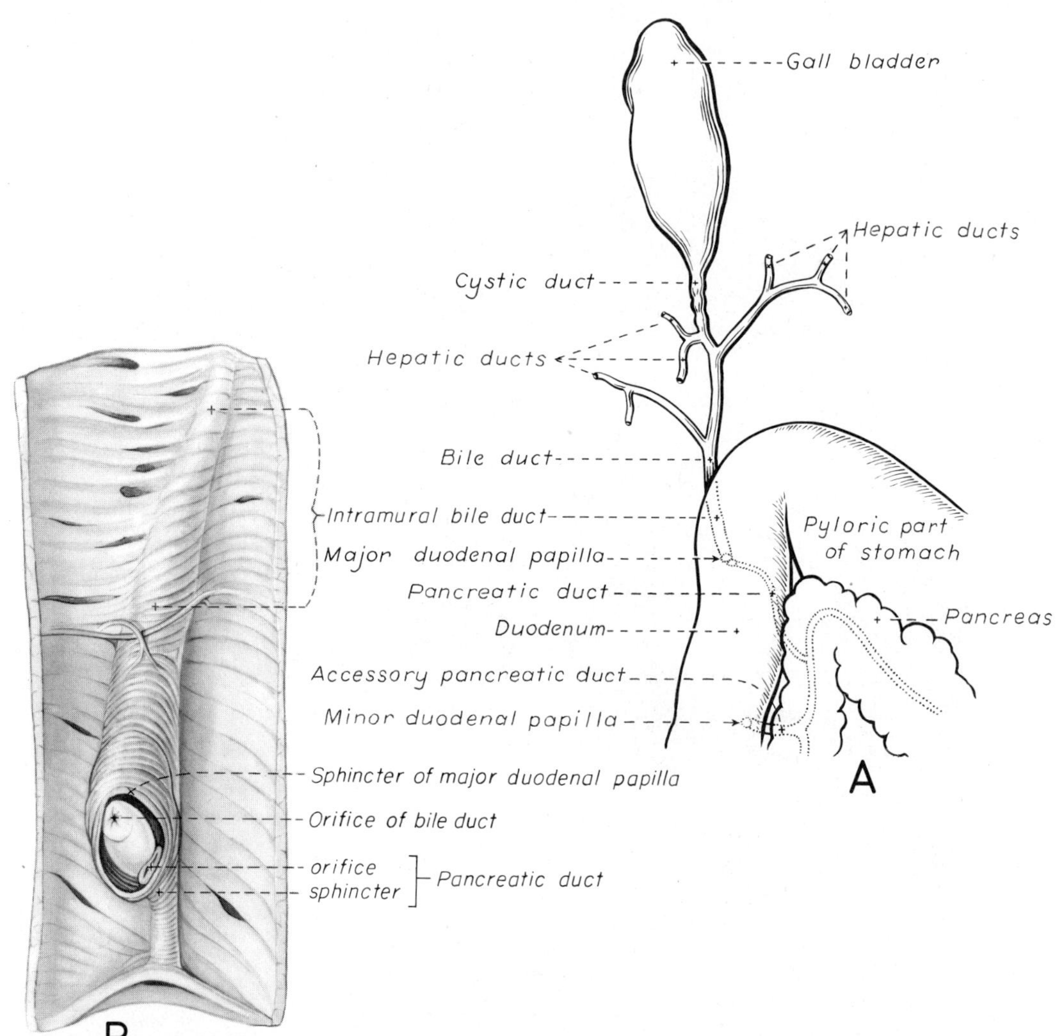

FIGURE 137. Biliary and pancreatic ducts. A, *Topographic relations, ventral view.* B, *Interior of the duodenum with the tunica mucosa removed to show musculus proprius in relation to the ducts and major duodenal papilla (after Eichorn and Boyden, 1955).*

the major duodenal papilla alongside the pancreatic duct. There are no valves in the biliary ducts, and bile may flow in either direction. Observe the duct system in your specimen.

The **stomach** (Figs. 130, 134, 138) is divided into parts that blend imperceptibly with one another. The **cardiac part** is the smallest part of the stomach and is situated nearest the esophagus. The **fundus** is dome-shaped and lies to the left of and dorsal to the cardia. The **body** of the stomach is the large middle portion. It extends from the fundus on the left to the pyloric part on the right. The body joins the pyloric part at the incisure, which is the relatively sharp bend on the lesser curvature. The **pyloric part** is the distal third of the stomach as measured along the lesser curvature. The initial thin-walled portion is the **pyloric antrum,** which narrows to a **pyloric canal** before joining the duodenum at the sphincter, the **pylorus.**

The stomach is bent so that its **greater curvature** faces mainly to the left and its **lesser curvature** mainly to the right; its parietal surface faces ventrally toward the liver, and its visceral surface dorsally faces the intestinal mass. Its position changes depending on its fullness.

The **empty stomach** is completely hidden from palpation and observation by the liver and diaphragm cranioventrally and the intestinal mass caudally. It lies to the left of the median plane. The empty stomach is cranial to the costal arch and sharply curved, so that it is more V-shaped than C-shaped. The greater curvature faces ventrally, caudally and to the left. This curvature lies above and to the left of the mass of the small intestine. The lesser curvature is strongly curved around the papillary process of the liver and faces craniodorsally and to the right. The left lobe of the pancreas and transverse colon are dorsocaudal to it.

The **full stomach** lies in contact with the ventral abdominal wall and protrudes beyond the costal arches. It displaces the intestinal mass.

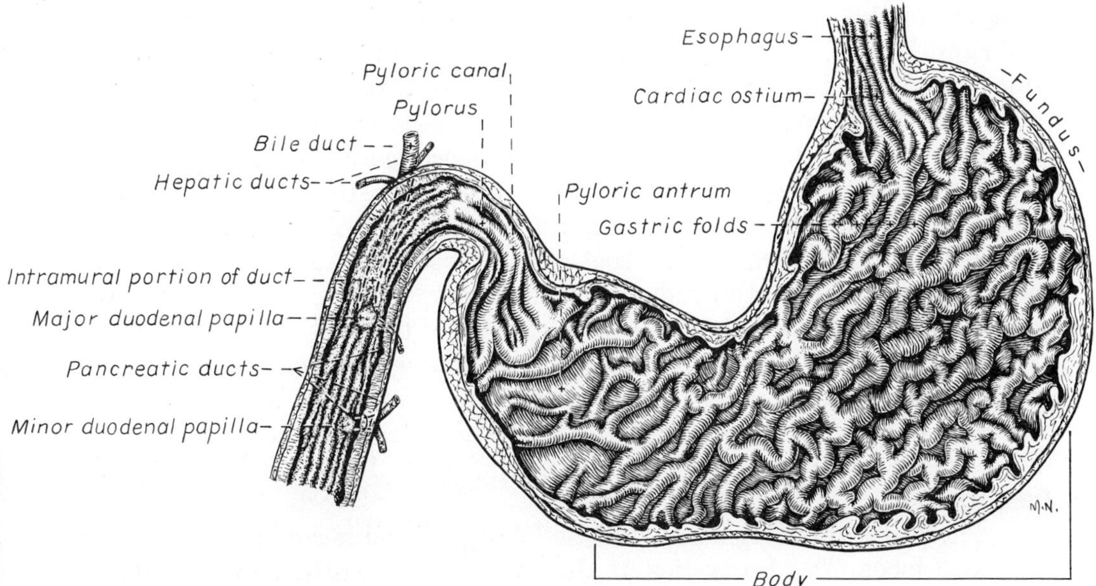

FIGURE 138. Longitudinal section of stomach and proximal duodenum.

Open the stomach along its parietal surface, remove the contents and observe the longitudinal folds of mucosa.

The **duodenum** (Figs. 130, 132, 134, 138) is the most fixed part of the small intestine. It is suspended by the mesoduodenum, which will be studied later. Reflect the greater omentum cranially and the jejunum to either side to expose the duodenum. The duodenum begins at the pylorus to the right of the median plane. After a short dorsocranial course, it turns as the **cranial duodenal flexure.** It continues caudally on the right as the **descending part,** where it is in contact with the parietal peritoneum. Farther caudally the duodenum turns, forming the **caudal duodenal flexure,** and continues cranially as the **ascending part.** The ascending part lies to the left of the root of the mesentery, where it forms the **duodenojejunal flexure.**

The **jejunum** forms the coils of the small intestine (Figs. 130, 131, 132, 134), which occupy the ventrocaudal part of the abdominal cavity. They receive their nutrition from the cranial mesenteric artery, which is in the root of the mesentery. The root of the mesentery attaches the jejunum and ileum to the dorsal body wall. The **mesenteric lymph nodes** lie along the vessels in the mesentery. The jejunum begins at the left of the root of the mesentery and is the longest portion of the small intestine. Trace it from the duodenojejunal flexure on the left to its termination at the ileum on the right side of the abdomen. The **ileum** is the terminal portion of the small intestine. It is short and passes cranially on the right side of the root of the mesentery and joins the ascending colon at the **ileocolic orifice.** There is no clear demarcation between jejunum and ileum. Note the vessel that courses on the antimesenteric side of the ileum from the cecum toward the jejunum. This approximates the length of the ileum (10 cm.).

The **cecum** (Figs. 132, 134, 139), a part of the large intestine, is an S-shaped, blind tube located to the right of the median plane at the junction of the ileum and colon. It is ventral to the caudal extremity of the right kidney, dorsal to the small intestines and medial to the descending part of the duodenum. The cecum communicates with the ascending colon at the **cecocolic orifice.** Open the cecum, terminal ileum and adjacent ascending colon and observe the ileocolic and cecocolic orifices.

The **colon** (Figs. 134, 139) is located dorsally in the abdomen, suspended by a mesocolon. It is divided into a short **ascending colon,** which lies on the right of the root of the mesentery; a transverse colon, which lies cranial to the root of the mesentery; and a long **descending colon,** which lies at its beginning on the left of the root of the mesentery. The bend between the ascending and transverse colons is known as the **right colic flexure** and that between the transverse and descending colons is known as the **left colic flexure.** The descending colon terminates at a transverse plane through the pelvic inlet. It is continued by the **rectum.**

The **pancreas** (Figs. 134, 137) is lobulated and is composed of a body and two lobes. The **body** lies at the pylorus. The **right lobe** lies dorsomedial to the descending part of the duodenum enclosed by the mesoduo-

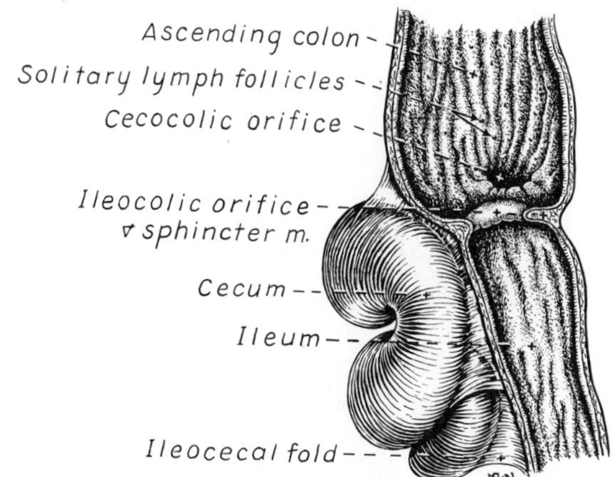

FIGURE 139. Longitudinal section through ileocolic junction showing cecocolic orifice.

denum. It is ventral to the right kidney. The **left lobe** of the pancreas lies between the peritoneal layers that form the deep leaf of the greater omentum. It is caudal to the stomach and liver and cranial to the transverse colon. Reflect the greater omentum cranially and the small intestine caudally to observe the pancreas.

The pancreatic duct system (Figs. 134, 137) is variable. Most dogs have two ducts; these open separately in the duodenum but communicate in the gland. The **pancreatic duct** is the smaller of the two ducts and is sometimes absent. It opens close to the bile duct on the **major duodenal papilla.** Make an incision through the free border of the descending part of the duodenum. Scrape away the mucosa with the scalpel handle and identify the major duodenal papilla. This is on the side where the mesoduodenum attaches. The larger **accessory pancreatic duct** opens into the duodenum on the **minor duodenal papilla** about 2 or 3 centimeters caudal to the major papilla. Locate the accessory duct by blunt dissection in the mesoduodenum between the right lobe of the pancreas and the descending duodenum.

The **adrenal glands** (Figs. 133, 140, 143) are light-colored and are located at the cranial aspect of each kidney. Each gland is crossed by the common trunk of the caudal phrenic and cranial abdominal veins, which leaves a deep groove on its ventral surface.

The **right adrenal gland** lies between the caudal vena cava and the caudate lobe of the liver ventrally and the sublumbar muscles dorsally. Expose the gland by dissection between the caudal vena cava and the kidney cranial to the renal vein. The **left adrenal gland** lies between the aorta and the left kidney. Transect each adrenal and note the lighter-colored cortex and darker medulla.

The **kidneys** (Figs. 130–133, 140, 141) are dark brown. They are partly surrounded by fat and are covered on their ventral surface by peritoneum. The lateral border is strongly convex, and the medial, nearly straight. At the middle of the medial border is an indention, the **hilus** of the kidney, where the renal vessels and nerves and the ureter communicate with the organ.

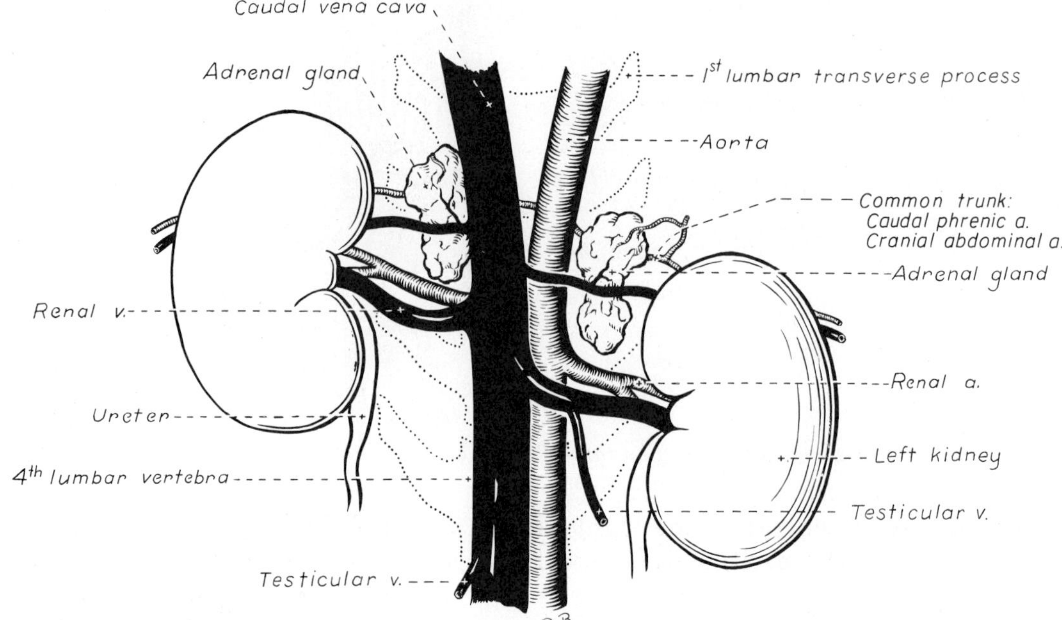

FIGURE 140. Kidneys and adrenal glands, ventral view.

The **right kidney** lies opposite the first three lumbar vertebrae. It is farther cranial than the left kidney by the length of half a kidney. The right kidney is more extensively related to the liver than to any other organ. Its cranial third is covered by the caudate process of the caudate lobe of the liver. The remaining ventral surface is related to the descending duodenum, the right lobe of the pancreas, the cecum and the ascending colon. The caudal vena cava is on the medial border of the right kidney.

The **left kidney** lies opposite the second, third and fourth lumbar vertebrae. It is related ventrally to the descending colon and the small intestine. The spleen is related to the cranial extremity of the kidney. The medial border is close to the aorta.

The expanded part of the **ureter** within the kidney is the **renal pelvis.** The ureter courses caudally in the sublumbar region. It opens into the dorsal part of the neck of the urinary bladder. Throughout this course it is enveloped by a fold of peritoneum from the dorsal body wall. Follow the course of the ureter. The **renal sinus** is the fat-filled space that contains the renal vessels and surrounds the renal pelvis.

Free the left kidney from its covering peritoneum and fascia. Do not cut its vascular attachment. Make a longitudinal section of the left kidney from its lateral border to the hilus, dividing it into dorsal and ventral halves. Note the granular appearance of the peripheral portion of the renal parenchyma. This is the **renal cortex,** which contains primarily the renal corpuscles and convoluted portions of the tubules. The more centrally positioned parenchyma is the **medulla.** It has a striated appearance due to numerous collecting ducts. The vessels that are apparent at the corticomedullary junction are the **arcuate branches** of the renal vessels. The longitudinal ridge projecting into the renal pelvis is the **renal**

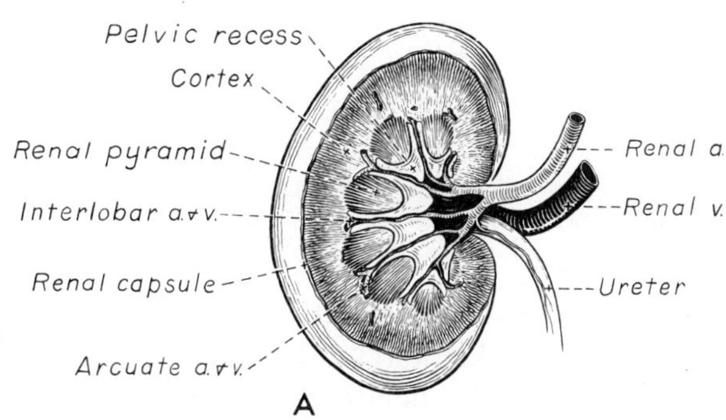

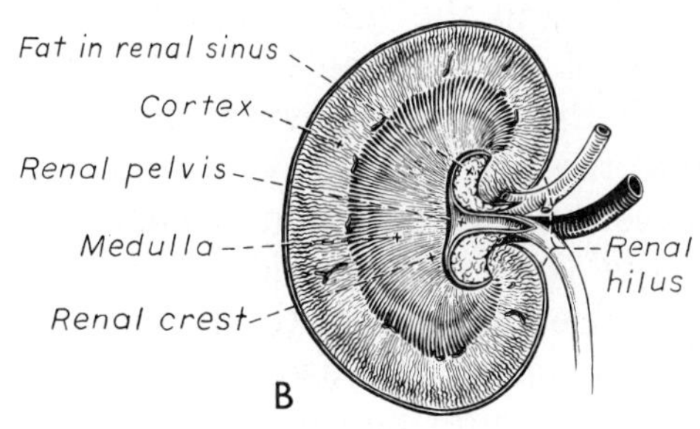

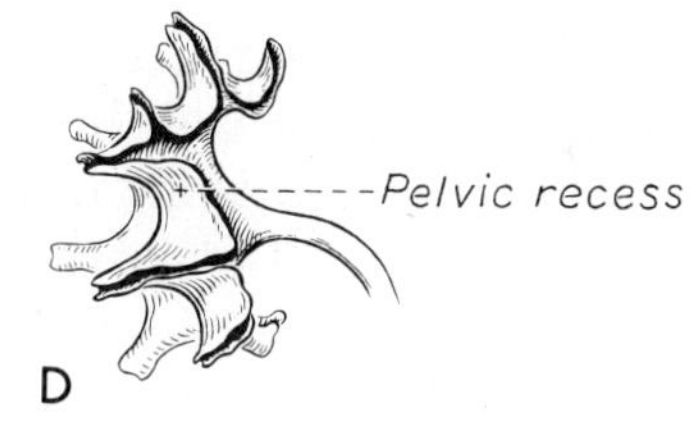

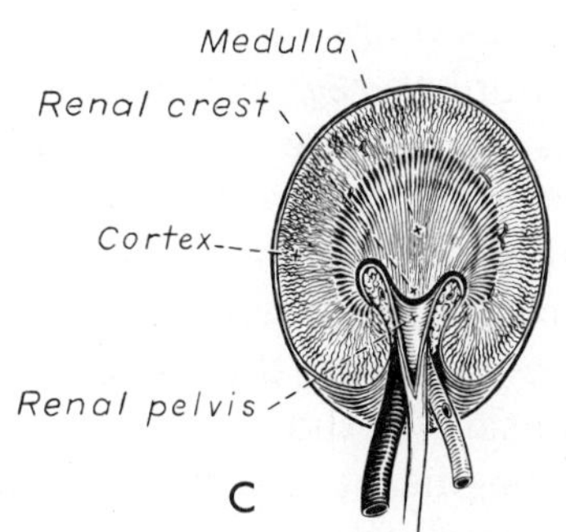

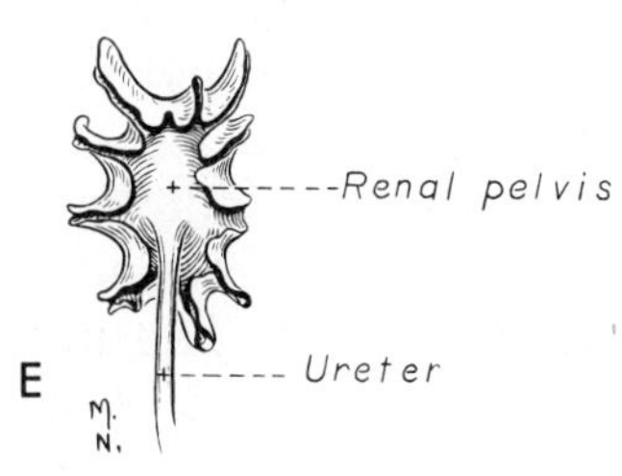

FIGURE 141. Details of kidney structure.
A) Sectioned in dorsal plane, off center.
B) Sectioned in middorsal plane.
C) Transverse section.
D) Cast of renal pelvis, dorsal view.
E) Cast of renal pelvis, medial view.

crest, through which collecting tubules of the kidney excrete urine into the renal pelvis. Make a second longitudinal section parallel to the first and note the **renal pyramids** formed by the medulla. The **pelvic recesses** of the renal pelvis project outward between the renal pyramids.

Free the right kidney from its peritoneum and fascia and make a transverse section through it. Note the renal cortex, medulla, crest and pelvis.

The **ovaries** are located near the caudal pole of the kidneys. The right ovary lies cranial to the left ovary and is dorsal to the descending duodenum. The left ovary is between the descending colon and the abdominal wall. Each ovary is enclosed in a thin-walled peritoneal sac, the **ovarian bursa** (Fig. 142), formed by the mesovarium and mesosalpinx. The ovarian bursa is open to the peritoneal cavity via a slitlike orifice on the medial surface.

The **uterine tube** courses cranially and then caudally through the lateral wall of the bursa on its way to the uterine horn. Examine the surface of the bursa and observe the small cordlike thickening within its wall. This is the uterine tube. Open the bursa by a lateral incision dorsal to the uterine tube and examine the ovary and the infundibulum. The **infundibulum** is the dilated ovarian end of the uterine tube. It has a fimbriated margin and functions to engulf the ova after ovulation. Note that several of the fimbriae protrude into the peritoneal cavity from the opening of the ovarian bursa. In life these fimbriae function to close the opening into the peritoneal cavity at the time of ovulation and thus prevent transperitoneal migration of ova. The entrance of the infundibulum into the uterine tube is spoken of as the abdominal ostium, and it is in this region that fertilization takes place. The uterine tube is short and slender. It opens into the much wider uterine horn at the tubouterine junction. This region is of physiological importance because it is here that sperm and ova are regulated in their transit.

The **broad ligaments** of the uterus (Fig. 142) are the peritoneal folds on each side that attach to the lateral sublumbar region. They suspend all the internal genitalia except the caudal part of the vagina, which is not covered by peritoneum. Each ligament is divided into three parts: The **mesometrium** arises from the lateral wall of the pelvis and the lateral part of the sublumbar region and attaches to the lateral part of the cranial end of the vagina, uterine cervix and uterine body, and the corresponding uterine horn. The **mesovarium,** a continuation of the mesometrium, is the cranial part of the broad ligament. It begins at a transverse plane through the cranial end of the uterine horn and attaches the ovary and the ligaments associated with the ovary to the lateral part of the sublumbar region. The **mesosalpinx** is the peritoneum that attaches the uterine tube to the mesovarium and forms with the mesovarium the wall of the ovarian bursa.

The **suspensory ligament of the ovary** joins the transversalis fascia medial to the dorsal end of the last rib (Fig. 142). It functions to hold the ovary in a relatively fixed position. In ovariohysterectomy this ligament is freed from its attachment to the body wall to facilitate removal of the

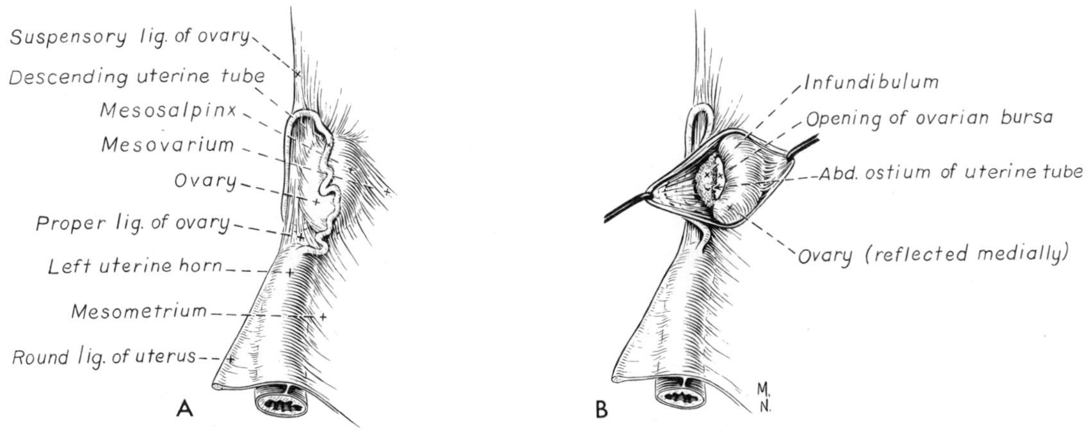

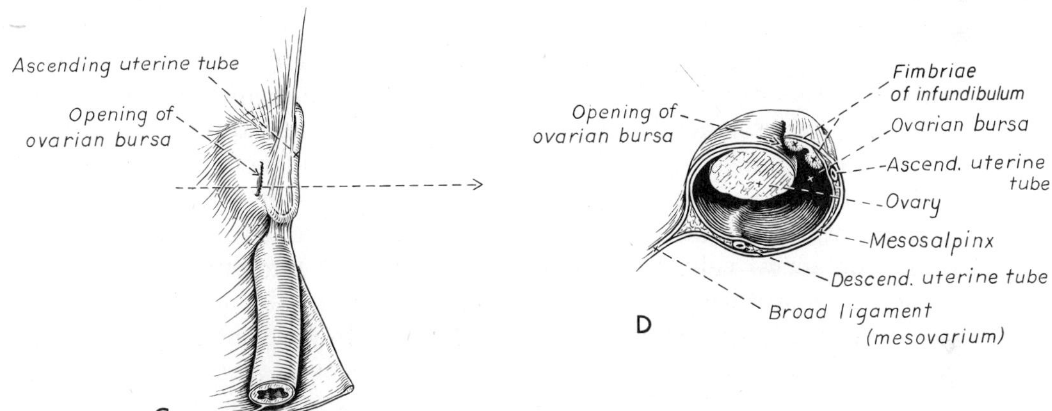

FIGURE 142. Relations of left ovary and ovarian bursa.
A) Lateral view.
B) Lateral view, bursa opened.
C) Medial view.
D) Section through ovary and bursa.

ovary. The **proper ligament of the ovary** is short and attaches the ovary to the cranial end of the uterine horn. From this point caudolaterally to the inguinal canal there is a fold from the lateral layer of the mesometrium which contains the **round ligament of the uterus** in its free border (see Fig. 168). The round ligament, a homologue of the embryonic gubernaculum, has no function in the adult. It passes through the inguinal canal and is wrapped by the vaginal process and adipose tissue (Fig. 127).

Peritoneum

The peritoneum lines the abdominal, pelvic and scrotal cavities and reflects around their contained organs. The **parietal peritoneum** covers the inner surface of the walls of the abdominal, pelvic and scrotal cavities; the **visceral peritoneum** covers the organs of these cavities.

In the embryo the dorsal common mesentery is a double layer of peritoneum that passes from the dorsal abdominal wall to the digestive tube. It serves as a route by which the nerves and vessels reach the organs. In the dog the dorsal common mesentery persists as the greater omentum, mesoduodenum, mesentery and mesocolon.

An **omentum** (epiploon) attaches the stomach to the body wall or other organs:

The **greater omentum** (see Figs. 129, 131, 132) attaches the greater curvature of the stomach to the dorsal body wall. From the greater curvature of the stomach it extends caudally as the superficial leaf over the floor of the abdomen. It turns dorsally on itself near the pelvic inlet and returns as the deep leaf dorsal to the stomach, where it contains the left lobe of the pancreas between its peritoneal layers. It attaches to the dorsal abdominal wall. Caudal and to the left of the fundus of the stomach is the spleen, which lies largely in an outpocketing of the superficial leaf of the greater omentum.

The **lesser omentum** loosely spans the distance from the lesser curvature of the stomach to the porta of the liver. Between the liver and the cardia of the stomach it attaches for a short distance to the diaphragm. The papillary process of the liver is loosely enveloped by the lesser omentum. On the right of the free edge of the lesser omentum is the **hepatoduodenal ligament,** which attaches the liver to the duodenum. It contains the portal vein, the hepatic artery and the bile duct.

The **omental bursa** is formed by the omenta and the adjacent organs. It has an **epiploic foramen** opening into the main peritoneal cavity. This opening lies dorsally to the right of the median plane at the level of the cranial duodenal flexure, caudomedial to the caudate lobe of the liver. It is bounded dorsally by the caudal vena cava, ventrally by the portal vein, caudally by the hepatic artery in the mesoduodenum and cranially by the liver. Find this foramen and insert a finger through it.

The **mesoduodenum** originates at the dorsal abdominal wall and the root of the mesentery and extends to the duodenum. On the right side it passes to the descending duodenum and encloses the right lobe of the pancreas between its layers. Cranially it is continuous with the greater omentum across the ventral surface of the portal vein. Caudally the mesoduodenum passes from the root of the mesentery to the caudal flexure of the duodenum. On the left it is attached to the ascending duodenum, and at the duodenojejunal flexure it is continuous with the mesentery of the jejunum. The ascending duodenum is attached to the mesocolon of the descending colon by the **duodenocolic fold.**

The **mesentery** (mesojejunoileum) attaches to the abdominal wall opposite the second lumbar vertebra by a short peritoneal attachment known as the **root of the mesentery.** Vessels and nerves pass in the mesentery to supply the large and small intestine. The peripheral border of the mesentery attaches to the jejunum and the ileum. At the ileocolic junction the mesentery is continuous with the ascending mesocolon.

The **ascending, transverse** and **descending mesocolons** connect the ascending, transverse and descending colons to the dorsal body wall. They are continuous with each other from right to left.

There are a few short peritoneal folds that serve more to fix organs in position than as channels for blood vessels. These are called ligaments:

The **right triangular ligament** extends from the right crus of the diaphragm above the central tendinous part to the right lateral lobe of the liver.

The **left triangular ligament** extends from the left crus of the diaphragm to the left lateral lobe of the liver.

The **coronary ligament** is a sheet of peritoneum that passes between the diaphragm and the liver around the caudal vena cava and hepatic veins. On the right it is continuous with the right triangular ligament and on the left it is continuous with the left triangular ligament. Ventrally, right and left parts of the coronary ligament converge to form the falciform ligament.

The **falciform ligament** extends from the liver to the diaphragm and ventral abdominal wall to the umbilicus. The round ligament of the liver, which is the remnant of the **umbilical vein** of the fetus, may be found in the young animal as a small fibrous cord lying in the free edge of the falciform ligament. It enters the fissure for the round ligament of the liver between the quadrate and left medial lobes. In adult dogs the falciform ligament is filled with fat and it persists only from the diaphragm to the umbilicus.

Vessels and Nerves of the Abdominal Viscera

Vagal Nerves

The **vagus nerves** (Figs. 105, 143) carry preganglionic parasympathetic efferent fibers to and visceral afferents from all of the thoracic and most of the abdominal organs. Right and left vagus nerves divide caudal to the roots of the lungs into **dorsal** and **ventral branches.** The dorsal branches unite near the diaphragm to form the **dorsal vagal trunk.** The ventral branches unite caudal to the roots of the lungs to form the **ventral vagal trunk.** These trunks lie on the dorsal and ventral surfaces of the terminal part of the esophagus. From the trunks, as well as from the dorsal and ventral branches of the vagi, nerves arise to supply the esophagus. The vagal trunks pass through the esophageal hiatus of the diaphragm and course along the lesser curvature of the stomach.

Transect the left crus of the diaphragm at the esophageal hiatus and reflect it to expose the vagal trunks. Observe the ventral vagal trunk. It supplies the liver, the parietal surface of the stomach and the pylorus. The terminal branches need not be dissected.

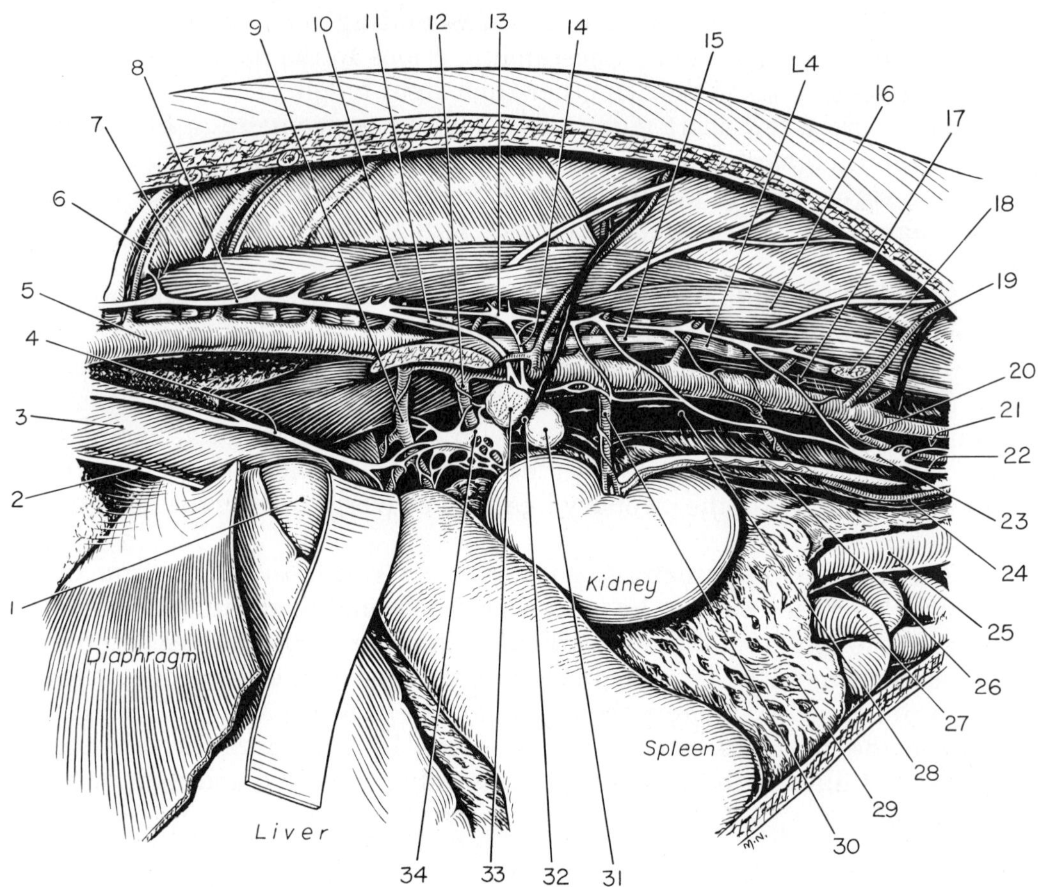

FIGURE 143. Exposure of abdominal autonomic nervous system on left side.

1. *Stomach*
2. *Ventral trunk of vagus n.*
3. *Esophagus*
4. *Dorsal trunk of vagus n.*
5. *Aorta*
6. *Intercostal a. and n.*
7. *Ramus communicans*
8. *Sympathetic trunk*
9. *Celiac a.*
10. *Quadratus lumborum*
11. *Major splanchnic n.*
12. *Cranial mesenteric a.*
13. *Lumbar sympathetic ganglion at L2*
14. *Minor splanchnic n.*
15. *Tendon of left crus of diaphragm*
16. *Psoas major*
17. *Lumbar splanchnic n.*
18. *Transected psoas minor*
19. *Deep circumflex iliac a.*
20. *Caudal mesenteric a.*
21. *Left hypogastric n.*
22. *Caudal mesenteric plexus*
23. *Caudal mesenteric ganglion*
24. *Testicular a. and v.*
25. *Descending colon*
26. *Cranial ureteral a.*
27. *Jejunum*
28. *Caudal vena cava*
29. *Greater omentum*
30. *Renal a. and plexus*
31. *Adrenal gland*
32. *Common trunk for caudal phrenic and cranial abdominal v.*
33. *Adrenal plexus*
34. *Celiac and cranial mesenteric ganglia and plexus.*

The dorsal vagal trunk (Fig. 143) gives off a **celiac branch** that passes dorsocaudally and contributes to the formation of the celiac and cranial mesenteric plexuses. These plexuses are nerve networks lying on, around and passing along the three abdominal vessels for which they are named. The sympathetic components of the plexuses have synapses in the associated ganglia. Parasympathetic axons are simply passing through. The branches in these plexuses follow the terminal branching of the respective blood vessels to the intestines at least as far caudally as the left colic flexure. The dorsal vagal trunk continues along the lesser curvature of the stomach to supply the visceral surface of the pylorus.

Sympathetic Trunk

To expose the caudal thoracic and lumbar parts of the sympathetic trunk on the left side, reflect the abdominal and caudal thoracic walls so that the kidney and the crura of the diaphragm are freely accessible. Reflect the kidney toward the median plane. The deep circumflex iliac artery and vein arise from the aorta and caudal vena cava in the caudal part of the lumbar region and may be transected and turned aside. These vessels will be described later with the branches of the abdominal aorta and caudal vena cava.

The psoas minor arises from the fascia of the muscle dorsal to it, the quadratus lumborum, and from the last thoracic and the first five lumbar vertebral bodies. It is ventral and medial to the quadratus lumborum and psoas major (Figs. 143, 172). It inserts on the ilium on the tubercle of the psoas minor dorsal to the iliopubic eminence. Transect the tendon of the psoas minor caudal to the deep circumflex iliac vessels. Remove the muscle from its vertebral origins. Examine the psoas major and its union with the iliacus to form the iliopsoas muscle.

Splanchnic Nerves

Identify the thoracic and lumbar parts of the sympathetic trunk as it passes along the vertebral bodies. Since most of the preganglionic axons in the sympathetic trunk at levels T10 to T13 pass into the major splanchnic nerve rather than into the lumbar region of the trunk, there is a distinct narrowing of the trunk caudal to the major splanchnic nerve. The trunk widens caudally as lumbar splanchnic components enter it.

The **splanchnic nerves** contain sympathetic neurons that run between the sympathetic trunk and the abdominal autonomic ganglia.

The **major splanchnic nerve** (Fig. 143) leaves the sympathetic trunk at the level of the twelfth or thirteenth thoracic sympathetic ganglion. It passes dorsal to the crus of the diaphragm, enters the abdominal cavity and courses to the adrenal and celiacomesenteric ganglia and plexuses.

The **minor splanchnic nerves** (Fig. 143), generally two, usually leave the last thoracic and first lumbar sympathetic ganglia. They supply nerves to the adrenal gland, ganglion and plexus, and they terminate in the celiacomesenteric ganglia and plexus. Dissect the origin of these nerves and their course to the adrenal gland.

The **lumbar splanchnic nerves** (Fig. 143) arise from the second to the fifth lumbar sympathetic ganglia. In general they are distributed to the aorticorenal, cranial mesenteric and caudal mesenteric ganglia and plexuses. Observe the origin of these nerves.

Abdominal Nerve Plexuses and Ganglia

In the abdomen, branches of the vagus and splanchnic nerves intermingle around the major abdominal arteries to form nerve plexuses. These plexuses are named according to the vessel upon which they are found. The plexuses supply the musculature of the artery and the viscera supplied by the branches of that artery.

Several sympathetic ganglia are located in the abdomen in close association with the plexuses. These ganglia are collections of cell bodies of postganglionic axons. Preganglionic axons in the splanchnic nerves must synapse in one of these ganglia. Preganglionic vagal axons (parasympathetic) do not synapse here but pass through the plexuses to the wall of the organ innervated, where they synapse on a cell body of a postganglionic axon.

The **celiac ganglia** (Fig. 143) lie on the right and left surfaces of the celiac artery close to its origin. They are often interconnected, and numerous nerves from the ganglia follow the terminal branches of the celiac artery as a plexus.

The **cranial mesenteric ganglion** (Fig. 143) is located caudal to the celiac ganglion on the sides and caudal surface of the cranial mesenteric artery, which it partly encircles. Most of its nerves continue distally on the cranial mesenteric artery as the cranial mesenteric plexus. Because of the close relationship of the celiac and cranial mesenteric plexuses and ganglia, they are referred to as the **celiacomesenteric ganglion and plexus.**

The caudal mesenteric ganglion (Fig. 143) is located ventral to the aorta around the caudal mesenteric artery, which is an unpaired branch of the aorta caudal to the kidneys that supplies the colon. Lumbar splanchnic nerves enter the ganglion on each side. Branches may also come from the aortic and celiacomesenteric plexuses. Some of the nerves leaving the ganglion continue along the artery as the **caudal mesenteric plexus.**

The **right** and **left hypogastric nerves** leave the ganglion and course caudally near the ureters. They run in the mesocolon, incline laterally and enter the pelvic canal. Their connections with the pelvic plexuses will be described later.

Branches of the Abdominal Aorta

Abdominal aorta
- Common trunk
  - Caudal phrenic a.
  - Cranial abdominal a.

- Lumbar aa.
- Celiac a.
  - Hepatic a.
    - Hepatic branches
      - Cystic a.
    - Right gastric a.
    - Gastroduodenal a.
      - Right gastroepiploic a.
      - Cranial pancreaticoduodenal a.
  - Left gastric a.
    - Esophageal branches
  - Splenic a.
    - Short gastric aa.
    - Pancreatic branches
    - Left gastroepiploic a.
- Cranial mesenteric a.
  - Common trunk
    - Middle colic a.
    - Right colic a.
    - Ileocolic a.
      - Colic br.
      - Mesenteric ileal br.
      - Cecal a.
        - Antimesenteric ileal br.
  - Caudal pancreaticoduodenal a.
  - Jejunal aa.
  - Ileal aa.
- Renal aa.
- Testicular and ovarian aa.
- Caudal mesenteric a.
  - Left colic a.
  - Cranial rectal a.
- Deep circumflex iliac aa.

1. The paired **lumbar arteries** (Figs. 144–146) leave the dorsal surface of the aorta. Each extends dorsally and terminates in a spinal and a dorsal branch. The spinal branches pass through the intervertebral foramina and anastomose with the ventral spinal artery that is within the vertebral canal and supplies part of the spinal cord. The dorsal branches supply the muscles and skin above the lumbar vertebrae.

2. The **celiac artery** (Figs. 143–145) is short and arises from the aorta between the crura of the diaphragm. It has three branches: the hepatic artery, the left gastric artery and the splenic artery. The celiac plexus of nerves covers the artery in its course through the mesentery.

The **hepatic artery** (Figs. 144, 145) is the first branch to leave the celiac. Find its origin from the celiac. Follow this vessel dorsal to the pylorus between the lesser curvature of the stomach and the liver. One to five branches leave the hepatic artery and enter the liver. (These branches are covered with nerves.) The **cystic artery** leaves the last hepatic branch and supplies the gall bladder. It need not be dissected. After giving off

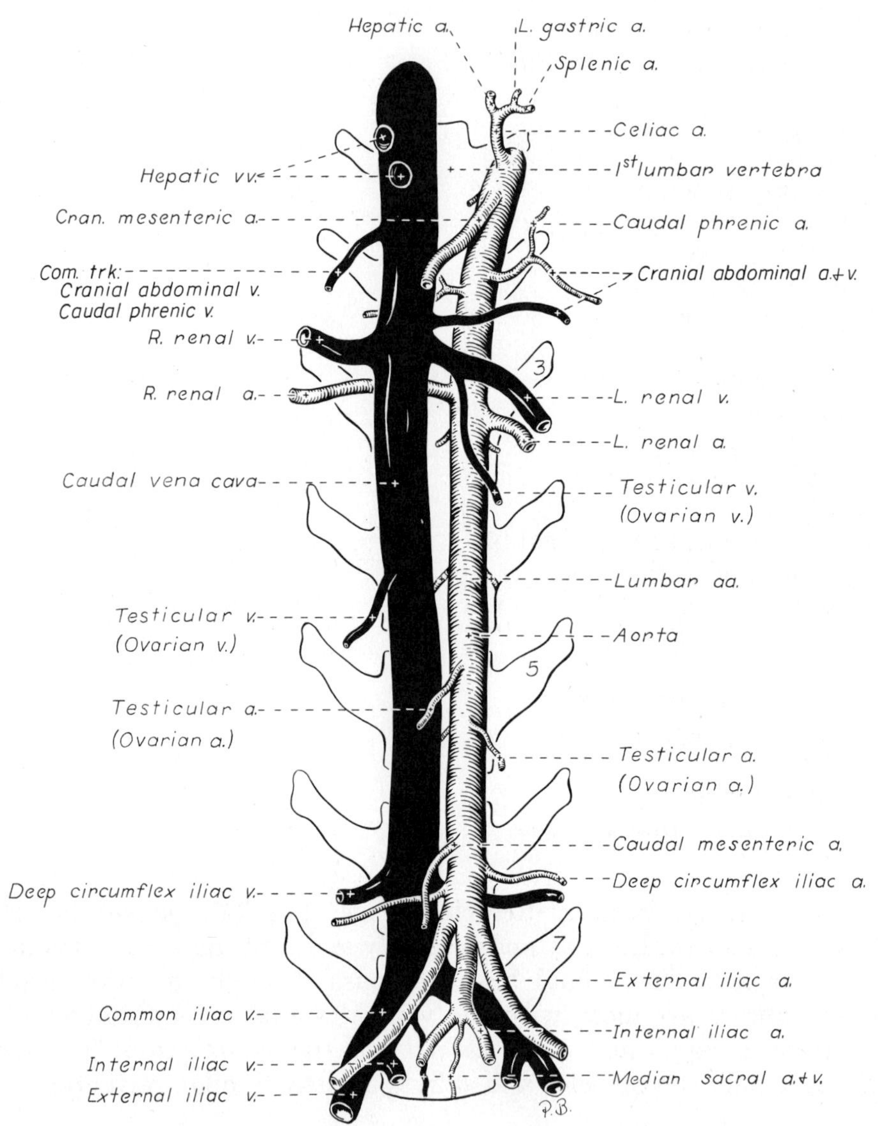

FIGURE 144. Branches of abdominal aorta and tributaries of the caudal vena cava, ventral view.

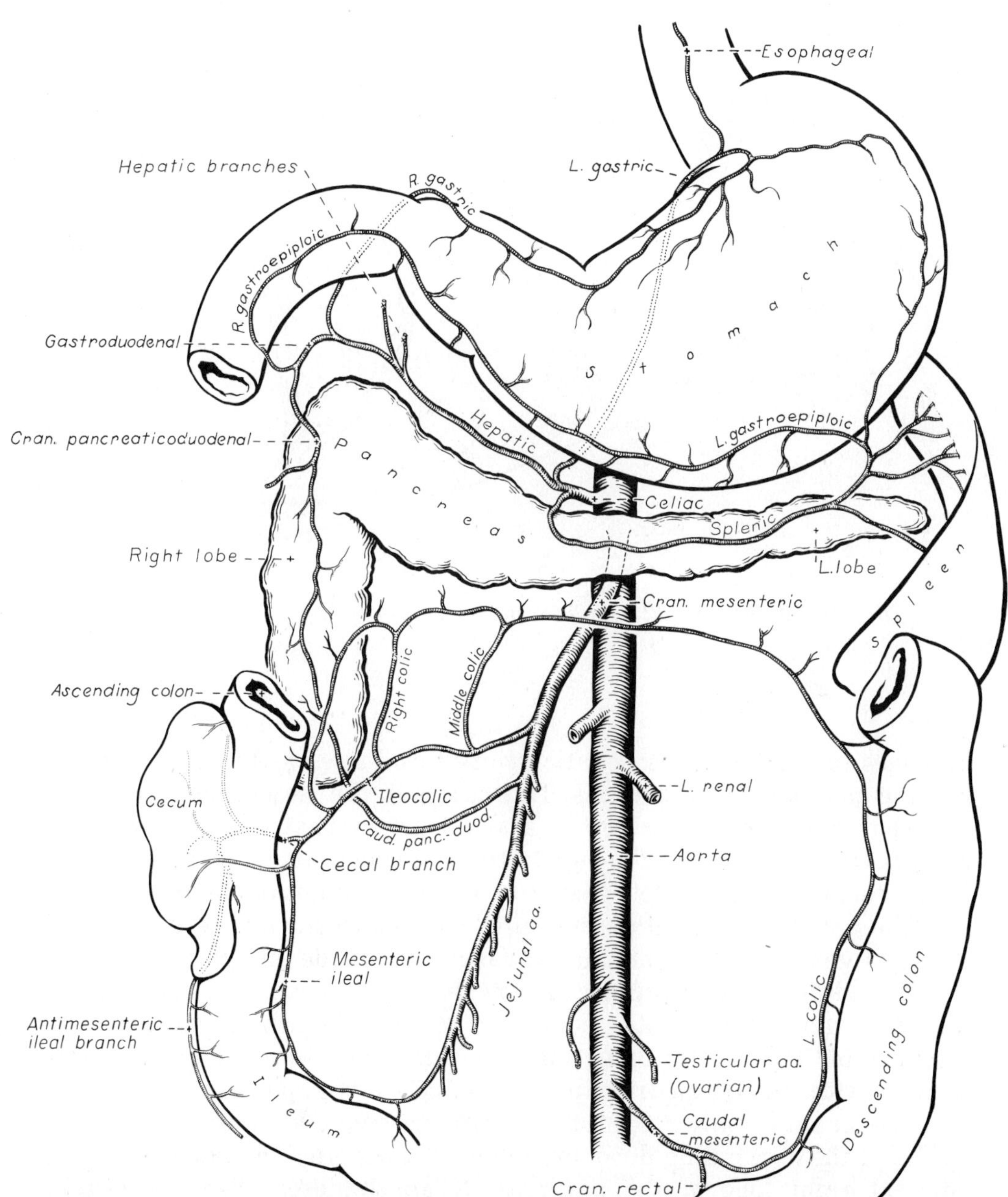

FIGURE 145. Branches of celiac and cranial mesenteric arteries with principal anastomoses.

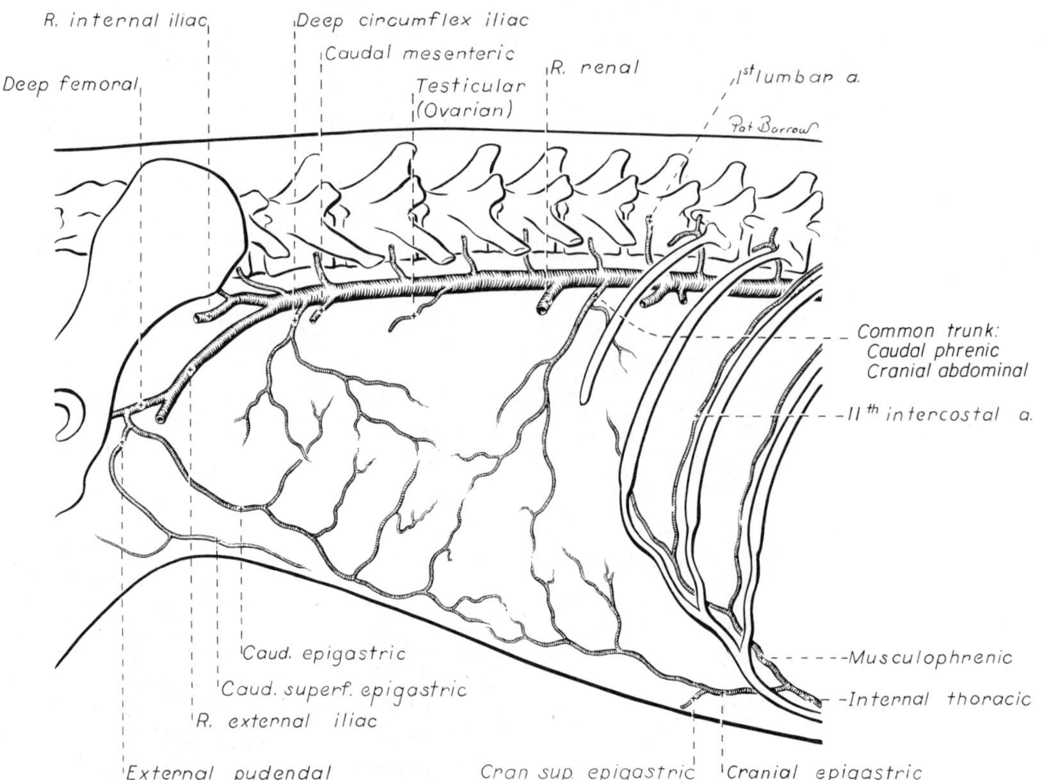

FIGURE 146. Abdominal aorta in relation to epigastric arteries, lateral view.

branches to the liver, the hepatic artery terminates as the right gastric and the gastroduodenal arteries. This occurs in the lesser omentum.

The **right gastric artery** is a small artery that extends from the pylorus toward the cardia to supply the lesser curvature of the stomach. It anastomoses with the left gastric artery. It need not be dissected.

The **gastroduodenal artery** supplies the pylorus and terminates as the right gastroepiploic and cranial pancreaticoduodenal arteries.

The **right gastroepiploic artery** enters and runs in the greater omentum along the greater curvature of the stomach. It supplies the stomach and the greater omentum. The right gastroepiploic artery anastomoses with the left gastroepiploic, a branch of the splenic artery.

The **cranial pancreaticoduodenal artery** follows the mesenteric border of the descending duodenum, where it supplies the duodenum and adjacent right lobe of the pancreas. It anastomoses with the caudal pancreaticoduodenal artery, which is a branch of the cranial mesenteric artery.

The **left gastric artery** (Figs. 144, 145) runs to the lesser curvature of the stomach near the cardia and supplies both surfaces of the stomach. One or more esophageal rami pass cranially on the esophagus. It extends toward the pylorus, where it anastomoses with the right gastric artery.

The **splenic artery,** one of the three main branches of the celiac artery, crosses the dorsal surface of the left lobe of the pancreas and approaches the hilus of the spleen near its middle. It sends major branches to each end of the organ and smaller branches to the pancreas and central portions of the spleen. Short gastric branches pass from the spleen to the greater curvature of the stomach in the gastrosplenic ligament. The splenic artery continues to the greater curvature of the stomach as the **left gastroepiploic artery.** This vessel supplies the greater omentum and passes towards the pyloric end of the stomach where it anastomoses with the right gastroepiploic, a branch of the hepatic artery.

3. The cranial **mesenteric artery** (Figs. 144, 145) leaves the aorta caudal to the celiac artery. It is surrounded proximally by the cranial mesenteric plexus of nerves and partly by the cranial mesenteric ganglion. Peripheral to the ganglion are the mesenteric lymph nodes and branches of the portal vein. Reflect these from the vessel. Observe the branches of the cranial mesenteric artery.

The middle, right and ileocolic arteries arise from a common trunk from the cranial mesenteric artery. Reflect the small intestine caudally and expose the colon cranial to the root of the great mesentery. Dissect its blood supply within the mesocolon.

The **middle colic artery,** the first branch from the common trunk, runs cranially in the mesocolon to the mesenteric border of the left colic flexure and descending part of the colon. It bifurcates near the left colic flexure. One branch runs distally in the descending mesocolon, supplies the descending colon and then anastomoses with the left colic artery, a branch of the caudal mesenteric artery. The other branch passes to the right and forms an arcade with the smaller right colic artery, and supplies the transverse colon.

The **right colic artery** runs in the right mesocolon toward the right colic flexure, giving off branches to the distal part of the ascending colon and the adjacent transverse colon. It forms arcades with the middle colic artery and the colic branch of the ileocolic artery.

The **ileocolic artery** (Fig. 147) supplies the ileum, cecum and ascending colon. The ascending colon is supplied by the **colic branch.** The **cecal artery** crosses the dorsal surface of the ileocolic junction and supplies the cecum and antimesenteric side of the ileum. The ileocolic artery continues as the mesenteric ileal branch to anastomose with ileal arteries of the cranial mesenteric artery.

The **caudal pancreaticoduodenal artery** (Fig. 145) arises from the cranial mesenteric artery close to the common trunk for the colon. It runs to the right in the mesoduodenum to the descending portion of the duodenum near the caudal flexure. It supplies the descending duodenum and the right lobe of the pancreas and anastomoses with the cranial pancreaticoduodenal artery.

The **jejunal arteries** arise from the caudal side of the cranial mesenteric artery. They form arcades in the mesentery close to the jejunum.

The cranial mesenteric artery is terminated by **ileal arteries,** the last of which anastomoses with a branch of the ileocolic artery.

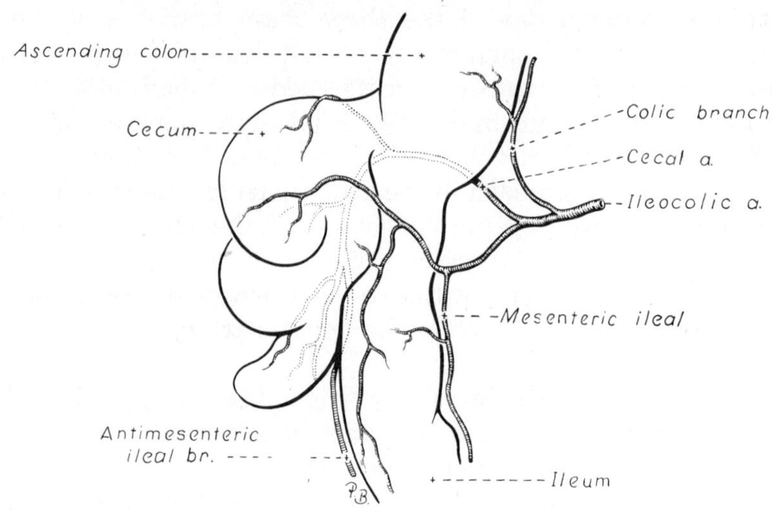

FIGURE 147. Terminations of ileocolic artery.

4. The common trunk of the caudal phrenic and cranial abdominal arteries (Fig. 146) is paired and arises from the aorta between the cranial mesenteric and renal arteries. This common trunk crosses the ventral surface of the psoas muscles dorsal to the adrenal gland. The **caudal phrenic artery** runs cranially to supply the diaphragm. The **cranial abdominal artery** continues into the abdominal wall and ramifies between the transversus abdominis and the internal abdominal oblique, where it was previously dissected.

The adrenal gland may receive branches from the aorta or caudal phrenic, renal or lumbar artery.

5. The **renal arteries** (Figs. 140, 144, 145) leave the aorta at different levels. The right one arises cranial to the left, in conformity with the more cranial position of the right kidney. It is longer than the left and lies dorsal to the caudal vena cava.

6. The **ovarian artery** (Figs. 133, 144, 146) of the female is homologous to the testicular artery of the male. This paired vessel arises from the aorta about halfway between the renal and external iliac arteries. The ovarian artery varies in size, position and tortuosity depending on the degree of development of the uterus. Each ovarian artery divides into two or more branches in the mesovarium just medial to the ovaries. Branches supply the ovary and its bursa and the uterine tube and horn. The branch to the uterine horn anastomoses with the uterine artery, which runs cranially in the mesometrium.

The **testicular artery** (Fig. 143) leaves the aorta in the midlumbar region and crosses the ventral surface of the ureter. The testicular artery, vein and nerve plexus lie in a peritoneal fold, the **mesorchium,** which can be followed to the level of the vaginal ring. Their course in the spermatic cord has been dissected.

The right testicular and ovarian veins enter the caudal vena cava near the origin of the artery from the aorta. However, the left testicular and ovarian veins usually enter the left renal vein. This is important surgically.

7. The **caudal mesenteric artery** (Figs. 144–146) is unpaired and arises near the termination of the aorta. It enters the descending mesocolon and runs caudoventrally to the mesenteric border of the descending colon, where it terminates in two branches of similar size. The **left colic artery** follows the mesenteric border of the descending colon cranially to anastomose with the middle colic artery. The **cranial rectal artery** descends along the rectum and anastomoses with the caudal rectal artery from the internal pudendal.

8. The **deep circumflex iliac artery** (Fig. 146) is paired and arises from the aorta close to the origin of the external iliac artery. It crosses the sublumbar muscles laterally, and at the lateral border of the psoas major it supplies the musculature of the caudodorsal portion of the abdominal wall. The deep circumflex iliac artery perforates the abdominal wall and becomes superficial ventral to the tuber coxae. It supplies the skin of the caudal abdominal area, the flank and the cranial thigh. This vessel was transected when the psoas minor muscle was removed.

Portal Venous System

The **portal vein** (Figs. 148, 149) carries venous blood to the liver from abdominal viscera: the stomach, the small intestine, the cecum, the colon, the pancreas and the spleen. Separate the caudate process of the liver from the descending part of the duodenum and find the portal vein in the ventral border of the epiploic foramen. Reflect the peritoneum and fat from the surface of the vein as far caudally as the root of the mesentery and expose its branches.

1. The **gastroduodenal vein** is a small, proximal branch of the portal vein. It enters the portal vein from the right side near the body of the pancreas and drains the pancreas, the stomach, the duodenum and the greater omentum.

2. The **splenic vein** enters the portal vein from the left side just caudal to the gastroduodenal branch. It is a large branch that receives blood from the spleen, the stomach, the pancreas and the greater omentum. It receives the **left gastric vein,** which drains the lesser curvature of the stomach.

3. The **cranial** and **caudal mesenteric veins** are the distal terminal branches of the portal vein. The cranial mesenteric vein arborizes in the mesentery and collects blood from the jejunum, the ileum, the caudal duodenum and the right lobe of the pancreas. The caudal mesenteric vein drains the cecum and the colon.

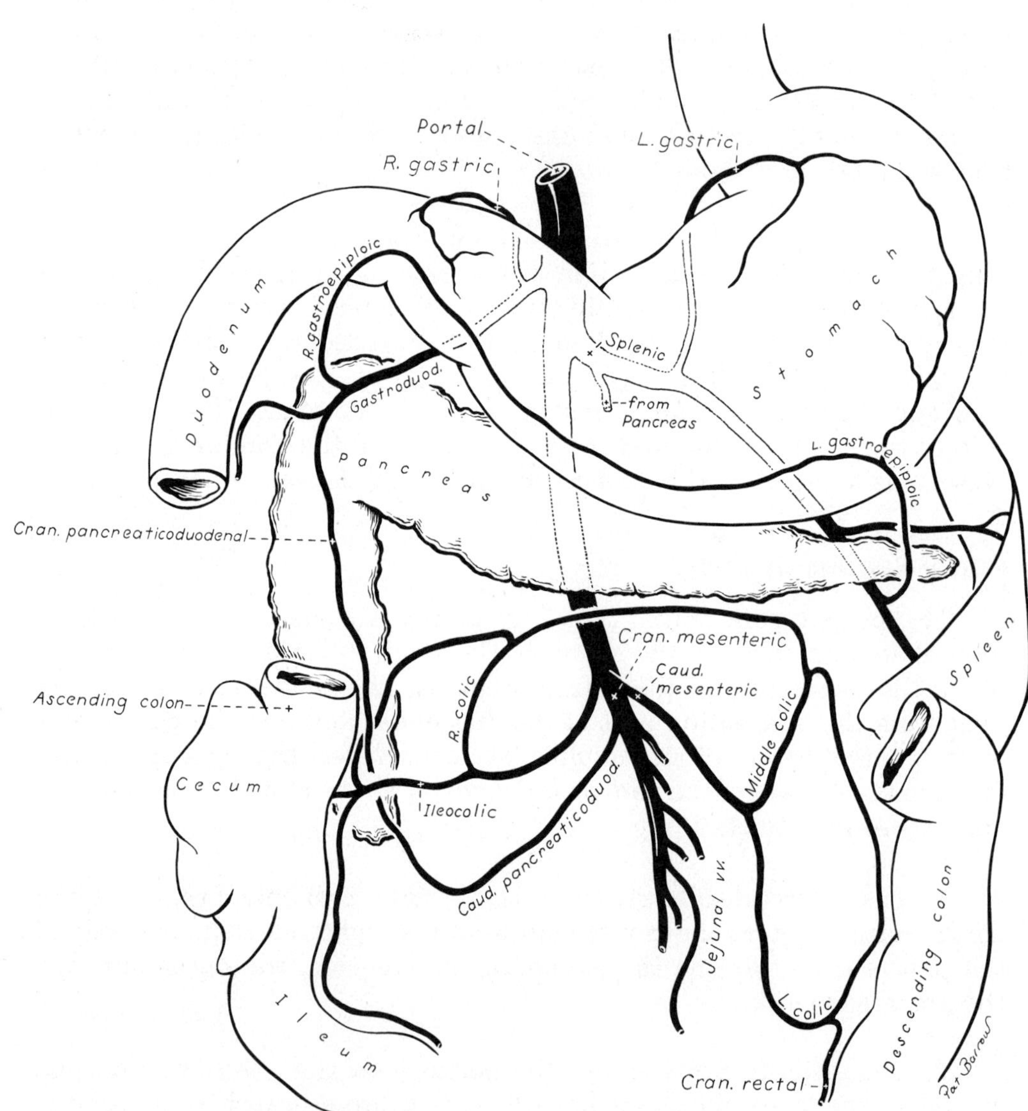

FIGURE 148. Tributaries of portal vein, ventral view.

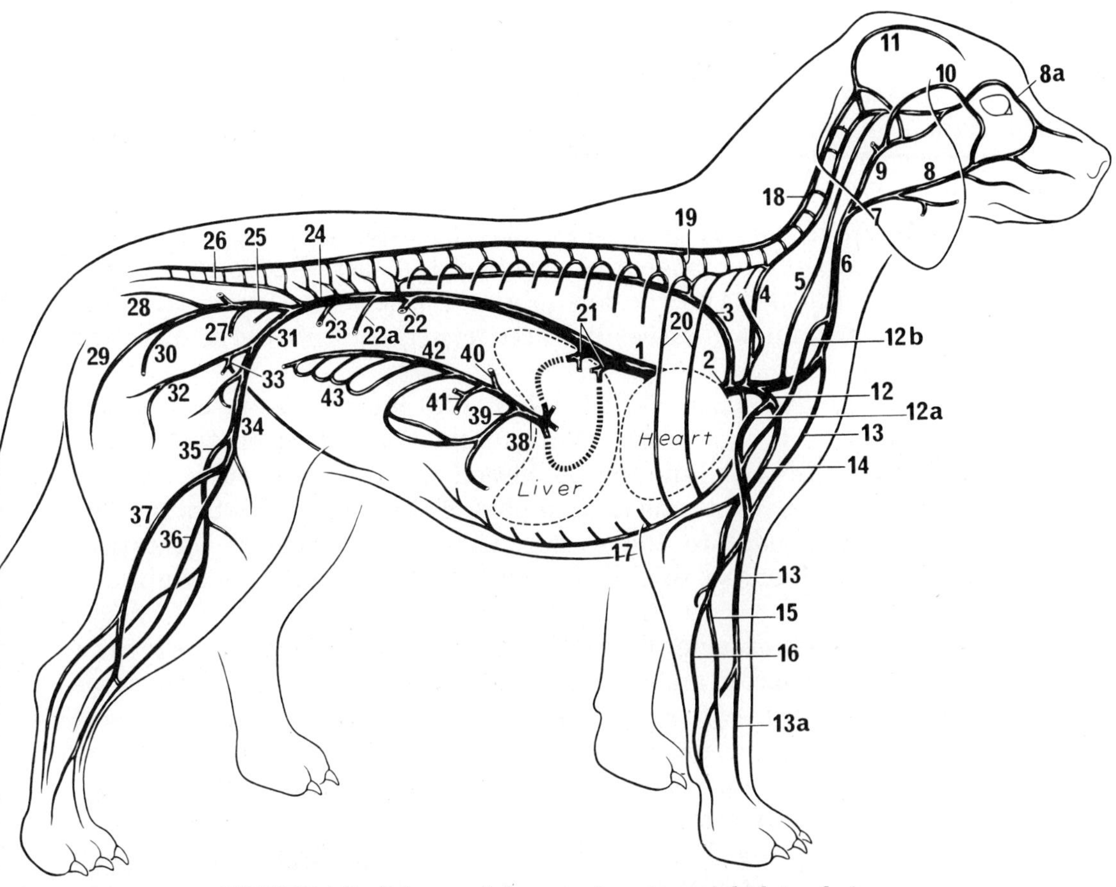

FIGURE 149. Schema of the venous system, right lateral view.

1. *Caudal vena cava*
2. *Cranial vena cava*
3. *Azygos*
4. *Vertebral*
5. *Internal jugular*
6. *External jugular*
7. *Linguofacial*
8. *Facial*
8a. *Angularis oculi*
9. *Maxillary*
10. *Superficial temporal*
11. *Dorsal sagittal sinus*
12. *Axillary*
12a. *Axillobrachial*
12b. *Omobrachial*
13. *Cephalic*
13a. *Accessory cephalic*
14. *Brachial*
15. *Median*
16. *Ulnar*
17. *Internal thoracic*
18. *Right internal vertebral venous plexus*
19. *Intervertebral*
20. *Intercostal*
21. *Hepatic*
22. *Renal*
22a. *Testicular or Ovarian*
23. *Deep circumflex iliac*
24. *Common iliac*
25. *Right internal iliac*
26. *Median sacral*
27. *Vaginal or Prostatic*
28. *Lateral caudal*
29. *Caudal gluteal*
30. *Internal pudendal*
31. *Right external iliac*
32. *Deep femoral*
33. *Pudendoepigastric trunk*
34. *Femoral*
35. *Medial saphenous*
36. *Cranial tibial*
37. *Lateral saphenous*
38. *Portal*
39. *Gastroduodenal*
40. *Splenic*
41. *Caudal mesenteric*
42. *Cranial mesenteric*
43. *Jejunal*

PELVIC VISCERA, VESSELS AND NERVES

To reflect the left pelvic limb, first reflect the penis and scrotum to the right. Cut through the pelvic symphysis with a cartilage knife, saw, or bone cutters. Locate the wing of the left ilium and sever all muscles that attach to its medial and ventral surfaces. Apply even but constant lateral pressure to the left os coxae by abducting the limb and at the same time cut through the cranioventral aspect of the sacroiliac joint. Cut the attachment of the penis to the left ischiatic tuberosity. Leave the limb attached. All structures more easily traced from the left should be dissected from this side.

The **levator ani** (Figs. 150, 157, 159, 172) muscle lies medial to the coccygeus muscle. It is a broad, thin muscle originating on the medial edge of the shaft of the ilium and the dorsal surface of the pubis and the pelvic symphysis. It covers the cranial part of the internal obturator. The muscle appears caudal to the coccygeus, where it inserts on caudal vertebrae 3 to 7. Transect this muscle on the left side near its origin and reflect it.

The **coccygeus muscle** (Figs. 150, 157, 159) lies lateral to the levator ani muscle. It is shorter and thicker and arises from the ischiatic spine and inserts on the transverse processes of caudal vertebrae 2 to 4. Transect the muscle at its origin and remove the left pelvic limb. The levator ani and coccygeus muscles of each side form a **pelvic diaphragm** through which the genitourinary and digestive tracts open to the outside.

The **pelvic plexus** (Fig. 150) lies caudal to a plane passing through the pelvic inlet and dorsal to the prostate. It is closely applied to the surface of the rectum and can be identified by tracing the left hypogastric nerve to it. Occasionally ganglia are large enough to be recognized in the plexus. The pelvic plexus contains sympathetic fibers from the hypogastric nerve and parasympathetic fibers from the pelvic nerve.

The **pelvic nerve** (Fig. 150) is formed by parasympathetic preganglionic axons that leave the sacral spinal nerves. Find the left pelvic nerve on the lateral wall of the distal portion of the rectum. Trace it proximally to its origin. It supplies branches to the urogenital organs, the rectum and the descending colon.

The extension of the peritoneal cavity dorsal to the rectum on either side of the mesorectum is the **pararectal fossa** (Figs. 129, 154). Caudally it extends approximately to the plane of the second caudal vertebra (see Fig. 154). The pararectal fossa is continuous ventrally with the common peritoneal space between the rectum and uterus or prostate. This is the **rectogenital pouch.** In the female the rectogenital pouch communicates ventrally on either side of the uterus with the **vesicogenital pouch** between the uterus and bladder. The vesicogenital pouch in the female and the rectogenital in the male communicate with the short **pubovesical pouch** between the bladder and ventral body wall and pubis. This is divided by the median ligament of the bladder.

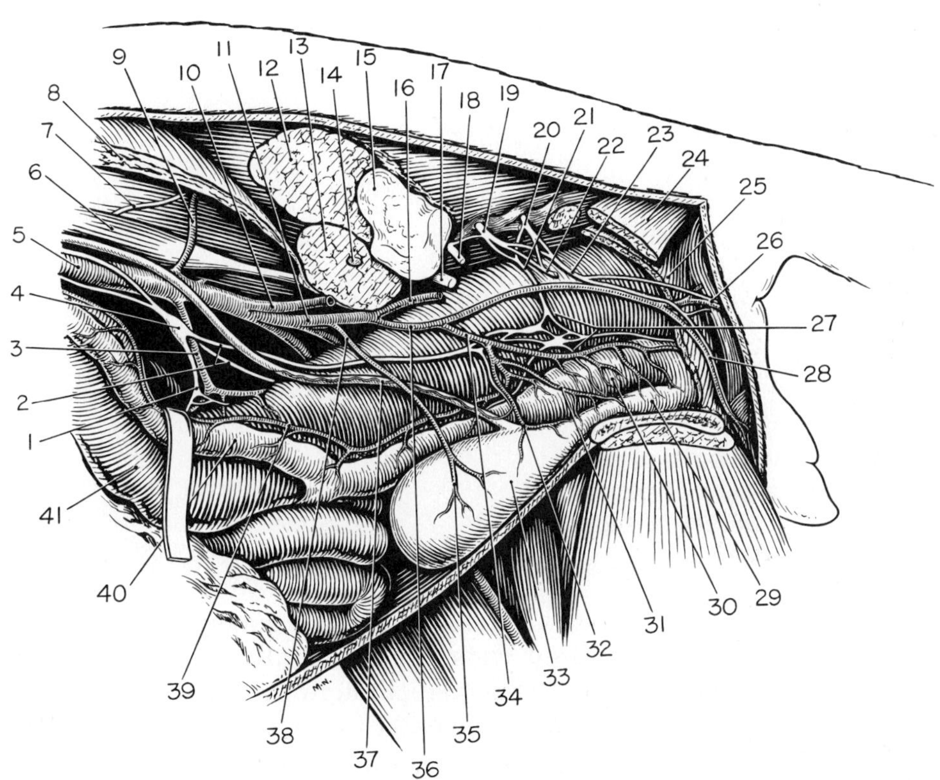

FIGURE 150. Autonomic nerves and vessels of pelvic region, left lateral view.

1. *Caudal mesenteric plexus*
2. *Right and left hypogastric nerves*
3. *Caudal mesenteric artery*
4. *Caudal mesenteric ganglion*
5. *Aorta*
6. *Psoas minor*
7. *Lateral cutaneous femoral nerve*
8. *Abdominal oblique muscles*
9. *Deep circumflex iliac artery*
10. *External iliac artery*
11. *Internal iliac artery*
12. *Quadratus lumborum*
13. *Iliopsoas*
14. *Femoral nerve*
15. *Sacroiliac articulation*
16. *Caudal gluteal artery*
17. *Lumbar nerves 6 and 7*
18. *First sacral nerve*
19. *Second sacral nerve*
20. *Third sacral nerve*
21. *Pelvic nerve*
22. *Caudal cutaneous femoral nerve*
23. *Pudendal nerve*
24. *Coccygeus*
25. *Levator ani*
26. *Perineal nerve and artery*
27. *Pelvic plexus*
28. *Artery and nerve to clitoris*
29. *Urethra*
30. *Vagina*
31. *Urethral branch of vaginal artery*
32. *Caudal vesical artery*
33. *Bladder*
34. *Vaginal artery*
35. *Cranial vesical artery*
36. *Internal pudendal artery*
37. *Ureter and ureteral branch of vaginal artery*
38. *Umbilical artery*
39. *Uterine artery*
40. *Uterine horn*
41. *Descending colon*

Iliac Arteries

- Iliac arteries
  - Internal iliac
    - Umbilical a.
    - Internal pudendal a.
      - Vaginal a.
        - Uterine a.
          - Caudal vesical a.
        - Middle rectal a.
      - Prostatic a.
        - Artery of ductus deferens
          - Caudal vesical a.
        - Middle rectal a.
      - Urethral a.
      - Ventral perineal a.
        - Caudal rectal a.
        - Dorsal scrotal or labial a.
      - Artery of penis or artery of clitoris
        - Artery of bulb of penis
        - Deep artery of penis
        - Dorsal artery of penis
    - Caudal gluteal a.

The paired **iliac arteries** (Figs. 144, 146, 150) supply the pelvis and pelvic limb. The **external iliac** runs ventrocaudally and becomes the femoral artery as it leaves the abdomen through the **vascular lacuna.** The **internal iliac** arises caudal to the external iliac and passes caudolaterally into the pelvis.

The internal iliac artery and the smaller, unpaired **median sacral artery** terminate the aorta. Find the origin of these vessels. The internal iliac artery gives off the rudimentary umbilical artery and terminates cranial to the sacroiliac joint as the caudal gluteal and internal pudendal arteries. The caudal gluteal primarily supplies muscles on the outside of the pelvis and in the caudal thigh. The internal pudendal is distributed to the pelvic viscera and external genitalia at the ischial arch. Dissect the following vessels on the left side.

In the fetus, the **umbilical artery** is a large, paired vessel that carries blood from the aorta to the placenta through the umbilicus. Find the remnant of this vessel. It arises near the origin of the internal iliac artery and courses to the apex of the bladder in its lateral ligament. In some specimens it remains patent this far and supplies the bladder with cranial vesical arteries. Distal to the bladder the vessel is obliterated.

Find the origin of the **internal pudendal artery** (Figs. 150–152) from the internal iliac and dissect its branches. It is the smaller, more ventral branch that runs caudally on the terminal tendon of the psoas minor. At the level of the sacroiliac joint the internal pudendal gives rise to the vaginal or prostatic artery.

The **vaginal** or **prostatic artery** forms an angle of about 45 degrees with the internal pudendal. It passes ventrally in an arch and terminates

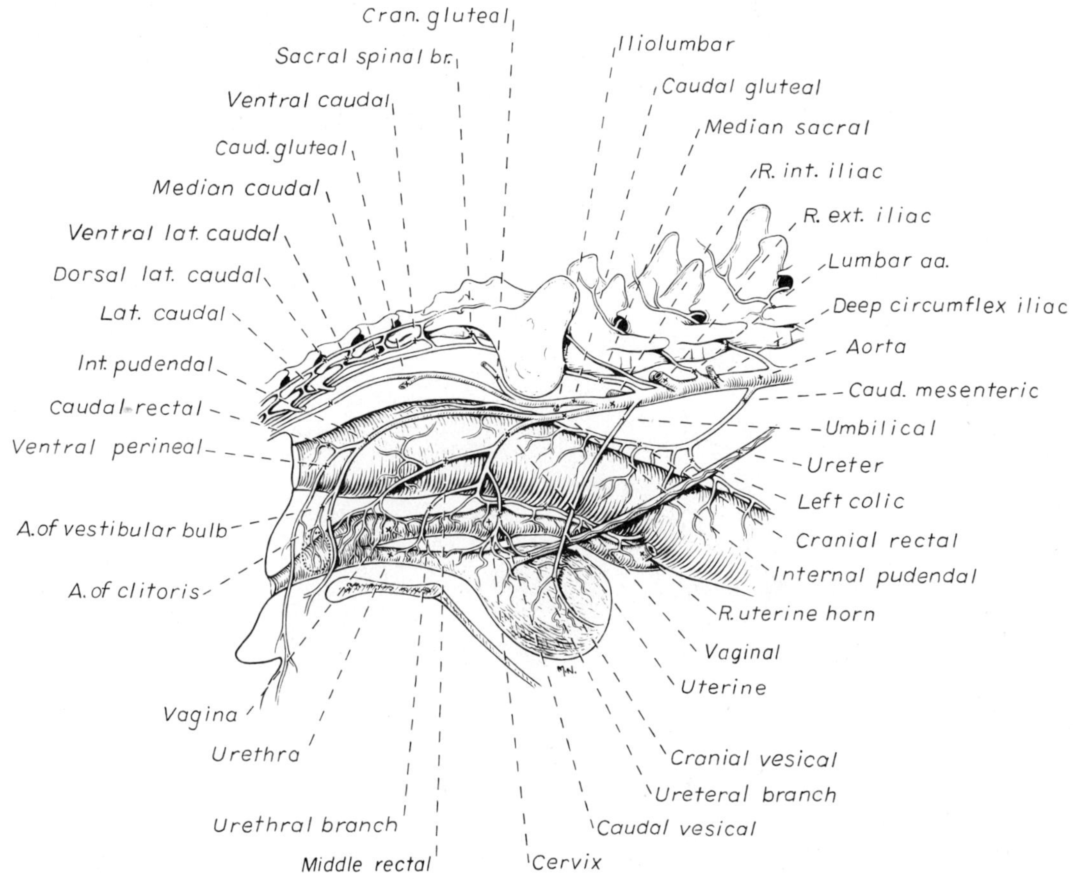

FIGURE 151. Arteries of female pelvic viscera, right lateral view.

in cranial and caudal branches. In the female the cranial branch is the **uterine artery.** The uterine artery supplies a **caudal vesical artery** to the bladder; this artery has ureteral and urethral branches. The uterine artery courses cranially along the body and horn of the uterus in the broad ligament and anastomoses with the uterine branch of the ovarian artery in the mesometrium. The caudal branch of the vaginal artery is the **middle rectal artery,** which supplies branches to the rectum and vagina. In the male the prostatic artery passes caudoventrally from the internal pudendal towards the prostate gland. Its cranial branch is the **artery of the ductus deferens,** which gives off a **caudal vesical artery** to the bladder, with ureteral and urethral branches. It then continues along the ductus deferens, which it supplies. The caudal branch is the **middle rectal artery,** which supplies the rectum, prostate and urethra.

Reflect the skin and fat from the right ischiorectal fossa. The internal pudendal artery (Figs. 150, 153, 174) passes obliquely across the greater ischiatic notch. It continues along the dorsal border of the ischiatic spine lateral to the coccygeus muscle and medial to the gluteal muscles and sacrotuberous ligament. It then courses medially into the ischiorectal fossa. Here it terminates as a ventral perineal artery, a variable urethral

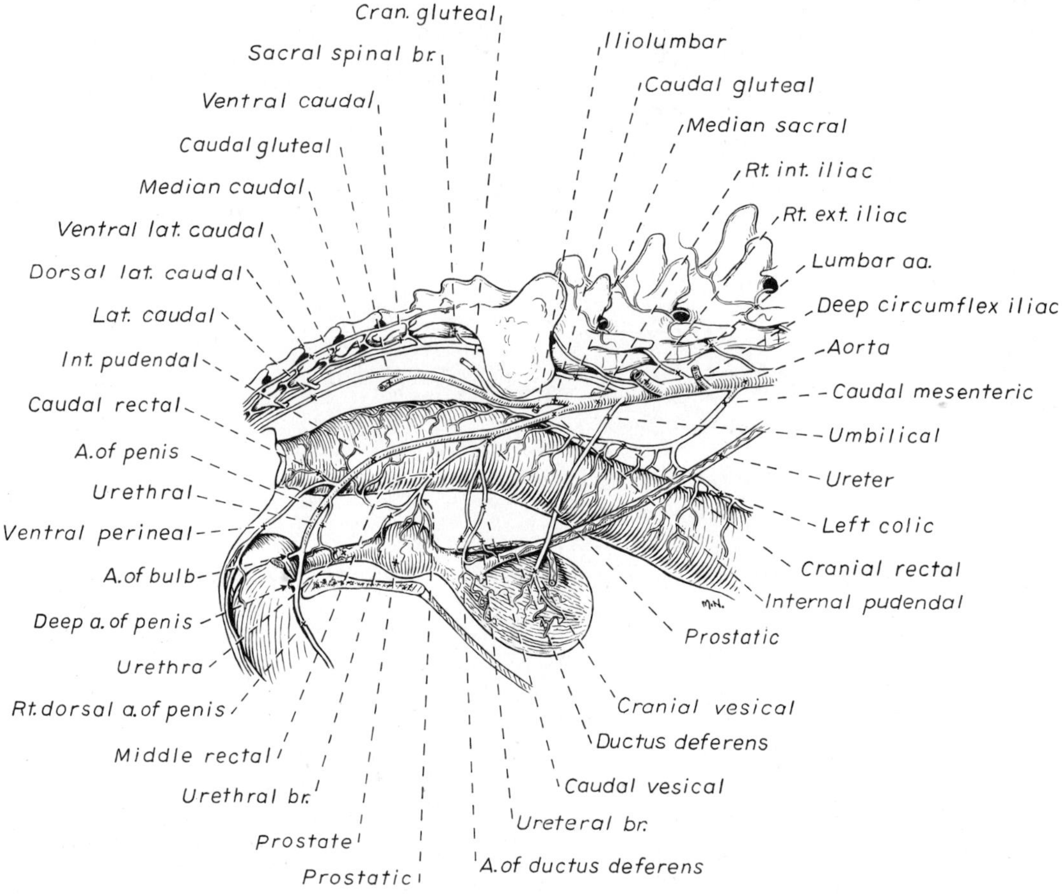

FIGURE 152. Arteries of male pelvic viscera, right lateral view.

artery, and an artery of the penis or clitoris. These vessels may be dissected on either side.

The **ventral perineal artery** may be seen passing caudally. It supplies a caudal rectal artery to the rectum and anus and terminates in the skin of the perineum and the scrotum or vulva.

The **artery of the penis** (Figs. 152, 153) courses caudoventrally and terminates at the level of the ischial arch as three branches: The **artery of the bulb of the penis** arborizes in the bulb and continues to supply the corpus spongiosum and penile urethra. Observe this artery as it enters the bulb. The **deep artery of the penis** arises close to the artery of the bulb and enters the corpus cavernosum at the crus. This is at the level of the ischial arch lateral to the penile bulb. The **dorsal artery of the penis** runs on the dorsal surface to the level of the bulbus glandis, where it divides and sends branches to the prepuce and pars longa glandis. The penile arteries are accompanied by veins that have an important role in the mechanism of erection (Fig. 153).

In the female the **artery of the clitoris** courses caudoventrally to supply the clitoris and vestibular bulb.

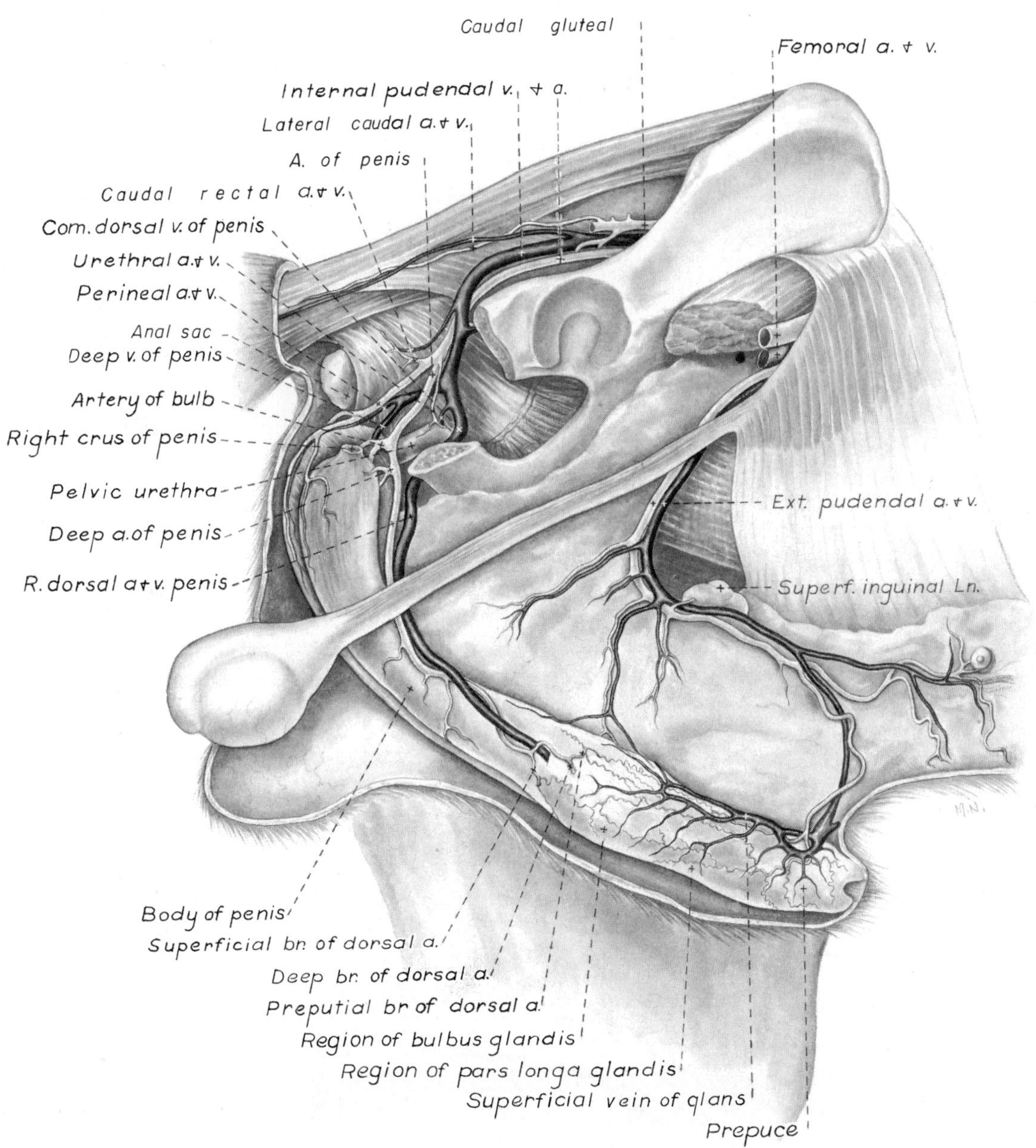

FIGURE 153. Vessels of penis and prepuce (modified from Christensen, 1954).

Pelvic Viscera

The **urinary bladder** (Figs. 154–156) has an apex, a body and a neck. Three peritoneal folds, ligaments, are reflected from the bladder upon the pelvic and abdominal walls. The **median ligament of the bladder** (Fig. 156) leaves the ventral surface of the bladder and attaches to the abdominal wall as far cranial as the umbilicus. In the fetus it contains the urachus and umbilical arteries. The **lateral ligament of the bladder** passes to the pelvic wall and often contains an accumulation of fat along with the ureter and umbilical artery.

Observe the pattern of the bundles of smooth muscle on the surface of the bladder. They pass obliquely across the neck of the bladder and the origin of the urethra. The muscle is innervated by the pelvic nerve (sacral parasympathetic neurons). No anatomical sphincter is present in the bladder.

The **urethral muscle** is striated and is confined to the pelvis, where it surrounds the pelvic urethra and serves as a voluntary sphincter to retain the urine. It is innervated by the pudendal nerve (sacral somatic efferent neurons). Smooth muscle in the urethra functions as a sphincter and is primarily innervated by sympathetic visceral efferent neurons. Make a midventral incision through the bladder wall and the urethra. In the male this should include the prostate gland.

Examine the mucosae of the bladder and urethra. If the bladder is contracted, its mucosa will be thrown into numerous folds, or **rugae,** as a result of its inelasticity.

Observe the entrance of the ureters into the bladder. These lie opposite each other near the neck of the organ. The **trigone of the bladder** is the

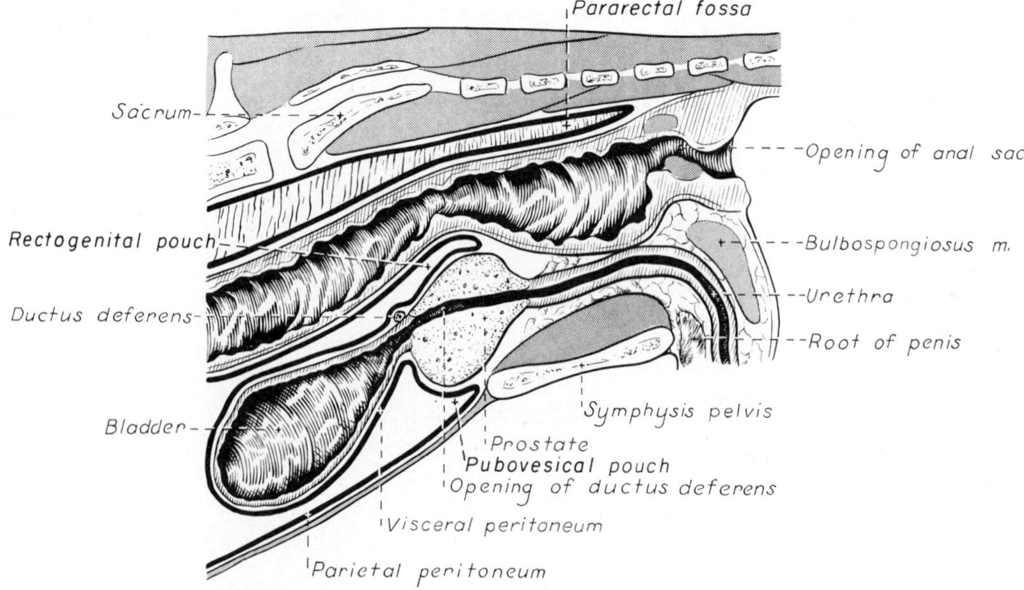

FIGURE 154. Median section through male pelvic region.

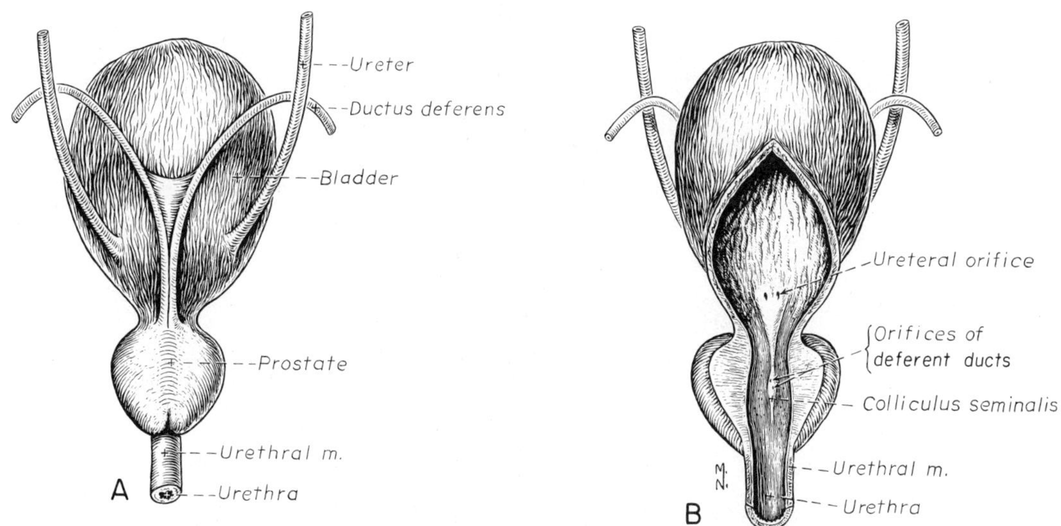

FIGURE 155. Bladder, prostate and associated structures.

dorsal triangular area located within lines connecting the ureteral openings into the bladder and the urethral exit from it.

The **rectum** (Figs. 154, 162) continues the descending colon through the pelvis. It begins at the pelvic inlet. The **anal canal** is a continuation of the rectum to the anus. It consists of three zones and begins with the columnar zone, where the mucosa of the rectum forms longitudinal folds. The longitudinal ridges are called **anal columns.** They terminate at the **intermediate zone,** or anocutaneous line, where small pockets, **anal sinuses,** are formed between the columns. Distal to this line is the larger **cutaneous zone** of the anal canal, which has fine hairs, microscopic circumanal glands and, on each side, the prominent ventrolateral opening of the **anal sac** (paranal sinus). The anal canal is surrounded by both a

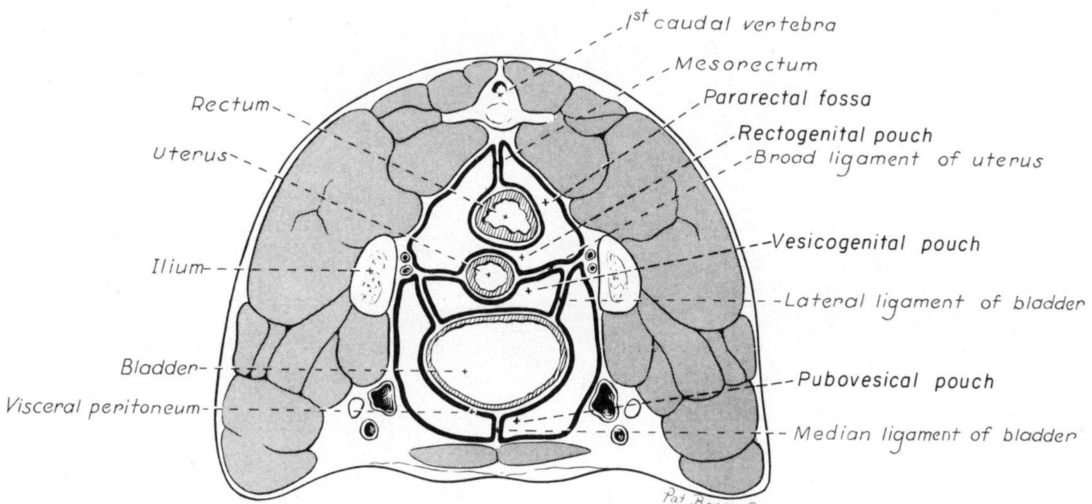

FIGURE 156. Schematic transection of female pelvic cavity.

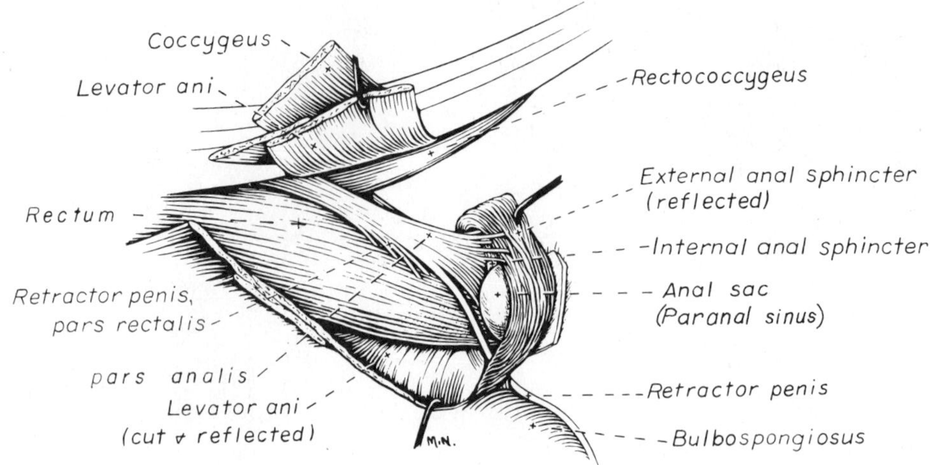

FIGURE 157. Muscles of the anal region, left lateral aspect.

smooth **internal** and a striated **external sphincter muscle** (Figs. 157, 158). The external opening of the anal canal is the **anus.** Transect the external sphincter on the left and reflect it from the anal sac. This muscle receives its nerve supply from the caudal rectal nerve (pudendal) and its blood supply from the ventral perineal artery.

Observe the anal sac (Figs. 157, 158), expose its duct and find the opening in the cutaneous zone of the anal canal. In the wall of this sac are microscopic glands, the secretion of which accumulates in the lumen of the sac. The secretion is discharged through the duct of the anal sac. Open the sac and examine its interior.

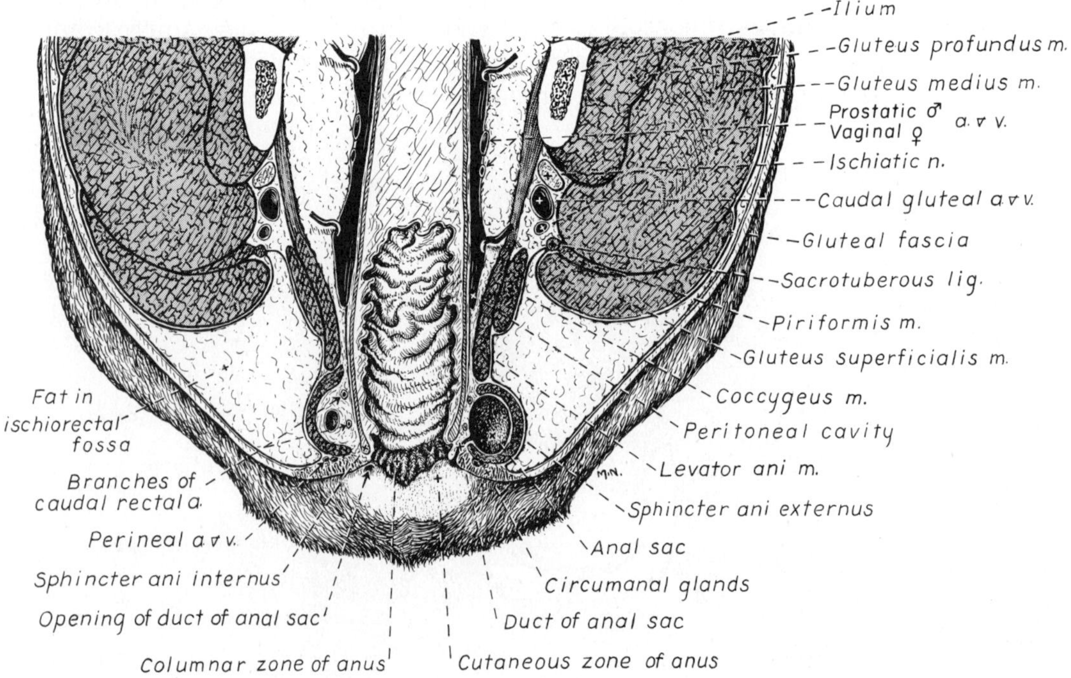

FIGURE 158. Dorsal section through anus. (Right side cut at lower level through duct of anal sac.)

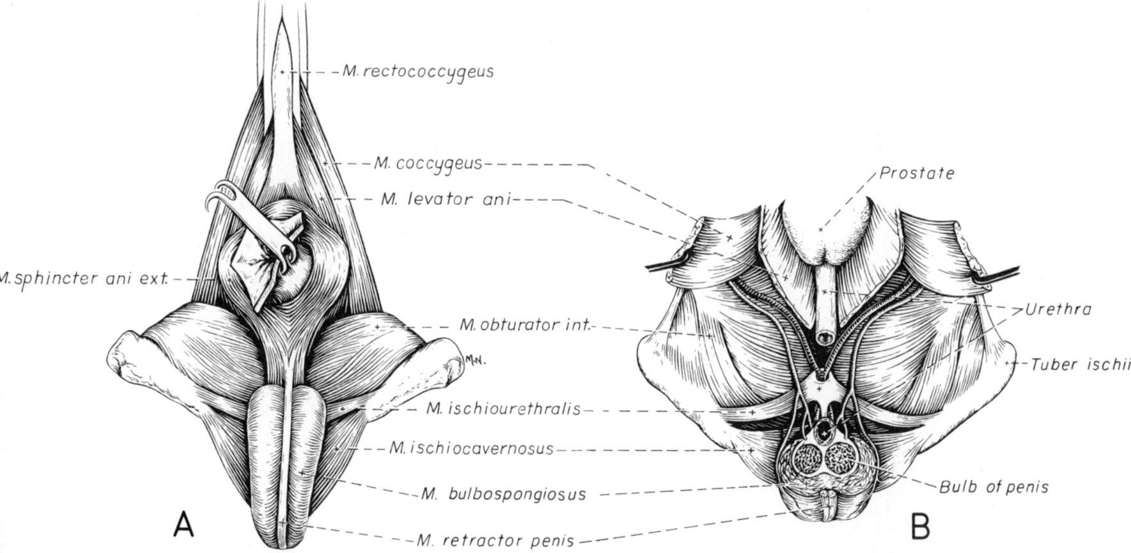

FIGURE 159. Male perineum. A, *Superficial muscles, caudal aspect.* B, *Dorsal section through pelvic cavity. The bilobed bulb of the penis is transected and the proximal portion removed.*

The **internal sphincter muscle** of the anus is an enlargement of the smooth circular muscle coat of the anal canal. It is not as distinct as the external sphincter.

The **rectococcygeus muscle** (Figs. 157, 159) continues the longitudinal coat of the rectum to the ventral surface of the tail. Reflect the levator ani and coccygeus muscles from the left side of the rectum. Observe the rectococcygeus muscle arising from the dorsal surface of the rectum cranial to the sphincter muscles. Trace it caudally to its insertion on the caudal vertebrae.

Male

The **prostate gland** (Figs. 154, 155) completely surrounds the neck of the bladder and the beginning of the urethra. Examine the surface, form, size and location of the prostate on several specimens. The normal size and weight of the prostate vary greatly. The organ generally lies at the pelvic inlet. It is larger and extends further into the abdomen in older dogs. The prostate is flattened dorsally and rounded ventrally and on the sides. It is heavily encapsulated. Muscle fibers from the bladder run caudally on its dorsal surface. A longitudinal septum leaves the ventral part of the capsule and reaches the urethra, thus partially dividing the gland ventrally into right and left lobes. This is indicated on the ventral surface by a shallow but distinct longitudinal furrow. Notice that the urethra runs through the center of the gland. Open the urethra and examine its lumen.

The male **urethra** is composed of a **pelvic part** within the pelvis and a **spongy part** within the penis. The **urethral crest** protrudes into the lumen from the dorsal wall of the prostatic part of the pelvic urethra.

Near its middle and protruding farthest into the lumen of the urethra is a hillock, the **colliculus seminalis.** On each side of this eminence the deferent ducts open. Many prostatic openings are found on both sides of the urethral crest and can usually be seen if the gland is compressed.

PENIS

The **penis** is composed of a root, a body and a glans (Figs. 153, 154, 160). The dorsal surface of the penis faces the pelvic symphysis and the

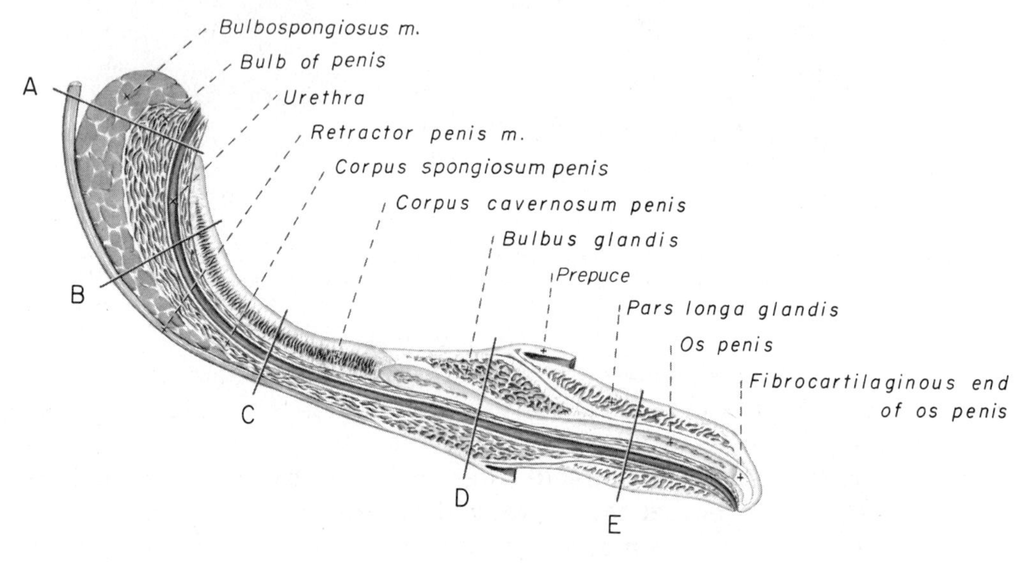

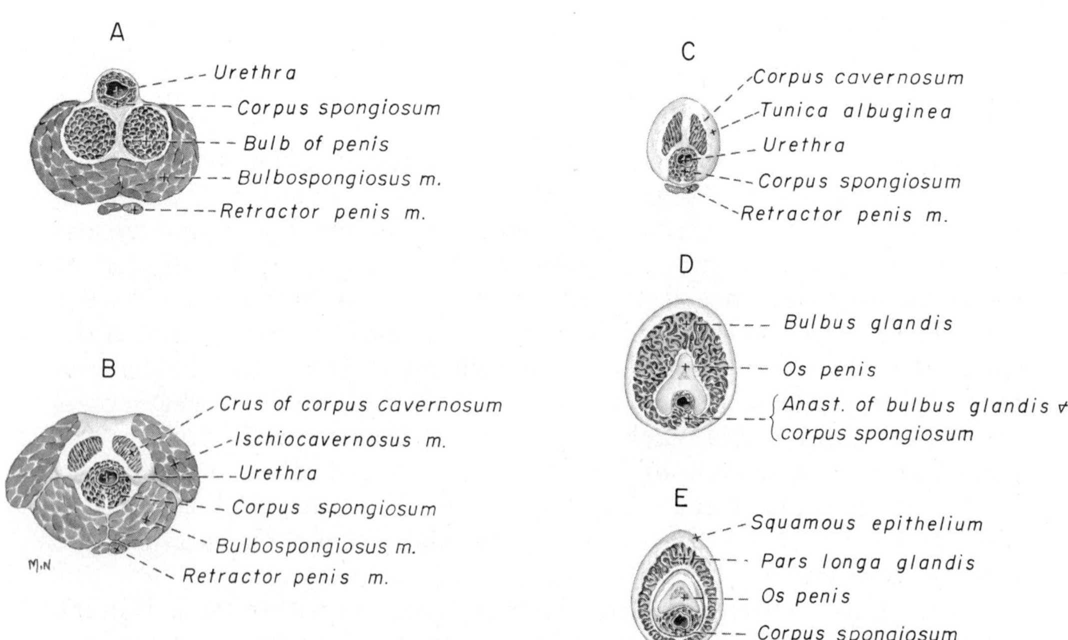

FIGURE 160. Median and transverse sections of the penis (from Christensen, 1954).

abdominal wall. In the nonerect state the glans is entirely withdrawn into the prepuce.

The **prepuce** is a tubular sheath or fold of integument that is continuous with the skin of the ventral abdominal wall and is reflected over the glans. It has a smooth internal layer and a haired external layer, which meet at the **preputial orifice.** At its deepest recess, the **fornix of the prepuce,** the internal layer is reflected onto the glans as the skin of the glans. In the erect state the fornix is eliminated since the internal layer of the prepuce is closely applied to the body of the penis. Open the prepuce by a midventral incision from the orifice to the fornix. Continue the midventral skin incision to the anus so as to expose the entire length of the penis.

The **root** of the penis is formed by right and left crura, which originate on the ischiatic tuberosity of each side. The root ends where the crura blend with each other on the midline to form the body. Each **crus** is composed of cavernous tissue, **corpus cavernosum penis,** supplied by the deep artery of the penis and surrounded by a thick fibrous tunic, **tunica albuginea.** Examine the root. The left crus was transected when the limb was reflected. The trabeculae and vascular spaces can be seen on the cut surface. Note the firm attachment of the right crus to the ischiatic tuberosity.

The **ischiocavernosus muscle** (Figs. 159, 160) arises from the ischiatic tuberosity, covers the origin of the crus and inserts distally on the crus.

The **retractor penis muscle** (Figs. 159, 160) originates from the ventral surface of the sacrum, or first two caudal vertebrae, blends with the external anal sphincter and extends distally on the ventral surface of the penis to the level of the glans, where it inserts. In the region of the anal sphincter there is muscle fiber exchange between the retractor penis muscle and the external anal sphincter. Observe the retractor muscle on the body of the penis.

The **bulbospongiosus muscle** (Figs. 159, 160) bulges between the ischiocavernosus muscles ventral to the external anal sphincter. The fibers of the bulbospongiosus are transverse proximally where they cover the **bulbs** of the penis and longitudinal distally where they pass onto the body of the penis.

Between the crura is the **bulb of the penis,** which is a bilobed dorsal expansion of the **corpus spongiosum penis** that surrounds the urethra. This expansion to form the bulb of the penis is located at the ischial arch. It is supplied by the artery of the bulb and is covered caudally by the **bulbospongiosus muscle.** Observe the penile bulb and its relationship to the urethra at the root of the penis.

The **body** of the penis extends from the root where the crura blend with each other to the glans covering the os penis in the caudal portion of the prepuce (Fig. 153). Note that the region at the beginning of the body of the penis is compressed from side to side and wrapped by a thick tunic. It is capable of being bent without twisting when the male dismounts during coitus and remains “locked” for a variable period.

The **corpus cavernosum penis** of each crus converges toward its fellow on the dorsal side of the body of the penis, and the two corpora extend side by side throughout the body to the os penis. A median septum completely separates the two corpora, and each is covered by a white capsule (tunica albuginea) throughout its length. The two corpora cavernosa form a groove ventrally that contains the urethra and the thin **corpus spongiosum** that surrounds the urethra. Make several incomplete transections through the body of the penis to study these structures.

The **glans** of the penis is composed of two parts, the proximal **bulbus glandis** and the distal, more elongate **pars longa glandis** (Figs. 153, 160). The bulbus glandis, surrounding the proximal end of the os penis, is an expansile vascular structure that is largely responsible for retaining the penis within the vagina during copulation. The pars longa glandis is a cavernous tissue structure that overlaps the distal half of the bulbus glandis and continues to the distal end of the penis, partially encircling the os penis and the urethra. The pars longa glandis has no vascular communication with the corpus spongiosum penis and is separated from the bulbus glandis by a layer of connective tissue. Venous channels drain the pars longa glandis into the bulbus glandis through this layer. Make a longitudinal incision on the dorsum of the penis through the glans to observe its structure.

The **os penis** (Figs. 160, 161) is a long, ventrally grooved bone that lies almost entirely within the glans penis. The expanded, rough, truncate base of the bone originates in the tunica albuginea at the distal end of the corpora cavernosa. It develops as an ossification of the fused distal ends of the corpora cavernosa. The body of the os penis extends through the glans penis. The base and body are grooved ventrally by the **urethral groove,** which surrounds the urethra and the corpus spongiosum on three sides. The bone ends as a long, pointed fibrocartilage in the tip of the glans, dorsal to the urethral opening.

At the level of the collarlike bulbus glandis of the glans of the penis, there is a communication between the **corpus spongiosum penis** and

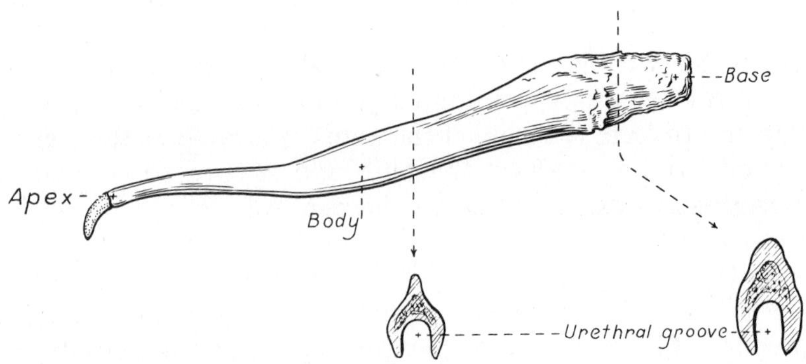

FIGURE 161. Os penis with transverse sections, left lateral view.

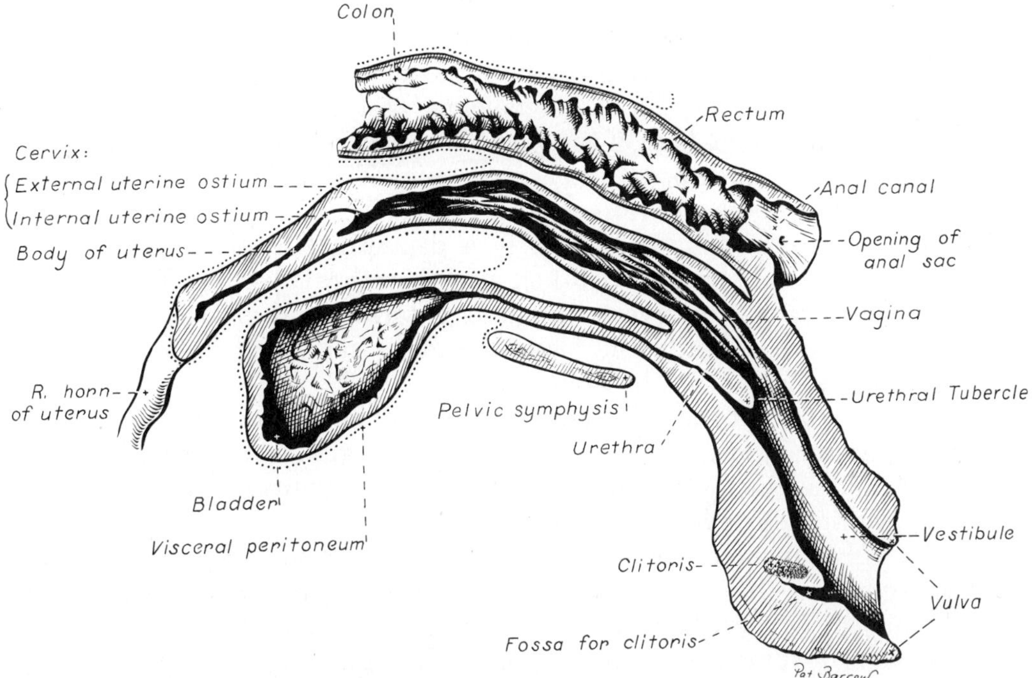

FIGURE 162. Female pelvic viscera, median section, left lateral view.

the **bulbus glandis.** The dorsal artery of the penis courses to the glans, where it supplies the prepuce, corpus spongiosum and pars longa glandis.

Female

The **cervix** (Figs 151, 162) is the constricted caudal portion of the uterus. The **cervical canal** is in a nearly vertical position, with the uterine opening (**internal uterine ostium**) dorsal and the vaginal opening (**external uterine ostium**) ventral in position.

The **vagina** (Fig. 162) is located between the uterine cervix and the vestibule. The most cranial part of the vagina is the **fornix,** which extends cranial to the cervix along its ventral margin. The mucosal lining of the remaining part of the vagina is thrown into longitudinal folds that have small transverse folds. These are evidence of its ability to enlarge in both diameter and length. The longitudinal folds end dorsally at the level of the urethral orifice, where the vagina joins the vestibule. A prominent dorsal longitudinal fold in the cranial vagina nearly obscures the external uterine os and makes catheterization difficult.

The **vestibule** (Fig. 162) is the cavity that extends from the vagina to the vulva. Open the vestibule and vagina by an incision through the dorsal wall. The **urethral tubercle** projects from the floor of the cranial part of the vestibule. The urethra opens on this tubercle. This tubercle is at the level of the ischial arch. Note its relationship to the more ventrally placed vulva.

In the floor of the vestibule, deep to the mucosa, are two elongate masses of erectile tissue, the **vestibular bulbs.** They are homologous to the bulbs of the penis of the male and lie in close proximity to the body of the clitoris. They are difficult to distinguish.

The **clitoris** is the female homologue of the penis. It is a small structure located in the floor of the vestibule near the vulva. It is composed of paired crura, a short body and a glans clitoridis, which are difficult to distinguish. The glans clitoridis is a very small erectile structure that lies in the **fossa clitoridis.** The fossa is a depression in the floor of the vestibule and should not be mistaken as the urethral opening. The dorsal wall of the fossa partly covers the glans clitoridis and is homologous to the male prepuce. Identify the fossa and glans clitoridis. Only rarely is an os clitoridis present in the glans.

The retractor clitoridis muscle (Fig. 163), a homologue of the retractor penis in the male, originates on the first two caudal vertebrae, joins the external anal sphincter, blends with the constrictor vulvae and continues onto the ventral surface of the clitoris.

The **vulva** includes the two **labia** and the external urogenital orifice that they bound, the **rima pudendi.** The labia fuse above and below the rima pudendi to form **dorsal** and **ventral commissures.** The dorsal commissure is ventral to a dorsal plane through the pelvic symphysis. The ventral commissure is directed caudoventrally.

Observe the course of the female urethra by passing a flexible probe through it. It extends from the bladder caudodorsally over the cranial edge of the pelvic symphysis to the genital tract caudal to the vaginovestibular junction. It ends at the external urethral orifice on the urethral tubercle, which is dorsal to the rima pudendi.

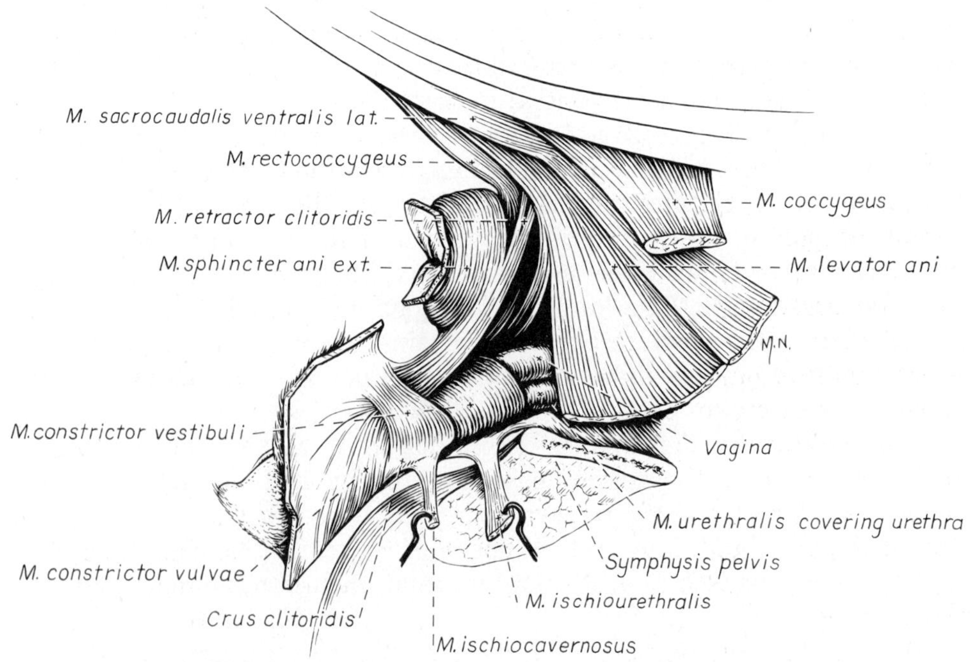

FIGURE 163. Muscles of female perineum, right lateral view.

Live Dog

Abdominal palpation is an art that takes considerable experience to develop. It is dependent on your knowledge of the topographic anatomy of abdominal organs. Not all organs are palpable and some cannot be felt in all dogs. In the standing dog, grasp the caudal abdomen gently with one hand and palpate the bladder ventrally and the descending colon above it. In the female the nonpregnant uterus can sometimes be felt between them. As the uterus enlarges with advanced gestation it will be felt in the ventral abdomen. Dorsal to the colon, lumbar and iliac lymph nodes that are enlarged from disease can be felt.

Stand over or beside the standing dog and palpate the two sides simultaneously, starting cranially at the costal arch and progressing caudally. The liver is not usually felt. On the left side the empty stomach is not palpable, but the full stomach will be felt. The spleen should be felt on the left behind the costal arch. The left kidney is deeper but usually palpable. Cranially on the right no specific organ is palpable. If pain is elicited, the organs to be considered as a source of irritation include the liver, pancreas, pylorus and right kidney. Sometimes the descending duodenum can be felt on the right. The right kidney is not usually felt. Remember its close relationship with the caudate process of the caudate lobe of the liver. The descending colon can be felt on the left. In the ventral midabdominal region, you should be able to feel the coils of small intestine slip through your fingers. The ileum and cecum are not usually palpable. Deep in the midabdominal region you may feel mesenteric lymph nodes, but only if they enlarged from disease.

In the male palpate the root of the penis. Feel the ischiocavernosus muscles covering the crura on either side of the bulbospongiosus muscle which covers the penile bulb. Feel the slightly compressed, firm body of the penis formed by the paired corpora cavernosa and the groove ventrally that contains the urethra and corpus spongiosum penis. Bend the body and note its flexibility. In normal canine mating, the male will step over the female with one hind limb and face in the opposite direction while still breeding the female. The penis bends without twisting at the level of the penile body. Palpate the junction of the corpora cavernosa of the penis with the os penis. This is sometimes the site of blockage of the urethra with calculi. Palpate the length of the os penis and the two parts of the glans penis that cover it. Palpate the superficial inguinal lymph nodes in the skin fold that suspends the penis at the level of the bulbus glandis. These are normally flat and difficult to feel. Palpate the spermatic cord from the superficial inguinal ring to the testes. Palpate the testes and epididymis in the scrotum.

In the female, open the vulva and observe the clitoral fossa. The urethral opening on its tubercle is dorsal to this at the level of the ischial arch and is not visible.

Gently extend the tail and observe the cutaneous zone of the anal canal. Find the openings of the anal sacs on either side of the cranial part of this zone.

VESSELS AND NERVES OF THE PELVIC LIMB (TABLE 3)

Internal Iliac Artery

Internal iliac artery
- Umbilical a.
- Internal pudendal a.
- Caudal gluteal a.
  - Iliolumbar a.
  - Cranial gluteal a.
  - Lateral caudal a.
  - Dorsal perineal a.

The **caudal gluteal artery** (Figs 148, 150–153, 164, 166, 174) is the larger of the two terminal branches of the internal iliac artery. It arises opposite the sacroiliac joint and passes caudally across the greater ischiatic notch and over the ischiatic spine lateral to the coccygeus muscle, parallel to the internal pudendal artery. The branches of the caudal gluteal are the iliolumbar, cranial gluteal, lateral caudal and dorsal perineal arteries (Fig. 164). The veins (Fig. 165) will not be dissected. Observe the origin of the caudal gluteal artery on the medial aspect of the right ilium at the pelvic inlet. Pull the caudal gluteal artery away from the ilium and observe the iliolumbar branch coursing cranial to the wing of the ilium and the cranial gluteal branch coursing caudal to the wing of the ilium across the greater ischiatic notch (Fig. 164).

Make a skin incision on the medial surface of right thigh to the stifle. Encircle the stifle and reflect the skin from the lateral pelvis, rump and thigh.

Expose the insertion of the superficial gluteal deep to the proximal edge of the biceps femoris. Transect the insertion at this level. Reflect the proximal portion of the superficial gluteal to its origin. Transect the middle gluteal muscle 1 centimeter from the crest of the ilium. Start at the cranial border of the bone and detach the muscle from the gluteal surface.

The **cranial gluteal artery** and **nerve** (Figs. 164, 166, 167) pass across the cranial part of the greater ischiatic notch of the ilium and between the middle and deep gluteal muscles, which they supply. The cranial gluteal nerve also continues into and innervates the tensor fasciae latae.

The **iliolumbar artery** (Figs. 151, 166, 174) arises close to the origin of the caudal gluteal artery or directly from the internal iliac. It passes across the cranioventral border of the ilium and supplies the psoas minor, iliopsoas, sartorius, tensor fasciae latae and middle gluteal muscles. On the lateral side, observe its terminal distribution to the deep surface of the cranial end of the middle gluteal.

Transect the biceps femoris midway between its origin and the stifle.

Table 3. *VESSELS AND NERVES OF THE PELVIC LIMB*

Area	Arterial Supply	Nerve Supply
Cranial Thigh Muscle		
Extensor of stifle:	Lateral circumflex femoral	Femoral
Quadriceps femoris		
Medial Thigh Muscles		
Adductors of pelvic limb:	Deep femoral	Obturator
Gracilis, Adductor,	Caudal femorals	
Pectineus		
Caudal Thigh Muscles		
Flexors and extensors of stifle:	Deep femoral	Sciatic
Biceps femoris,	Caudal gluteal	
Semimembranosus,	Caudal femorals	
Semitendinosus		
Cranial Muscles of Crus		
Flexors of tarsus:	Cranial tibial	Peroneal
Cranial tibial,		
Peroneus longus		
Extensor of digits:		
Long digital extensor		
Caudal Muscles of Crus:		
Flexor of stifle:	Popliteal	Tibial
Popliteus	Distal caudal femoral	
Extensor of tarsus:		
Gastrocnemius		
Flexors of digits:		
Superficial digital flexor		
Deep digital flexors		
Dorsal Surface of Paw		
Superficial	Saphenous	Peroneal
Deep	Dorsal pedal	
Plantar Surface of Paw		
Superficial	Saphenous	Tibial
Deep	Dorsal pedal (perforating ramus)	

Transect the semitendinosus 1 centimeter distal to the transection through the biceps. Turn both muscles toward their origins. The caudal gluteal artery lies on the ventrocranial side of the sacrotuberous ligament and in this location gives off several small branches to adjacent muscles: the lateral caudal artery to the tail and the dorsal perineal artery to the perineum. These need not be dissected.

Follow the caudal gluteal artery as it passes over the ischiatic spine with the sciatic nerve ventral to the sacrotuberous ligament (Fig. 167). Here the artery supplies the superficial and middle gluteals, the rotators of the hip, and the adductor muscle. It divides into several branches, which supply the biceps femoris, the semitendinosus and semimembranosus muscles. Turn the biceps femoris caudally to expose the caudal gluteal artery lying deep to it, close to the sacrotuberous ligament and ischiatic tuberosity.

Text continued on page 238

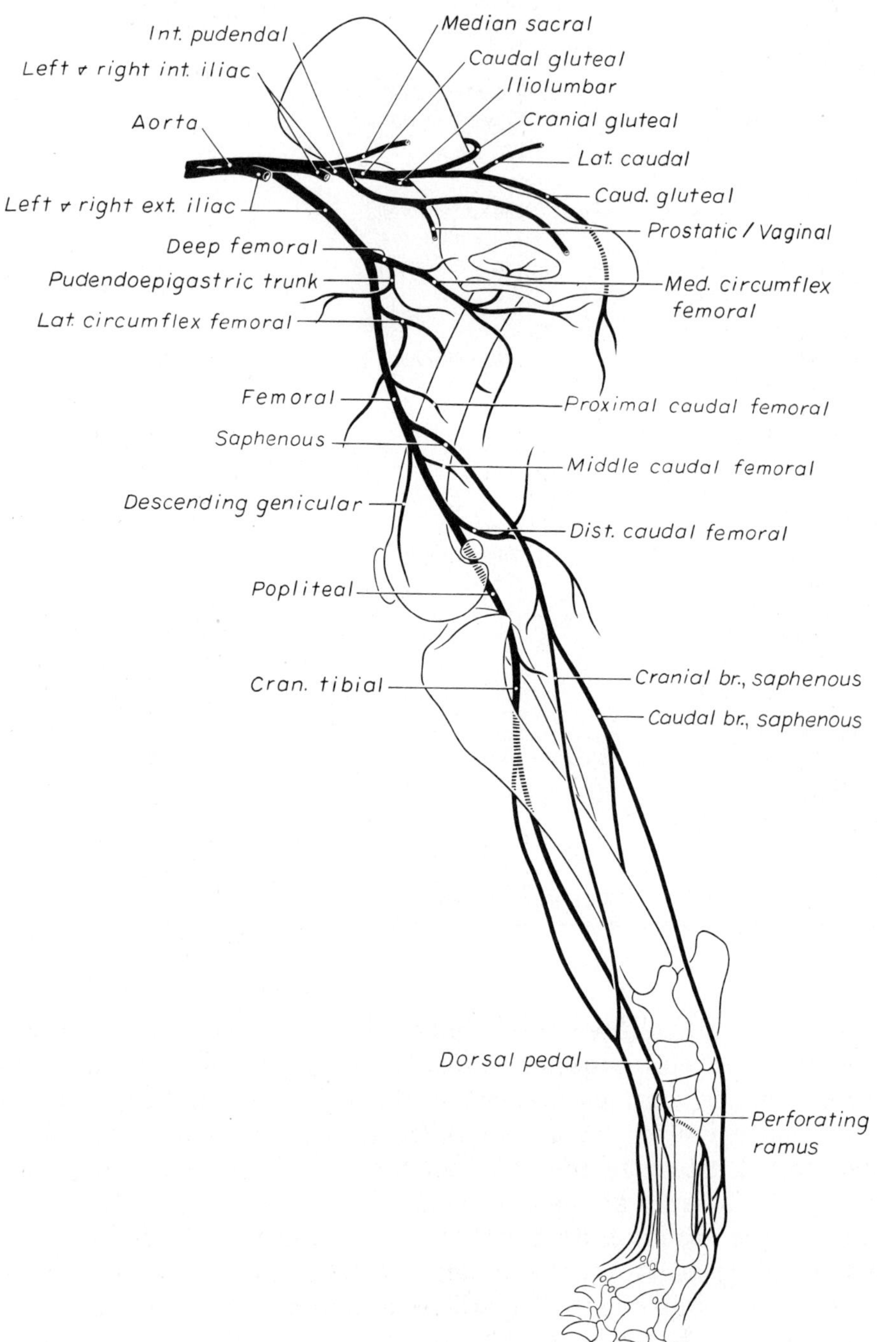

FIGURE 164. Arteries of right pelvic limb, schematic medial view.

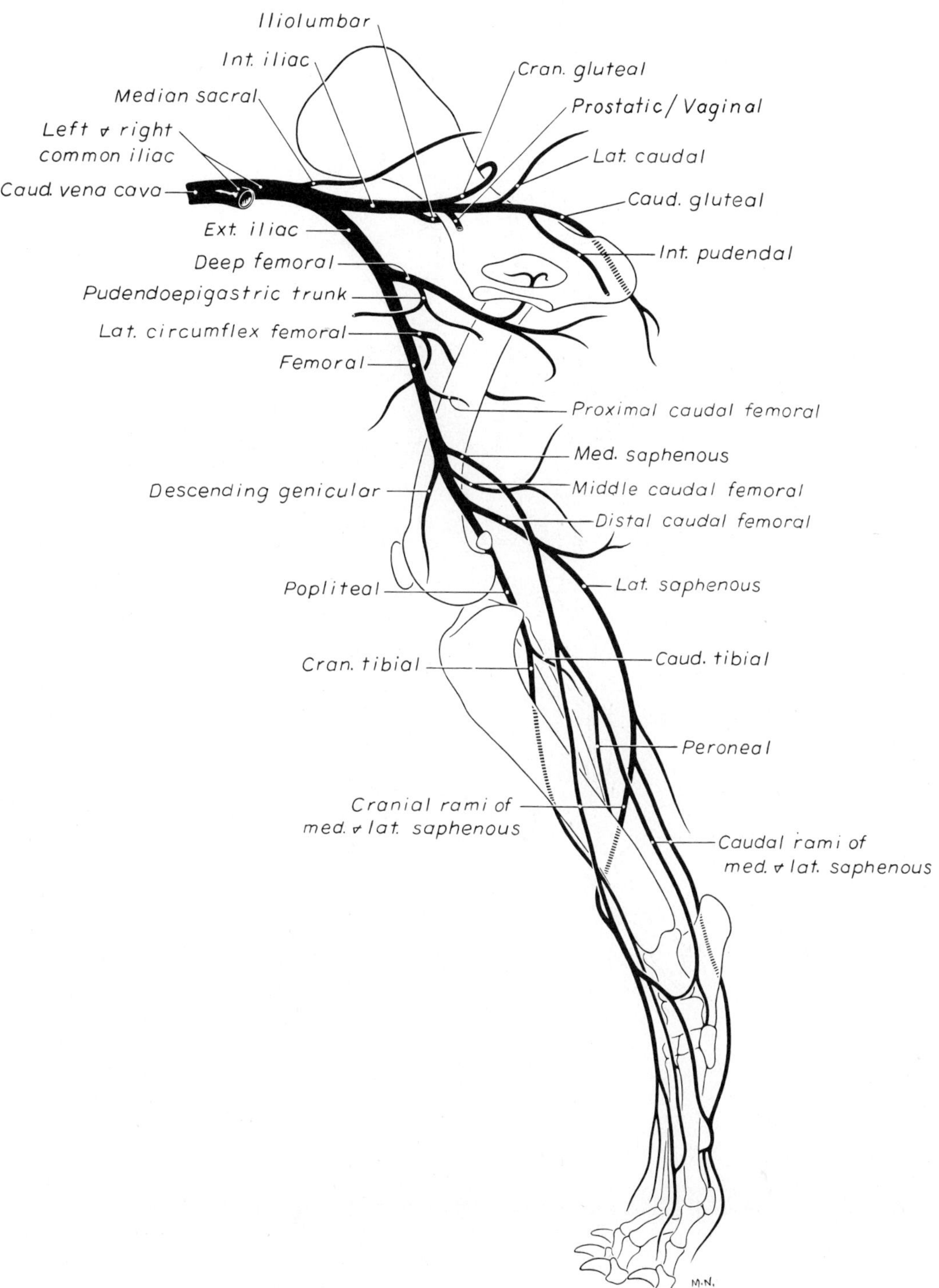

FIGURE 165. Veins of right pelvic limb, schematic medial view.

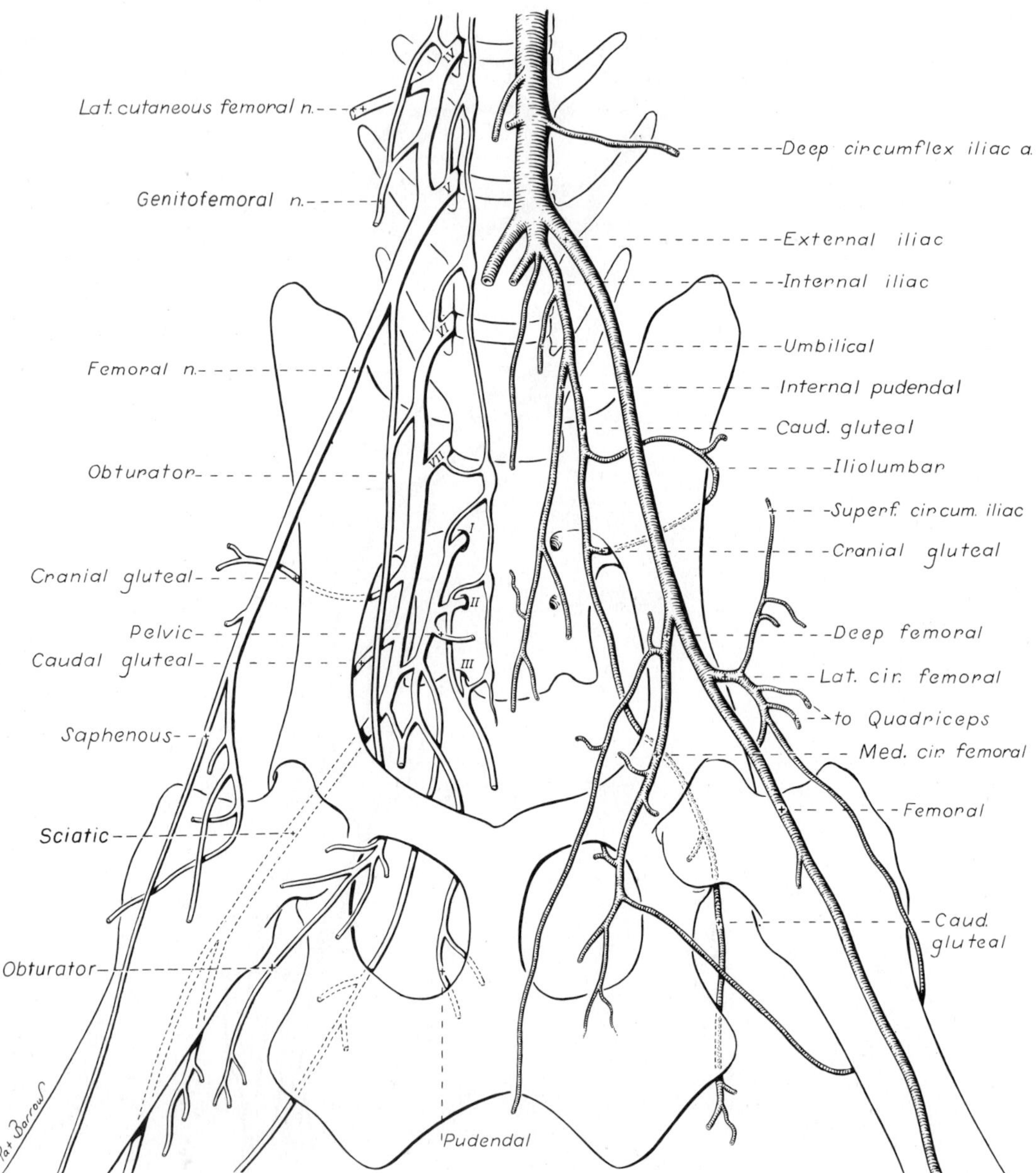

FIGURE 166. Lumbosacral nerves (left) *and arteries* (right), *ventral view.*

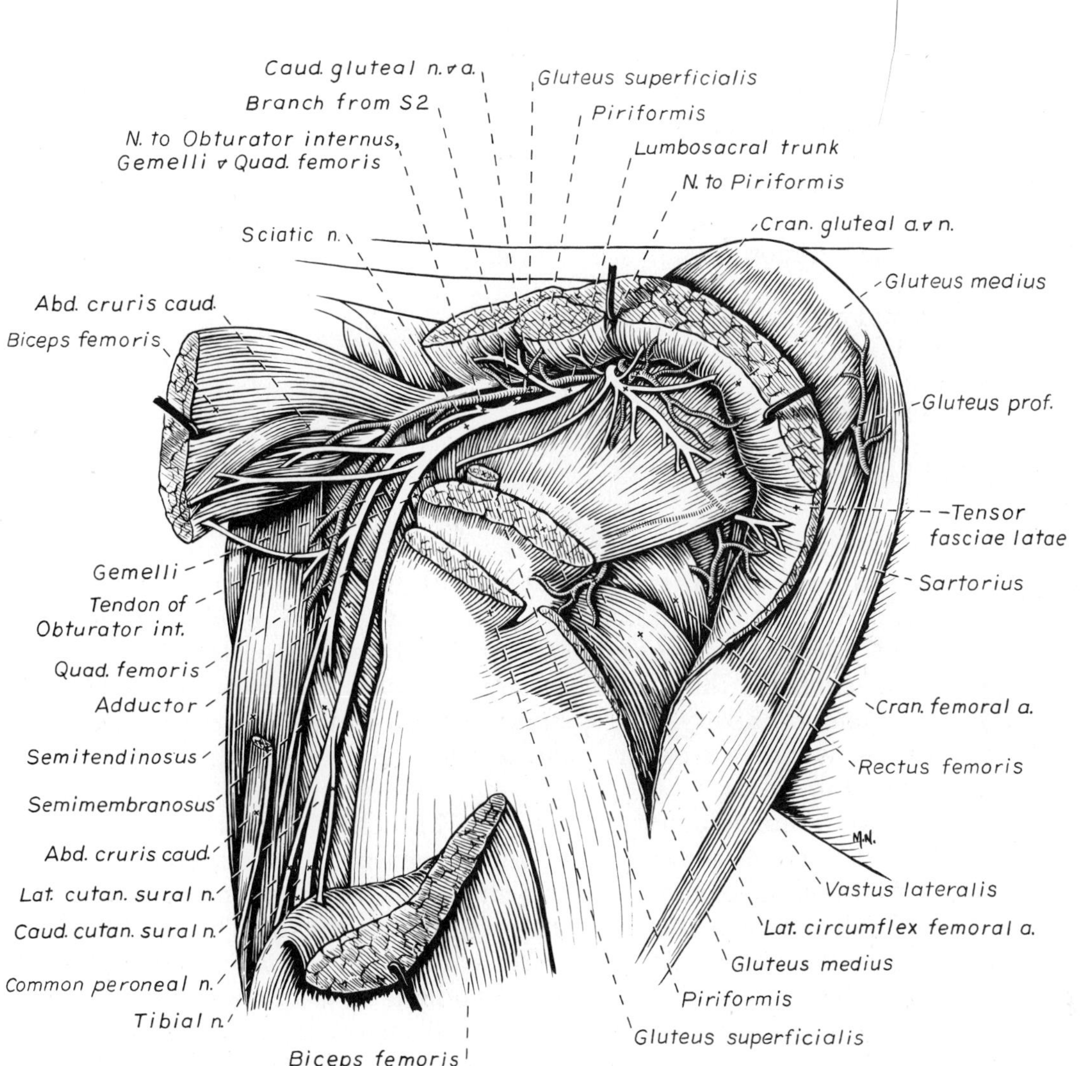

FIGURE 167. Nerves, arteries and muscles of the right hip, lateral aspect.

External Iliac Artery and Primary Branches

External Iliac a.
  Deep femoral a.
    Pudendoepigastric trunk
      Caudal epigastric a.
      External pudendal a.
    Medial circumflex femoral a.
Femoral a.
  Superficial circumflex iliac a.
  Lateral circumflex femoral a.
  Proximal caudal femoral a.
  Saphenous a.
  Descending genicular a.
  Middle caudal femoral a.
  Distal caudal femoral a.
Popliteal a.
Cranial tibial a.
Dorsal pedal a.
  Arcuate a.
    Dorsal metatarsal aa.
      Perforating br.
Caudal tibial a.

The right **external iliac artery** (Figs. 144, 146, 164, 166) arises from the aorta on a level with the sixth and seventh lumbar vertebrae. It runs caudoventrally and is related laterally near its origin to the common iliac vein and the psoas minor muscle. Farther distally it lies on the iliopsoas muscle. The external iliac upon passing through the abdominal wall becomes the femoral artery. The opening through which the external iliac artery passes is the **vascular lacuna,** located between the inguinal ligament and the pelvis.

The **deep femoral artery** is the only branch of the external iliac artery and arises inside the abdomen near the vascular lacuna and courses caudally. Two vessels leave the ventral surface of the deep femoral artery within the abdomen by a short **pudendoepigastric trunk.** These are the external pudendal and caudal epigastric arteries. The external pudendal artery passes through the inguinal canal and has already been dissected.

The **caudal epigastric artery** (Fig. 168) arises from the pudendoepigastric trunk and passes cranially on the dorsal surface of the rectus abdominis. It supplies the caudal half of the rectus abdominis and the ventral parts of the oblique and transverse muscles.

Expose the femoral artery and vein and the saphenous nerve in the **femoral triangle.** This triangle is bounded cranially by the sartorius, laterally by the vastus medialis and rectus femoris and caudally by the pectineus and adductor.

After giving off the pudendoepigastric trunk, the deep femoral artery continues as the **medial circumflex femoral artery** (Figs. 166, 168), which leaves the abdomen through the vascular lacuna. It continues

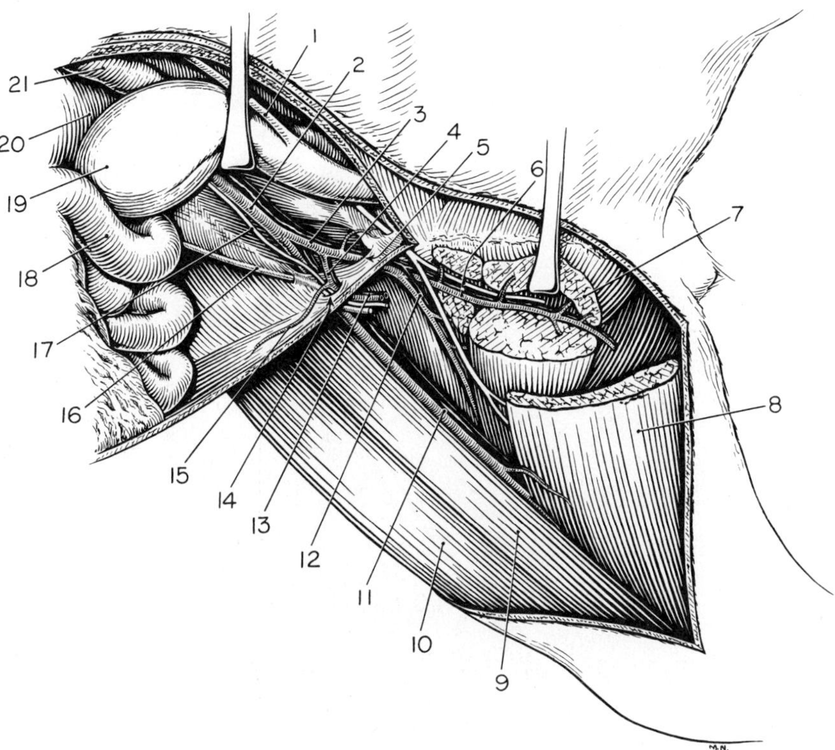

FIGURE 168. Deep femoral artery, medial view, pectineus muscle removed and adductor transected.

1. *Left ureter*
2. *External iliac artery*
3. *Deep femoral a. and v.*
4. *Pudendoepigastric trunk*
5. *Femoral lacuna*
6. *Transverse branch of medial circumflex femoral a. and v. and obturator n.*
7. *Adductor*
8. *Gracilis*
9. *Caudal part sartorius*
10. *Cranial part sartorius*
11. *Femoral a. and v.*
12. *Deep branch of medial circumflex femoral a. and v.*
13. *External pudendal a. and v.*
14. *Deep inguinal ring*
15. *Caudal epigastric artery*
16. *Round ligament of uterus*
17. *Genitofemoral nerve*
18. *Small intestine*
19. *Bladder*
20. *Rectum*
21. *Uterine horn*

caudally between the quadriceps femoris and pectineus muscles and enters the adductor. Transect the pectineus at its origin and reflect it. Transect the origin of the gracilis and reflect the muscle caudally. Transect the origin of the adductor. Spare the branches of the medial circumflex femoral artery and obturator nerve that enter its cranial aspect. Remove portions of the adductor muscle to follow the distribution of the medial circumflex femoral artery.

As the medial circumflex femoral artery approaches the large adductor muscle, it gives off a **deep branch** that descends distally between the adductor and vastus medialis muscles, both of which it supplies. Small branches of the medial circumflex femoral supply the obturator muscles and the hip joint capsule. The **transverse branch** passes caudally through

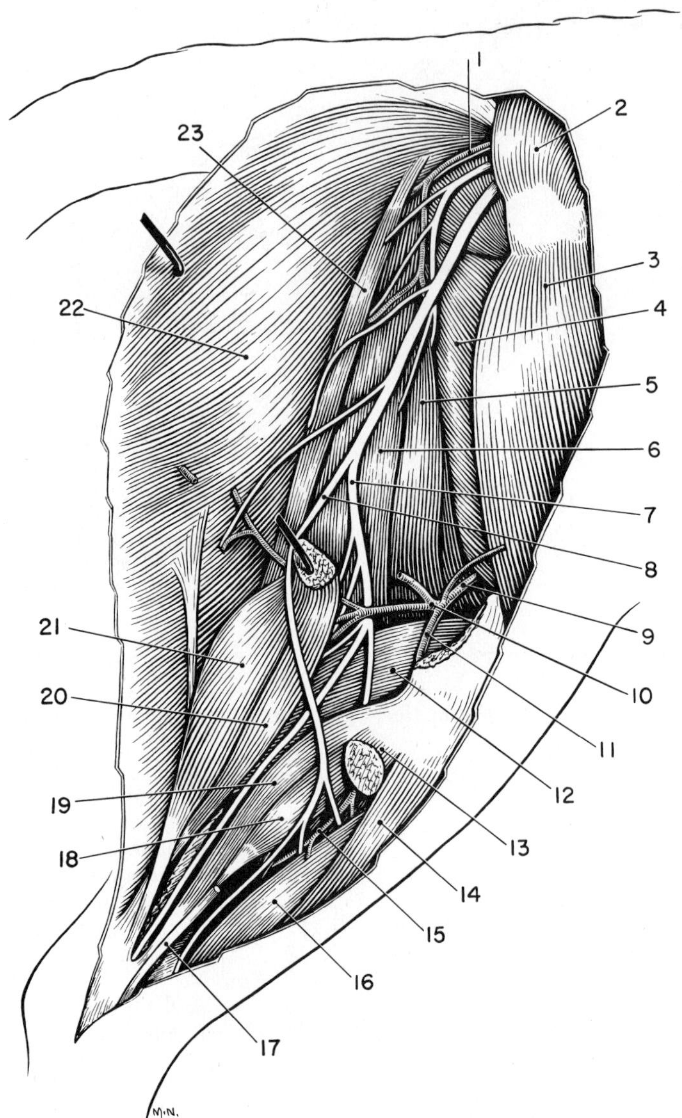

FIGURE 169. Arteries and nerves of right thigh and leg, lateral view.

1. *Caudal gluteal a.*
2. *Superficial gluteal a.*
3. *Vastus lateralis*
4. *Adductor*
5. *Semimembranosus*
6. *Semitendinosus*
7. *Tibial n.*
8. *Peroneal n.*
9. *Femoral a.*
10. *Distal caudal femoral a.*
11. *Popliteal a.*
12. *Gastrocnemius, medial head*
13. *Peroneus longus*
14. *Cranial tibial*
15. *Cranial tibial a.*
16. *Long digital extensor*
17. *Peroneus longus*
18. *Lateral digital extensor*
19. *Lateral digital flexor*
20. *Superficial digital flexor*
21. *Gastrocnemius, lateral head*
22. *Biceps femoris (reflected)*
23. *Caudal crural abductor*

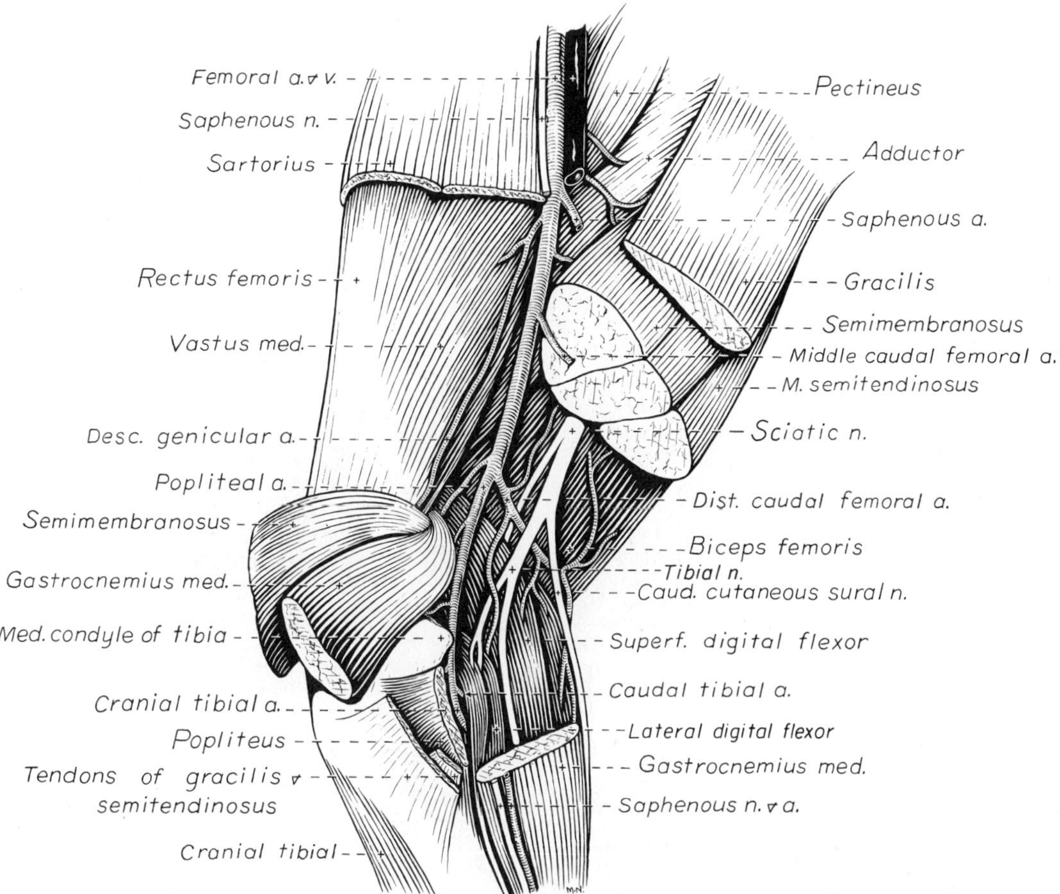

FIGURE 170. Arteries of right popliteal region, medial view.

the adductor muscle, which it supplies and terminates in the semimembranosus muscle.

The **femoral artery** (Figs. 164, 166–171) is the continuation of the external iliac artery beyond the level of the vascular lacuna. The branches of the femoral artery in the order that they arise are: superficial circumflex iliac, lateral circumflex femoral, proximal caudal femoral, saphenous, descending genicular and middle and distal caudal femoral.

The **superficial circumflex iliac artery** is a small branch that arises from the lateral side of the femoral artery near or with the lateral circumflex femoral artery. The superficial circumflex iliac artery courses cranially and supplies both parts of the sartorius, the tensor fascia lata and the rectus femoris. It becomes superficial at the cranial ventral iliac spine of the tuber coxae. Transect both parts of the sartorius over the vessel and observe its branches.

The **lateral circumflex femoral artery** (Figs. 164, 166) is a large branch that passes between the rectus femoris and the vastus medialis. Although most of the vessel arborizes in the quadriceps, it also supplies the tensor fasciae latae, the superficial and middle gluteals and the hip joint capsule.

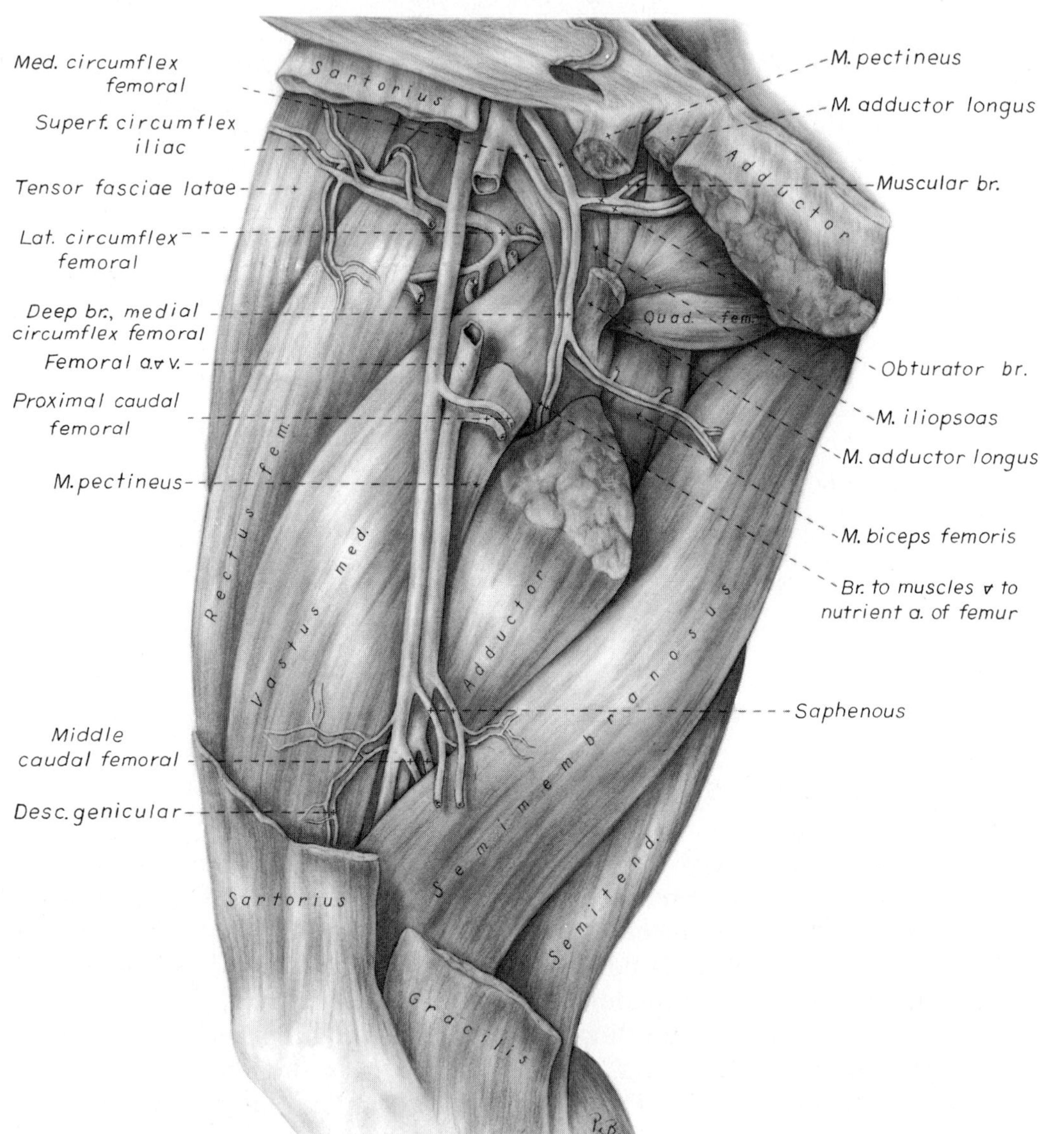

FIGURE 171. Deep structures of the right thigh, medial view.

The **proximal caudal femoral artery** (Fig. 164) leaves the caudal surface of the femoral distal to the origin of the lateral circumflex femoral in the midthigh region. It extends distocaudally over the pectineus and adductor muscles, which it supplies, and enters the deep surface of the gracilis.

Make a skin incision from the stifle to the claw of the second digit. Remove the skin as far distally as the metatarsal pad. Try to leave the subcutaneous vessels on the limb.

The **saphenous artery** (Figs. 164, 170), vein and nerve continue distally between the converging borders of the caudal part of the sartorius and gracilis. Observe that the saphenous artery arises from the femoral, proximal to the stifle. The saphenous artery supplies the skin on the medial side of the stifle and terminates in a cranial and a caudal branch.

The **cranial branch** (Figs. 164, 170, 176) of the saphenous artery arises opposite the proximal end of the tibia, obliquely crosses the medial surface of the tibia and passes distally on the cranial tibial muscle. It crosses the flexor surface of the tarsus with this muscle. At the proximal part of the metatarsus it terminates as the dorsal common digital arteries.

The **caudal branch** (Figs. 164, 178) of the saphenous artery arises at the proximal end of the tibia. It lies between the medial head of the gastrocnemius and the tibia. Distally it is related to the flexors of the digits, and with the tibial nerve, it crosses the medial plantar surface of the tarsus to enter the metatarsus. The vessel supplies branches to the tarsus and the deep structures of the proximal end of the metatarsus. In the metatarsus it supplies the paw by deep branches that contribute to a deep plantar arch that is the source of plantar metatarsal arteries and terminate in the plantar common digital arteries. These branches need not be dissected

In some specimens the saphenous (Fig. 165) veins may be congested enough to be identified. The **medial saphenous vein** is similar to the artery in its origin in the paw and termination in the femoral vein. The **lateral saphenous vein** has no comparable artery. It is formed by cranial and caudal branches in the leg that arise from venous arcades in the paw. It terminates in the distal caudal femoral vein. The **cranial branch** of the lateral saphenous vein is often used for venipuncture. It arises from an anastomosis with the cranial branch of the medial saphenous vein on the cranial aspect of the tarsus and courses proximocaudally across the lateral surface of the leg.

After the saphenous artery arises from the femoral, the latter disappears lateral to the semimembranosus. Transect and turn the distal end of the semimembranosus craniomedially and trace the femoral artery to the gastrocnemius muscle.

The **descending genicular artery** (Figs. 164, 170) arises from the femoral distal to the origin of the saphenous and supplies the medial surface of the stifle.

The **middle caudal femoral artery** (Figs. 164, 170) arises distal to the descending genicular and saphenous arteries and ramifies in the distal parts of the adductor and semimembranosus muscles.

The **distal caudal femoral artery** (Figs. 164, 169, 170) is a large vessel that arises from the caudal surface of the last centimeter of the femoral. The femoral is continued by the popliteal artery on entering the gastrocnemius. Reflect the insertions of the gracilis, semimembranosus and semitendinosus to uncover the medial head of the gastrocnemius muscle. Transect the medial head of the gastrocnemius and reflect it. This will expose the distal caudal femoral artery and its branches, which supply the biceps femoris, the semimembranosus, the semitendinosus, the gastrocnemius and the digital flexors.

The **popliteal artery** (Figs. 164, 169, 170), a continuation of the femoral, passes between the two heads of the gastrocnemius muscle, crosses the medial surface of the superficial digital flexor muscle and courses over the flexor surface of the stifle and through the popliteal notch of the tibia. It inclines laterally under the popliteus muscle and perforates the lateral digital flexor to reach the interosseous space. The popliteal artery supplies the stifle, gastrocnemius and popliteus muscles and terminates as cranial and caudal tibial arteries. The caudal tibial artery is a small vessel that leaves the caudal surface of the popliteal in the interosseous space. It need not be dissected.

Transect the popliteus where it covers the popliteal artery and follow the artery to the interosseous space.

The **cranial tibial artery** (Figs. 164, 169, 170, 177) passes between the tibia and the fibula. Reflect the fascia on the cranial aspect of the stifle and leg where it serves for the insertion of the biceps femoris. Separate the cranial tibial and long digital extensor muscles throughout their length. Observe the cranial tibial artery between these two muscles. Transect the peroneus longus muscle at its origin and reflect it to expose the cranial tibial artery emerging from the interosseous space between the tibia and fibula. It supplies the peroneus longus, long digital extensor and cranial tibial muscles. The termination of the artery will be dissected with the peroneal nerves.

The **lumbosacral plexus** (Fig. 172) consists of the ventral branches of the lumbar and sacral nerves. Of the nerves that arise from this plexus, the cranial and caudal iliohypogastric, the ilioinguinal, the lateral cutaneous femoral and the genitofemoral have been dissected. The remainder will be dissected in the order of their accessibility.

1. The **obturator nerve** (Figs. 168, 172, 173) arises from the fourth, fifth and sixth lumbar nerves. It is formed within the caudomedial portion of the iliopsoas muscle. It leaves the muscle dorsomedially, runs caudoventrally along the body of the ilium, penetrates the medial side of the levator ani muscle and leaves the pelvis by passing through the cranial part of the obturator foramen. It supplies the adductor muscles of the limb: the external obturator, the pectineus, the gracilis and the adductor. Locate this nerve on the medial side of the right ilium. Find it as it emerges ventrally from the obturator foramen and arborizes in the adductor muscles with the branches of the medial circumflex femoral artery.

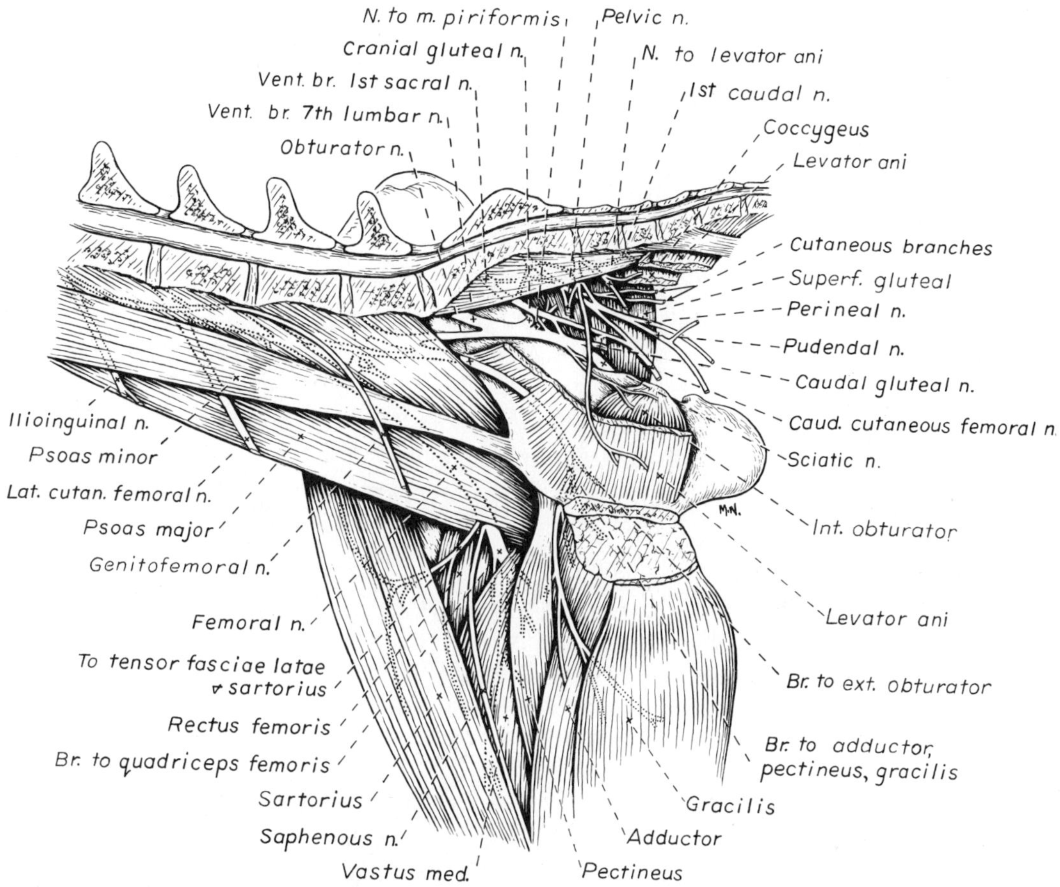

FIGURE 172. Lumbosacral plexus, left medial view.

2. The **femoral nerve** (Figs. 172, 173) arises primarily from the fourth, fifth and sixth lumbar nerves. Find the femoral nerve with the lateral circumflex femoral artery. Observe its emergence from the iliopsoas muscle. Within the iliopsoas the **saphenous nerve** arises from the cranial side of the femoral nerve. The saphenous or femoral innervates the sartorius muscle. The cutaneous portion of the saphenous supplies the skin on the medial side of the thigh, the stifle, leg, tarsus and paw. Follow this nerve as far distally as the tarsus.

The femoral nerve supplies branches to the iliopsoas muscle. It enters the quadriceps muscle between the rectus femoris and vastus medialis and supplies all four heads of the quadriceps.

In the right ischiorectal fossa, identify the coccygeus muscle. Transect the origins of the superficial gluteal muscle and the middle gluteal muscle and reflect them to their attachments to uncover the greater ischiatic notch. Transect and reflect the sacral attachment of the sacrotuberous ligament. This exposes the caudal gluteal artery and the sciatic nerve. Deep to these are the internal pudendal artery and pudendal nerve and the ventral branches of the sacral nerves. These ventral branches emerge from the two pelvic sacral foramina and the sacrocaudal intervertebral foramen.

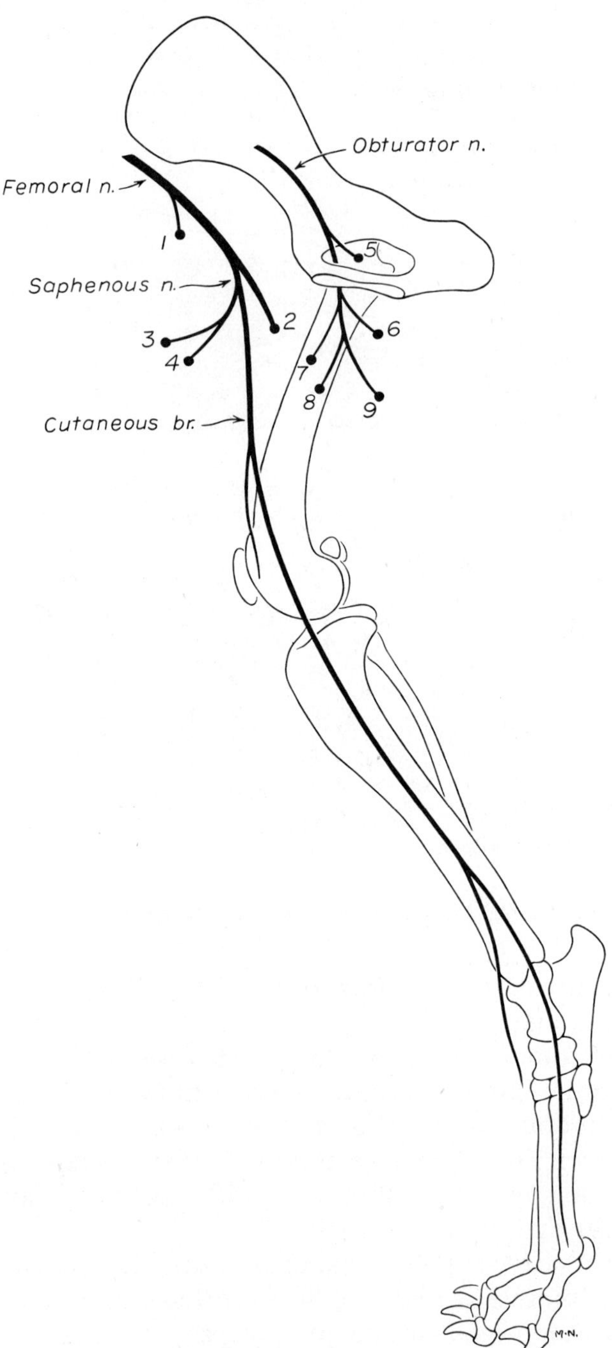

FIGURE 173. Distribution of saphenous, femoral and obturator nerves of right pelvic limb, schematic medial aspect.

Muscles innervated by numbered nerves.

Femoral nerve
1. *Iliopsoas*
2. *Quadriceps femoris*

Saphenous nerve
3. *Sartorius, cranial part*
4. *Sartorius, caudal part*

Obturator nerve
5. *External obturator*
6. *Adductor longus*
7. *Pectineus*
8. *Adductor magnus and brevis*
9. *Gracilis*

3. The **pudendal nerve** (Figs. 172, 174) arises from all three sacral nerves. It passes caudolaterally, where it lies lateral to the levator ani and coccygeus muscles, medial to the superficial gluteal muscle and dorsal to the internal pudendal vessels. It appears superficially in the ischiorectal fossa after emerging from the medial side of the superficial gluteal muscle and courses caudomedially towards the pelvic symphysis at the ischial arch. The following branches arise from the pudendal nerve:

a. The **caudal rectal nerve** may arise from sacral nerves or may

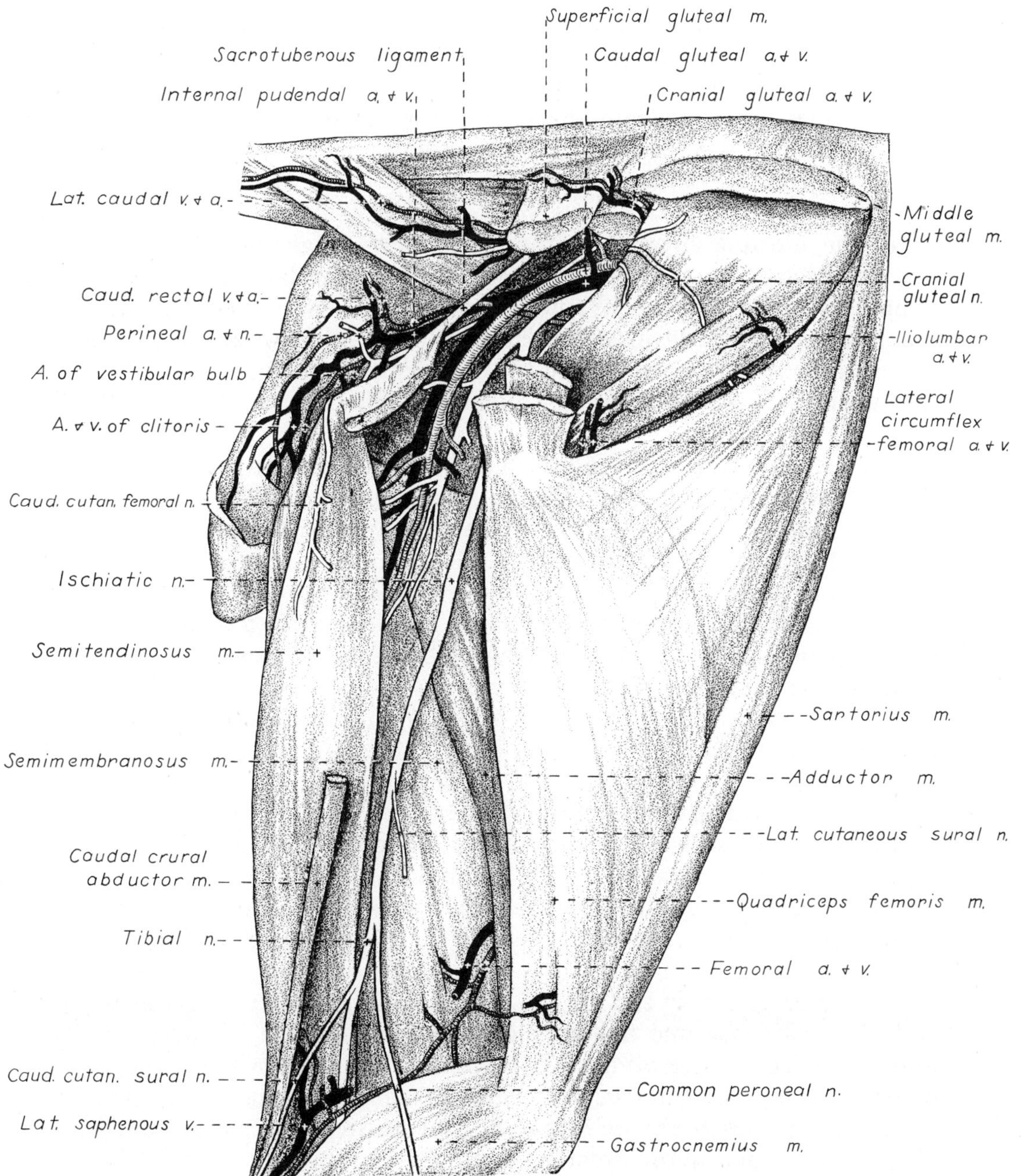

FIGURE 174. Vessels and nerves of right thigh and perineum, lateral view.

leave the pudendal nerve at the caudal border of the levator ani muscle. It innervates the external anal sphincter. It need not be dissected.

b. The **perineal nerves** arise from the dorsal surface of the pudendal nerve. They supply the skin of the anus and the perineum and continue to the scrotum or labium. Short nerves from the pudendal or perineal nerves supply the muscles of the penis or the vestibule and vulva.

c. The **dorsal nerve of the penis** in the male (or of the clitoris in the female) curves around the ischial arch and reaches the dorsal surface of the penis where it courses cranially. It continues through the glans penis and ends in the skin covering the apex of the glans. It provides sensory nerves to the skin of the glans. In the female the smaller dorsal nerve of the clitoris runs ventrally to the ventral commissure of the vulva, where it terminates in the clitoris.

4. The **caudal cutaneous femoral nerve** (Figs. 172, 174) arises from the sacral plexus and is united to the pudendal for most of its intrapelvic course. The caudal cutaneous femoral nerve follows the caudal gluteal artery to the level of the ischiatic tuberosity, where it becomes superficial near the attachment of the sacrotuberous ligament and terminates in the skin on the proximal caudal half of the thigh.

The ventral branches of the sixth and seventh lumbar nerves and the first two sacral nerves unite to form a lumbosacral trunk adjacent to the greater ischiatic notch. The nerves that arise from this trunk are the caudal gluteal, cranial gluteal and sciatic.

5. The **caudal gluteal nerve** (Figs. 166, 167, 172) passes over the ischiatic notch medial to the middle gluteal muscle and enters the medial surface of the superficial gluteal muscle. It is the sole innervation to the superficial gluteal. It has a variable origin from the seventh lumbar and first two sacral nerves.

6. The **cranial gluteal nerve** (Figs. 166, 172) passes over the greater ischiatic notch, crosses the lateral surface of the ilium at the origin of the deep gluteal muscle and innervates the middle and deep gluteal muscles and the tensor fasciae latae. It arises from the sixth and seventh lumbar and first sacral nerves. It was dissected with the cranial gluteal artery.

7. The **sciatic nerve** (Figs 166, 167, 170, 172, 174, 175) arises from the last two lumbar and first two sacral nerves. Isolate the nerve as it passes over the greater ischiatic notch. Small branches leave within the pelvis to supply the internal obturator, gemelli and quadratus femoris muscles. Do not dissect these branches. The sciatic nerve passes caudally over the hip medial to the greater trochanter and then distally, caudal to the femur on the lateral side of the adductor muscle. A branch leaves the

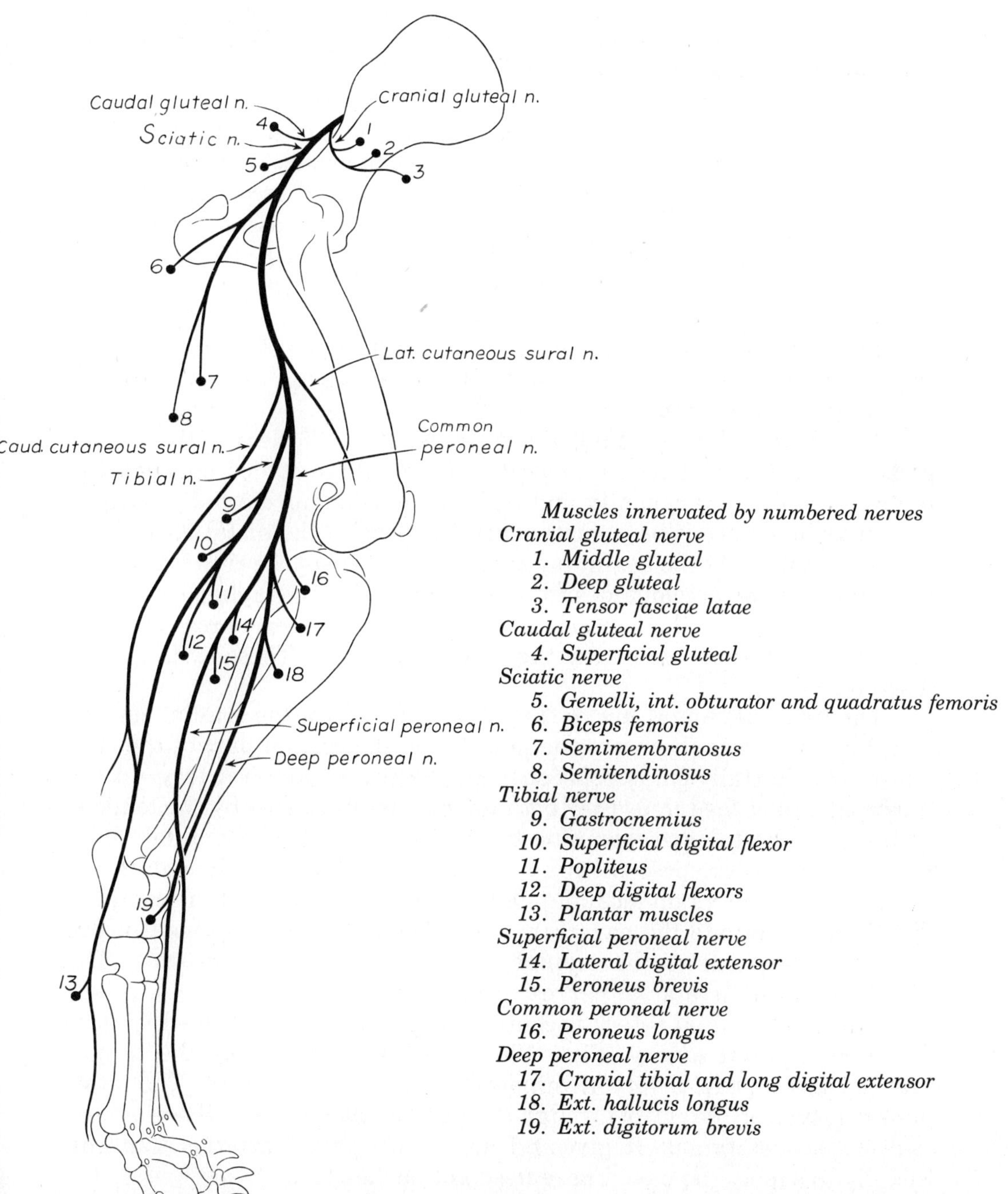

Muscles innervated by numbered nerves

Cranial gluteal nerve
1. *Middle gluteal*
2. *Deep gluteal*
3. *Tensor fasciae latae*

Caudal gluteal nerve

4. *Superficial gluteal*

Sciatic nerve

5. *Gemelli, int. obturator and quadratus femoris*
6. *Biceps femoris*
7. *Semimembranosus*
8. *Semitendinosus*

Tibial nerve

9. *Gastrocnemius*
10. *Superficial digital flexor*
11. *Popliteus*
12. *Deep digital flexors*
13. *Plantar muscles*

Superficial peroneal nerve

14. *Lateral digital extensor*
15. *Peroneus brevis*

Common peroneal nerve

16. *Peroneus longus*

Deep peroneal nerve

17. *Cranial tibial and long digital extensor*
18. *Ext. hallucis longus*
19. *Ext. digitorum brevis*

FIGURE 175. Distribution of cranial and caudal gluteal nerves and sciatic nerve of right pelvic limb, schematic lateral view.

nerve at the level of the hip and innervates the biceps femoris, semitendinosus and semimembranosus muscles.

There are **lateral** and **caudal cutaneous sural nerves** (Fig. 167) that arise from the peroneal and tibial components, respectively, of the sciatic nerve in the thigh and supply the skin on the lateral and caudal surfaces of the crus. These need not be dissected.

The sciatic nerve terminates in the thigh as the common peroneal and tibial nerves. The **common peroneal nerve** arises primarily from L6 and L7 (Figs. 169, 174, 175). It is smaller and passes laterally, deep to the thin terminal part of the biceps femoris muscle. It crosses the lateral head of the gastrocnemius muscle and the fibula and passes between the lateral digital flexor muscle caudally and the peroneus longus muscle cranially to enter the muscles on the cranial side of the crus. Here the common peroneal divides into superficial and deep peroneal nerves. These nerves innervate the flexor muscles of the tarsus and extensor muscles of the digits. These include the cranial tibial, peroneus longus and long digital extensor muscles.

The **superficial peroneal nerve** (Figs. 175, 176) leaves the lateral portion of the parent nerve below the stifle, where it lies between the lateral digital flexor caudally and the peroneus longus cranially. Expose the nerve and trace it as it curves distally, deep to the distal part of the peroneus longus. At the beginning of the distal third of the crus it becomes subcutaneous and accompanies the cranial branch of the saphenous artery. Distal to the tarsus the superficial peroneal nerve forms dorsal common digital nerves that innervate the paw (Fig. 176). These will not be dissected.

The **deep peroneal nerve** (Fig. 175) arises from the cranial surface of the parent nerve. It enters the muscles on the cranial part of the crus and courses distally in association with the cranial tibial artery. In the proximal half of the tarsus they both lie in a groove formed by the tendons of the long digital extensor muscle laterally and by the cranial tibial muscle medially. Expose them by cutting the extensor retinaculum. At the tarsus the nerve divides into dorsal metatarsal nerves that continue distally to innervate the paw (Fig. 176). Follow the deep peroneal nerve as the termination of the cranial tibial artery is now dissected. The terminal dorsal metatarsal nerves will not be dissected.

The cranial tibial artery is continued opposite the talocrural joint as the **dorsal pedal artery** (Figs. 176, 177). Branches supply the tarsus, and the dorsal pedal artery terminates in the **arcuate artery.** The latter runs transversely laterally through the ligamentous tissue at the proximal end of the metatarsus. It gives off dorsal metatarsal arteries that run distally to supply the paw. These need not be dissected.

A **perforating branch** leaves dorsal metatarsal artery II, a branch of the arcuate artery, and courses distally in the space between the second and third metatarsal bones. This perforating branch passes from the dorsal to the plantar surface of the metatarsus in the proximal end of this space. It anastomoses with the branches of the caudal branch of the

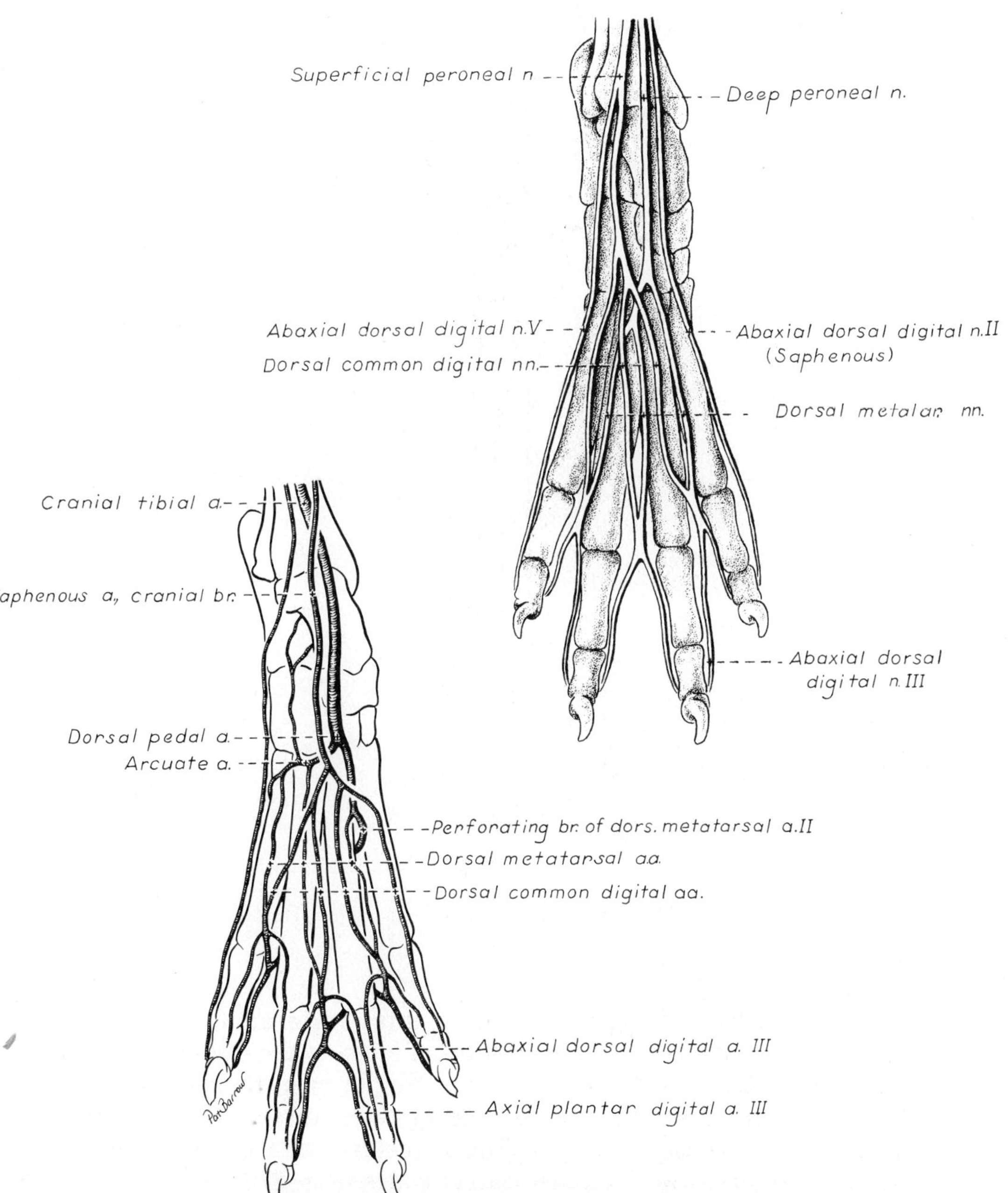

FIGURE 176. Arteries and nerves of right hind paw, dorsal view.

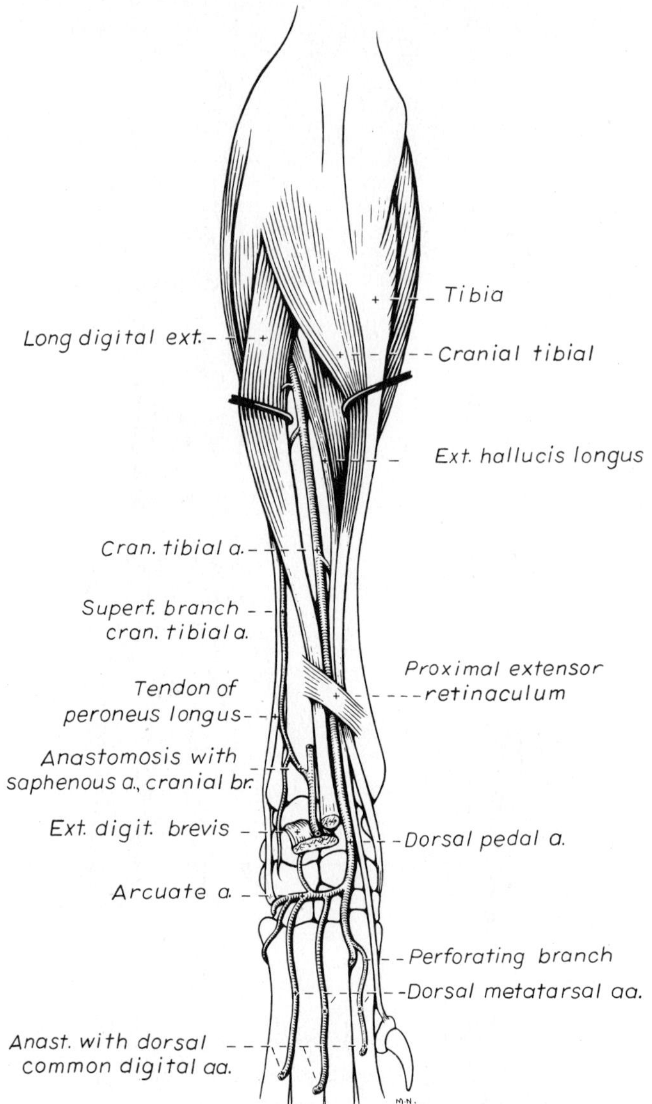

FIGURE 177. Cranial tibial artery of right pelvic limb, cranial view.

saphenous artery to contribute to the plantar metatarsal arteries that supply the paw. This is the largest source of blood to the digits of the paw. Expose the arcuate artery and the perforating branch. Remove the proximal half of the interosseous muscle covering the plantar side of the second metatarsal bone to observe the perforating metatarsal artery that emerges between the second and third metatarsal bones (Fig. 178).

The **tibial nerve** arises primarily from L7 and S1 and is the caudal portion of the sciatic nerve (Figs. 169, 179, 174, 175). It separates from the common peroneal nerve in the thigh. At the stifle it passes between the two heads of the gastrocnemius muscle. The tibial nerve supplies the muscles caudal to the tibia and fibula, which include the extensors of the tarsus and flexors of the digits, and sends branches to the stifle. It

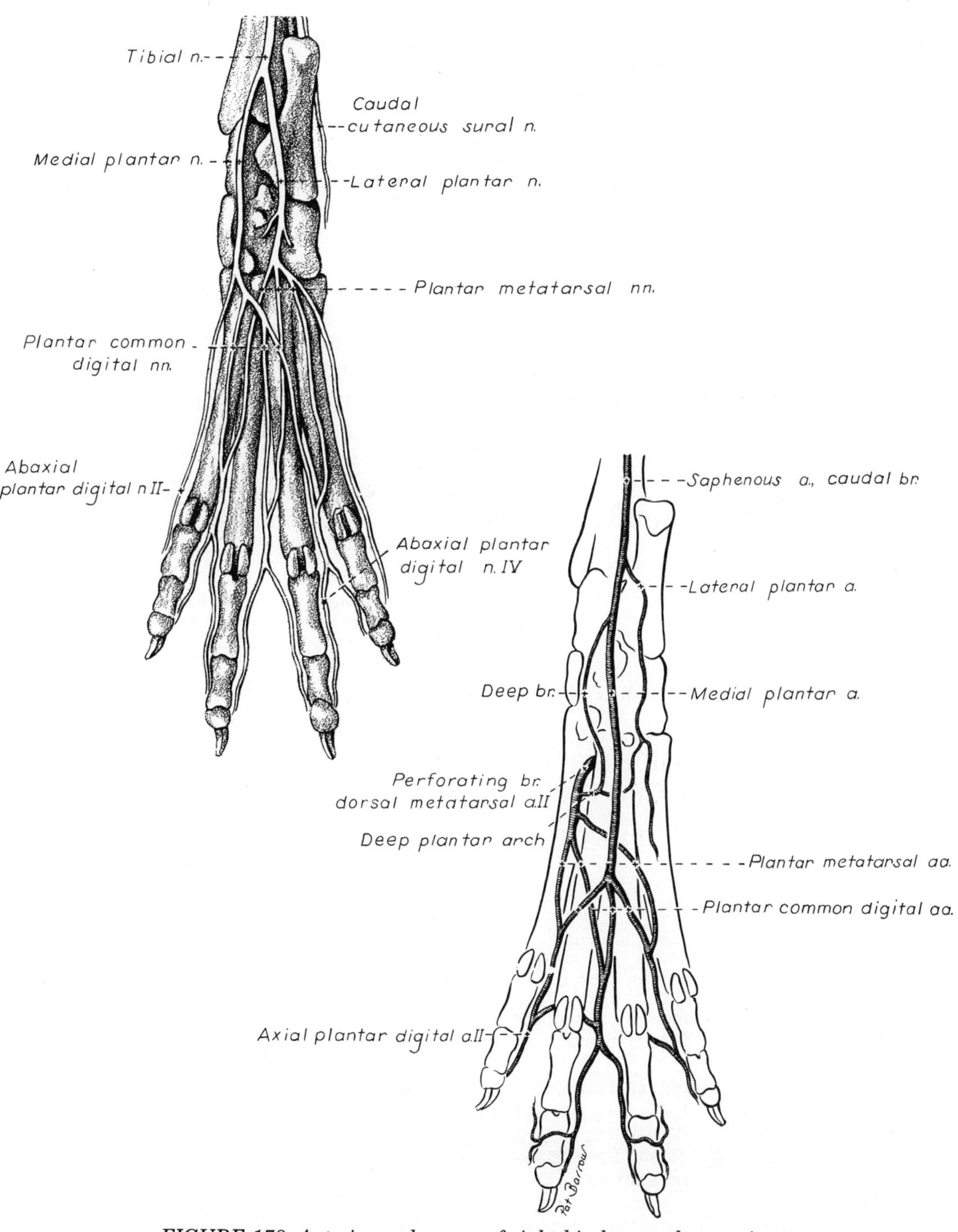

FIGURE 178. Arteries and nerves of right hind paw, plantar view.

innervates both heads of the gastrocnemius muscle and the superficial digital flexor, the popliteus and both digital flexor muscles. The tibial nerve is continued beyond these branches on the lateral digital flexor muscle. It emerges from the deep surface of the medial head of the gastrocnemius and continues distally along the medial side of the caudal surface of the tibia. Proximal to the tarsocrural joint the tibial nerve divides into the medial and lateral plantar nerves. These cross the tarsus medial to the tuber calcanei and terminate as the plantar common digital and plantar metatarsal nerves, which are sensory to the paw.

Blood supply and innervation of the digits of the pelvic limb are summarized in Table 4.

Live Dog

Place the palm of your hand over the cranial thigh with your fingers on the medial side and palpate the borders of the femoral triangle. Feel the pulse in the femoral artery. This is the most common site to determine the pulse rate and quality in a physical examination. The pulse can also be felt at two other sites in the pelvic limb. One is where the cranial branch of the saphenous artery crosses the medial side of the middle third of the tibia. Both the bone and the artery are subcutaneous here. The

Table 4. *BLOOD SUPPLY AND INNERVATION OF THE DIGITS OF THE PELVIC LIMB*

	Dorsal Surface		
Vessels			
(Superficial)	Cranial br. saphenous	Dorsal common digitals	Axial or abaxial dorsal
(Deep)	Cranial tibial Dorsal pedal—Arcuate	Dorsal metatarsals	digital a., v.
Nerves			
(Superficial)	Superficial peroneal	Dorsal common digitals	Axial or abaxial dorsal
(Deep)	Deep peroneal	Dorsal Metatarsals	digital n.
	Plantar Surface		
Vessels			
(Superficial)	Caudal br. saphenous Medial plantar	Plantar common digitals	Axial or abaxial plantar
(Deep)	Deep plantar arch Perforating ramus from dorsal pedal; lateral plantar from caudal br. saphenous	Plantar metatarsals	digital a., v.
Nerves			
(Superficial)	Tibial–Medial plantar	Plantar common digitals	Axial or abaxial plantar
(Deep)	Tibial–Lateral plantar	Plantar metatarsals	digital n.

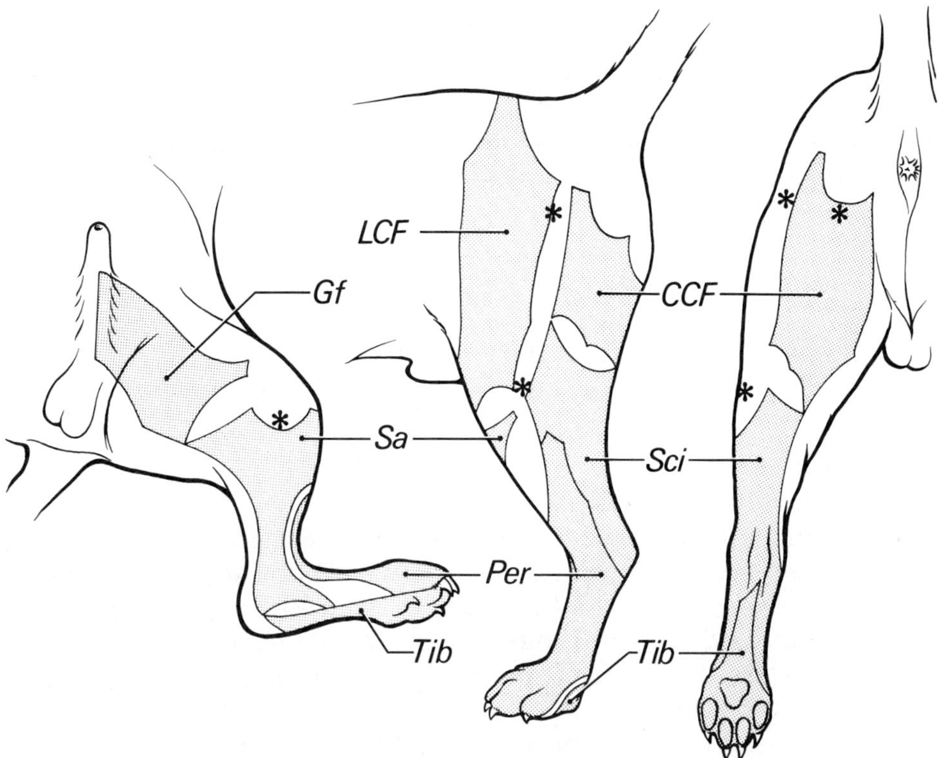

FIGURE 179. Autonomous zones of cutaneous innervation of the pelvic limb. Medial, lateral and caudal aspects. Caudal cutaneous femoral (CCF), genitofemoral (Gf), lateral cutaneous femoral (LCF), peroneal (Per), saphenous (Sa), sciatic (Sci), tibial (Tib). Asterisks indicate palpable bony landmarks—medial and lateral tibial condyles, greater trochanter and lateral end of tuber ischiadicum (from R. L. Kitchell). The sciatic nerve autonomous zone is for lesions proximal to the greater trochanter and includes the zones for the peroneal and tibial nerves. For sciatic nerve lesions caudal to the femur, the autonomous zone varies, depending on how many of its cutaneous branches are affected.

other is from the dorsal pedal artery where it crosses the dorsal surface of the tarsus.

Trace the course of the sciatic nerve and appreciate where the nerve is close to bones that can fracture and injure it. The common peroneal nerve is palpable where it crosses the proximal end of the fibula. Feel the head of the fibula and stroke the skin from proximal to distal to roll the nerve on the bone here. The tibial nerve can be felt proximal to the tarsus between the skin layers cranial to the common calcanean tendon. It is accompanied by saphenous vessels here.

Study Figure 179 and identify the autonomous zones of the nerves in the pelvic limb on the live dog.

THE HEAD

THE SKULL

The **skull,** or cranium, is a complex of bones formed in both membrane and cartilage around the brain, sense organs and entrances to the digestive and respiratory systems. The braincase is formed by a dermal-bone roof, the **calvaria,** fused with cartilage-bone walls and floor. Articulated with the braincase are the bones of the face, jaws and palate, the ear ossicles and the hyoid apparatus.

Various skull elements fuse with each other during development or have been lost phylogenetically, with the result that each species has distinctive features that may not be present in others. The dog's skull varies more in shape among the different breeds than does the skull of other species of domestic animals.

Dorsal and Lateral Surfaces of the Skull (Figs. 180–182)

Braincase. The paired frontal and parietal bones form the dorsum of the braincase or calvaria. The **parietal bone** joins the frontal bone rostrally and its fellow medially. Caudally the parietal bone meets the occipital bone, which forms the caudal surface of the skull. The unpaired interparietal bone fuses with the occipital bone prenatally and appears as a process extending rostrally. The ventral border of the parietal bone joins the squamous temporal and basisphenoid bones. Rostral to the parietal bone is the **frontal bone,** which forms the dorsomedial part of the orbit.

The **sagittal crest** is a median ridge formed by the parietal and interparietal bones. It varies in height and may be absent.

In most brachycephalic breeds the sagittal crest is replaced by a pair of sagittal **temporal lines.** Each extends from the **external occipital protuberance** to the zygomatic process of the frontal bone. The **nuchal crest** is a transverse ridge that marks the transition between the dorsal and caudal surfaces of the skull. The external occipital protuberance is median in position at the caudal end of the sagittal crest.

On each side of the dorsum of the skull is the **temporal fossa.** The fossa is convex. It is bounded medially by the sagittal crest or the temporal line, caudally by the nuchal crest and ventrally by the zygomatic process of the temporal bone. The temporal fossa is continuous rostrally with the orbit. The temporal muscle arises from this temporal fossa on the frontal and parietal bones.

Facial Bones. The facial part of the dorsal surface of the skull is formed by parts of the frontal, nasal, maxillary and incisive bones. All are paired. The **nasal bone** meets its fellow on the midline, the frontal bone caudally and the maxilla and incisive bone laterally. The **maxilla** contains the upper cheek teeth. The **incisive bone** bears the three incisor teeth and has a long nasal process that articulates with the maxilla and nasal bone. The **nasal aperture** is bounded by the incisive and nasal

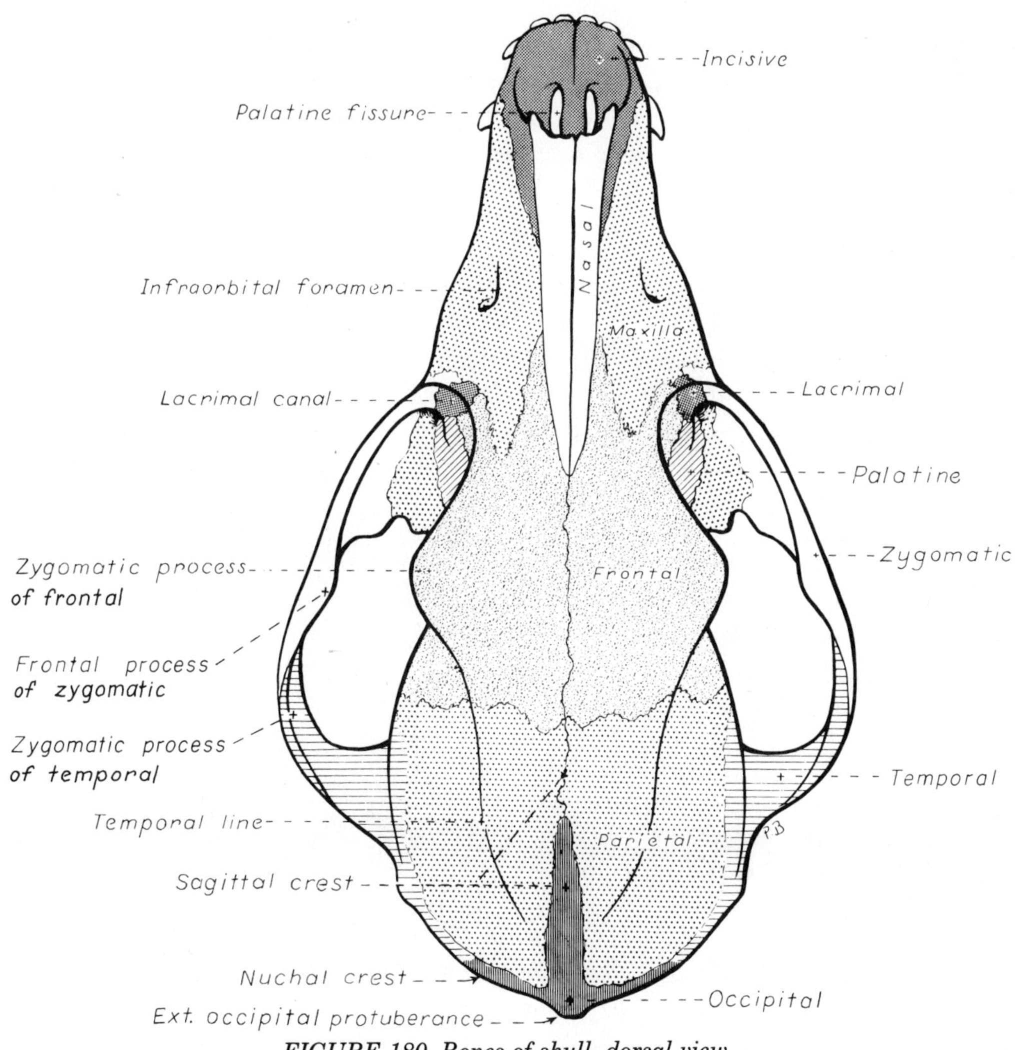

FIGURE 180. Bones of skull, dorsal view.

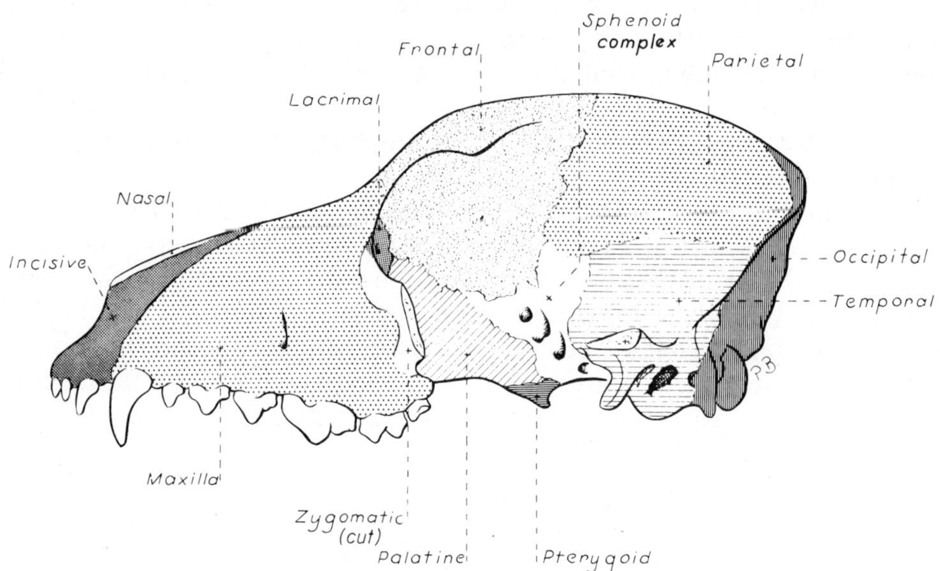

FIGURE 181. Bones of skull, lateral view, zygomatic arch removed.

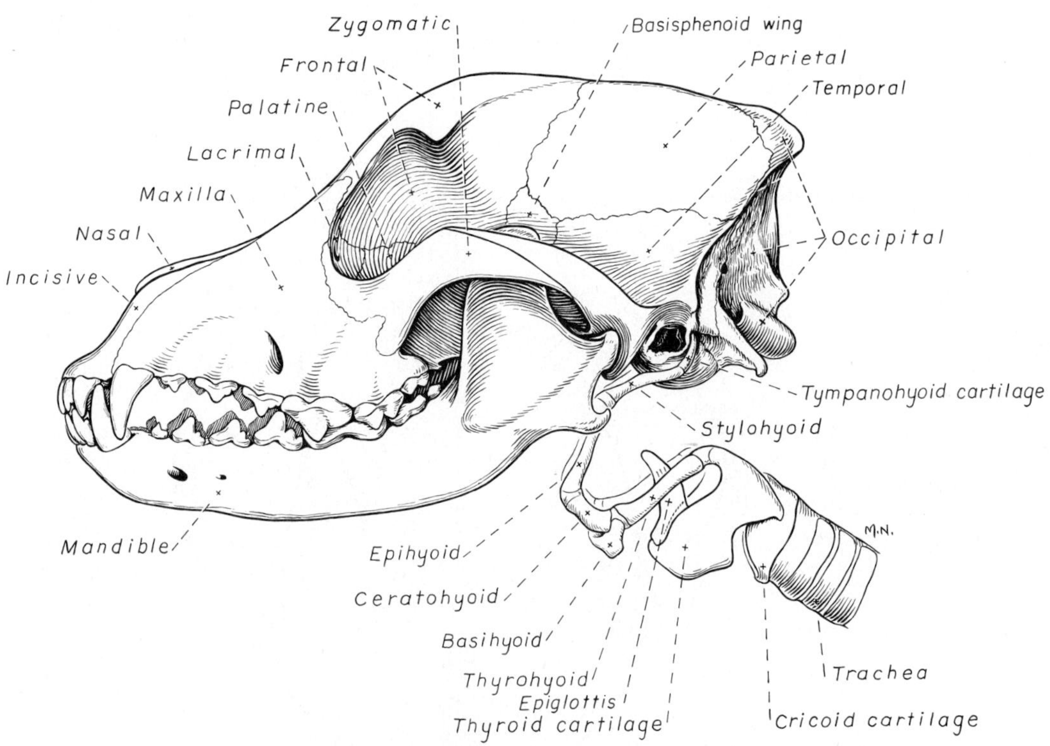

FIGURE 182. Skull, hyoid apparatus and larynx, lateral view.

bones. It is nearly circular in brachycephalic breeds and is oval in the dolichocephalic breeds.

Lateral Surface of the Skull

Braincase. Lateral portions of the frontal and parietal bones form the lateral surface (Figs. 181, 182) of the braincase. The caudoventral part of the lateral surface of the skull is composed largely of the **temporal bone.** This compound bone is composed of squamous, tympanic and petrous parts.

The **squamous part** forms the ventral portion of the temporal fossa and bears a **zygomatic process,** which forms the caudal part of the zygomatic arch. It articulates dorsally with the parietal bone, rostrally with the basisphenoid wing and caudally with the occipital bone.

Rostral to the squamous temporal bone the ventrolateral surface of the braincase is formed by the wings of the **basisphenoid** and **presphenoid.** The wing of the basisphenoid articulates caudally with the squamous part of the temporal bone, dorsally with the parietal and frontal bones and rostrally with the wing of the presphenoid.

Facial Bones. The **orbit** is the cavity in which the eye is situated. A portion of it is bony. The **orbital margin** is formed by the frontal, lacrimal and zygomatic bones. The lateral margin of the orbit is formed by the **orbital ligament,** which extends from the frontal process of the zygomatic bone to the zygomatic process of the frontal bone. The medial wall of the orbit is formed by the orbital surfaces of the frontal, lacrimal, presphenoid and palatine bones.

The **zygomatic arch** is formed by the zygomatic process of the maxilla, the zygomatic bone and the zygomatic process of the temporal bone. The arch forms the cheekbone and serves as an origin for the masseter muscle, which closes the jaws.

The **pterygopalatine fossa** is located ventral to the orbit. The maxilla, palatine bone and zygomatic bone bound the rostral part. The caudal part is bounded by the palatine and pterygoid bones and the wings of the sphenoid bones. The pterygoid muscles arise from this fossa.

The three openings in the caudal part of the orbit are, from rostral to caudal, the **optic canal,** the **orbital fissure** and the **rostral alar foramen.** The optic canal passes through the presphenoid, and the rostral alar foramen passes through the basisphenoid. The orbital fissure is formed in the articulation between the basisphenoid and presphenoid bones. The optic nerve passes through the optic canal. Passing through the orbital fissure are the oculomotor, trochlear, abducent and ophthalmic nerves and some vessels. Emerging from the rostral alar foramen are the maxillary artery and the maxillary nerve.

In the rostral part of the pterygopalatine fossa are several foramina. The **caudal palatine foramen** and **sphenopalatine foramen** are closely related openings of about equal size located on the rostromedial wall of the fossa. The sphenopalatine foramen is dorsal to the caudal palatine.

The major palatine artery, vein and nerve enter the palatine canal through the caudal palatine foramen and together course to the hard palate. The sphenopalatine artery and vein and the caudal nasal nerve enter the nasal cavity via the sphenopalatine foramen. Rostrolateral to these is the **maxillary foramen,** the caudal opening of the infraorbital canal. The infraorbital artery, vein and nerve course rostrally through this canal. A small part of the rostromedial wall of the pterygopalatine fossa, just caudal to the maxillary foramen, often presents an opening that is normally occupied by a thin plate of bone, which serves as the origin of the ventral oblique eye muscle. Caudal to the maxillary foramen are a number of small openings, most of them for the small nerves and vessels that pass through their respective **alveolar canals** to the roots of the last two cheek teeth and the caudal root of the last premolar. Above the maxillary foramen in the lacrimal bone is the shallow **fossa for the lacrimal sac.** The fossa is continued by the nasolacrimal canal for the nasolacrimal duct.

The facial part of the lateral surface of the skull rostral to the orbit includes the lateral surface of the maxilla and the incisive bone. Dorsal to the third premolar tooth is the **infraorbital foramen,** the rostral opening of the infraorbital canal. The roots of the cheek teeth produce lateral elevations, the **alveolar juga.**

Ventral Surface of the Skull (Figs. 183, 184)

Braincase. The ventral aspect of the braincase consists of the basioccipital bone, tympanic and petrosal parts of the temporal bone, the basisphenoid bone and the presphenoid bone. The basioccipital bone forms the caudal third of the cranial base. It articulates laterally with the tympanic and petrous parts of the temporal bone and rostrally with the body of the basisphenoid. Caudally the **occipital condyle** articulates with the atlas. The **paracondylar process,** a ventral projection of the occipital bone, articulates with the caudolateral part of the tympanic bulla. The digastricus muscle arises from the paracondylar process.

The **tympanic part** of the temporal bone has a bulbous enlargement, the **tympanic bulla,** which encloses the middle ear cavity (Fig. 185). On the lateral side of the bulla is the **external acoustic meatus.** In life the tympanic membrane closes this opening, and the annular cartilage of the external ear attaches around its periphery.

The **petrosal part** of the temporal bone contains the membranous and bony labyrinths of the inner ear (Fig. 185). The major portion of this bone is visible inside the cranial cavity. The tympanic bulla has been removed on one side. This exposes a barrel-shaped eminence, the **promontory,** on the ventral surface of the petrosal bone. The promontory contains the **cochlear window,** which is closed in life by a membrane. The **vestibular window** lies dorsal to the promontory and contains the foot-plate of the **stapes.** The stapes articulates with the **incus,** which in turn articulates with the **malleus.** The malleus is attached to the medial side of the tympanic membrane. The **mastoid process** is the only part of the petrosal bone to reach the exterior. It is small and lies caudal to the external acoustic meatus lateral and dorsal to the root of the prominent paracondylar process. The cleidomastoideus and sternomastoideus muscles terminate on the mastoid process.

The **basisphenoid** articulates caudally with the basioccipital and rostrally with the presphenoid and pterygoid bones. The oval foramen, round foramen and alar canal pass through the basisphenoid bone. The **presphenoid** articulates caudally with the basisphenoid and pterygoid, laterally with the perpendicular part of the palatine and rostrally with the vomer. Only a small median portion of the presphenoid is exposed on the ventral surface of the braincase. The optic canals pass through the orbital wing of the presphenoid.

The **rostral alar foramen,** caudoventral to the orbital fissure, is the rostral opening of the **alar canal.** The caudal opening of this short canal

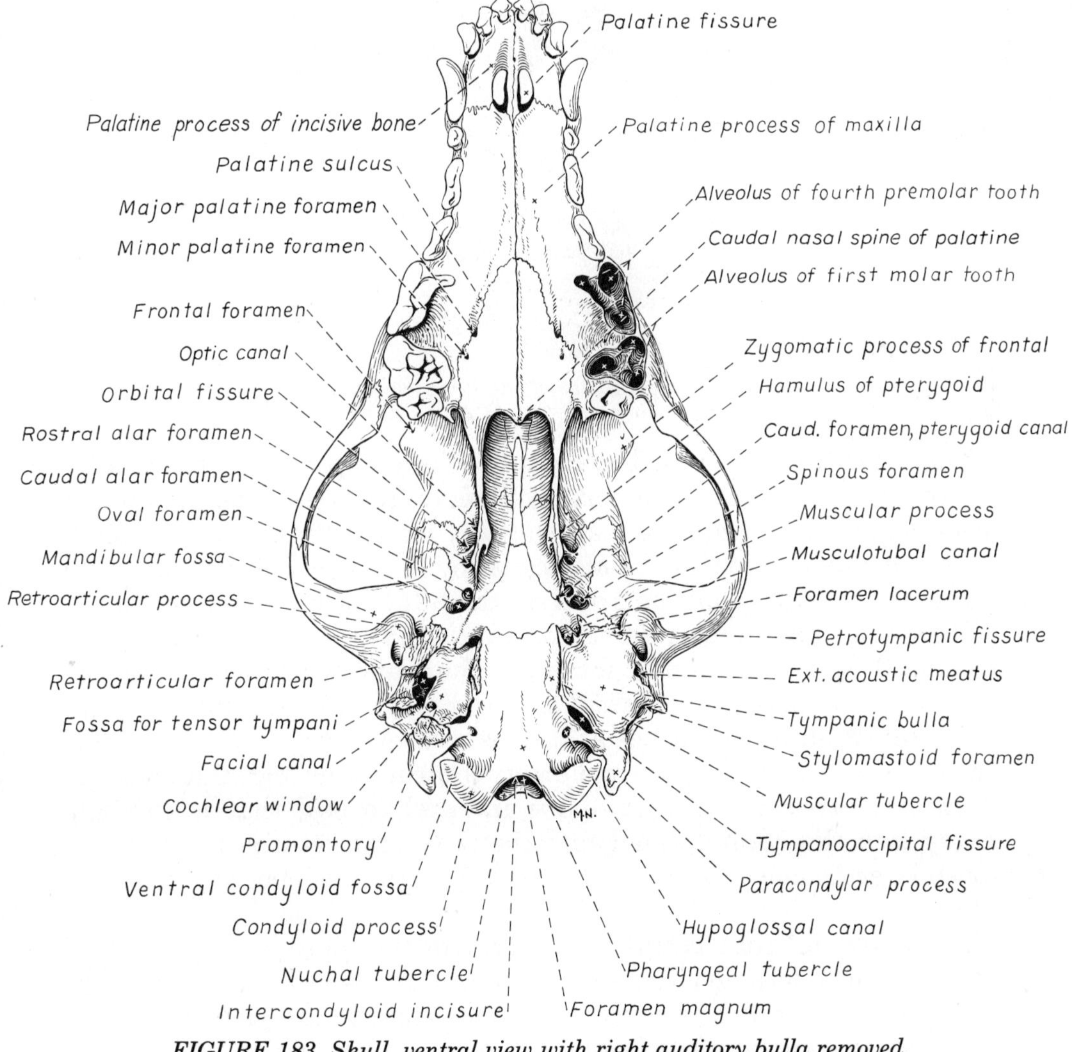

FIGURE 183. Skull, ventral view with right auditory bulla removed.

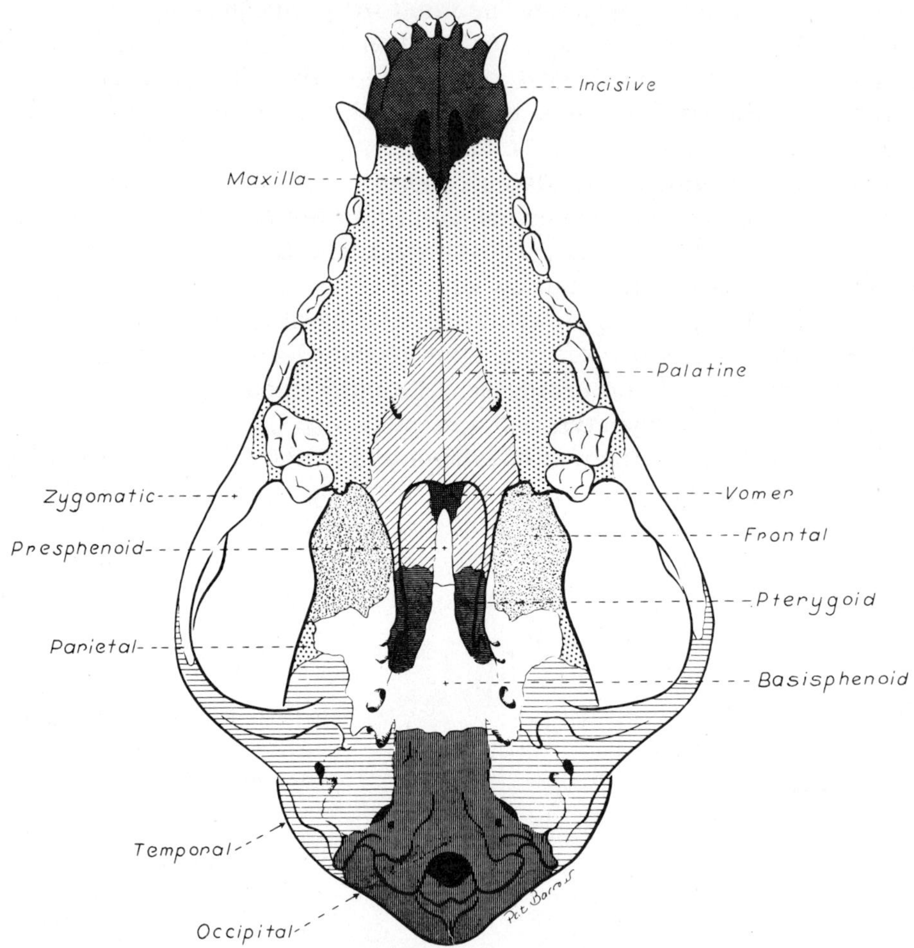

FIGURE 184. Bones of skull, ventral view.

is the **caudal alar foramen.** The **round foramen** opens from the cranial cavity into the alar canal. The maxillary nerve from the trigeminal nerve enters the alar canal from the cranial cavity via this foramen. The nerve courses rostrally and leaves the alar canal by the rostral alar foramen. In addition, the maxillary artery traverses the full length of the alar canal.

The **oval foramen,** a direct opening into the cranial cavity, is caudolateral to the caudal alar foramen. The mandibular nerve from the trigeminal nerve leaves the cranial cavity through this opening.

The **foramen lacerum** lies at the rostromedial edge of the tympanic bulla. A loop of the internal carotid artery protrudes through this opening. This loop is between the part of the internal carotid that is coursing rostrally into the carotid canal and the part that returns through the foramen lacerum and enters the cavernous sinus on the floor of the cranial cavity.

The **musculotubal canal** lies lateral to the foramen lacerum and caudal to the oval foramen. It is the bony enclosure of a tubular connection, the auditory tube, which runs from the middle ear to the pharynx.

The **tympano-occipital fissure** is an oblong depression between the basilar part of the occipital bone and the tympanic part of the temporal bone. The petro-occipital canal and the carotid canal leave the depths of the fissure at about the same place. The carotid canal transmits the internal carotid artery. The petro-occipital canal transmits the ventral petrosal venous sinus. Neither canal can be adequately demonstrated on an articulated skull. The glossopharyngeal, vagus and accessory nerves course peripherally from the jugular foramen through the tympano-occipital fissure. Also passing through this fissure are the internal carotid artery, venous radicles of the vertebral and internal jugular veins and sympathetic postganglionic axons from the cranial cervical ganglion.

The **hypoglossal canal,** for the passage of the hypoglossal nerve, lies caudal to the petro-occipital fissure in the occipital bone.

The **mandibular fossa** of the zygomatic process of the temporal bone articulates with the condyle of the mandible to form the temporomandibular joint. The **retroarticular process** forms the caudal wall of the mandibular fossa. The **retroarticular foramen** caudal to this process conducts the retroarticular vein from the temporal venous sinus.

Between the tympanic bulla and the mastoid process of the temporal bone is the **stylomastoid foramen.** This is the opening of the facial canal that conducts the facial nerve peripherally through the petrosal bone.

Facial Bones. The ventral surface of the facial part of the skull is characterized by the teeth and hard palate. There is an upper (maxillary) and a lower (mandibular) dental arch. Each tooth lies in an alveolus or socket. Interalveolar septa separate the alveoli of adjacent teeth. An

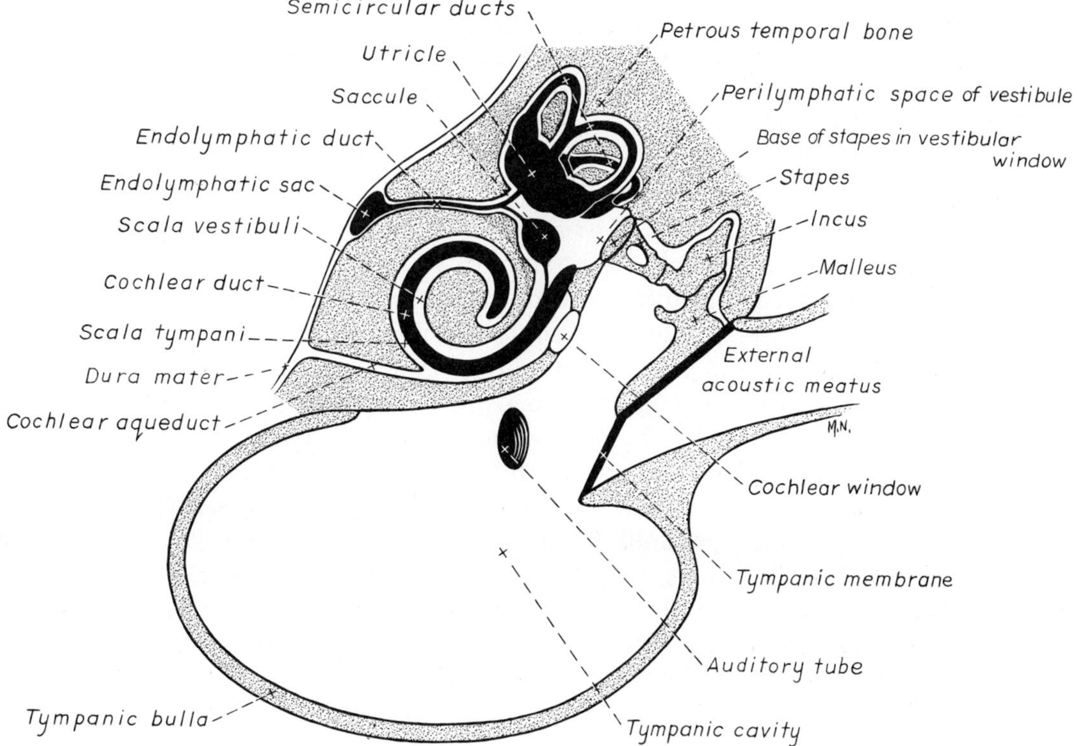

FIGURE 185. Diagram of middle and inner ear.

alveolus is subdivided by interradicular septae for those teeth with more than one root. In the upper dental arch single alveoli are present in the incisive bones for the incisor teeth and in the maxilla for the canine and first premolar tooth. There are two alveoli in the maxilla for the second premolar and two for the third. The fourth premolar and the two molars each have three alveoli in the maxilla. The **hard palate** is composed of the horizontal parts of the palatine, the maxillary and the incisive bones. An opening, the **palatine fissure,** is located on each side of the midline between the canine teeth.

The palatine bones form the caudal third of the hard palate. The **major palatine foramen** is medial to the fourth cheek tooth. Caudal to this is the **minor palatine foramen.** The major palatine artery, vein and nerve and their branches emerge through these foramina. The **choanae** are the openings of the right and left nasal cavities into the nasopharynx. They are located at the caudal end of the hard palate, where the vomer articulates with the palatine bones.

Caudal Surface of the Skull (Figs. 182, 183)

During development the occipital bone is formed by paired exoccipitals (which bear the condyles), a supraoccipital and a basioccipital. The lateral borders form a **nuchal crest** where the occipital meets the parietal and squamous temporal bones. Middorsally an **external occipital protuberance** is formed where the interparietal bone is fused to the occipital at the caudal end of the sagittal crest. The **foramen magnum** is the large opening into the cranial cavity through which the spinal cord is continued as the brain stem. The **mastoid foramen** is located in the occipitotemporal suture, dorsolateral to the occipital condyle. It transmits the caudal meningeal artery and vein. The rest of the caudal surface of the skull is roughened for muscular attachment.

If a disarticulated skull is available for study, try to locate each of the bones in the whole skull. Being able to visualize one bone in relation to another aids in radiographic interpretation of normal features. Several of the skull bones overlap each other to a considerable degree, so that it is not always possible to see the boundaries of an individual bone in situ. The presphenoid bone, through whose orbital wings the optic nerves pass, is a good example. Identify the optic canal of an intact skull and note the outline of the presphenoid sutures with adjacent bones. Orient an individual presphenoid bone in a similar position. Do the same for the more caudally located basisphenoid bone with its temporal wings, and the basioccipital bone which is part of the occipital ring. These three bones form the basicranial axis of the skull upon which the brain rests. The cribriform plate with the attached ethmoturbinates would be located at the rostral end of the basicranial axis.

With the aid of the diagram of the disarticulated skull (Fig. 186), locate the other bones on the intact skull.

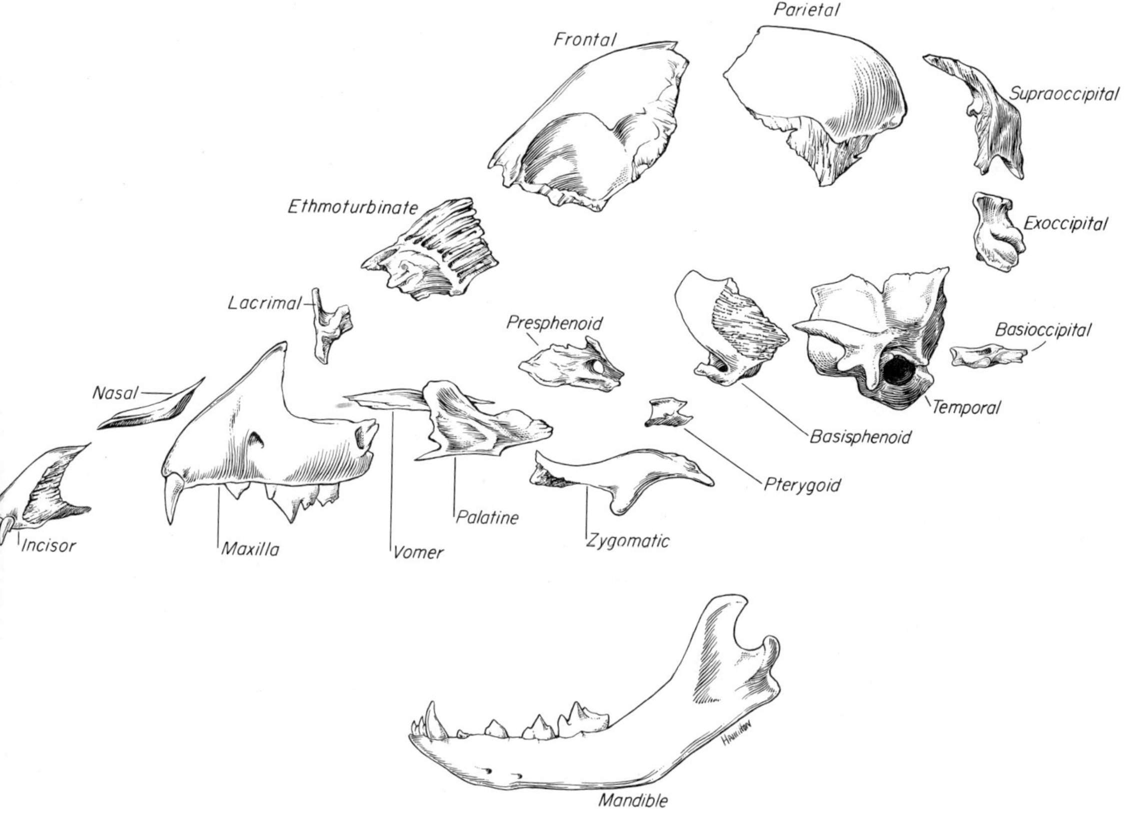

FIGURE 186. Disarticulated bones of a puppy skull, left lateral view.

Mandible

The mandible (Fig. 187), or lower jaw, bears the lower teeth and articulates with the temporal bone. The two mandibles join rostrally at the symphysis. Each mandible can be divided into a **body,** or horizontal part, and a **ramus,** or perpendicular part. The **alveolar border** of the mandible contains alveoli for the roots of the teeth. The incisors, canine, first premolar and third molar have one root each. The last three premolars and first two molars have two roots each.

On the lateral surface of the ramus of the mandible is the triangular **masseteric fossa** for the insertion of the masseter muscle. The dorsal half of the ramus is the **coronoid process.** Its medial surface has a shallow depression for insertion of the temporal muscle. Ventral to this is the **mandibular foramen.** This foramen is the caudal opening of the **mandibular canal,** which is located in the ramus and body of the mandible. It transmits the inferior alveolar artery and vein and the inferior alveolar nerve. It opens rostrally at the three **mental foramina,** where mental nerves supply sensory innervation to the adjacent lower lip and chin. The pterygoid muscles insert on the medial surface of the mandible and on the angular process, ventral to the insertion of the temporal muscle.

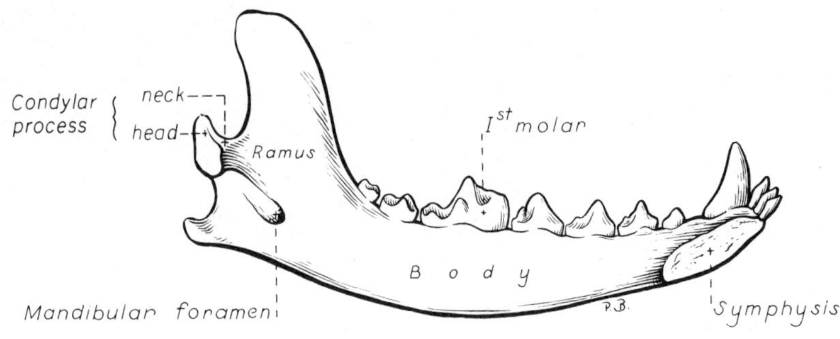

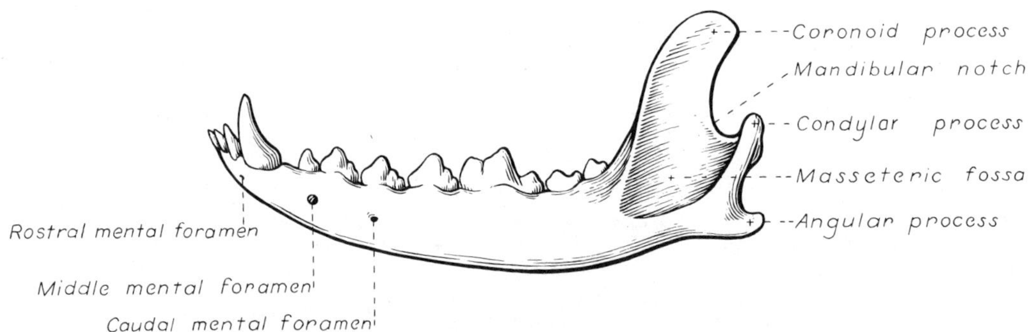

FIGURE 187. Left mandible, medial and lateral views.

The **condylar process** enters into the formation of the **temporomandibular joint.** Between the condylar process and the coronoid process is a U-shaped depression, the **mandibular notch.** Motor branches of the mandibular nerve pass across this notch to innervate the masseter muscle.

The **angular process** is a hooked eminence ventral to the condylar process. It serves for the attachment of the pterygoids medially and the masseter laterally.

Hyoid Bones

The **hyoid apparatus** (Figs. 188, 197) is composed of the hyoid bones, which stabilize the tongue and the larynx. This apparatus extends from the mastoid process of the skull to the thyroid cartilage of the larynx. It consists of the short tympanohyoid cartilage and the following articulated bones: the stylohyoid, epihyoid, ceratohyoid, basihyoid and thyrohyoid. All of the bones are paired except for the basihyoid, which unites the elements of the two sides in the root of the tongue. Examine these on the wet specimens provided, as well as on your own specimen.

Teeth

The teeth are arranged in (upper) and (lower) arches that face each other. The lower arch is narrower than the upper (Figs. 181, 183, 187).

The upper teeth are contained in the incisive and maxillary bones. Those whose roots are embedded in the incisive bone are known as the **incisor teeth.** Caudal to these and separated from them by a space is the **canine tooth.** Behind this are the **cheek teeth,** which are divided into **premolars** and **molars.** The lower teeth are generally similar to the upper. There is one more molar tooth in each mandible than in the corresponding maxilla. Some of the teeth—the incisors, the fourth premolar and the molars—usually meet those of the opposite arch when the jaws are closed. The upper fourth premolar and first molar shear along the first lower molar. The first three premolars fail to meet during normal closure, and the opening between the teeth is called the premolar carrying space. In dogs with long muzzles, there may be a considerable interval between teeth, supernumerary teeth may be present and some premolars do not occlude. In short-muzzle breeds the teeth are usually crowded, have shallow roots and are all in wear. In addition, most of the teeth are set obliquely.

The dental formula for the permanent teeth of the dog is:

$$\begin{array}{r} \\ \textit{Upper} \\ \textit{Lower} \end{array} \begin{array}{cccc} \mathbf{I} & \mathbf{C} & \mathbf{Pm} & \mathbf{M} \\ \hline 3 & 1 & 4 & 2 \\ 3 & 1 & 4 & 3 \end{array} = \frac{10}{11} \text{ 42 total right and left.}$$

The temporary or deciduous dentition ("milk teeth") can be expressed by the formula:

$$\begin{array}{r} \\ \textit{Upper} \\ \textit{Lower} \end{array} \begin{array}{ccc} \mathbf{I} & \mathbf{C} & \mathbf{Pm} \\ \hline 3 & 1 & 3 \\ 3 & 1 & 3 \end{array} = \frac{7}{7} \text{ 28 total right and left.}$$

The first incisor tooth (central) is next to the median plane and is followed by the second incisor (intermediate) and the third incisor (corner). In the permanent dentition the premolar and molar teeth are numbered from rostral to caudal; thus the tooth nearest the canine is number one. The fourth premolar is the largest cheek tooth of the maxilla; the largest

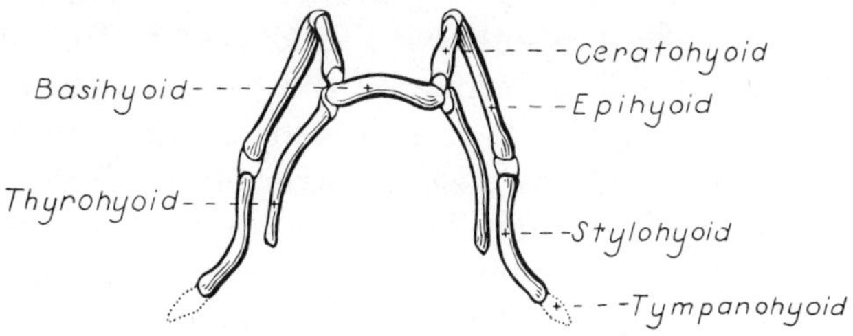

FIGURE 188. Hyoid bones, ventral view.

cheek tooth of the mandible is the first molar. These are known as the sectorial or shearing teeth.

The first premolar in the dog has no deciduous predecessor. The teeth of the permanent set are much larger than those of the deciduous set. The last permanent tooth erupts at six or seven months.

Each tooth possesses a **crown** and a **root** (or roots), which is embedded in the alveoli of the jaws. The junction of root and crown is the **neck** of the tooth.

The roots of the teeth are fairly constant. The incisors and canines of both jaws have one each. In the upper jaw the first premolar has one root; the second and third have two each; the fourth premolar and the first and second molars have three each. The lower cheek teeth have two roots each, except the first premolar and third molar, which have one. The **upper shearing tooth,** which is the fourth premolar, has two rostral roots in a transverse plane and a large caudal root. Notice that the lateral roots of the fourth premolar lie ventrolateral to the infraorbital canal and that the medial root rostral pair lies ventromedial to the infraorbital foramen.

The outer surface of the teeth is the **vestibular surface,** and the inner surface is the **lingual surface.** The sides of a tooth that lie in contact with or face an adjacent tooth are the **contact surfaces.** The surface of the tooth that faces the opposite dental arch is known as the **occlusal** or **masticating surface.**

Cavities of the Skull

Cranial Cavity

The cranial cavity (Figs. 189, 190) contains the brain and its coverings and vessels. The roof of the braincase, the calvaria, is formed by the parietal and frontal bones. The rostral two-thirds of the base of the cranium is formed by the sphenoid bones. The caudal third is formed by the occipital and temporal bones. The caudal wall is the occipital bone, and the rostral wall is the cribriform plate of the ethmoid bone. The lateral walls are formed by the temporal, parietal, frontal and sphenoid bones. The interior of the cranial cavity contains impressions formed by the gyri and sulci of the brain. Arteries leave grooves on the cerebral surface of the bones, while many of the veins lie in the diploë between the outer and inner tables of the bones. The base of the cranial cavity is divided into rostral, middle and caudal fossae.

The **rostral fossa** is located in front of the optic canals. The floor of this fossa is formed by the presphenoid bone and the cribriform plate of the ethmoid. It is bounded laterally by the frontal bone. The olfactory bulbs and the rostral parts of the frontal lobes of the brain lie in this fossa. The numerous foramina in the **cribriform plate** transmit blood vessels and olfactory nerves from the olfactory epithelium of the nasal cavity to the olfactory bulbs of the brain. The much perforated cribriform

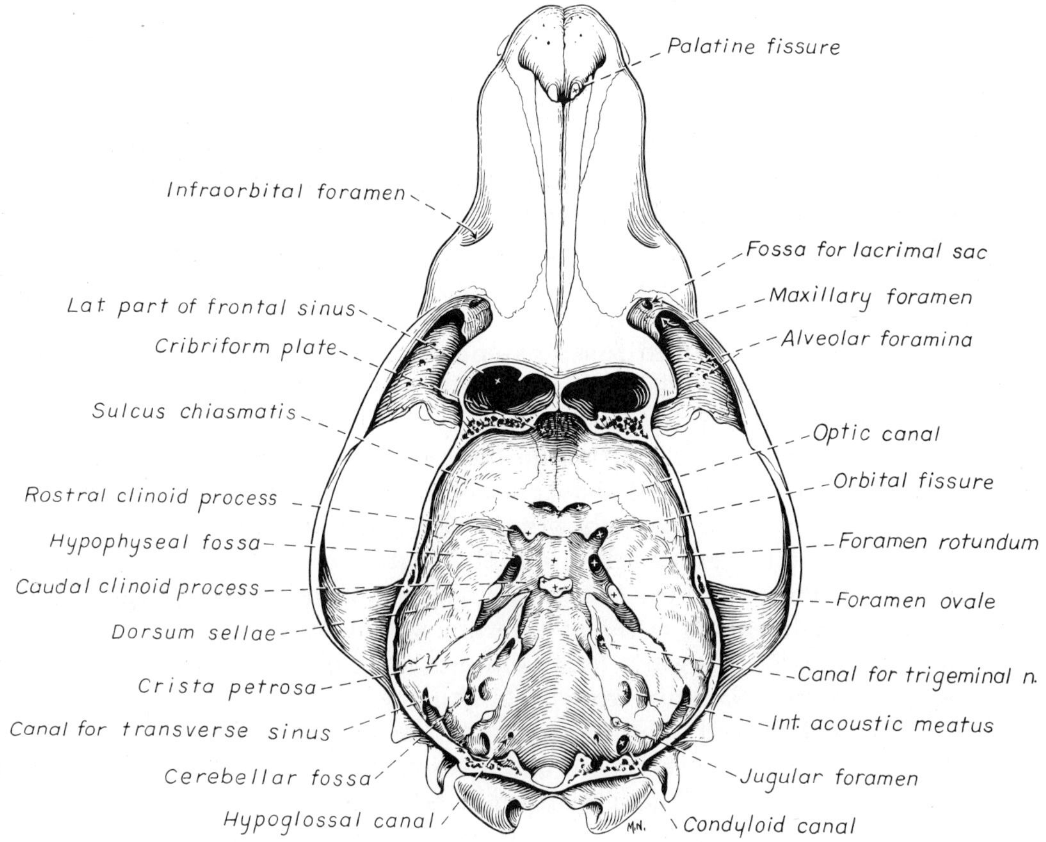

Palatine fissure
Infraorbital foramen
Fossa for lacrimal sac
Maxillary foramen
Alveolar foramina
Lat. part of frontal sinus
Cribriform plate
Sulcus chiasmatis
Optic canal
Orbital fissure
Rostral clinoid process
Hypophyseal fossa
Foramen rotundum
Caudal clinoid process
Foramen ovale
Dorsum sellae
Canal for trigeminal n.
Crista petrosa
Int. acoustic meatus
Canal for transverse sinus
Jugular foramen
Cerebellar fossa
Hypoglossal canal
Condyloid canal
M.N.

FIGURE 189. Skull with calvaria removed, dorsal view.

Tentorium osseum
Groove for middle meningeal a.
Transverse groove
Lat. part of frontal sinus
Internal table of frontal bone
Transverse canal
Medial part of frontal sinus
Cerebellar fossa
Cribriform plate
Dorsal nasal concha
Mastoid foramen
Ventral nasal concha
Condyloid canal
Jugular foramen
Hypoglossal canal
Int. acoustic meatus
Petrooccipital fissure
Canal for trigeminal n.
Dorsum sellae
Foramen ovale
Foramen rotundum
Orbital fissure
Optic canal
Ethmoidal foramina
Sphenopalatine foramen
Entrance to maxillary recess
M.N.

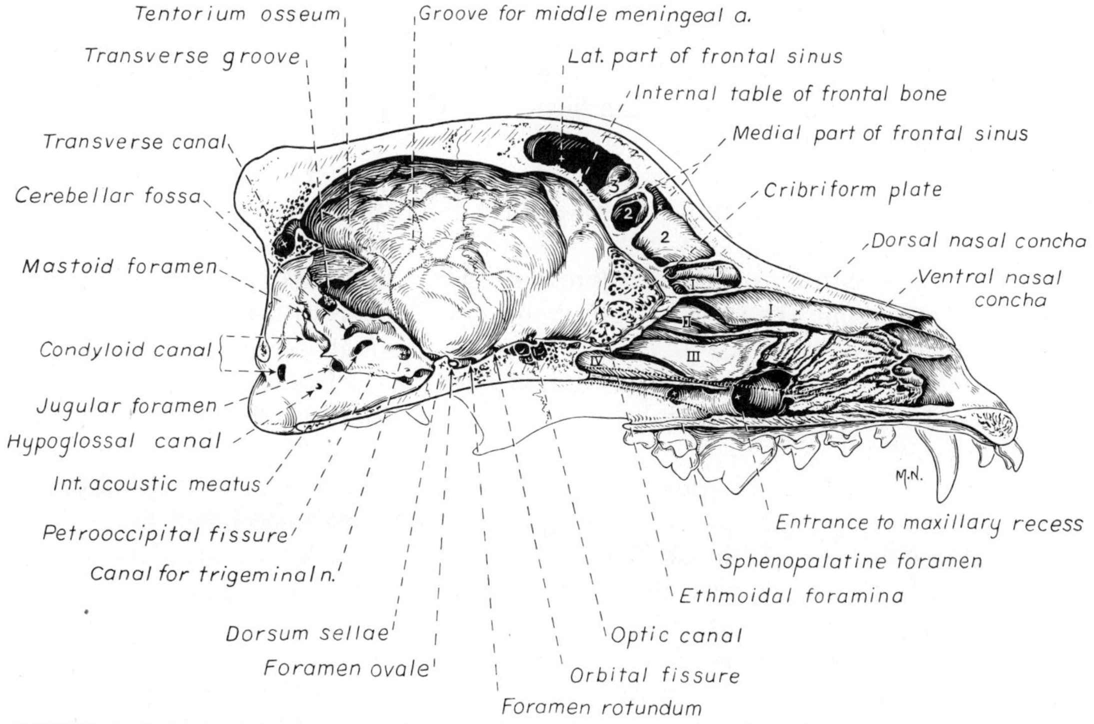

FIGURE 190. Sagittal section of skull, medial view. Roman numerals indicate endoturbinates, arabic numerals ectoturbinates.

plate of the ethmoid bone separates the braincase from the nasal cavity. (It can be the route of invasive organisms via the nasal cavity.) It is concave on its caudal surface and can be seen in a skull through the foramen magnum. On its rostral side it has many ethmoturbinate scrolls which project into and fill the caudal part of the nasal cavity. In life these thin ossified scrolls are covered by olfactory mucosa. The **optic canal** is a short passage in each orbital wing of the presphenoid bone through which the optic nerve courses.

The **middle cranial fossa** extends caudally from the optic canals to the petrosal crests and the dorsum sellae. It is situated on a lower level than the rostral cranial fossa. Several paired foramina are found on the floor of this fossa. Caudal to the optic canal is the orbital fissure. The oculomotor, trochlear and abducent nerves and the ophthalmic nerve from the trigeminal nerve leave the cranial cavity through this opening. Caudal and lateral to the orbital fissure is the round foramen that transmits the maxillary nerve from the trigeminal nerve to the alar canal. Caudolateral to the round foramen is the oval foramen that transmits the mandibular nerve from the trigeminal nerve and the middle meningeal artery, which enters the cranial cavity from the maxillary artery.

The **sella turcica,** on the dorsal surface of the basisphenoid, contains the hypophysis. It is composed of a shallow **hypophyseal fossa,** which is limited rostrally by the presphenoid bone and bounded caudally by a raised quadrilateral process, the **dorsum sellae.** The caudal part of the middle fossa is the widest part of the cranial cavity. The parietal and temporal lobes of the cerebrum are located here.

The **caudal cranial fossa** contains the cerebellum, the pons, the medulla and part of the occipital lobes of the cerebrum. It extends from the petrosal crests and dorsum sellae to the foramen magnum. The floor of this fossa is formed by the basioccipital bone and petrosal parts of the temporal bones.

Study the foramina and canals on the floor and sides of the braincase (Fig. 190). The opening of the carotid canal is located under the rostral tip of the petrosal bone.

The **canal for the trigeminal nerve** is in the rostral end of the petrosal bone. The trigeminal ganglion is located in the canal. Caudal to this is the **internal acoustic meatus,** through which the facial and vestibulocochlear nerves pass. Dorsocaudal to the internal acoustic meatus is the **cerebellar fossa** that contains a small lateral portion of the cerebellum.

The **jugular foramen** is located between the petrosal and the occipital bones. It opens to the outside through the tympano-occipital fissure and transmits the glossopharyngeal, vagus and accessory cranial nerves as well as the sigmoid venous sinus. Caudomedial to the jugular foramen is the **hypoglossal canal** for the hypoglossal nerve. Dorsal to this foramen is the **condyloid canal,** which transmits a venous sinus.

Projecting rostroventrally from the caudal wall of the cranial cavity is the **tentorium osseum.** This is composed of processes from the parietal and occipital bones. The dural membrane, the **tentorium cerebelli,** attaches to the petrosal crests and the tentorium osseum, separating the cerebrum from the cerebellum. A relatively median **foramen for the dorsal sagittal sinus** is located on the rostral surface of the occipital bone dorsal to the tentorium osseum. It opens into the paired **transverse canals.** This foramen transmits the dorsal sagittal venous sinus to the transverse sinus in the transverse canal. The transverse sinus continues ventrolaterally through the **transverse groove** of the occipital bone. Then as the temporal sinus it passes through the temporal bone lateral to the petrous portion. At the retroarticular foramen the temporal sinus communicates with the maxillary vein.

Nasal Cavity

The nasal cavity is the facial part of the respiratory passages. It begins at the bony **nasal aperture** and is composed of two symmetrical halves separated by a median nasal septum. The caudal openings of the nasal cavities are the choanae. In a medisected skull, study the bony scrolls, the conchae, which lie in the nasal fossa. Compare them with the mucosa-covered conchae of a hemisected embalmed head.

The conchae (Figs. 190, 191) project into each half of the nasal cavity and, with their mucosa, act as baffles to warm and cleanse inspired air. Their caudal portions also contain olfactory neurons, whose axons course to the olfactory bulbs of the brain through the cribriform plate.

The **dorsal nasal concha** originates as the most dorsal scroll on the cribriform plate and extends rostrally as a shelf attached along the medial surface of the nasal bone.

The **ventral nasal concha** consists of several elongate scrolls that attach to a crest on the medial surface of the maxilla. It lies in the middle of the nasal cavity but does not contact the median nasal septum.

The **ethmoidal labyrinth** is composed of many delicate scrolls that

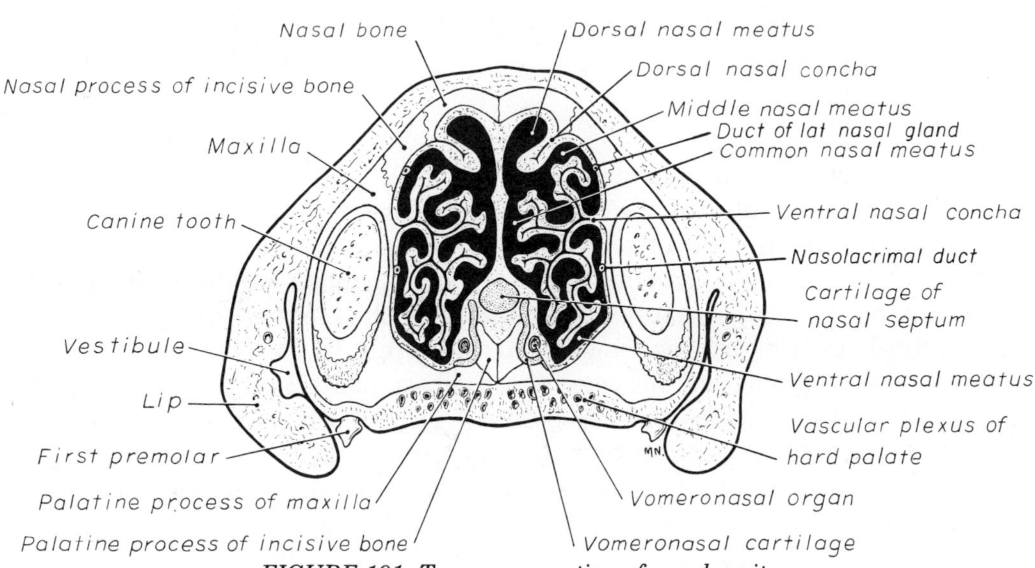

FIGURE 191. Transverse section of nasal cavity.

attach to the cribriform plate caudally and occupy the fundus of the nasal cavity. Dorsally the scrolls extend as ectoturbinates into the rostral portion of the frontal sinus. Ventrally, as endoturbinates, the scrolls attach to the vomer, which separates the entire ethmoidal labyrinth from the nasopharynx.

The **ethmoid bone** complex is located between the braincase and the facial part of the skull. It consists of the ethmoidal labyrinth, the cribriform plate and the median bony perpendicular plate of the nasal septum. The ethmoid bone complex is surrounded by the frontal bone dorsally, the maxilla laterally and the vomer and the palatine ventrally. The ethmoidal labyrinth, consisting of ecto- and endoturbinates attached to the cribriform plate, has an orbital lamina on each side that forms the medial wall of the maxillary recess.

The **nasal septum** separates the right and left nasal cavities. It is composed of cartilage and bone. The cartilaginous part, the **septal cartilage,** forms the rostral two-thirds of this median partition. It articulates with other cartilages at the nares, which prevent collapse of the nostrils. Ventrally the septal cartilage fits into a groove formed by the vomer, and dorsally it articulates with the nasal bones where they meet at the midline. The osseous part of the nasal septum is formed by the perpendicular plate of the ethmoid bone, the septal processes of the frontal and nasal bones and the sagittal portion of the vomer.

In each nasal cavity the shelflike dorsal nasal concha and the scrolls of the ventral nasal concha divide the cavity into four primary passages known as meatuses (Fig. 191). The **dorsal nasal meatus** lies between the nasal bone and the dorsal nasal concha. The small **middle nasal meatus** lies between the dorsal nasal concha and the ventral nasal concha, while the **ventral nasal meatus** is dorsal to the hard palate. Since the conchae do not reach the nasal septum, a vertical space, the **common nasal meatus,** is formed on each side of the nasal septum. This space extends from the nasal aperture to the choanae in a longitudinal direction and from the nasal bone to the hard palate in a vertical direction.

Paranasal Sinuses

There are three frontal sinuses (Figs. 189, 190) located between the outer and inner tables of the frontal bone. They are designated lateral, rostral and medial. The **lateral frontal sinus** is much larger than the others. It occupies the zygomatic process and extends caudally, bounded laterally by the temporal line and medially by the median septum. It may be partially divided by bony septa extending into it. The **rostral frontal sinus** is small and lies between the median plane and the orbit. The ethmoid labyrinth bulges into this sinus. The **medial frontal sinus** lies between the median septum and the walls of the other two sinuses. It is very small and may be absent. All three sinuses communicate with the nasal cavity.

The **maxillary recess** (Fig. 190) communicates with the nasal cavity. Its opening lies in a transverse plane through the rostral roots of the upper fourth premolar tooth. The recess continues caudally to a plane through the last molar tooth. The walls of the maxillary recess are formed laterally and ventrally by the maxilla and medially by the orbital lamina of the ethmoid bone. The **lateral nasal gland** occupies this recess.

Tympanic Cavity

The tympanic cavity is the cavity of the middle ear. It houses the auditory ossicles and communicates with the nasopharynx via the auditory tube. It is bounded ventrally by the tympanic bulla and dorsally by the petrosal bone. Laterally the external acoustic meatus is closed by the tympanic membrane (Figs. 183, 185).

Joints of the Head

The atlanto-occipital joint was described with the vertebrae. The **temporomandibular joint** is between the condyloid process of the mandible and the mandibular fossa of the temporal bone. The articulation is elongated transversely. There is a thin cartilaginous meniscus, the **articular disk,** which separates the articular surfaces of each bone and divides the joint capsule into two compartments. Lateral and caudal ligaments strengthen the joint capsule.

The **mandibular symphysis** is a synchondrosis with an interdigitating surface that persists throughout life in the dog.

Live Dog

Palpate the features of the skull of the dog's head. The widest part of the head is the palpable zygomatic arch. Palpate the zygomatic process of the frontal bone (its widest point) and the orbital ligament between it and the frontal process of the zygomatic bone. This ligament forms the lateral border of the rostral part of the orbit. Follow the frontal bone caudally to the temporal line and sagittal crest. Feel the external occipital protuberance and follow the nuchal crest ventrally on each side to the mastoid process of the temporal bone. Flex and extend the atlanto-occipital joint. Follow the frontal bones rostrally to the nose and maxillary bones. Feel the infraorbital foramen on the side of the maxilla at the level of the rostral roots of the fourth premolar. Reflect the lip and study the crowns and necks of the teeth of each dental arch. Find the shearing teeth and note how they contact each other.

Palpate the coronoid process of the mandible medial to the zygomatic arch and the angular process ventrally. Feel the body of the mandible and move the temporomandibular joint by opening the mouth.

STRUCTURES OF THE HEAD

To facilitate the dissection of the head, it should be removed and divided on the median plane. Using a hand-saw, make a complete transection of the neck at the level of the fourth cervical vertebra.

Section the head and attached portion of the neck on the median plane using a band saw. Wash the cut surface to remove bone dust and hair.

Skin both halves, leaving the muscles in place. Leave a narrow rim of skin around the margin of the eyelids and at the edge of the lips. Preserve the nose and the **philtrum,** which is the median groove separating the right and left parts of the upper lip. Skin only the base of the ear.

The right side of the head will be used for the dissection of vessels and nerves, while the left side will be used for muscles. At the end of each dissection period, wrap the head in cheesecloth moistened with 1 per cent phenol water or 2 per cent phenoxyethanol before covering your specimen. A plastic bag is useful to prevent desiccation.

Muscles of the Face

The muscles of the face function to open, close or move the lips, eyelids, nose and ear. They are all innervated by the facial (seventh cranial) nerve.

Cheek, Lips and Nose. The **platysma** (Fig. 192) is a cutaneous muscle that passes from the dorsal median raphe of the neck to the angle of the mouth, where it radiates into the orbicularis oris. It is the most superficial muscle, covering the ventrolateral surface of the face. Transect this muscle and reflect it.

The **orbicularis oris** lies near the free borders of the lips and extends from one lip to the other around the angle of the mouth. The fibers of each side end at the median plane in the incisor region of both jaws.

The **buccinator muscle** is a thin, wide muscle that forms the foundation of the cheek. It attaches to the alveolar margins of the mandible and maxilla and the adjacent buccal mucosa. It may be found between the rostral border of the masseter muscle and the orbicularis oris. Place your finger within the cheek and press outward. This will help define the buccinator. A portion of the buccinator lies deep to the orbicularis oris muscle and is difficult to separate from it. It functions to return food from the vestibule to the occlusal surface of the teeth.

The **levator nasolabialis** is a flat muscle lying beneath the skin on the lateral surface of the maxillary bone. It arises from the maxillary bone, courses rostroventrally and attaches to the edge of the upper lip and on the external naris. It dilates the nostril and raises the upper lip.

Eyelids. Before dissecting the eyelids, **palpebrae,** observe their external features. The **superior and inferior palpebrae** (upper and lower eyelids) border the **palpebral fissure.** They join at either end of the fissure to form the **medial** and **lateral palpebral commissures.**

Each commissure is attached by ligaments to adjacent bone. The medial palpebral ligament is well developed and attaches the medial commissure to the frontal bone near the nasomaxillary suture. The lateral palpebral ligament is poorly developed and attaches to the zygomatic bone at the orbital ligament.

The upper eyelid bears cilia (eyelashes) on its free border. The lower eyelid lacks cilia. The cutaneous or external surface of the eyelid is covered by hair. The posterior or inner surface is covered by a mucous membrane, the **palpebral conjunctiva** (see Fig. 206). Follow the palpebral conjunctiva posteriorly to its reflection from the eyelids onto the globe of the eye, which is the **bulbar conjunctiva.** The angle formed by this reflection is called the **fornix.** The potential cavity thus formed, the

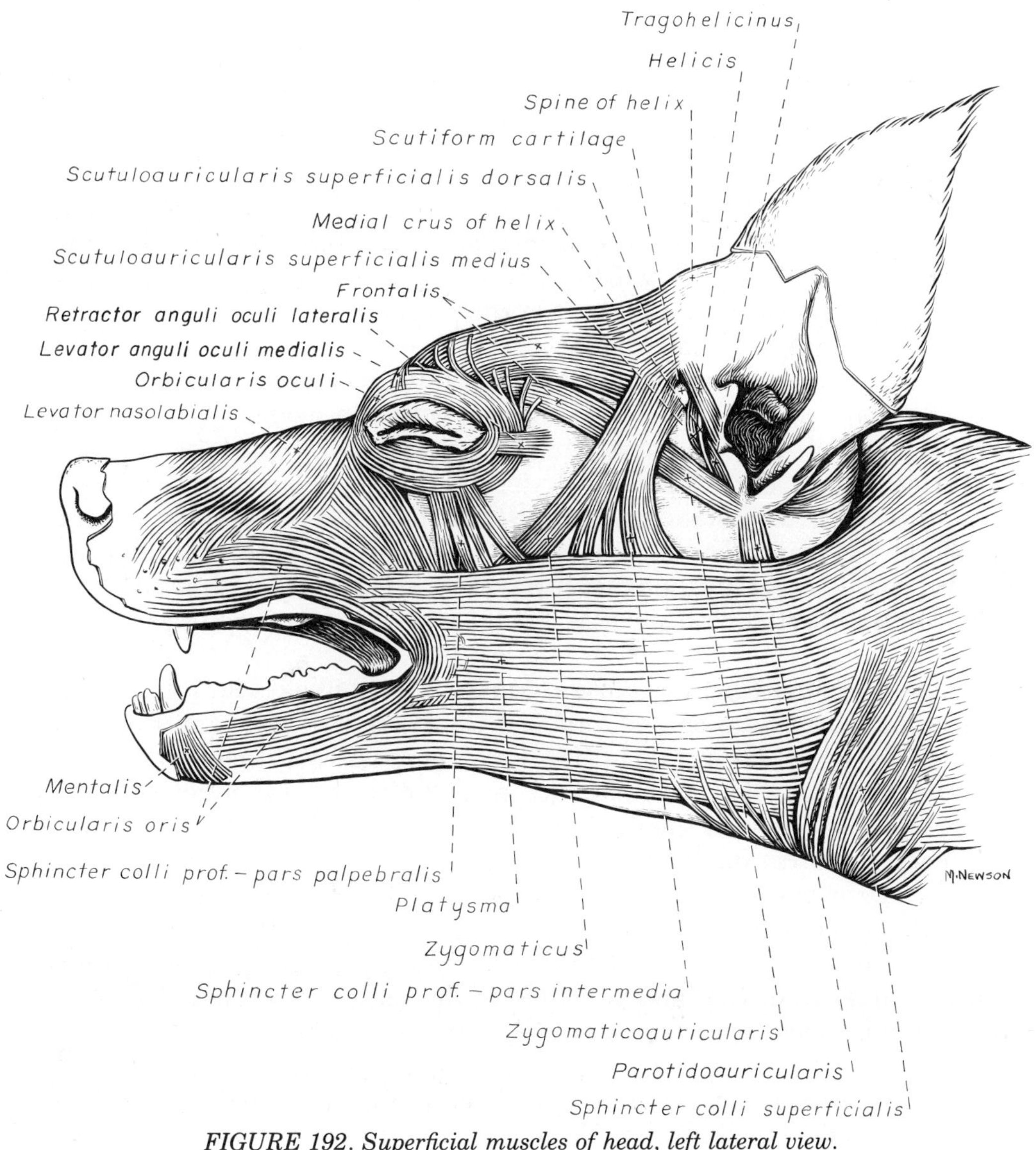

FIGURE 192. Superficial muscles of head, left lateral view.

conjunctival sac, is bounded posteriorly by the bulbar conjunctiva and the cornea and anteriorly by the palpebral conjunctiva.

At the medial commissure observe the triangular-shaped prominence of finely-haired skin, the **lacrimal caruncle.** The **lacrimal punctum** (see Fig. 203) of each lid is the beginning of the dorsal and ventral lacrimal ducts. Each is a small opening on the conjunctival margin of the lid a few millimeters from the medial commissure. The puncta may be difficult to see without the aid of magnification. The **lacrimal gland,** located ventral to the zygomatic process of the frontal bone, secretes through many duct openings into the dorsolateral part of the conjunctival sac. After this serous fluid has passed across the cornea, it is collected by the puncta and passes in succession through the **lacrimal duct** of each lid, the **lacrimal sac** and the **nasolacrimal duct** to the ventral nasal meatus of the nasal cavity. There, evaporation of the lacrimal secretion takes place. A significant contribution to tear secretion comes from the gland of the third eyelid, conjunctival goblet cells and tarsal glands of the upper eyelid. The lacrimal gland and the rostral opening of the nasolacrimal duct will be seen later.

The **plica semilunaris,** or **third eyelid** (see Fig. 203), is a concave fold of palpebral conjunctiva and cartilage that protrudes from the medial angle of the eye. The cartilage extends ventrally into the orbit and is surrounded by a body of fat and glandular tissue, the **superficial gland of the third eyelid.** This will be dissected shortly. Its serous secretion enters the conjunctival sac under the third eyelid at the medial commissure. Lift the third eyelid from the bulbar conjunctiva and examine its medial surface. Note the slightly raised lymphoid tissue.

There are several muscles associated with the eyelids. The **orbicularis oculi** (Fig. 192) lies partly in the eyelids and is attached medially to the medial palpebral ligament. Laterally the fibers of the muscle blend with those of the **retractor anguli oculi,** which covers the lateral palpebral ligament. The action of the muscle is to close the palpebral fissure. The **levator palpebrae superioris** arises deep within the orbit and will be dissected with the muscles of the eyeball. It elevates the upper lid.

The External Ear. The **rostral auricular muscles** (Fig. 192) include those muscles that lie on the forehead caudal to the orbit and converge towards the auricular cartilage. Transect the muscles at their origins and turn them towards the auricular cartilage. Notice that the middorsal part arises from its fellow of the opposite side.

The **scutiform cartilage** is a small, boot-shaped, cartilaginous plate located in the muscles rostral and medial to the external ear. It is an isolated cartilage interposed in the auricular muscles.

The **caudal auricular muscles** are the largest group. Most of these muscles arise from the median raphe of the neck and attach directly to the auricular cartilage. Transect the caudal auricular muscles and turn the external ear ventrally to expose the temporal muscle.

The other superficial muscles of the face, all innervated by the facial nerve, will not be dissected. They are collectively referred to as the mimetic muscles or muscles of facial expression.

Oral Cavity

The oral cavity, or mouth, is divided into the vestibule and the oral cavity proper. The **vestibule** is the cavity lying outside the teeth and gums and inside the lips and cheeks. The ducts of the parotid and zygomatic salivary glands open into the dorsocaudal part of the vestibule. The **parotid duct** opens through the cheek on a small papilla located opposite the caudal end of the upper fourth premolar or shearing tooth. The **ducts of the zygomatic gland** open into the vestibule lateral to the last upper molar tooth.

The **oral cavity proper** is bounded dorsally by the hard palate and a small part of the adjacent soft palate; laterally and rostrally by the dental arches; and ventrally by the tongue and adjacent mucosa. Its caudal boundary ventrally is the body of the tongue at the palatoglossal arch. Pull the tongue away from one mandible and note the fold of tissue that extends from the body of the tongue to the beginning of the soft palate; this is the **palatoglossal arch** (Fig. 193). It freely communicates with the vestibule by numerous interdental spaces and is continued caudally by the oropharynx.

Tongue

Examine the tongue (Fig. 193). It is a muscular organ composed of the interwoven bundles of intrinsic and extrinsic muscles. These will be dissected later. It is divided into a **root,** which composes its caudal third; a **body,** which is the long, slender, rostral part of the tongue; and a free extremity, the **apex.** The mucosa covering the dorsum of the tongue is modified to form various types of papillae. A dissecting microscope facili-

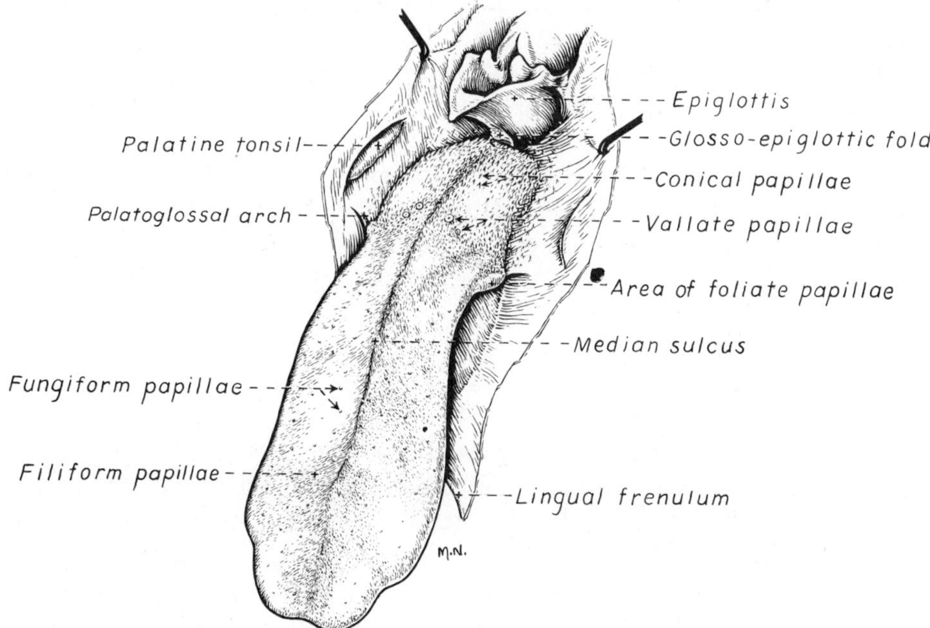

FIGURE 193. The tongue, dorsal aspect.

tates examination of these structures. Five types of papillae are recognized by their shape. The **filiform papillae** are found predominantly on the body and apex of the tongue. They are arranged in rows like shingles, with their multiple pointed tips directed caudally. At the root of the tongue the filiform papillae are replaced by **conical papillae,** which have only one pointed tip. The **fungiform papillae** have a smooth, rounded surface and are fewer in number. They are located among the filiform papillae. A few may be scattered caudally among the conical papillae. The **foliate papillae** are found on the lateral margins of the root of the tongue, rostral to the palatoglossal arch. They are leaflike but appear as a row of parallel grooves in the fixed specimen. The **vallate papillae** are located at the junction of the body and root of the tongue. There are four to six in the dog, and they are arranged in the form of a V with the apex directed caudally. They are larger than the others, have a circular surface and are surrounded by a sulcus. There are taste buds on vallate, foliate and fungiform papillae.

The tongue is attached rostrally to the floor of the oral cavity by a ventral median fold of mucosa, the **lingual frenulum** (Fig. 194). Examine the medial cut surface of the apex of the tongue. On the midline, just under the mucosa, is the **lyssa** (Fig. 195). This fusiform fibrous spicule extends from the apex to the level of the attachment of the frenulum. Expose the lyssa.

Salivary Glands

Turn the tongue medially and observe the slightly raised elevation that is closely adjacent to the frenulum laterally and protruding from the floor of the oral cavity. This is the **sublingual caruncle.** Extending caudally from the caruncle is the **sublingual fold.** The **mandibular duct** and **major sublingual duct** (Fig. 194) are found in this fold. They course rostrally to open on or beside the sublingual caruncle, separately or through a common opening. Carefully incise the mucosa over these ducts from the caruncle to the root of the tongue at the palatoglossal arch. Bluntly reflect the mucosa to expose the ducts and associated salivary tissue. The major sublingual duct is connected caudally with the monostomatic sublingual gland. This is closely associated with the mandibular salivary gland. There are also sublingual gland lobules (the polystomatic sublingual gland) deep to the mucosa of the sublingual fold. These have independent microscopic ducts opening into the oral cavity. Follow the mandibular and major sublingual ducts to the root of the tongue. Their origin will be seen when these glands are dissected from the lateral side of the head.

Expose the **mandibular salivary gland** (Fig. 194) on the lateral side of the head just caudal to the angle of the mandible, where it lies between the maxillary and linguofacial veins. It is covered by a thick capsule that also includes the caudal part of the monostomatic **sublingual gland** (Figs. 194, 196). Incise the capsule that surrounds these two glands and free the caudal portion by elevating it from the capsule. Locate the division

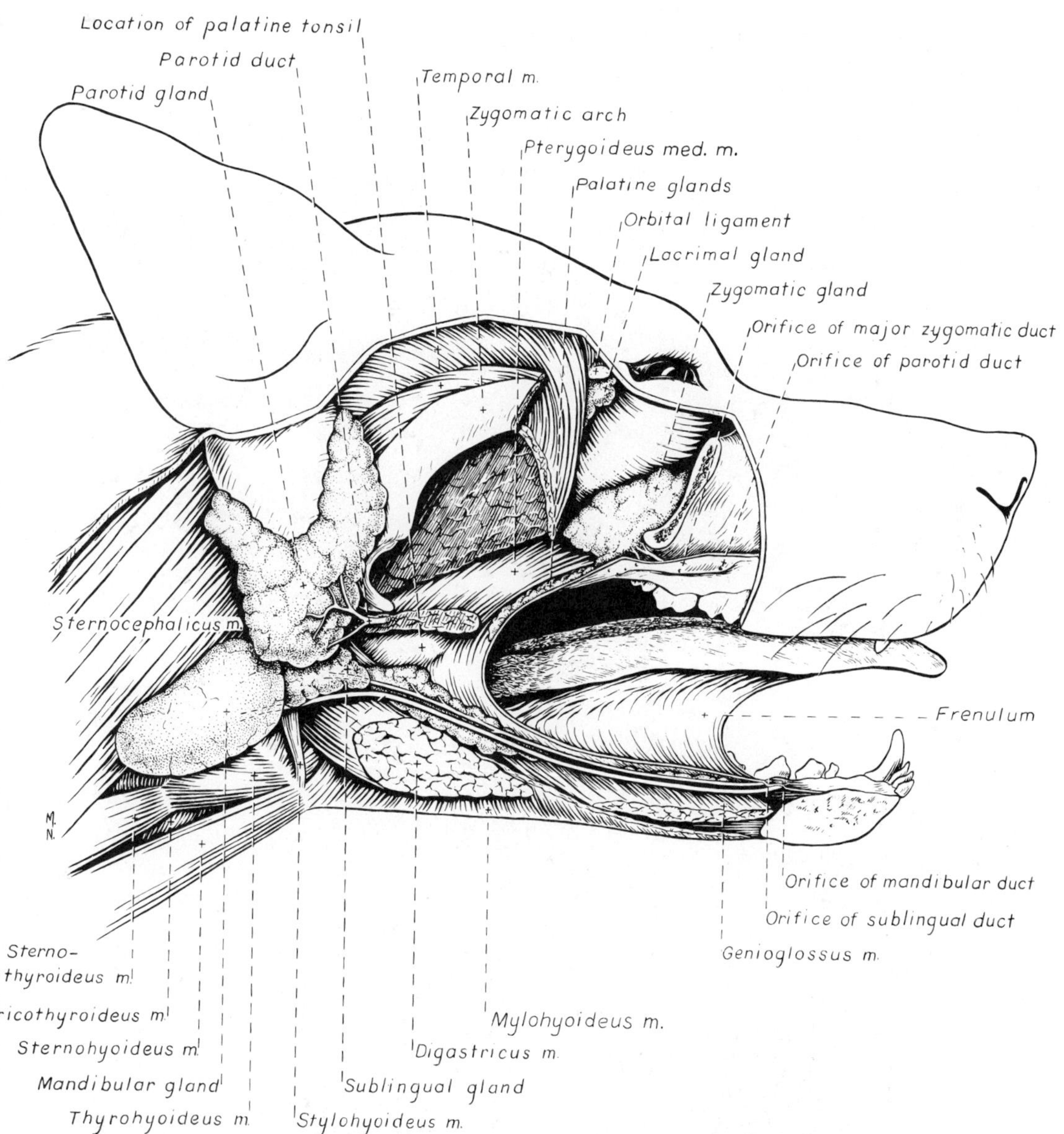

FIGURE 194. Salivary glands: parotid, mandibular, sublingual and zygomatic.

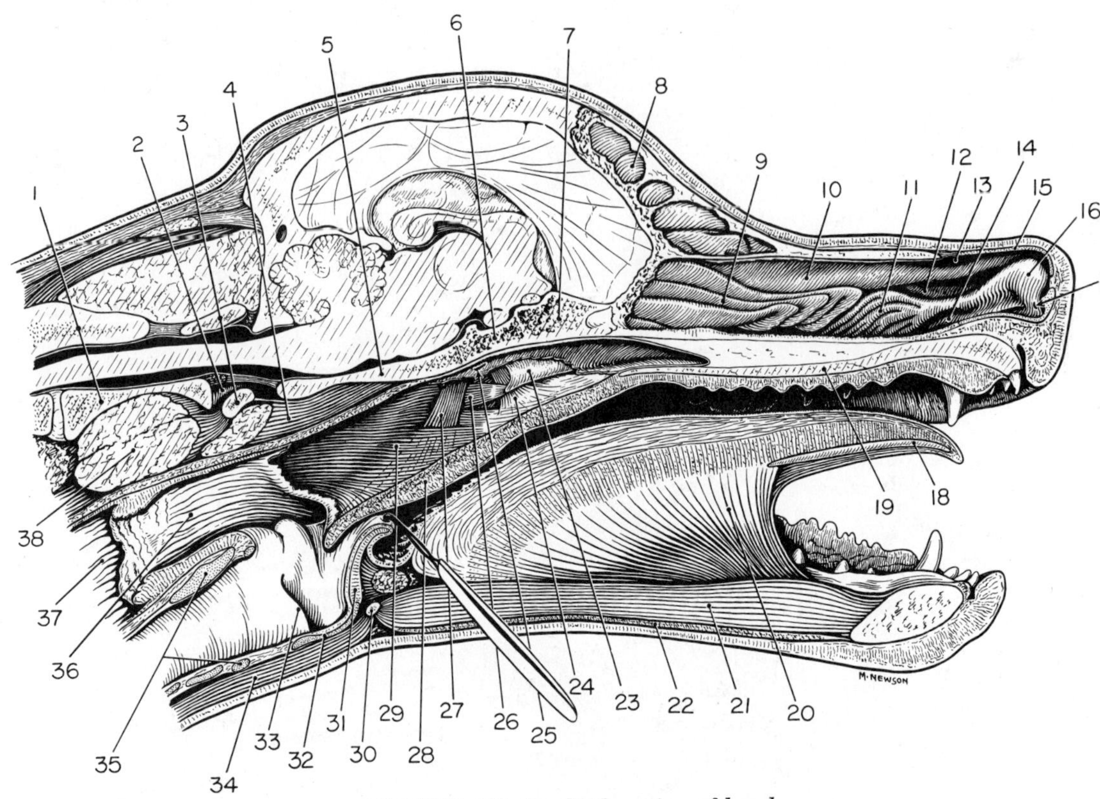

FIGURE 195. Sagittal section of head.

1. *Axis*
2. *Dens*
3. *Atlas*
4. *Longus capitis*
5. *Basioccipital*
6. *Basisphenoid*
7. *Presphenoid*
8. *Frontal sinus*
9. *Ethmoid labyrinth*
10. *Dorsal nasal concha*
11. *Ventral nasal concha*
12. *Middle nasal meatus*
13. *Dorsal nasal meatus*
14. *Ventral nasal meatus*
15. *Dorsal lateral nasal cartilage*
16. *Alar fold*
17. *Nasolacrimal duct orifice*
18. *Lyssa*
19. *Hard palate*
20. *Genioglossus*
21. *Geniohyoideus*
22. *Mylohyoideus*
23. *Pterygoid bone*
24. *Tensor veli palatini*
25. *Pharyngeal orifice of auditory tube*
26. *Pterygopharyngeus*
27. *Levator veli palatini*
28. *Soft palate*
29. *Palatopharyngeus*
30. *Basihyoid*
31. *Epiglottis*
32. *Thyroid cartilage*
33. *Vocal fold*
34. *Sternohyoideus*
35. *Cricoid cartilage*
36. *Laryngopharynx*
37. *Esophagus*
38. *Longus colli*

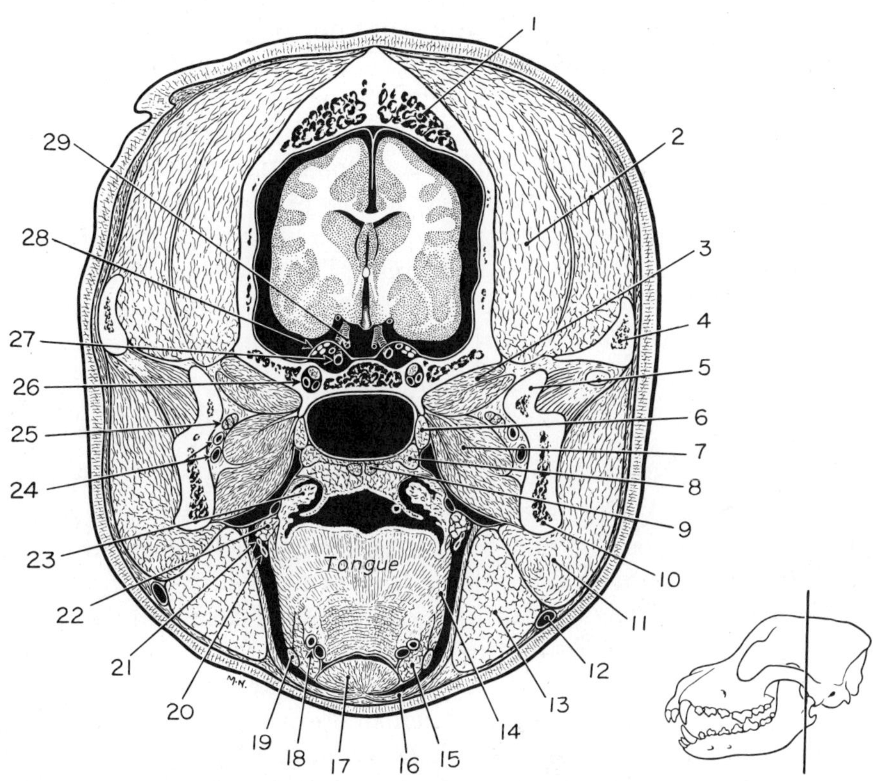

FIGURE 196. Transection of head through palatine tonsil.

1. *Diploe*
2. *Temporal m.*
3. *Lateral pterygoid m.*
4. *Zygomatic process of temporal bone*
5. *Condylar process*
6. *Tensor veli palatini*
7. *Medial pterygoid m.*
8. *Pterygopharyngeus m.*
9. *Palatinus m.*
10. *Mandible*
11. *Masseter m.*
12. *Facial vein*
13. *Digastricus m.*
14. *Styloglossus m.*
15. *Hyoglossus m.*
16. *Mylohyoideus m.*
17. *Geniohyoideus m.*
18. *Lingual a. and v.*
19. *Hypoglossal n.*
20. *Mandibular duct*
21. *Major sublingual duct*
22. *Sublingual salivary gland*
23. *Palatine tonsil in tonsillar fossa*
24. *Inferior alveolar a. and v.*
25. *Mylohyoid, inferior alveolar and lingual nn.*
26. *Maxillary a., v. and n. in alar canal*
27. *Internal carotid a. in cavernous sinus*
28. *Cranial nerves III, IV and VI and ophthalmic division of V*
29. *Cerebral arterial circle—caudal communicating a.*

between the rostrally located sublingual gland, which is roughly triangular, and the larger ovoid mandibular gland. The ducts of both glands leave the rostromedial surface and enter the space between the masseter and the digastricus muscles. Lobules of the sublingual gland continue into the oral cavity where they can be seen beneath the oral mucosa.

The **parotid salivary gland** (Fig. 194) lies between the mandibular gland and the ear. It is closely applied to the base of the auricular cartilage of the ear. The **parotid duct** is formed by two or three converging radicles, which unite and leave the rostral border of the gland. It grooves the lateral surface of the masseter muscle as it passes to the cheek. It opens into the vestibule on a small papilla at the level of the caudal margin of the fourth upper premolar. Evert the upper lip near the commissure and find the small opening into the vestibule. The ducts of the zygomatic salivary gland open into the vestibule near the last molar, caudal to the parotid duct. The gland will be dissected later (see p. 288).

Palate

Examine the roof of the oral cavity. The hard palate is crossed by approximately eight transverse ridges (Fig. 195). A small eminence, the **incisive papilla,** is located just caudal to the central incisor teeth. The fissure on either side of this papilla is the oral opening of the **incisive duct.** The incisive duct passes through the palatine fissure and opens into the ventral nasal meatus. Extending caudally from the incisive duct, close to its entrance into the nasal cavity, is the **vomeronasal organ** (Fig. 191). This tubular structure about 2 centimeters long lies at the base of the nasal septum dorsal to the hard palate and is an olfactory receptor of sexual stimuli. It can sometimes be seen on the sagittal section of the head.

Pharynx

The pharynx (Figs. 195, 196) is a passage that is common, in part, to both the respiratory and digestive systems. It is divided into oral, nasal and laryngeal parts. The **oropharynx** extends from the level of the palatoglossal arches to the caudal border of the soft palate and the base of the epiglottis at the caudal end of the root of the tongue. The dorsal and ventral boundaries of the oropharynx are the soft palate and the root of the tongue. The lateral wall of the oropharynx contains the palatine tonsil in the tonsillar fossa.

The **palatine tonsil** (Figs. 193, 194, 196) is elongate and is located caudal to the palatoglossal arch. The medial wall of the fossa, which partially covers the tonsil, is the **semilunar fold.** The tonsil is attached laterally throughout its entire length. Reflect the tonsil from the fossa.

The nasal cavity extends from the nostrils to the choanae. It is divided into right and left halves by the nasal septum. Each half is divided into four meatuses. These were described with the bones of the nasal cavity and should be reviewed now. On the floor of the rostrolateral end of the ventral meatus, find the **opening of the nasolacrimal duct.** It may be necessary to remove the rostral end of the cartilaginous septum to see the opening. This duct comes from the lacrimal sac at the medial commissure of the eye.

The **nasopharynx** extends from the choanae to the junction of the palatopharyngeal arches. A **palatopharyngeal arch** extends caudally on each side from the caudal border of the soft palate to the dorsolateral wall of the nasopharynx. It is a fold of mucosa that covers the palatopharyngeus muscle. On the lateral wall of the nasopharynx, dorsal to the middle of the soft palate, is an oblique, slitlike opening, the pharyngeal opening of the **auditory tube** (Fig. 195).

The **laryngopharynx** is dorsal to the larynx. It extends from the palatopharyngeal arches to the beginning of the esophagus. The esophagus begins at an annular constriction at the level of the cricoid cartilage, the **pharyngoesophageal limen.**

Larynx

Cartilages of the Larynx

Study the cartilages of the larynx (Fig. 197) on specimens from which the muscles have been removed. Visualize their topography in your bisected specimen and palpate them through the laryngeal mucosa. The laryngeal cartilages that will be dissected include the paired arytenoid and the unpaired epiglottic, thyroid and cricoid cartilages.

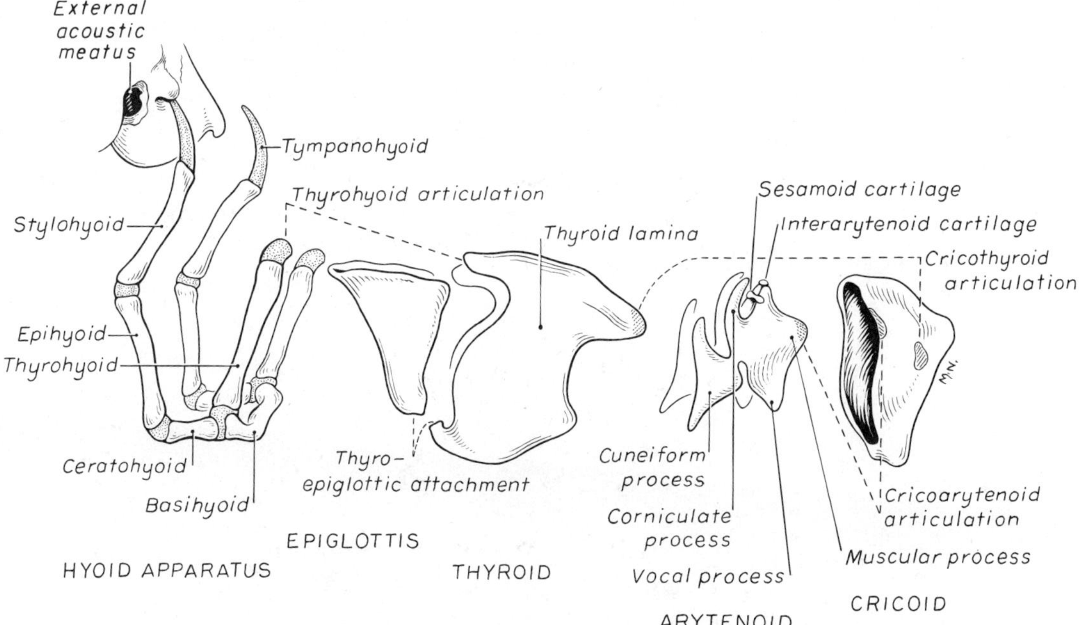

FIGURE 197. Cartilages of disarticulated larynx with hyoid apparatus intact, left lateral view.

The **epiglottic cartilage** lies at the entrance to the larynx. Its lingual surface is attached to the basihyoid bone and faces the oropharynx. The apex lies just dorsal to the edge of the soft palate. The lateral margin is attached by mucosa to the cuneiform process of the arytenoid to form the **aryepiglottic fold.** Caudally the epiglottis attaches to the body of the thyroid cartilage.

The **thyroid cartilage** forms a deep trough, which is open dorsally. The **rostral cornu** articulates with the thyrohyoid bone; the **caudal cornu** articulates with the cricoid cartilage. Ventrally the caudal border is notched by a median **caudal thyroid incisure.** The **cricothyroid ligament** attaches the caudal border to the ventral arch of the cricoid cartilage.

The **cricoid cartilage** forms a complete ring that lies partially within the trough of the thyroid cartilage. It has a wide dorsal plate, or lamina, and a narrow ventral arch. Near the caudal border at the junction of the lamina and the arch, there is a lateral facet for articulation with the caudal cornu of the thyroid cartilage. On the cranial border of the lamina, there is a prominent pair of lateral facets for articulation with the arytenoid cartilages.

The **arytenoid cartilage** is paired, irregular in shape and located in a sagittal plane. Each articulates medially with a facet on the rostral border of the cricoid lamina and has a lateral **muscular process** and a ventrally directed **vocal process.** The **vocal fold** is attached between the vocal process of the arytenoid and the thyroid cartilage. The arytenoid cartilage bears a **corniculate process** dorsally. Rostral to this process the **cuneiform process** is attached to the arytenoid (Fig. 197). The **vestibular fold** is attached to the ventral portion of the cuneiform process and forms the rostral boundary of the laryngeal ventricle. The **laryngeal ventricle** is a diverticulum of the laryngeal mucosa bounded laterally by the thyroid cartilage and medially by the arytenoid cartilage. It opens into the larynx between the vestibular fold rostrally and vocal fold caudally. Observe these folds and the laryngeal ventricle on the medial side of your specimen (Figs. 198–200).

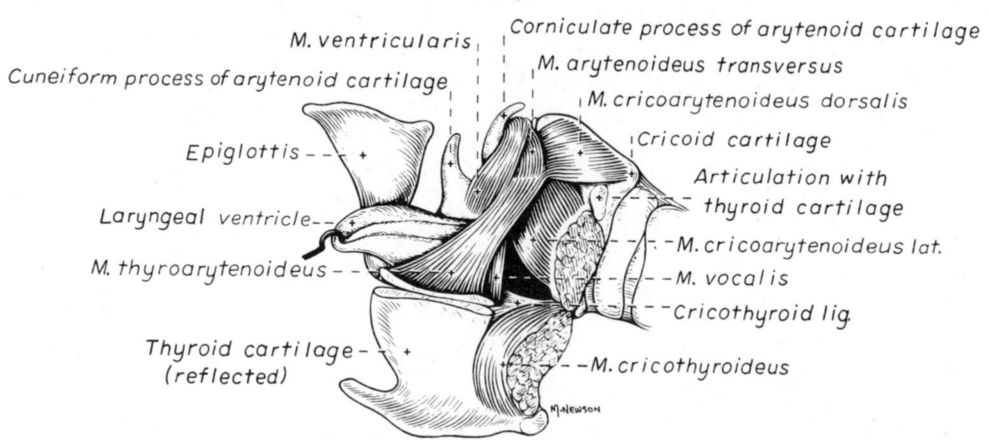

FIGURE 198. Laryngeal muscles, with thyroid cartilage reflected, left lateral view.

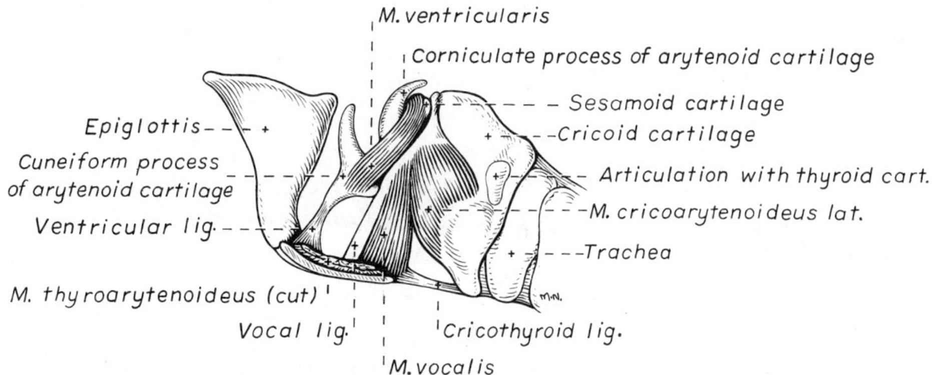

FIGURE 199. Laryngeal muscles, left lateral view after removal of thyroid cartilage, thyroarytenoideus, arytenoideus transversus and cricoarytenoideus dorsalis.

The **glottis** consists of the vocal folds, the vocal processes of the arytenoid cartilages and the **rima glottidis,** which is the narrow passageway through the glottis.

Muscles of the Larynx (Figs. 198–200)

The **cricothyroid muscle** lies ventral to the insertion of the sternothyroideus muscle and passes from the cricoid cartilage to the thyroid lamina. It tenses the vocal fold indirectly by drawing the ventral parts of the cricoid and thyroid cartilages together. It is innervated by the cranial laryngeal nerve, a branch of the vagus. Observe this on the lateral side.

On the medial side of the specimen reflect the mucosa of the laryngopharynx from the dorsal aspect of the larynx. Separate the thyroid lamina from the cricoid and arytenoid cartilages to expose the muscles.

The **cricoarytenoideus dorsalis** arises from the dorsolateral surface of the cricoid cartilage and inserts on the muscular process on the lateral surface of the arytenoid cartilage. It rotates the arytenoid so that the vocal process moves laterally, opening the glottis. It is the only laryngeal muscle that functions primarily to open the glottis. Transect the muscle

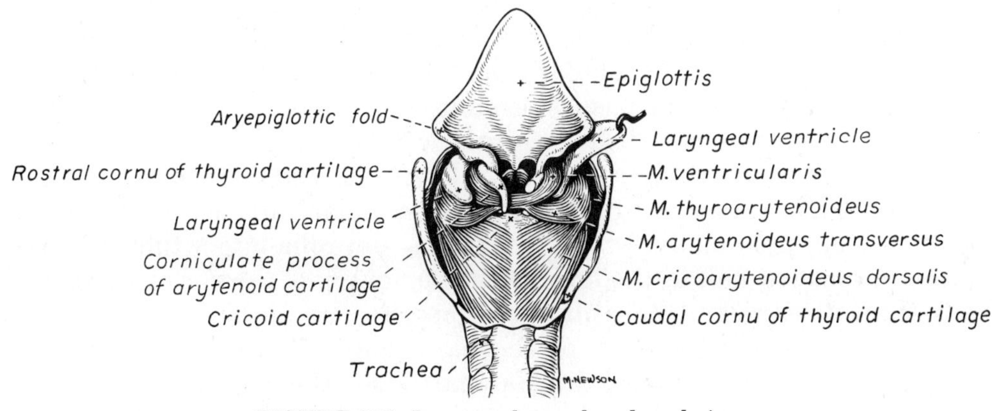

FIGURE 200. Laryngeal muscles, dorsal view.

and examine the articular surface of the cricoarytenoid joint. On the intact specimen of laryngeal cartilages grasp the muscular process of the arytenoid with forceps and pull it caudomedially, as it would be pulled by the functioning cricoarytenoideus dorsalis muscle. Observe the abduction of the vocal process and fold which widens the glottal opening.

The **cricoarytenoideus lateralis** arises from the lateral surface of the cricoid cartilage and inserts on the arytenoid cartilage between the cricoarytenoideus dorsalis and the vocalis. It acts to close the glottis. Expose this muscle by reflecting the cricoid medially away from the thyroid lamina.

The **thyroarytenoideus** is the parent muscle that gives rise to the vocalis muscle medially and ventricularis muscle rostrally. It arises along the internal midline of the thyroid cartilage and inserts on the arytenoid cartilage. Its function is to relax the vocal fold and to constrict the glottis.

The **vocalis** is a medial division of the thyroartenoid muscle. It arises on the internal midline of the thyroid cartilage and inserts on the vocal process of the arytenoid cartilage. Cut the laryngeal mucosa of the vocal fold and observe the medial side of this muscle. Attached along its rostral border is the **vocal ligament.**

The cricoarytenoideus dorsalis and lateralis and the thyroarytenoideus muscles are innervated by the caudal laryngeal nerve from the recurrent laryngeal nerve.

The ventricularis and arytenoideus transversus will not be dissected.

Notice the relationship of laryngeal ventricle to the laryngeal muscles. From its laryngeal opening it runs rostrally between the thyroid lamina and thyroarytenoideus laterally and the vestibular fold, the cuneiform process and the ventricularis medially.

The External Ear

The external ear (Fig. 201) consists of the auricle and the external ear canal. The external ear canal is mostly cartilaginous but has a short osseous part. It makes a right-angle turn in the deeper part of its course and extends to the tympanic membrane.

Except for one small annular cartilage adjacent to the skull, the external ear consists of a single auricular cartilage that is rolled into a tube ventromedially. The tubular part becomes the external ear canal. Remove the skin from the base of the auricular cartilage. The **auricular cartilage** is funnel-shaped. Its external, convex surface faces caudally; the internal, concave surface faces rostrally. It has a slightly folded medial and lateral margin called the **helix.** The auricular cartilage is thin and pliable except proximally, where it thickens and rolls into a tube. On its internal, concave wall at the level of the beginning of the ear canal there is a transverse ridge, the **anthelix.** Opposite the anthelix the rostral boundary of the initial part of the ear canal is formed by a thick, quadrangular plate of the auricular cartilage, the **tragus.** Projecting caudally from the tragus and completing the lateral boundary of the external ear canal is a thin, elongate piece of cartilage, the **antitragus.** The **intertragic incisure** separates these two parts of the auricular cartilage.

The lateral portion of the helix is indented proximally by an incisure. At this point the skin forms a pouch, the **marginal cutaneous sac.**

The medial helix is nearly straight. An abrupt angle in this border at its proximal end forms the **spine of the helix.** Between the spine of the helix and the tragus the medial border of the ear canal is formed by two curved portions of the cartilage, the **medial** and **lateral crura** of the helix. They both end laterally in a free border separated from the tragus by the **pretragic incisure.**

Incise the lateral wall of the ear canal by two parallel incisions starting at the intertragic and pretragic incisures. Reflect the isolated piece of the lateral wall to observe the course of the ear canal to the tympanic membrane.

Interposed between the auricular cartilage and the external acoustic meatus of the temporal bone is the **annular cartilage.** This is a band of cartilage that overlaps the bony projection of the meatus.

Muscles of Mastication and Related Muscles

On the left half of the head, cut all attachments of the temporal and masseter muscles to the zygomatic arch and remove the arch.

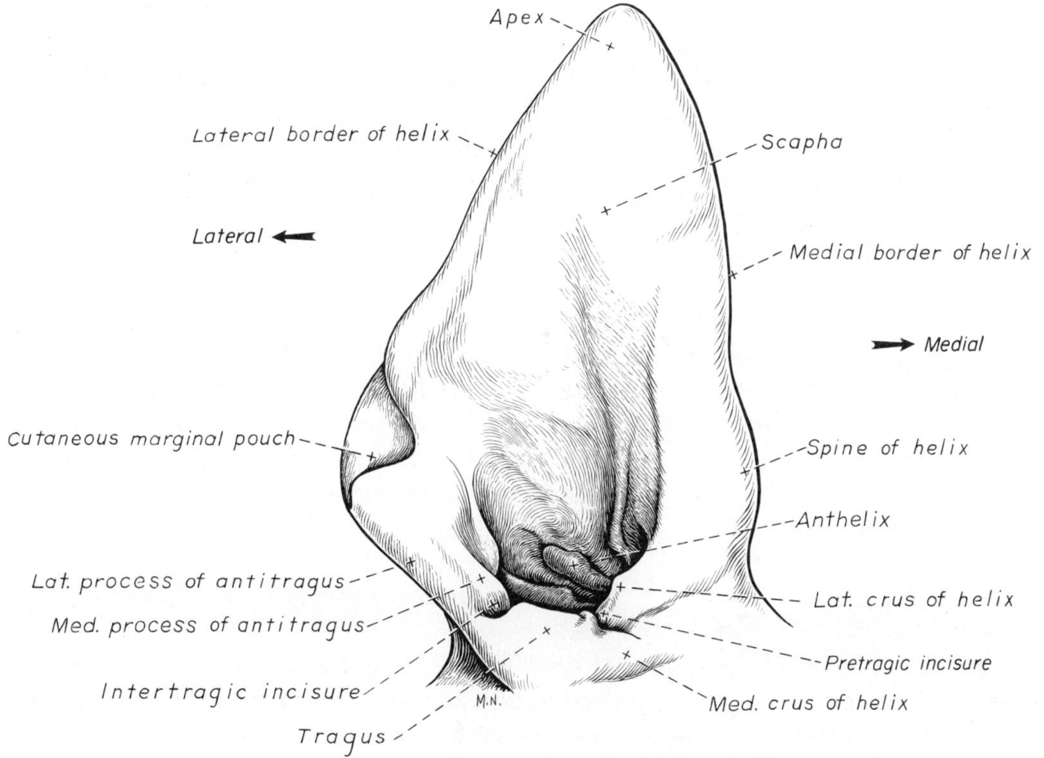

FIGURE 201. Right external ear.

The **temporalis muscle** (Figs. 194, 196) arises from the temporal fossa and inserts on the coronoid process of the mandible. Remove the muscle from its wide origin by scraping it off the bone with a blunt instrument, such as the scalpel handle, and reflect it. The temporal and masseter muscles fuse between the zygomatic arch and the coronoid process.

The **masseter muscle** (Fig. 196) arises from the zygomatic arch, where its deep portion is intermingled with the fibers of the temporal muscle. It inserts in the masseteric fossa, the ventrolateral surface of the ramus of the mandible and the angular process. The muscle is covered by a strong, glistening aponeurosis and contains many tendinous intermuscular strands.

Transect the temporal muscle as close to its insertion as possible. With bone cutters, remove the coronoid process and the remaining attached muscles. Observe the dorsal surface of the pterygoid muscles, which are now exposed in the ventral orbit.

Between the eyeball and the pterygoid muscles is the **zygomatic salivary gland** (Figs. 196, 212). It is hidden laterally by the zygomatic bone. The gland opens into the vestibule by one main and several minor ducts lateral to the last upper molar tooth.

The **medial** and **lateral pterygoid muscles** (Figs. 196, 212) arise from the pterygopalatine fossa and insert on the medial surface and caudal margin of the ramus of the mandible and angular process, ventral to the insertion of the temporal muscle. The muscles need not be distinguished from each other. The bulk of the muscle mass is the medial pterygoid. On the medial side of your specimen, cut the mucosa of the oropharynx from the rostral end of the palatine tonsil to the midline at the junction of the soft and hard palates. Reflect the cut edges to expose the ventral surface of the medial pterygoid muscle.

The temporal, masseter and pterygoid muscles function to close the jaws. They are innervated by the mandibular nerve, a branch of the trigeminal nerve (cranial nerve V).

Reflect the mandibular and parotid salivary glands to expose the digastricus muscle.

The **digastricus** (Figs. 194, 196) arises from the paracondylar process of the occipital bone and is inserted on the body of the mandible. A tendinous intersection crosses its belly and divides it into rostral and caudal parts. Transect it and expose its attachments. It acts to open the jaws. The rostral portion is innervated by the mandibular nerve (trigeminal), while the caudal belly receives facial nerve (cranial nerve VII) innervation.

Lingual Muscles

The muscles of the tongue (Figs 195, 202) may be divided into extrinsic and intrinsic groups. Three paired extrinsic muscles enter the tongue. The styloglossus and hyoglossus muscles are best exposed on the lateral side, the genioglossus on the medial side.

The **styloglossus** arises from the stylohyoid bone, passes rostroventrally lateral to the palatine tonsil and inserts in the middle of the tongue. It retracts and elevates the tongue.

The **hyoglossus** arises from the thyrohyoid and the basihyoid and passes into the root of the tongue. It lies medial to the styloglossus and retracts and depresses the tongue.

The **genioglossus** arises from the symphysis and adjacent surface of the body of the mandible. It joins its fellow at the median plane and is bounded laterally by the geniohyoideus and the hyoglossus. Its caudal fibers protrude the tongue and its rostral ones retract the apex. It lies partly in the frenulum. These muscles are all innervated by the hypoglossal nerve (cranial nerve XII).

Hyoid Muscles

The hyoid muscles (Fig. 202) are associated with the hyoid apparatus, which suspends the larynx and anchors the tongue. They function in swallowing, lolling, lapping and retching. All muscles of this group have names with the suffix-*hyoideus.* The prefixes of the hyoid muscles designate the bone or part from which they arise. Dissect the following hyoid muscles from the lateral side.

The **sternohyoideus,** from its origin on the sternum and first costal cartilage, is fused to the deeper sternothyroideus for the first third of its length. It then separates from this muscle and runs an independent course to insert on the basihyoid bone. Its origin was previously dissected.

The **thyrohyoideus** is a short muscle that lies dorsal to the sterno-

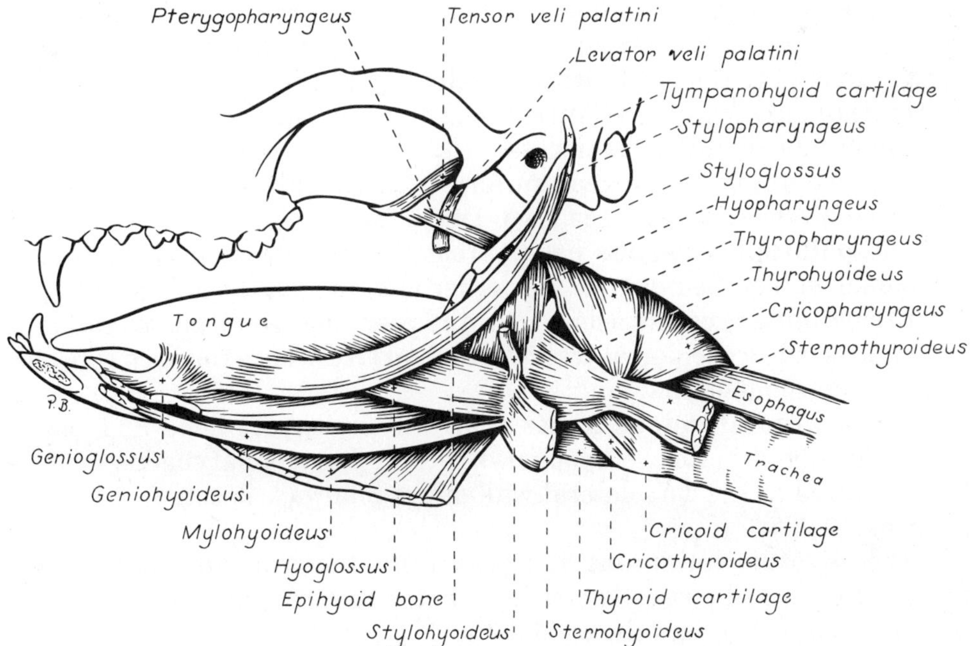

FIGURE 202. Muscles of pharynx and tongue, left lateral view, left mandible removed.

hyoideus. It extends from the thyroid cartilage of the larynx to the thyrohyoid bone.

The sternohyoideus and thyrohyoideus muscles are innervated by ventral branches of cervical nerves and the accessory nerve (cranial nerve XI).

The **mylohyoideus** spans the intermandibular space. It arises as a thin sheet of transverse fibers from the medial surface of the body of the mandible. It is inserted on its fellow muscle at the midventral raphe. Caudally it inserts on the basihyoid. It forms a sling that aids in the support of the tongue. It is innervated by the mandibular nerve of the trigeminal.

The **geniohyoideus** lies deep to the mylohyoideus. It is a muscular strap that arises on and adjacent to the mandibular symphysis. It parallels its fellow along the median plane and attaches to the basihyoid. Contraction of the geniohyoideus draws the hyoid apparatus and larynx rostrally. It is innervated by the hypoglossal nerve.

Pharyngeal Muscles

The pharyngeal muscles (Figs. 195, 202) aid directly in swallowing. The **cricopharyngeus** arises from the lateral surface of the cricoid cartilage. Its fibers are inserted on the median dorsal raphe of the laryngopharynx. Caudally its fibers blend with the esophagus.

The **thyropharyngeus** arises from the lateral side of the thyroid lamina and is inserted on the median dorsal raphe of the pharynx. This muscle is rostral to the cricopharyngeus and caudal to the hyopharyngeus.

The **hyopharyngeus** is in two parts as it arises from the lateral surface of the thyrohyoid bone and the ceratohyoid bone. This origin was previously transected. The fibers of both parts form a muscle plate that passes upward over the larynx and pharynx to be inserted on the median dorsal raphe of the pharynx with its fellow from the opposite side. These pharyngeal muscles are all innervated by pharyngeal branches of the glossopharyngeal and vagus nerves.

The remaining pharyngeal muscles and muscles of the palate listed here need not be dissected (see Figs. 195, 202).

The **palatopharyngeus** passes from the soft palate into the lateral and dorsal wall of the pharynx. Its border is in the palatopharyngeal arch. The **pterygopharyngeus** arises from the pterygoid bone, passes caudally and is inserted in the dorsal wall of the pharynx. These muscles constrict and shorten the pharynx.

The **stylopharyngeus** arises from the stylohyoid bone and passes caudolaterally, deep to the hyopharyngeus and thyropharyngeus muscles to be inserted in the dorsolateral wall of the pharynx. It acts to dilate the pharynx.

The **levator veli palatini** arises from the tympanic part of the temporal bone and passes ventrally to enter the soft palate caudal to the pterygoid bone. It raises the caudal end of the soft palate (Fig. 195).

The **tensor veli palatini** arises mainly from the cartilaginous wall of the auditory tube and is inserted on the pterygoid bone and medially on the soft palate.

The Eye and Related Structures

The **orbit** is a conical cavity containing the eyeball and ocular adnexa. The orbital margin is formed by the frontal, maxillary, lacrimal and zygomatic bones and the orbital ligament laterally. The medial wall of the orbit is formed by parts of the frontal, presphenoid and lacrimal bones. The ventral wall includes the zygomatic salivary gland and the pterygoid muscles. The dorsal and lateral walls are formed primarily by temporal muscle. To study the periorbita and the extraocular muscles, all of the zyomatic arch and the coronoid process of the mandible with the attached muscles must be removed. This is best done with a Stryker saw or bone cutters.

The **periorbita** is a cone-shaped sheath of connective tissue that encloses the eyeball and its muscles, vessels and nerves. Where the periorbita contacts the bone medially, it is the periosteum of the orbit. Its apex is caudal where it attaches to the bony margin of the optic canal and the orbital fissure. Rostrally it widens to blend with the periosteum of the face. Orbital fat may be observed on both sides of the periorbita.

Reflect the orbital ligament and the periorbita from the dorsolateral surface of the eyeball. The **lacrimal gland** (see Figs. 194, 214) is a small, flat lobular structure lying on the medial side of the orbital ligament within the periorbita. Small ducts that cannot be seen without a microscope empty their secretion into the conjunctival sac at the dorsal fornix.

Incise the periorbita longitudinally on its lateral side, and reflect it to expose the eyeball muscles and the levator of the upper eyelid (Fig. 203).

The **levator palpebrae superioris** (Fig. 203) is narrow and superficial. It begins at the apex of the orbit, extends over the dorsal rectus and widens to insert as a flat tendon in the upper eyelid.

There are seven extrinsic muscles of the eyeball (Figs 203, 204, 214): two **obliquus muscles,** four **rectus muscles** and the **retractor bulbi.** All of these extrinsic muscles insert on the fibrous coat of the eyeball, the sclera, near the equator of the eyeball. The rectus muscles insert closer to the corneoscleral junction than the retractor muscles.

The four rectus muscles are the **dorsal rectus, medial rectus, ventral rectus** and **lateral rectus.** As they course rostrally from their small area of origin around the optic canal and orbital fissure, they diverge and attach to the sclera on an imaginary line encircling the eyeball at its equator. In the spaces between the rectus muscles, parts of the **retractor bulbi** can be seen. The retractor bulbi muscle consists of four fascicles that surround the optic nerve, a dorsal pair and a ventral pair.

The **dorsal oblique** ascends on the dorsomedial side of the extraocular muscles dorsal to the medial rectus. Roll the dorsal aspect of the eyeball laterally to expose these muscles. The dorsal oblique is a narrow muscle

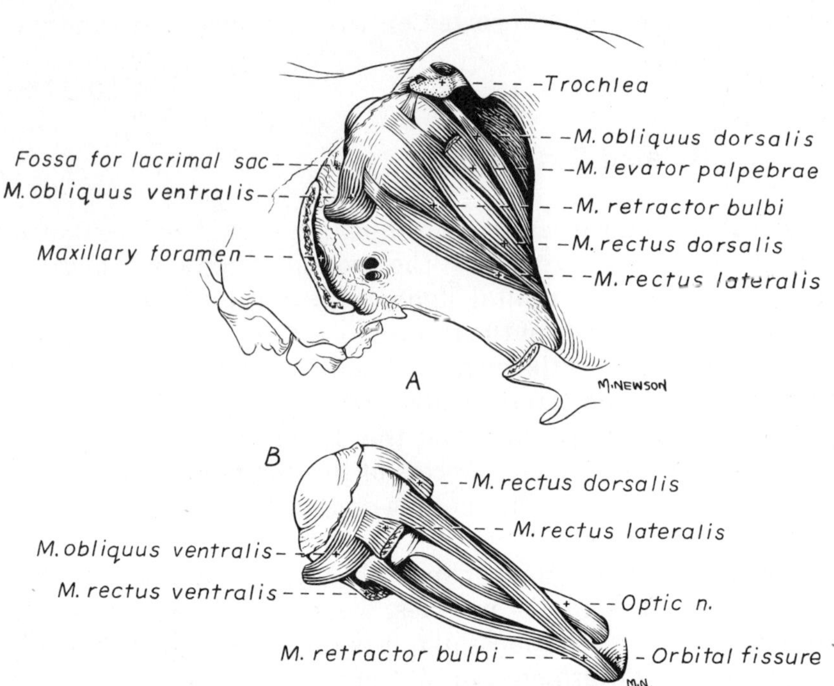

FIGURE 203. A, *Extrinsic muscles of left eyeball, dorsolateral view.* B, *Retractor bulbi muscle exposed, lateral view.*

that forms a long tendon rostrally that passes through a groove in the trochlea. The **trochlea** is a cartilaginous plaque attached at the level of the medial angle of the eye to the wall of the orbit. The tendon of the dorsal oblique muscle turns and courses laterally after passing around the trochlea and attaches to the sclera under the tendon of insertion of the dorsal rectus muscle. This insertion is difficult to distinguish and need not be dissected (Fig. 203a).

The **ventral oblique** arises from the rostral border of the palatine bone at the level of the maxillary foramen, passes ventral to the ventral rectus and is inserted on the sclera at the insertion of the lateral rectus.

To understand the action of these individual muscles, consider the eyeball as having three imaginary axes that cross in the center of the globe (Fig. 204). The dorsal and ventral rectus muscles would rotate the eyeball around a horizontal axis through the equator from medial to lateral. The medial and lateral rectus muscles would rotate it around a vertical axis through the equator from dorsal to ventral. The oblique muscles would rotate the eyeball around a longitudinal axis passing from rostral to caudal through the center of the eyeball.

Enlarge the opening of the palpebral fissure by cutting through the lateral commissure and the underlying conjunctiva to the eyeball. This will facilitate exposure of the third eyelid in the medial angle. Elevate the third eyelid and observe the lymph nodules on its bulbar conjunctival surface and the **superficial gland of the third eyelid** (Fig. 205).

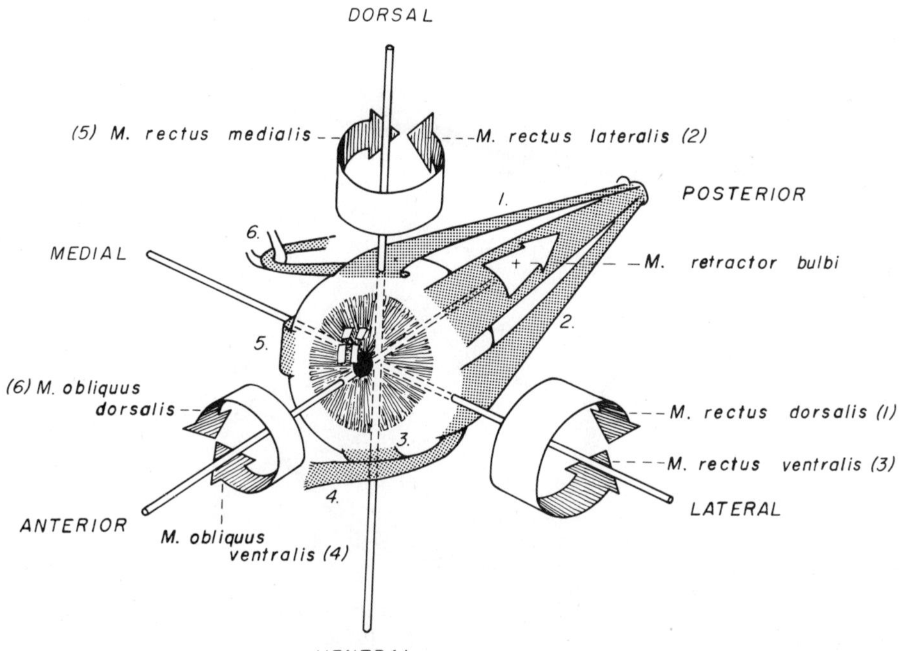

FIGURE 204. Schema of extrinsic ocular muscles and their action on the eyeball.

Bulbus Oculi

In the following dissection, leave the bulbus oculi, or eyeball, in the orbit. Many of these structures are best observed in a fresh eye under a dissecting microscope. Freeze-dried preparations of the opened eyeball are also instructive. Directional terms used with reference to the eyeball are anterior and posterior, and superior and inferior. The wall is composed of three layers: a fibrous coat, a vascular coat and an internal coat that includes the retina (Figs. 206, 207).

1. The **external fibrous coat** is composed of the **cornea,** which forms the anterior one-quarter, and the **sclera,** which forms the posterior three-quarters. The cornea is transparent and circular. It meets the dense, opaque sclera peripherally at the corneoscleral junction, or **limbus,** of the cornea. The sclera is dull gray-white. Anteriorly it is covered by bulbar conjunctiva. Posterior to this the extrinsic ocular muscles insert in its

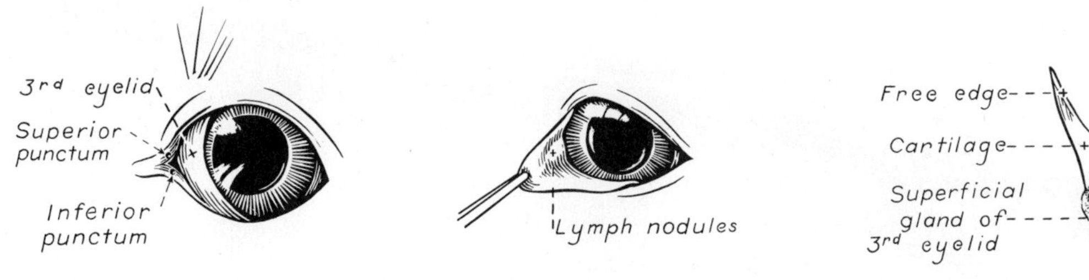

FIGURE 205. Third eyelid of left eye.

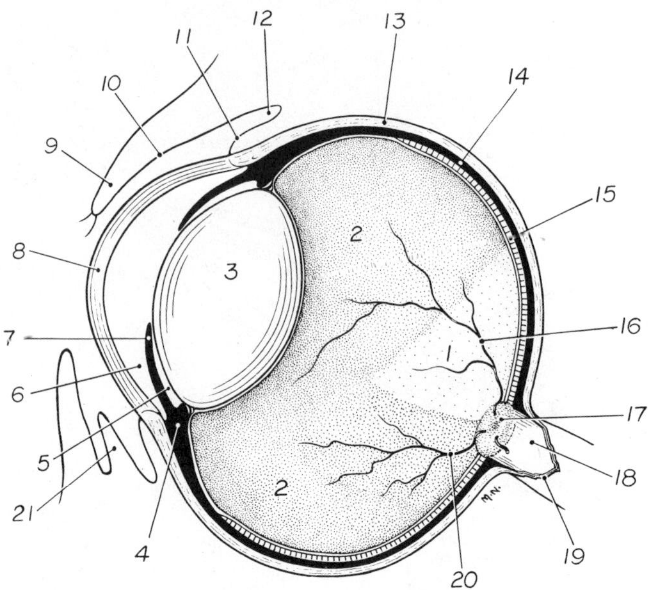

FIGURE 206. Sagittal section of the eyeball.

1. *Tapetum lucidum*
2. *Nontapetal nigrum*
3. *Lens*
4. *Ciliary body*
5. *Posterior chamber*
6. *Anterior chamber*
7. *Iris*
8. *Cornea*
9. *Superior eyelid*
10. *Palpebral conjunctiva*
11. *Bulbar conjunctiva*
12. *Fornix*
13. *Sclera*
14. *Choroid*
15. *Retina*
16. *Superior retinal vein*
17. *Optic disc*
18. *Optic nerve*
19. *Dura-arachnoid*
20. *Inferior medial retinal vein*
21. *Third eyelid*

wall, and it is penetrated by blood vessels and nerves, including the large optic nerve posteriorly. Transect and reflect the attachments of the lateral rectus and retractor bulbi muscles on the eyeball and observe the optic nerve on the posterior surface of the globe.

2. The **middle vascular coat** (the **uvea**) is deep to the sclera and consists of three continuous parts from posterior to anterior: the choroid,

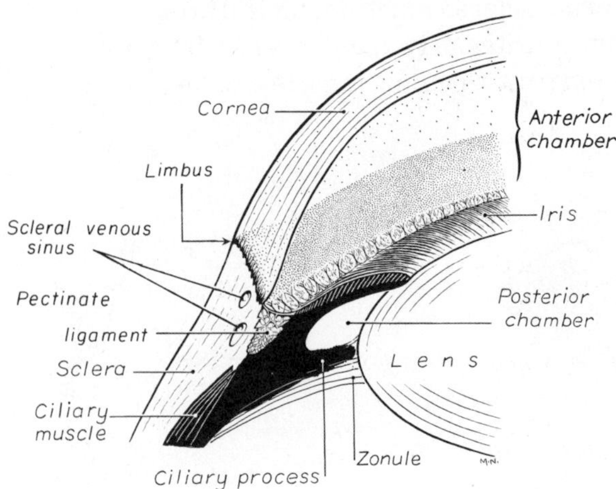

FIGURE 207a. Section of eyeball at iridocorneal angle.

FIGURE 207b. Inner surface of a ciliary body segment.

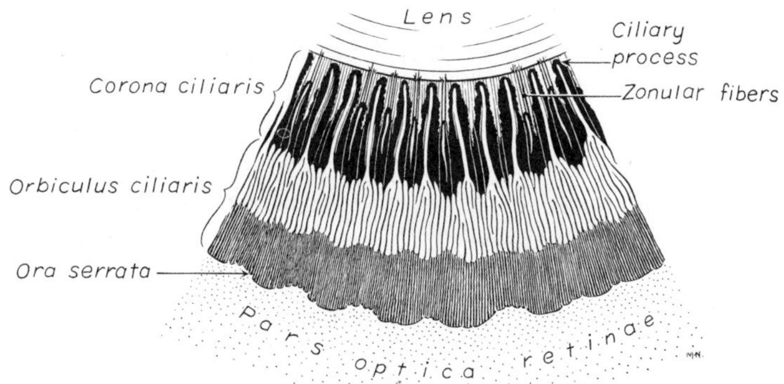

ciliary body and iris. The **iris,** which has circular and radial smooth muscle, can be seen through the cornea as a pigmented diaphragm with a central opening, the **pupil.** While the eyeball is in the orbit, use a sharp scalpel or razor blade to make a sagittal cut through the eyeball from anterior to posterior pole and remove the lateral half of the eyeball.

The **choroid** is the posterior portion of the vascular coat and is firmly attached to the sclera. It is pigmented and lines the internal surface of the sclera as far anterior as the ciliary body posterior to the lens. The junction of the choroid and the ciliary body is seen as an undulating line in the overlying retina called the **ora serrata** (Fig. 207). The fundus is the posterior or deep portion of the eyeball. The light-colored, reflective area in the dorsal part of the fundus is the **tapetum lucidum** (Fig. 206) of the choroid. This is a specialized layer of cells that reflects light rays.

The vascular coat forms a thick circular mound at the level of the limbus. This is the **ciliary body** (Fig. 207). The ciliary body, located

FIGURE 207c. Zonular fibers passing along ciliary processes before attaching to lens, anterior view.

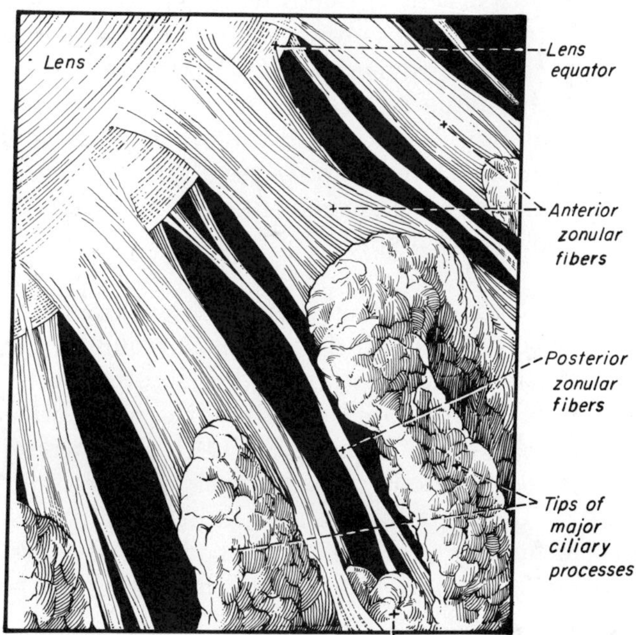

between the iris and choroid, contains numerous muscle bundles that function in the regulation of the shape of the lens. The internal surface of the ciliary body is marked by longitudinal folds, the **ciliary processes.** Observe these on the lateral portion of the eyeball that was removed. These processes surround but do not attach to the lens at its equator. They consist of several hundred pigmented folds alternating in length, which are small at their posterior border near the ora serrata, but enlarge in size anteriorly as they approach the lens.

The **zonule** is the suspensory apparatus of the lens. It is composed of numerous fine strands, **zonular fibers,** which pass from the region of the ora serrata along the ciliary processes to the equator of the lens. The lens is suspended from the ciliary processes by the zonular fibers. Contraction of the ciliary muscles pulls the ciliary body and processes towards the lens and therefore relaxes the tension on the zonular fibers attached to the lens. This allows the elastic lens to become more spherical and accommodate for near vision. (In other words contraction of the ciliary muscle is required for near vision.)

The **lens** in a fixed specimen is firm and opaque. In life it is transparent and elastic. It is bounded posteriorly by the transparent jellylike **vitreous body,** which fills the **vitreous chamber** posterior to the lens. Anteriorly it is bounded by the iris and by aqueous humor. The **aqueous humor** fills the space between the cornea and the lens. This space is divided by the iris into two chambers. The **anterior chamber** is the space between the cornea and the iris. The **posterior chamber** is a narrow cavity between the iris and the lens. Aqueous humor, produced by the ciliary epithelium covering the ciliary processes, circulates through the zonular fibers into the posterior chamber. It then passes through the pupil into the anterior chamber and is drained through the **trabecular meshwork** at the **iridocorneal angle,** where it passes into the venous system via the **venous scleral sinus.** The iridocorneal angle (Fig. 207) is traversed by a meshwork of fibers with intervening spaces. This meshwork of the angle is referred to as the **pectinate ligament.** The integrity of this angle is critical in the drainage of aqueous humor from the eyeball. Failure in the drainage results in increased intraocular pressure, which is known as glaucoma. Remove the lens. Note the zonular fibers as they stretch and break.

3. The internal coat of the eye consists of the retina and its associated blood vessels and the nerves surrounding the vitreous body. That portion of the retina containing the light-sensitive rods and cones, the bipolar cells and the ganglion cells is the pars optica retinae. It covers the internal surface of the choroid from the point where the optic nerve enters to the level of the ciliary body. From this boundary, called the ora serrata (Fig. 207), there is a thin non–light-receptive portion of the retina that passes anteriorly over the ciliary body as the pars ciliaris retinae and continues on the posterior surface of the iris as the pars iridica retinae. The ciliary portion of the retina is two layers thick and forms the blood-aqueous barrier through which aqueous fluid is secreted into the posterior chamber. The iridial portion of the retina is also a double cell layer consisting of

pigment cells, which give the iris its color, and myoepithelial cells, which form the dilator pupillae. (The ciliary and iridial portions of the retina are referred to collectively as the pars ceca retinae.)

In the embalmed specimen the nervous layer of the pars optica retinae has a gray-white appearance, and it readily peels away from the single layer of pigmented epithelial cells on its posterior surface, which remains attached to the choroid. This pigmented layer of the retina and the pigment of the choroid give the interior of the eyeball a brown to black appearance except where the specialized layer of the choroid, the **tapetum lucidum,** is located. This area exhibits a variety of brilliant colors from silver to blue to green or orange. In this area there is no pigment in the pigment epithelial cells of the retina covering the tapetum lucidum. The brown-black portion of the interior of the eyeball is sometimes referred to as the nontapetal area or nontapetal nigrum.

Notice the entrance of the optic nerve into the posterior aspect of the eyeball. This is the **optic disk.** With careful observation of the disk, you may see the retinal vessels that enter with it to supply the internal surface of the retina. The posterior portion of the eyeball that includes the area of the optic disk, tapetum lucidum and adjacent nontapetal nigrum is referred to as the **fundus** of the eyeball. The optic disk may be found in the inferior region of the tapetum lucidum or at or below its inferior border. This varies with the breed of dog.

Superficial Veins of the Head

The **external jugular vein** (Figs. 208, 209) is formed by the confluence of the linguofacial and maxillary veins caudal to the mandibular salivary gland, which lies between these two veins.

The **lingual vein** (Fig. 209) is the first large tributary that enters the **linguofacial vein** ventrally. Its radicles drain blood from the tongue, the larynx and part of the pharynx. These radicles will not be dissected.

The **facial vein** is the other tributary of the linguofacial. The radicles that form the facial vein lie on the dorsal surface of the muzzle. One of these, the **dorsal nasal,** runs caudally from the nares, while another, the **angularis oculi,** passes rostrally from the medial aspect of the orbit. Blood may drain from the face in either direction through the angularis oculi. Identify these veins. The remaining branches will not be dissected for to do so would sacrifice nerves and arteries. These branches drain the lateral muzzle and the upper and lower lips.

The **maxillary vein** drains the ear, orbit, palate, nasal cavity, cheek and mandible as well as the cranial cavity. It will not be dissected.

Facial Nerve

The **facial** or **seventh cranial nerve** (Fig. 210) supplies all of the superficial muscles of the head and face as well as the caudal belly of the digastricus and the platysma of the neck. The nerve enters the petrosal

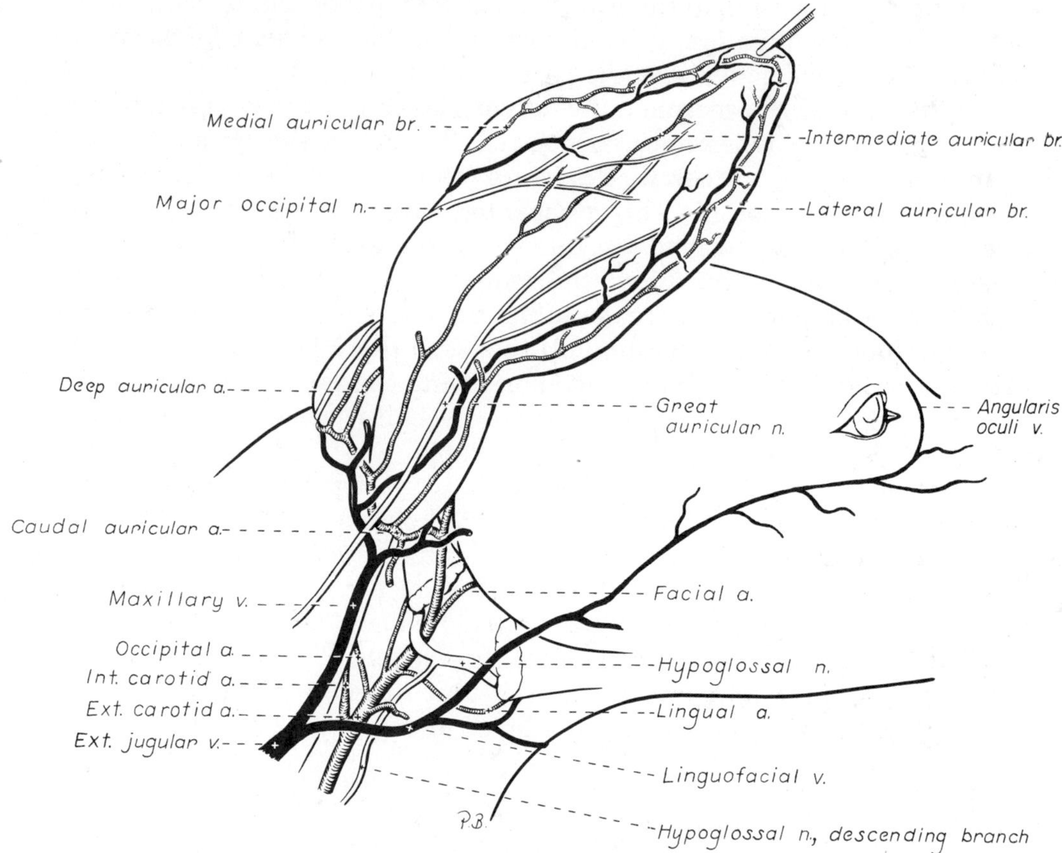

FIGURE 208. Vessels and nerves of external ear.

part of the temporal bone via the internal acoustic meatus, courses through the facial canal of that bone and leaves the skull at the stylomastoid foramen just caudal to the external acoustic meatus, where it divides into several branches.

Reflect the parotid gland. Dissect deep between the parotid and sublingual glands to expose the facial nerve arising from the stylomastoid foramen caudal to the horizontal portion of the ear canal. The maxillary vein may be transected as it crosses the lateral surface of the nerve at this site. Dissect the following branches of the facial nerve.

The **auriculopalpebral nerve** arises as the facial nerve curves rostrally ventral to the base of the ear. Dissect deep to the rostral edge of the parotid gland to locate the nerve. **Rostral auricular branches** course through the parotid gland and are distributed to the rostral auricular muscles. The auriculopalpebral nerve crosses the zygomatic arch, supplies branches to the rostral auricular plexus and continues to the orbit to supply **palpebral branches** to the orbicularis oculi. Beyond this, a branch passes medial to the orbit and continues rostrally on the nose to supply the muscles of the nose and upper lip.

Two **buccal branches** course across the masseter muscle to innervate the muscles of the cheek, the upper and lower lip and the lateral surface of the nose. One is dorsal to the parotid duct. The other, ventral to the parotid duct, is near the ventral border of the masseter. Identify these two branches and the parotid duct.

A branch of the mandibular nerve from the **trigeminal** or **fifth cranial nerve,** known as the **auriculotemporal nerve,** is apparent in the dissection field of the auriculopalpebral branch of the facial nerve. The auriculotemporal nerve emerges between the caudal border of the masseter muscle and the base of the external ear below the zygomatic arch. It may be found deep to the origin of the auriculopalpebral nerve. The auriculotemporal nerve supplies sensory branches to the skin of the external ear and the temporal, zygomatic and masseteric regions.

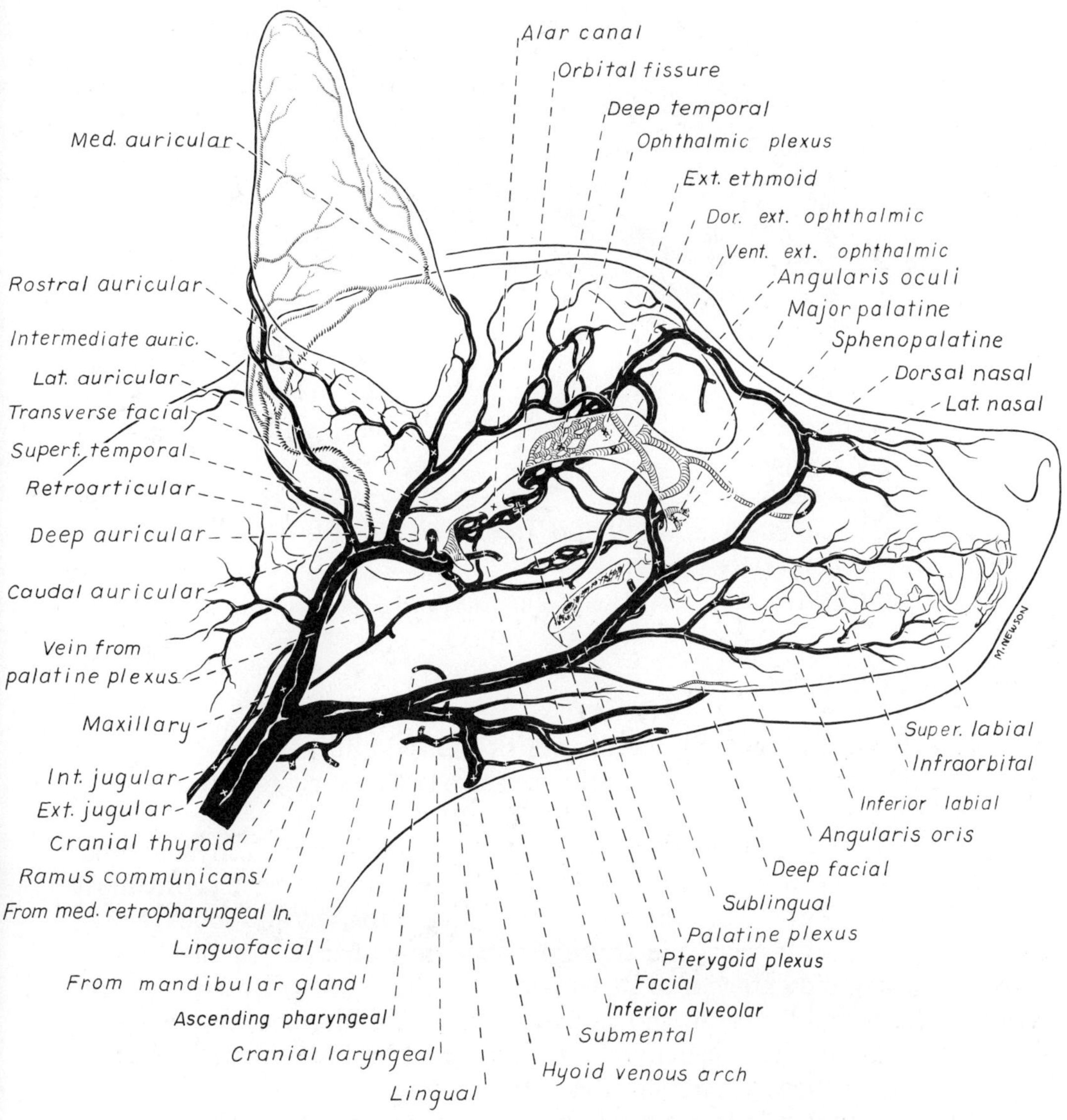

FIGURE 209. Superficial veins of the head, right lateral aspect.

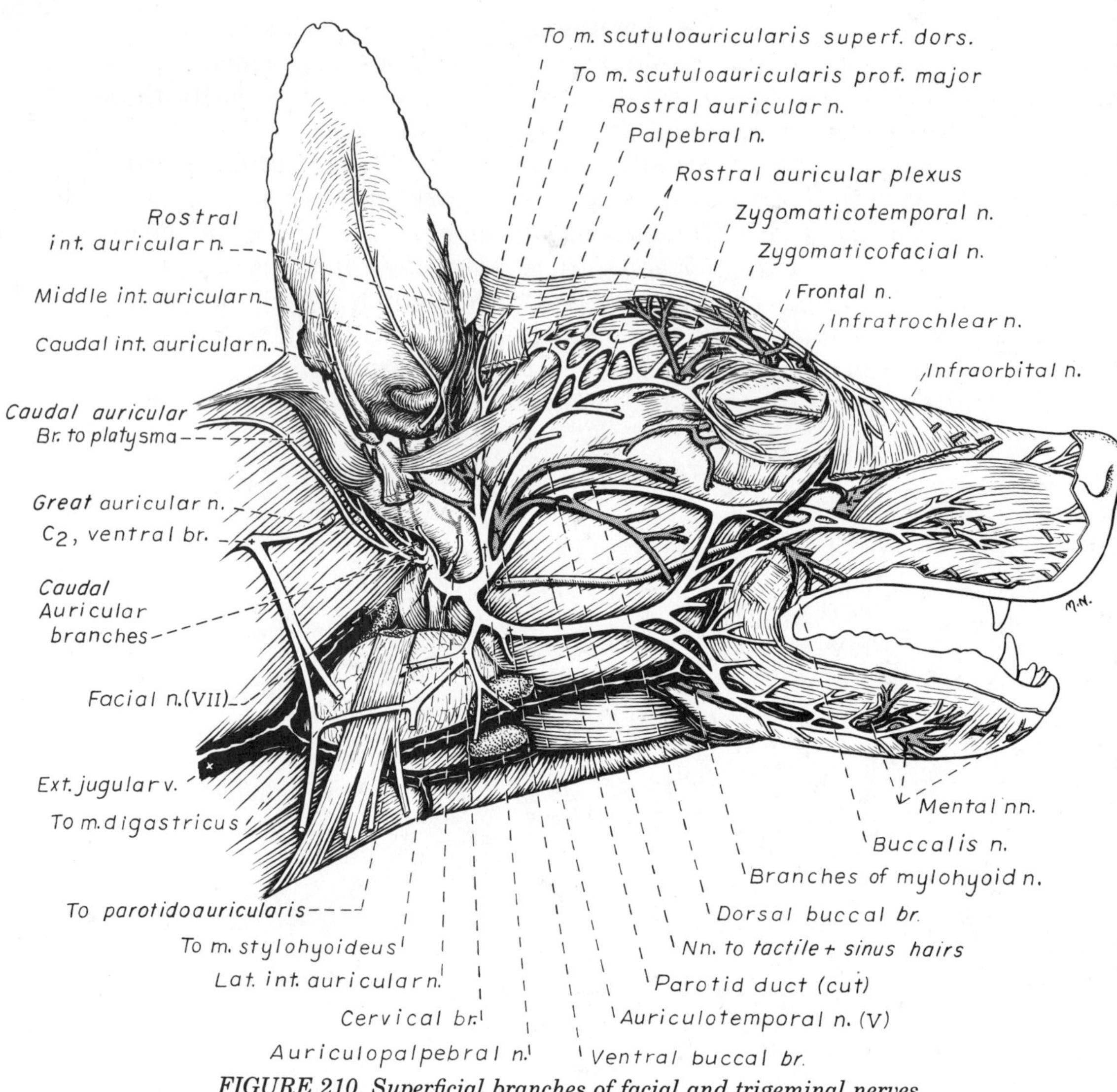

FIGURE 210. Superficial branches of facial and trigeminal nerves.

Cervical Structures

The **thyroid gland** (Fig. 216) is dark-colored and usually consists of two separate lobes lying lateral to the first five tracheal rings. Occasionally a connecting isthmus is present.

There are two **parathyroid glands** associated with each thyroid lobe. They are light-colored, spherical bodies. The **external parathyroid** most commonly lies in the fascia at the cranial pole of the thyroid lobe. It may be entirely separate from the thyroid tissue or embedded in the cranial pole of the thyroid external to its capsule. The **internal parathyroid** lies deep to the thyroid capsule on the medial aspect of the lobe. Occasionally it is embedded in the parenchyma of the thyroid and is difficult to locate. The location of these glands is subject to variation.

The **cervical portion of the esophagus** extends from the pharynx to the thoracic portion of the esophagus at the thoracic inlet. It begins opposite the middle of the axis dorsally and opposite the caudal border of the cricoid cartilage ventrally. A plicated ridge of mucosa, the **pharyngoesophageal limen,** marks the boundary between the laryngopharynx and the esophagus. The esophagus inclines to the left, so that at the thoracic inlet it usually lies to the left of the trachea.

The **trachea** extends from a transverse plane through the middle of the axis to a plane between the fourth and fifth thoracic vertebrae. It is composed of approximately 35 C-shaped **tracheal cartilages.** They are open dorsally and the space is bridged by the tracheal muscle.

Common Carotid Artery

- Common carotid a.
  - Caudal thyroid a.
  - Cranial thyroid a.
  - Internal carotid a.
  - External carotid a.
    - Occipital a.
    - Cranial laryngeal a.
    - Lingual a.
    - Facial a.
      - Sublingual a.
    - Caudal auricular a.
    - Superficial temporal a.
    - Maxillary a.

On the right side, expose the **common carotid artery** in the carotid sheath and observe the following branches.

1. The **caudal thyroid artery** has a variable origin from the major arterial branches in the thoracic inlet. It passes cranially on the trachea supplying the trachea, esophagus and thyroid gland.

2. The **cranial thyroid artery** (Fig. 211) arises from the ventral surface of the common carotid at the level of the larynx. It supplies the thyroid and parathyroid glands, the pharyngeal muscles, the laryngeal muscles and mucosa, the cervical parts of the trachea and esophagus and the adjacent portions of the sternocephalicus and cleidomastoideus. Clean the origin of this vessel.

The large **medial retropharyngeal lymph node** (see Fig. 13) is dorsal to the common carotid artery at the larynx and ventral to the wing of the atlas. Afferent vessels arise from the tongue, the nasal cavity, the pharynx, the salivary glands, the external ear, the larynx and the esophagus. The tracheal duct of each side arises from this lymph node.

Identify the internal and external carotid arteries, which are the terminal branches of the common carotid artery.

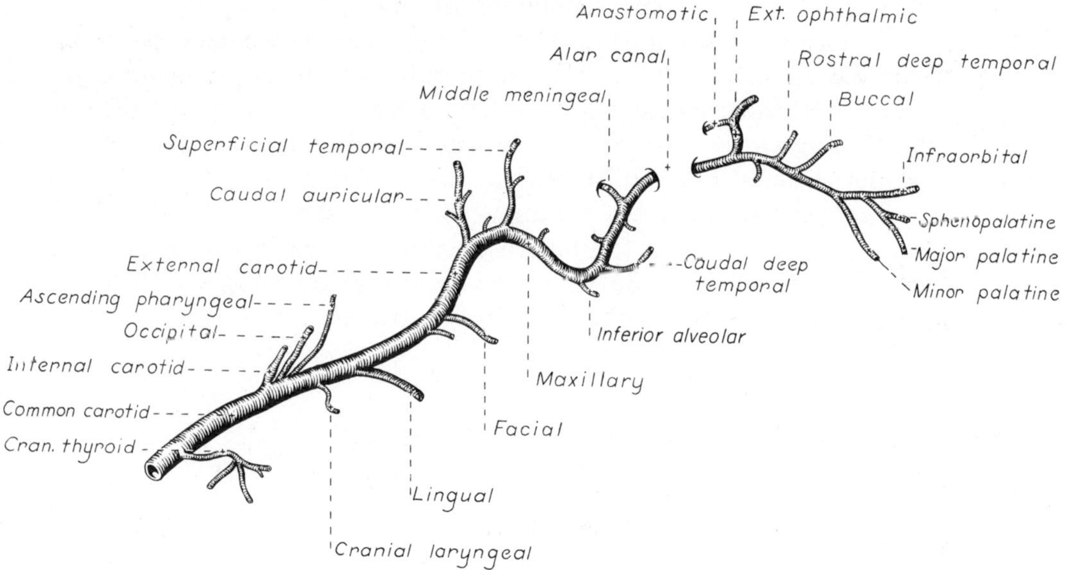

FIGURE 211. Branches of right common carotid artery.

3. The **internal carotid artery** is closely associated with the occipital artery, the first branch of the external carotid. A bulbous enlargement at the origin of the internal carotid artery is the **carotid sinus,** a baroreceptor. (The carotid body, a chemoreceptor, lies at the bifurcation of the carotid arteries.) Beyond this the internal carotid artery ascends across the lateral surface of the pharynx medial to the occipital artery. No branches leave the internal carotid in its extracranial course. It enters the carotid canal deep in the tympano-occipital fissure. Its course from here into the cranial cavity has been described. Its branches to the brain will be dissected later.

4. The **external carotid artery** (Fig. 211) passes cranially, medial to the digastricus. At the caudal border of the mandible, rostroventral to the annular cartilage of the ear, the vessel terminates by dividing into the **superficial temporal** and **maxillary** arteries. The maxillary is the direct continuation of the external carotid. Dissect the following branches of the external carotid. Transect the digastricus muscle and remove the caudal portion.

a. The **occipital artery** leaves the external carotid adjacent to the internal carotid and passes dorsally to supply the muscles on the caudal aspect of the skull and the meninges. Do not trace this artery.
b. The **cranial laryngeal artery** is a ventral branch that supplies the adjacent sternomastoideus and pharyngeal muscles. It enters the larynx between the thyrohyoid bone and the thyroid cartilage to supply the mucosa and laryngeal muscles.
c. The **lingual artery** leaves the ventral surface of the external carotid and passes rostrally to supply the tonsil and tongue.

d. The **facial artery** leaves the external carotid beyond the lingual. A branch, the **sublingual artery,** lies medial to the digastric muscle and is accompanied by the mylohyoid nerve. It runs rostrally into the tongue.

   The facial artery courses rostrolaterally between the digastric and masseter muscles to reach the cheek lateral to the mandible, where it supplies the lips and nose.

e. The **caudal auricular artery** usually arises from the external carotid at the base of the ear and ascends under the caudal auricular muscles. Reflect the caudal limb of the parotid gland to expose the caudal auricular artery and its branches, which need not be dissected. Lateral, intermediate and medial auricular arteries course distally on the convex surface of the ear. Occasionally this artery arises closer to the origin of the external carotid.

f. The **superficial temporal artery** arises rostral to the base of the auricular cartilage at the caudodorsal border of the mandible and courses dorsally. It supplies the parotid gland, the masseter and temporal muscles, the rostral auricular muscles and the eyelids.

g. The **maxillary artery** is the larger terminal branch of the external carotid artery. It is deeply placed and is closely associated with a number of cranial nerves. From its origin with the superficial temporal it passes rostromedially beneath the temporomandibular joint medial to the retroarticular process in its course to the alar canal.

Remove the auricular muscles and reflect the ear caudally. Cut through the origin of the temporal muscle along its margin. With a blunt instrument, remove it from the temporal fossa. Cut the attachments of the temporal and masseter muscles to both sides of the zygomatic arch. Sever the orbital ligament at the arch. With bone cutters, detach each end of the zygomatic arch and remove it. Transect the temporal muscle as close to its insertion on the coronoid process as possible. Cut off the coronoid process below the level of the ventral border of the zygomatic arch with bone cutters. Remove all the temporal muscle to expose the periorbital tissues and the pterygoid muscles. Vessels and nerves entering the temporal muscle must be severed. Loosen the temporomandibular joint by forcing the symphyseal end of the mandible medially. Rotate the mandible so that the stump of the coronoid process is forced laterally.

In the oral cavity, reflect the mucosa from the level of the root of the tongue to the frenulum along the sublingual fold.

Mandibular Nerve

The branches of the mandibular nerve from the **trigeminal** or **fifth cranial nerve** have been exposed by this dissection (Fig. 212). The mandibular nerve leaves the cranial cavity via the oval foramen (Fig.

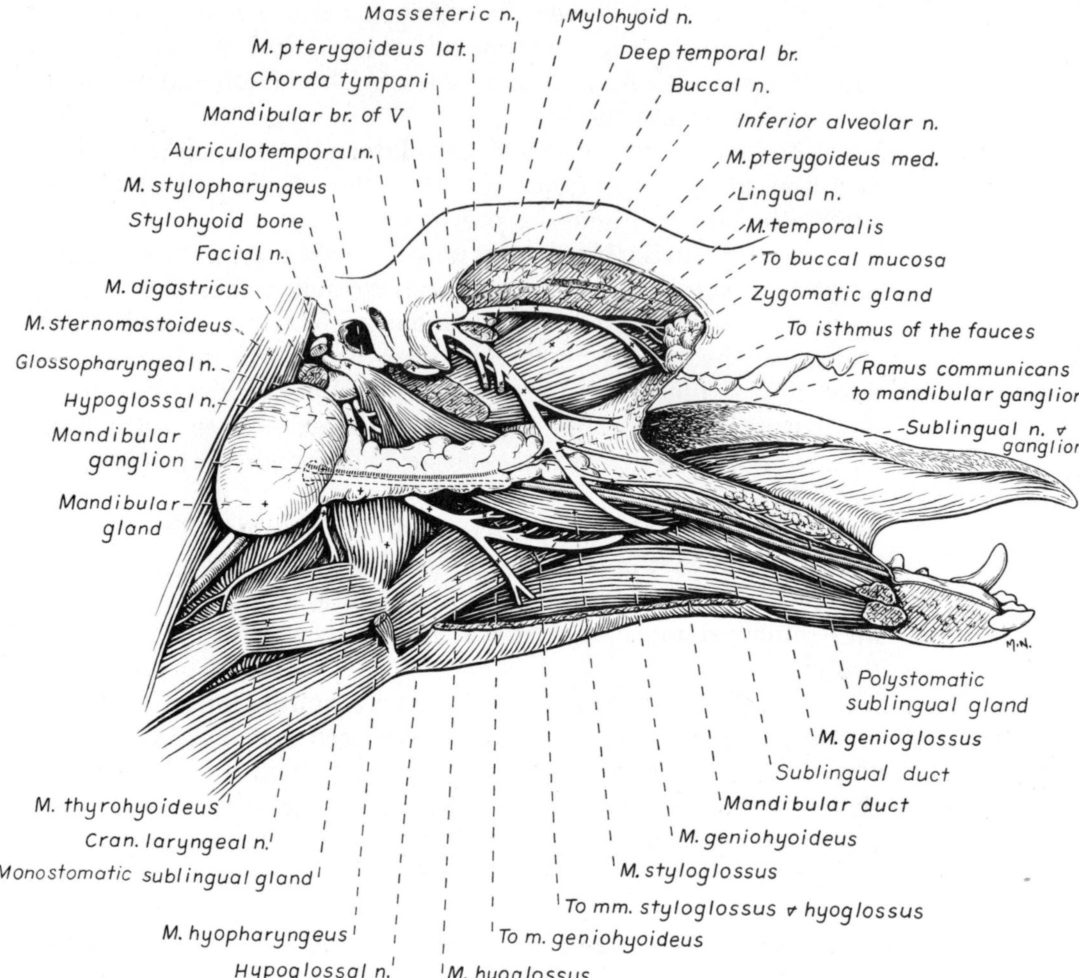

FIGURE 212. Muscles, nerves and salivary glands medial to right mandible, lateral view.

213). Branches arise on the surface of the pterygoid muscles ventral and lateral to the apex of the periorbita. These include the pterygoid, deep temporal and masseteric nerves that contain somatic motor neurons that innervate the muscles of mastication. Most of these branches have been severed by the dissection. The **buccal nerve** crosses the pterygoid muscles and enters the cheek lateral to the zygomatic salivary gland. Remove the zygomatic gland to get better exposure. This nerve is sensory to the mucosa and skin of the cheek.

Rotate the stump of the coronoid process laterally to observe the lingual, inferior alveolar and mylohyoid nerves, which cross the dorsal surface of the medial pterygoid muscle.

The **lingual nerve** (sensory) is the largest and most rostral of the three. It can be observed coursing across the pterygoid muscles and passing between the styloglossus and mylohyoideus. On the medial side of the specimen pull the tongue medially and observe the nerve between these muscles and the place where it crosses the lateral side of the mandibular and sublingual ducts and enters the tongue. It is sensory to the rostral two-thirds of the tongue.

The **inferior alveolar nerve** (sensory) enters the mandibular foramen on the medial side of the ramus of the mandible. It courses through the mandibular canal, supplying sensory nerves to the teeth. The mental nerves that emerge through the mental foramina and supply the lower lip are branches from this nerve.

The **mylohyoid nerve** (motor) is a caudal branch of the inferior alveolar. It reaches the ventral border of the mandible, supplies a branch to the rostral belly of the digastricus muscle and continues into the mylohyoid muscle. Observe this nerve emerging on the medial side of the angle of the mandible lateral to the mylohyoideus. It is motor to the mylohyoideus and sensory to the skin between the mandibles.

The **auriculotemporal nerve** (sensory) leaves the mandibular nerve at the oval foramen, passes caudal to the retroarticular process of the temporal bone and emerges between the base of the auricular cartilage and the masseter muscle, where it was previously seen.

Maxillary Artery

Maxillary artery
- Inferior alveolar a.
- Caudal deep temporal a.
- Middle meningeal a.

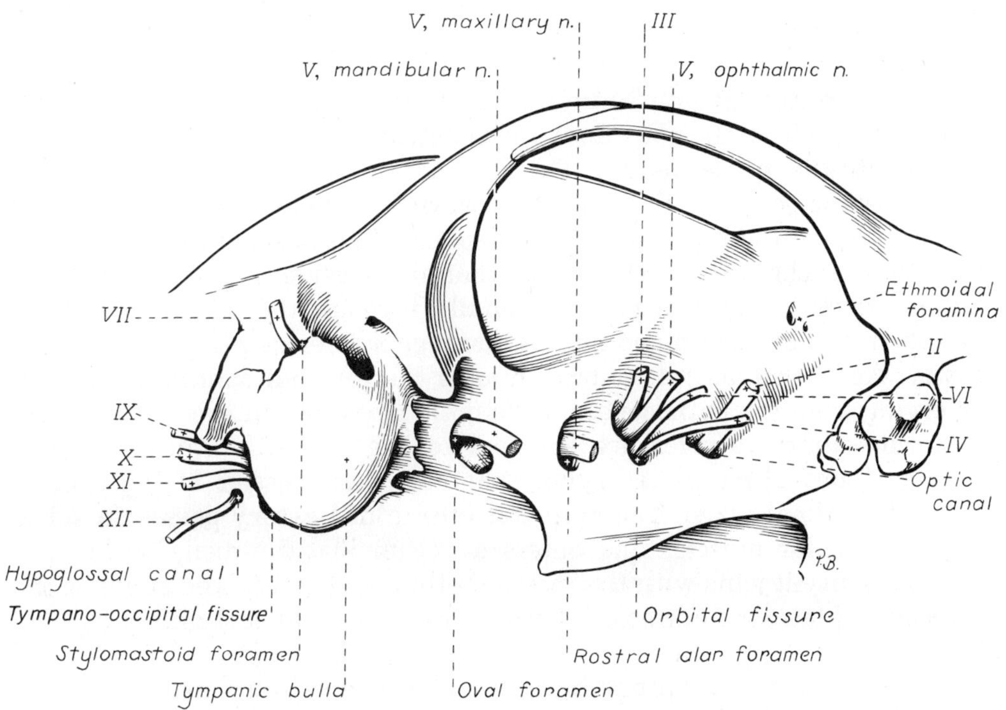

FIGURE 213. Cranial nerves leaving skull, ventrolateral view.

External ophthalmic a.
  Anastomotic rami
  Muscular rami
  External ethmoidal a.
Descending palatine
  Minor palatine a.
  Major palatine a.
  Sphenopalatine a.
Infraorbital a.

Complete the disarticulation of the temporomandibular joint and remove the lateral pterygoid muscle. Reflect the ramus of the mandible laterally and identify the following branches of the maxillary artery (Figs. 211, 214). The first three arise before the maxillary artery enters the alar canal.

1. The **inferior alveolar artery** (Fig. 211) enters the mandibular foramen with the inferior alveolar nerve and courses through the mandibular canal. It supplies branches to the roots of the teeth in the lower jaw.

2. The **caudal deep temporal artery** arises near the inferior alveolar artery and enters the temporal muscle. Only the origin of this artery may be seen.

3. The **middle meningeal artery** passes through the oval foramen and courses dorsally in a groove on the inside of the calvaria. It will be followed in a later dissection to the dura over the cerebral hemispheres. Do not dissect its origin.

4. The **external ophthalmic artery** arises from the maxillary upon its emergence from the alar canal and penetrates the apex of the periorbita adjacent to the orbital fissure. The external ophthalmic artery gives rise to the vessels that supply the structures within the periorbita. Incise the periorbita longitudinally along its dorsolateral border and reflect it.

The branches of the external ophthalmic artery need not be dissected. One anastomotic branch passes caudally through the orbital fissure to join the internal carotid and middle meningeal arteries within the cranial cavity. Another anastomotic branch joins the internal ophthalmic artery emerging from the optic canal on the optic nerve. From this anastomosis posterior ciliary arteries are supplied to the eyeball. Branches of the external ophthalmic artery supply the extrinsic muscles of the eyeball and the lacrimal gland. The **external ethmoidal artery** passes dorsal to the extraocular muscles and enters an ethmoidal foramen. Within the cranial cavity it joins with the internal ethmoidal artery and supplies the cribriform plate, the ethmoid labyrinth and the nasal septum.

5. Among the terminal branches of the maxillary artery are the minor and major palatine and sphenopalatine arteries. They usually arise from the descending palatine artery as it courses rostroventrally over the

medial pterygoid muscle. Only the origin of these vessels need be observed. Their origin can be seen at the rostral border of the medial pterygoid muscle deep to the zygomatic salivary gland, which should be removed.

The **minor palatine artery** passes ventrally, caudal to the hard palate, and is distributed to the adjacent soft and hard palates. Clean the mucosa from the palate just medial to the last molar and see the branches of this vessel.

The **major palatine artery** enters the caudal palatine foramen and passes through the major palatine canal to supply the hard palate.

The **sphenopalatine artery** passes through the sphenopalatine foramen to the interior of the nasal cavity.

6. The maxillary artery terminates as the **infraorbital artery,** which supplies dental branches to the caudal cheek teeth through alveolar canals, enters the maxillary foramen and passes through the infraorbital canal. Within the canal, dental branches arise. These supply the premolars, the canine teeth and incisor teeth. The infraorbital artery emerges from the infraorbital foramen and terminates as the lateral and rostral dorsal nasal arteries, which supply the nose and the upper lip.

Cranial Nerves

There are twelve pairs of cranial nerves (Figs. 213–216). Each pair is both numbered and named. The numbers indicate the rostrocaudal order

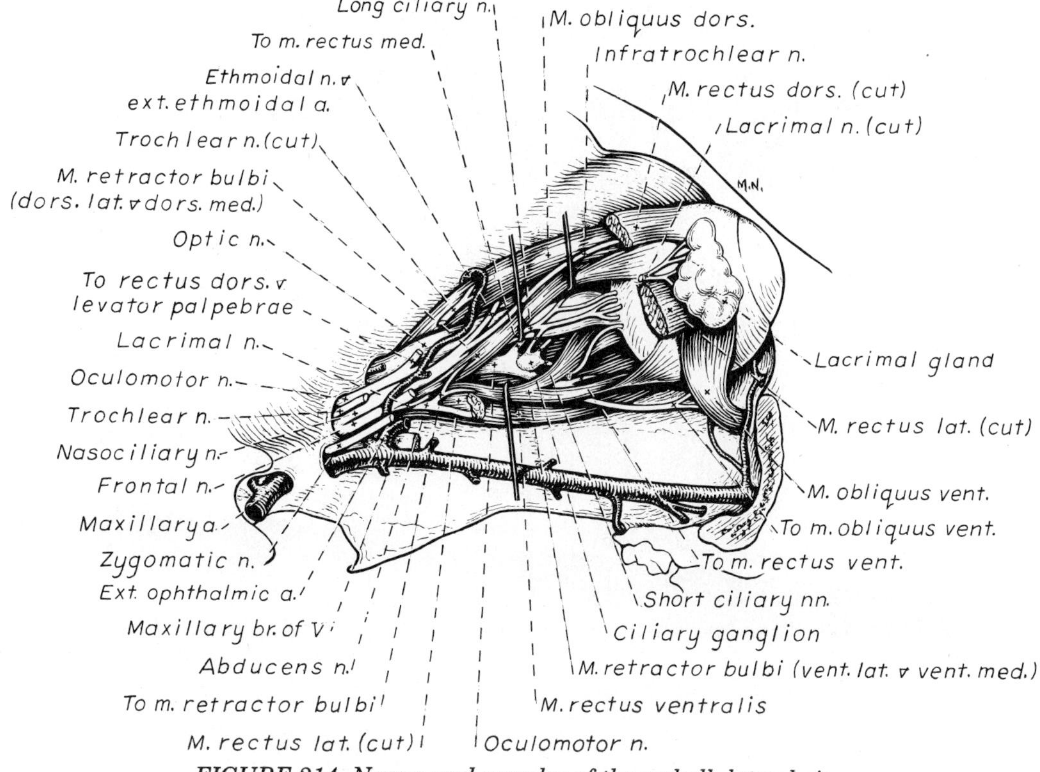

FIGURE 214. Nerves and muscles of the eyeball, lateral view.

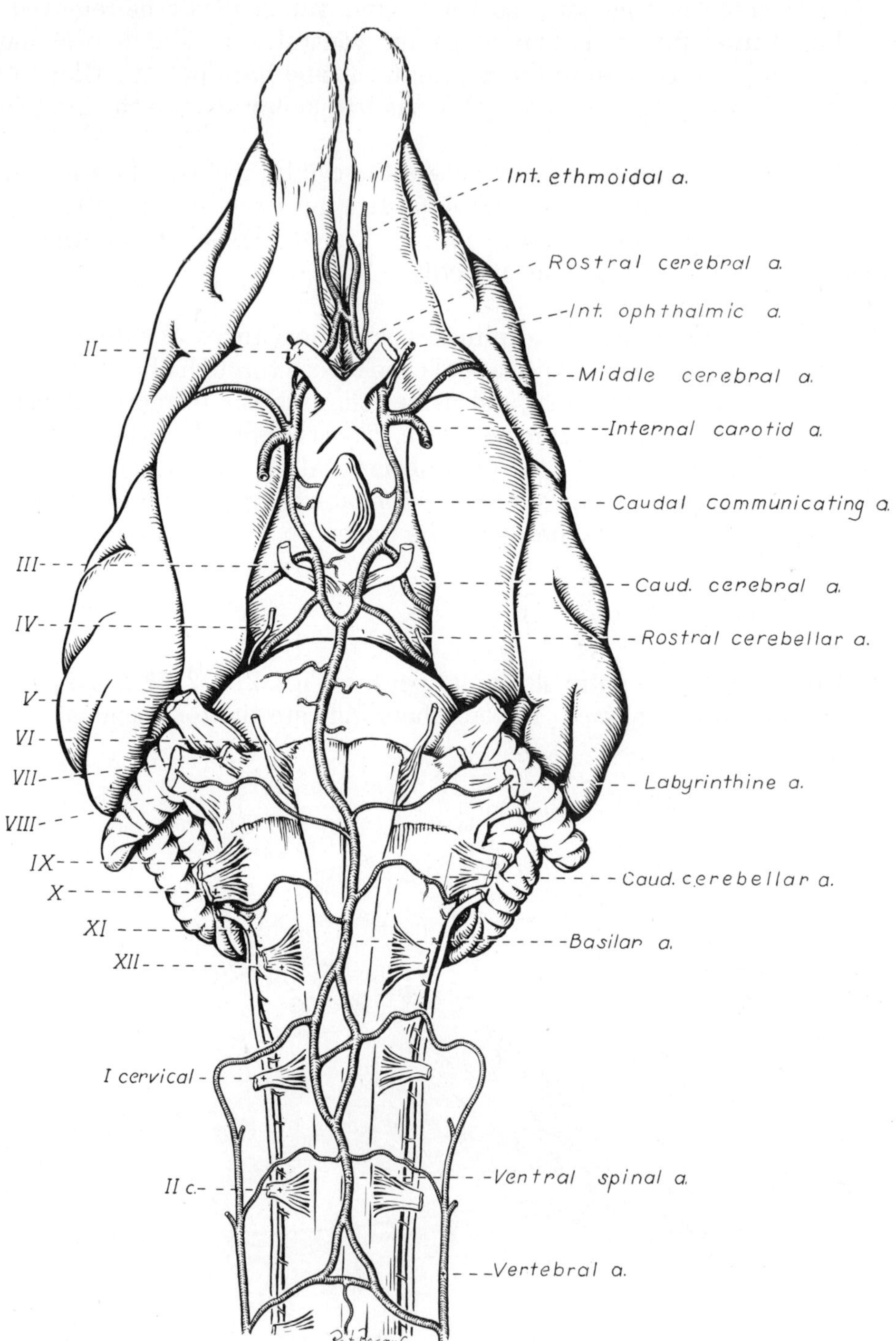

FIGURE 215. Vessels and nerves of the brain and first two cervical spinal cord segments, ventral view.

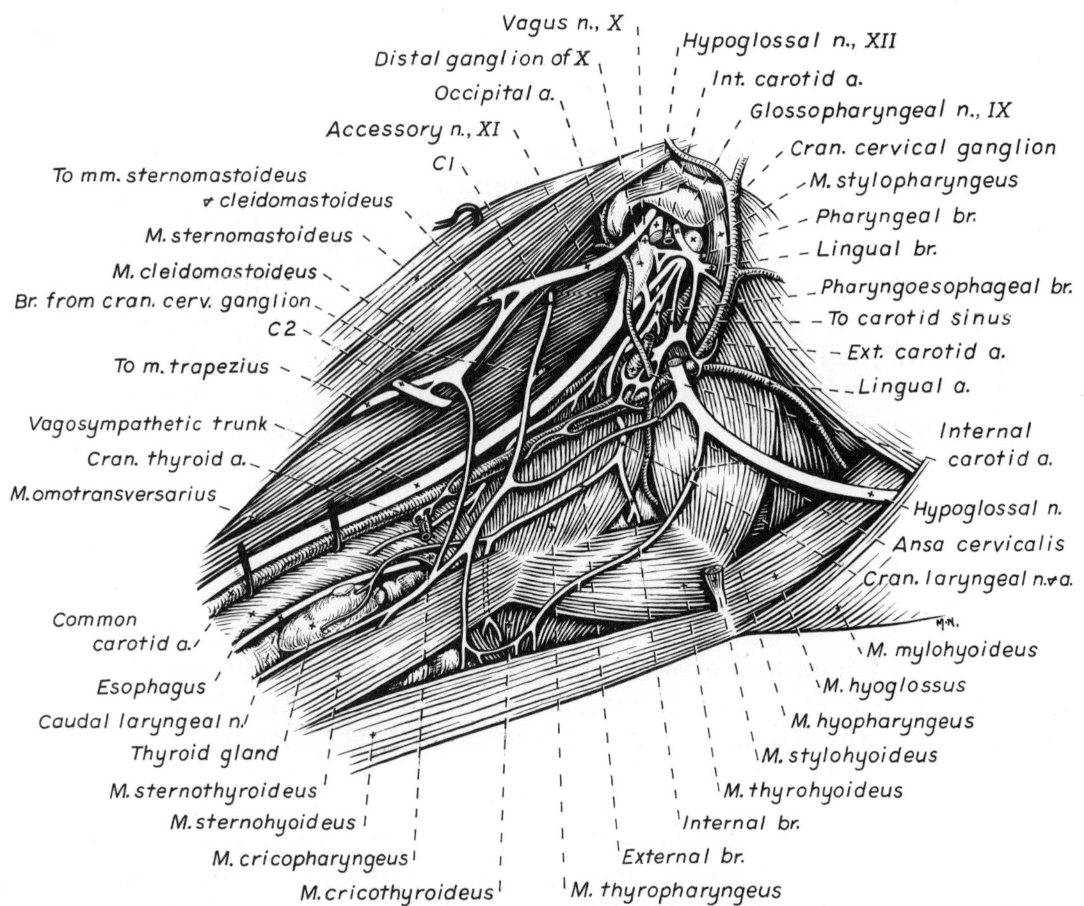

FIGURE 216. Nerves emerging from the tympano-occipital fissure, right lateral view; digastricus muscle removed.

in which they arise from the brain; their names are descriptive. Some of these have already been dissected, others will be dissected now.

1. The **olfactory** or **first cranial nerve** consists of numerous axons that arise in the olfactory epithelium of the caudal nasal mucosa and pass through the cribriform foramina to the olfactory bulbs. These include axons from the vomeronasal organ that course along the nasal septum. No dissection is necessary.

2. The **optic** or **second cranial nerve** (Fig. 214) is surrounded by the retractor bulbi muscle within the periorbita. Observe the nerve in the periorbita and as it enters the optic canal. It is surrounded by an extension of meninges from the cranial cavity.

3. The **oculomotor** or **third cranial nerve** (Fig. 214) passes through the orbital fissure and enters the periorbita with the optic nerve. Expose the proximal half of the optic nerve. Lift it gently and observe the oculomotor nerve on its lateroventral aspect. Do not dissect its individual branches. The oculomotor nerve innervates the dorsal, medial and ventral rectus, the ventral oblique and the levator palpebrae superioris. The **ciliary ganglion** is an irregular enlargement at the termination of the oculomotor nerve on the ventral surface of the middle of the optic nerve. This ganglion contains parasympathetic cell bodies of postganglionic axons that innervate the sphincter pupillae of the iris.

4. The **trochlear** or **fourth cranial nerve** enters the periorbita through the orbital fissure. It innervates the dorsal oblique; it need not be dissected.

5. The **trigeminal** or **fifth cranial nerve** divides into three nerves as it emerges from the trigeminal canal in the petrosal bone: the ophthalmic, the maxillary and the mandibular. The mandibular nerve has been dissected (see p. 303).

The **ophthalmic nerve** (sensory) passes through the orbital fissure and supplies sensory nerves that enter the periorbita. These need not be dissected. The **frontal** and **infratrochlear nerves** pass rostrally between the dorsal oblique and dorsal rectus muscles to innervate the medial aspect of the upper and lower eyelids. Long **ciliary nerves** follow the optic nerve and innervate the eyeball. The **ethmoidal nerve** passes through an ethmoidal foramen and the cribriform plate to innervate the nasal mucosa and skin of the nose.

The **maxillary nerve** (sensory) (Fig. 214) enters the alar canal via the round foramen. It emerges from the rostral alar foramen and crosses the pterygopalatine fossa dorsal to the pterygoid muscles and ventral to the periorbita, accompanied by the maxillary artery. Dissect the following branches.

a. The **zygomatic nerve** (sensory) enters the periorbita and divides into two branches that pass rostrally on the inner surface of the lateral part of the periorbita to innervate the lacrimal gland and the lateral portion of the upper and lower eyelids.
b. The **pterygopalatine ganglion** is dorsal to the maxillary nerve on the surface of the medial pterygoid muscle. It contains cell bodies of postganglionic parasympathetic axons that supply the lacrimal, nasal and palatine glands. The postganglionic axons course with the branches of the maxillary nerve to their terminations. Reflect the periorbita dorsally and the maxillary nerve ventrally to see this small, flat ganglion on the pterygoid muscle.
c. The **pterygopalatine nerve** arises beyond the level of the pterygopalatine ganglion from the ventral surface of the maxillary nerve and divides into three nerves: the **minor** and **major palatine nerves** of the palate and the **caudal nasal nerve** of the nasal mucosa. Dissect their origin only.
d. The **infraorbital nerve** (sensory) is the continuation of the maxillary nerve in the pterygopalatine fossa. It enters the infraorbital canal via the maxillary foramen. Along its course through the infraorbital canal it gives off **superior alveolar branches,** which supply the roots of the teeth via the alveolar canals. As the infraorbital nerve emerges from the infraorbital foramen, it divides into a number of fasciculi, which are distributed to the skin and adjacent structures of the upper lip and nose. Dissect these branches as they emerge from the infraorbital foramen.

6. The **abducent** or **sixth cranial nerve** (Fig. 214) passes through the orbital fissure and enters the periorbita. It innervates the retractor bulbi and the lateral rectus. Observe this small nerve entering the dorsal border of the lateral rectus near its origin.

7. The **facial** or **seventh cranial nerve** enters the internal acoustic meatus of the petrosal bone. It courses through the facial canal and emerges through the stylomastoid foramen, where its motor branches to the facial muscles have been dissected (see p. 297).

8. The **vestibulocochlear or eighth cranial nerve** enters the internal acoustic meatus and terminates in the membranous labyrinth of the inner ear. It is the nerve involved in balance and hearing. It will be dissected with the brain.

Observation of the ninth, tenth and eleventh cranial nerves and the sympathetic trunk ganglion at the base of the skull is facilitated by removal of the origin of the digastricus muscle for a lateral view. Likewise, removal of the insertion of the longus capitis muscle provides a medial view. These neural structures are found just medial and ventral to the tympanic bulla near the tympano-occipital fissure.

Locate the sympathetic trunk where it is joined with the vagus. Follow it cranially to a level ventral and medial to the tympanic bulla, where the two nerves separate. The sympathetic trunk is ventral to the vagus. The distal ganglion of the vagus is located dorsal to this separation and caudal to the cranial cervical ganglion. Trace the sympathetic trunk cranial to the separation and note an enlargement, the **cranial cervical ganglion.** This is the most cranial group of cell bodies of sympathetic postganglionic axons. These axons are distributed to the smooth muscles and glands of the head via blood vessels and other nerves. On the lateral side, observe the internal carotid artery coursing dorsocranially over the lateral surface of the cranial cervical ganglion. Notice the dense plexus that these nerves form in the immediate vicinity of the ganglion.

9. The **glossopharyngeal** or **ninth cranial nerve** (Fig. 216) passes through the jugular foramen and the tympano-occipital fissure. Beyond the fissure the glossopharyngeal crosses the lateral surface of the cranial cervical ganglion and divides into pharyngeal and lingual branches which are sensory to the pharyngeal mucosa and motor to the stylopharyngeus and other pharyngeal muscles. In addition, some branches course to the carotid sinus and others contribute to the pharyngeal plexus along with branches of the vagus nerve. Observe the nerve where it crosses the ganglion.

10. The **vagus** or **tenth cranial nerve** (Fig. 216) passes through the jugular foramen and the tympano-occipital fissure. It courses along the common carotid artery in the neck with the sympathetic trunk and through the thorax on the esophagus to terminal branches in the thorax and abdomen (which have already been dissected). The following branches are distributed to cranial cervical structures.

There are two ganglia associated with this nerve. The **proximal ganglion** of the vagus lies in the jugular foramen and cannot be seen. The **distal ganglion** of the vagus is found outside the tympano-occipital fissure, ventral and medial to the tympanic bulla. These are sensory ganglia. Find the distal ganglion with the vagus nerve caudal to the cranial cervical ganglion on the sympathetic trunk. The distal ganglion contains the cell bodies of the visceral afferent neurons that are distributed to most of the viscera of the body. Caudal to the distal ganglion the vagus joins the sympathetic trunk, with which it remains associated throughout its cervical course to the thoracic inlet. Branches from the vagus and glossopharyngeal nerves and the cranial cervical ganglion form a **pharyngeal plexus,** which innervates the caudal pharyngeal muscles and the cranial esophagus. The **cranial laryngeal nerve** leaves the vagus nerve at the distal ganglion and passes ventrally to the larynx, where it supplies the cricothyroid muscle and the laryngeal mucosa. Identify this nerve. The origin of the recurrent laryngeal nerve was seen in the thoracic inlet. It innervates the cervical esophagus in its course up the neck and terminates as the **caudal laryngeal nerve,** which enters the larynx under the caudal edge of the cricopharyngeus muscle. It innervates all the muscles of the larynx except the cricothyroideus.

11. The **accessory** or **eleventh cranial nerve** (Fig. 216) passes through the jugular foramen and the tympano-occipital fissure along with the ninth and tenth cranial nerves. It passes caudally and by a ventral branch innervates the sternomastoideus and cleidomastoideus muscles. The dorsal branch supplies the cleidocervicalis and the trapezius.

12. The **hypoglossal** or **twelfth cranial nerve** (Fig 216) passes through the hypoglossal canal. It passes ventrorostrally, lateral to the vagosympathetic trunk and the carotid arteries. It lies medial to the mandibular salivary gland, digastricus and mandible. The hypoglossal nerve is closely associated with the lingual artery. Deep to the mylohyoideus it innervates the extrinsic and intrinsic muscles of the tongue.

Live Dog

Examine the lips and cheeks and the vestibule they border. Evert the caudal part of the upper lip and find the opening of the parotid duct at the level of the fourth upper premolar. Try to feel the parotid duct through the platysma as the duct crosses the middle of the masseter muscle. Palpate the ventral portion of the external ear canal. This is covered by the parotid salivary gland, but the gland is difficult to feel. Palpate the firm ovoid mandibular salivary gland ventral to the parotid. The monostomatic sublingual gland is at the rostral end of the mandibular but cannot be identified by palpation. Open the mouth and examine the oral cavity. Observe the frenulum of the tongue and the sublingual caruncle where the mandibular and sublingual salivary ducts open. Examine the surface of the body and apex of the tongue. Recognize and feel the filiform papillae. Observe the prominent lingual vein on the ventral surface. In the anesthetized dog this vein can be used for intravenous injections. Pull the tongue to one side to see the palatoglossal arch where the oral cavity is continued by the oropharynx. Pull the tongue forward and see the root of the tongue in the floor of the oropharynx, the epiglottis caudal to it, the palatine tonsils partly covered by the semilunar folds on each side of the oropharynx and the soft palate dorsally.

Observe the philtrum on the nose. Feel the cartilaginous parts of the nose rostral to the incisive and nasal bones.

Examine the eyelids and observe the caruncle at the medial commissure. Evert the eyelids slightly and look for the lacrimal puncta along their borders near the medial commissure. Consider lacrimal flow from its origin at the lacrimal gland, across the cornea and along the conjunctival sacs, and through its collecting system to the ventral meatus where it is usually evaporated. Observe the margin of the third eyelid. Gently press the eyeball caudally into the periorbita in the orbit by pushing on it through the superior eyelid. Observe the passive protrusion of the third eyelid across the cornea. Recognize the palpebral and bulbar conjunctivae, the sclera, limbus, cornea, iris, pupil and anterior chamber. Move the head side to side and watch the involuntary quick horizontal movements of the eyeballs. The quick adduction towards the nose is a function of the medial rectus and its oculomotor (III) nerve innervation. The quick abduction away from the nose is a function of the lateral rectus and its abducent (VI) nerve innervation. Palpate along the zygomatic arch and try to feel the palpebral branch of the auriculopalpebral (VII) nerve where it crosses the arch to innervate the orbicularis oculi.

Palpate the temporal and masseter muscles above and below the zygomatic arch respectively. Palpate caudal and ventral to the masseter to feel the digastricus.

Palpate the scutiform cartilage in the auricular muscles medial to the external ear. Examine the external ear. Follow the helix and identify the cutaneous pouch laterally. At the opening into the external ear canal, identify the crura of the helix medially, the pretragic incisure, the tragus rostrally, the intertragic incisure and the antitragus laterally.

Recognize the anthelix on the caudal wall of the external ear canal opposite the tragus.

Palpate the trachea from the thoracic inlet to the larynx. Feel the cricoid and thyroid cartilages of the larynx. Palpate the basihyoid bone and try to feel the other hyoid bones. Normally the thyroid gland and medial retropharyngeal lymph node cannot be palpated. Palpate the small flat subcutaneous mandibular lymph nodes at the angle of the mandible.

THE NERVOUS SYSTEM

The nervous system may be divided into the **central nervous system,** consisting of the brain and spinal cord, and the **peripheral nervous system,** composed of cranial, spinal and peripheral nerves.

CEREBRAL MENINGES

The brain and spinal cord are covered by three membranes of connective tissue, the **meninges** (Fig. 217). The **dura mater,** or pachymeninx, is the thickest of these and the most external. Throughout most of the vertebral canal the dura is separated from the periosteum of the bony canal by the loose connective tissue of the epidural space, which often contains fat.

As the spinal cord approaches the brain stem, the dura adheres to the periosteum in the first two cervical vertebrae and to the atlanto-occipital membranes. Inside the cranial cavity the dura and periosteum are fused. Starting at the dorsal margin, free the hemisectioned brain from the skull of the sagittally split head. On one-half of the head a fold of dura will be found extending ventrally from the midline in the **longitudinal cerebral fissure** between the two cerebral hemispheres. This is the **falx cerebri,** which must be removed to allow reflection of the cerebral hemisphere. Remove the dura that adheres to the inside of the frontal, parietal and temporal bones.

The **pia mater** and **arachnoid** (the **leptomeninges**) are the other two connective tissue coverings of the central nervous system. The pia mater adheres to the external surface of the nervous tissue. The arachnoid in the live animal lies adjacent to the dura and sends delicate trabeculae to the pia. These trabeculae closely invest the blood vessels that course on the surface of the pia. The space between the pia and the arachnoid is the subarachnoid space, which is filled with cerebrospinal fluid. There is a closed potential space between the arachnoid and dura in life. In the embalmed specimen the arachnoid is collapsed onto the pia of the central nervous system.

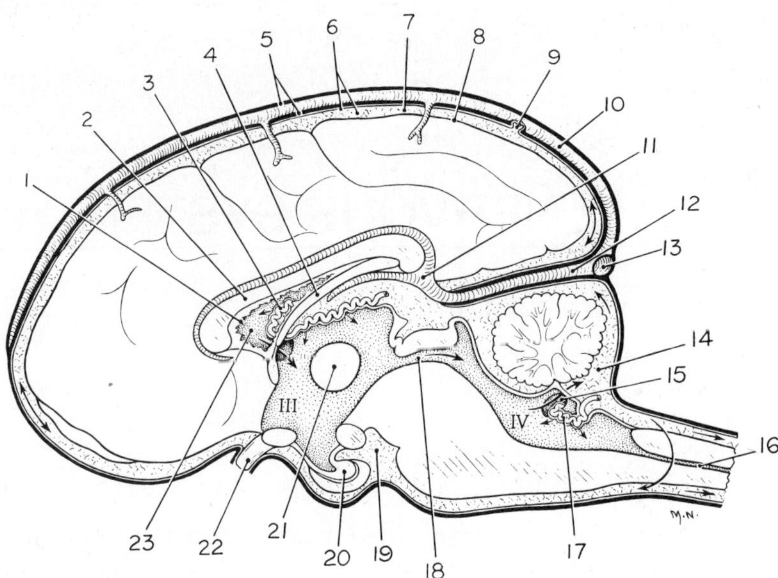

FIGURE 217. Meninges and ventricles of brain, median plane. (Arrows indicate flow of cerebrospinal fluid.)

1. *Cut edge of septum pellucidum*
2. *Corpus callosum*
3. *Choroid plexus, lateral ventricle*
4. *Fornix of hippocampus*
5. *Dura*
6. *Arachnoid membrane and trabeculae*
7. *Subarachnoid space*
8. *Pia*
9. *Arachnoid villus*
10. *Dorsal sagittal sinus*
11. *Great cerebral vein*
12. *Straight sinus*
13. *Transverse sinus*
14. *Cerebellomedullary cistern*
15. *Lateral aperture of fourth ventricle*
16. *Central canal*
17. *Choroid plexus, fourth ventricle*
18. *Mesencephalic aqueduct*
19. *Intercrural cistern*
20. *Hypophysis*
21. *Interthalamic adhesion*
22. *Optic nerve*
23. *Lateral ventricle*

Subarachnoid cisterns occur in areas where the arachnoid and pia are separated. The largest cistern is the **cerebellomedullary cistern** located in the angle between the cerebellum and medulla. Cerebrospinal fluid may be obtained from this cistern (Fig. 217) via a needle puncture through the atlanto-occipital space.

ARTERIES

Examine the arteries to the brain on the latex-injected specimen (Fig. 215). The arteries to the cerebrum and cerebellum are branches from the vessels on the ventral surface of the brain. The **basilar artery** is formed by the terminal branches of the **vertebral arteries,** which enter the floor of the vertebral canal through the lateral vertebral foramina of the atlas. It is continuous caudally with the **ventral spinal artery** of the spinal cord. The basilar artery courses along the midline of the ventral surface of the medulla and pons and then divides into two branches that form the caudal portion of the arterial circle of the brain (Fig. 215).

The **internal carotid arteries** are the other main source of blood to the arterial circle of the brain. After traversing the carotid canal in the tympanic part of the temporal bone and forming a loop at the foramen lacerum, each internal carotid artery enters the middle cranial fossa under the rostral end of the petrous temporal bone. It courses rostrally through the cavernous venous sinus beside the hypophyseal fossa and hypophysis. Between the hypophysis and the optic chiasm it emerges through the dura over the sinus and divides into a middle cerebral, rostral cerebral and caudal communicating artery. The small **caudal communicating arteries** course caudally and join the terminal branches of the basilar artery. Rostrally the two rostral cerebral arteries anastomose, completing the arterial circle on the ventral surface of the brain.

The **cerebral arterial circle** surrounds the pituitary gland, which receives small branches from the circle as well as directly from the internal carotid artery. Put the two halves of the latex-injected brain together to observe this arterial circle. Using the hemisectioned latex-injected brain, identify and trace the following vessels.

The **rostral cerebral artery** is a terminal branch of the internal carotid artery at the rostral aspect of the circle. It courses dorsally, lateral to the optic chiasm, and continues dorsally between the two frontal lobes in the longitudinal fissure. It courses over the fibers that connect the two cerebral hemispheres (corpus callosum) and caudally along the dorsal surface of the corpus callosum adjacent to the cerebral gyri.

Internal ethmoidal and internal ophthalmic arteries branch from the rostral cerebral artery. The internal ethmoidal anastomoses with the external ethmoidal and leaves the cranial cavity through the ethmoidal bone to supply structures in the nasal cavity. The internal ophthalmic artery anastomoses with a branch of the external ophthalmic on the optic nerve. This is the source of the long ciliary arteries that follow the optic nerve to the eyeball, which they supply. Do not dissect these arteries.

The **middle cerebral artery** arises from the arterial circle at the level of the rostral aspect of the pituitary gland. It courses laterally, rostral to the piriform lobe on the ventral surface of the olfactory peduncle. It continues dorsolaterally over the cerebral hemisphere, where it branches to supply the lateral surface.

The **caudal cerebral artery** arises from the caudal communicating artery at the level of the caudal aspect of the pituitary gland, rostral to the oculomotor nerve. The artery courses caudodorsally, following the optic tract over the lateral aspect of the thalamus to the longitudinal fissure. It passes rostrally on the corpus callosum to supply the medial surface of the caudal portion of the cerebral hemisphere. It also supplies the diencephalon and the rostral mesencephalon.

The **rostral cerebellar artery** leaves the caudal third of the arterial circle and courses dorsocaudally along the pons and the middle cerebellar peduncle to the lateral cerebellar hemisphere. It supplies the caudal midbrain and the rostral half of the cerebellum.

The **caudal cerebellar artery** is a branch of the basilar artery near the middle of the medulla. It courses dorsally to supply the caudal portion of the cerebellum.

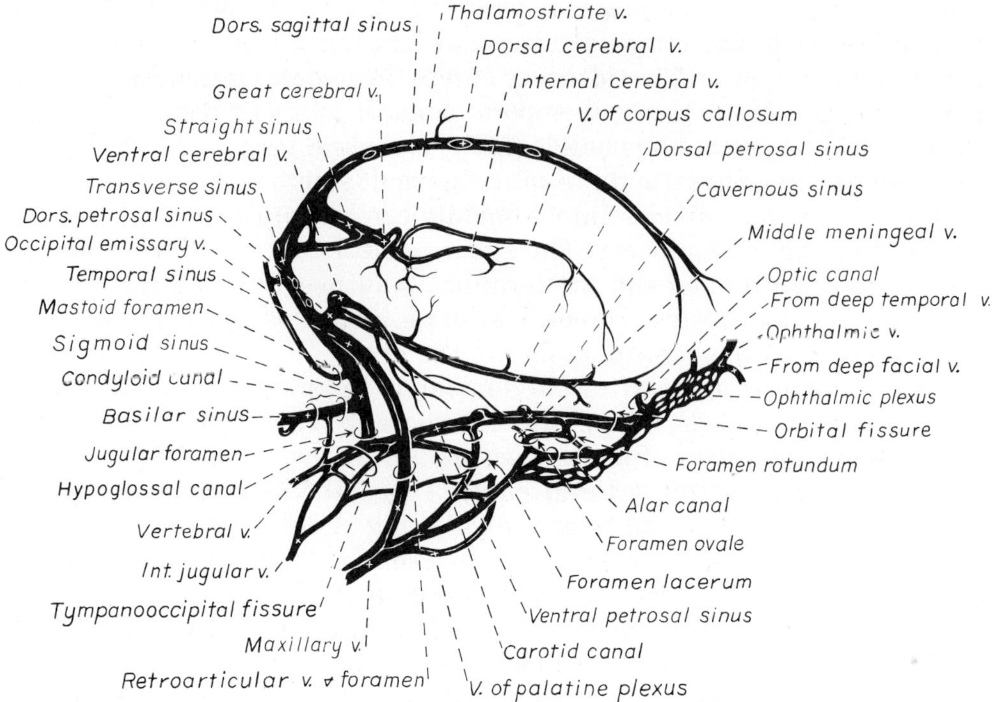

FIGURE 218. Cranial venous sinuses, right lateral aspect (modified from Reinhard, Miller and Evans, 1962).

VEINS

The **venous sinuses** (Figs. 218, 219) of the cranial dura mater are venous passageways located within the dura or within bony canals in the skull. These sinuses receive the veins draining the brain and the bones of the skull. They convey venous blood to the paired maxillary, internal jugular and vertebral veins and to the ventral internal vertebral venous plexuses. The following venous sinuses should be located.

The **dorsal sagittal sinus** is located in the attached edge of the falx cerebri, which is a fold of dura extending ventrally into the longitudinal fissure between the two cerebral hemispheres. Caudally this sinus enters the foramen for the dorsal sagittal sinus in the occipital bone where the tentorium attaches. There it joins the right and left transverse sinuses.

Each **transverse sinus** runs laterally through the transverse canal and sulcus. At the distal end of the sulcus, at the dorsal border of the petrosal bone, the sinus divides into a temporal and a sigmoid sinus. Coursing caudolateral to the petrosal bone, the **temporal sinus** extends to the retroarticular foramen, where it emerges as the retroarticular vein and joins the maxillary vein.

Each **sigmoid sinus** forms an S-shaped curve as it courses over the dorsomedial side of the petrous temporal bone. It passes through the jugular foramen into the tympano-occipital fissure. Within the fissure the ventral petrosal sinus enters rostrally from the petro-occipital canal and joins the sigmoid sinus. From this anastomosis the vertebral and internal jugular veins arise and leave the tympano-occipital fissure and course caudally. The vertebral vein descends the neck through the transverse foramina of the cervical vertebrae. The internal jugular vein was seen previously in the carotid sheath. A branch of the sigmoid sinus continues caudally through the condyloid canal to the ventral internal vertebral venous plexus in the vertebral canal.

The **cavernous sinus** lies on each side of the floor of the middle cranial fossa from the orbital fissure to the petro-occipital canal. Emissary veins connect each cavernous sinus with the ophthalmic plexus of veins rostrally and with the maxillary vein laterally. These sinuses are each continued caudally by the **ventral petrosal sinus,** which lies in the petro-occipital canal. Two or three intercavernous sinuses connect the left and right cavernous sinuses rostral and caudal to the pituitary gland. The internal carotid artery was seen coursing through the cavernous sinus.

The **ventral internal vertebral venous plexuses** are paired vessels that lie on the floor of the vertebral canal in the epidural connective tissue. They extend from the venous sinuses of the cranium throughout the vertebral canal. At each intervertebral foramen **intervertebral veins** connect the vertebral venous plexus with the vertebral veins of the neck, intercostal veins of the thorax (azygos and costocervical veins) and lumbar veins (caudal vena cava) in the abdomen. This plexus will be seen later when the spinal cord is removed from the vertebral canal.

There is a continuous venous pathway from the angularis oculi vein, through the external ophthalmic vein, ophthalmic plexus, cavernous sinus, ventral petrosal sinus, sigmoid sinus and occipital emissary vein, to the ventral internal vertebral venous plexus. This pathway can be demon-

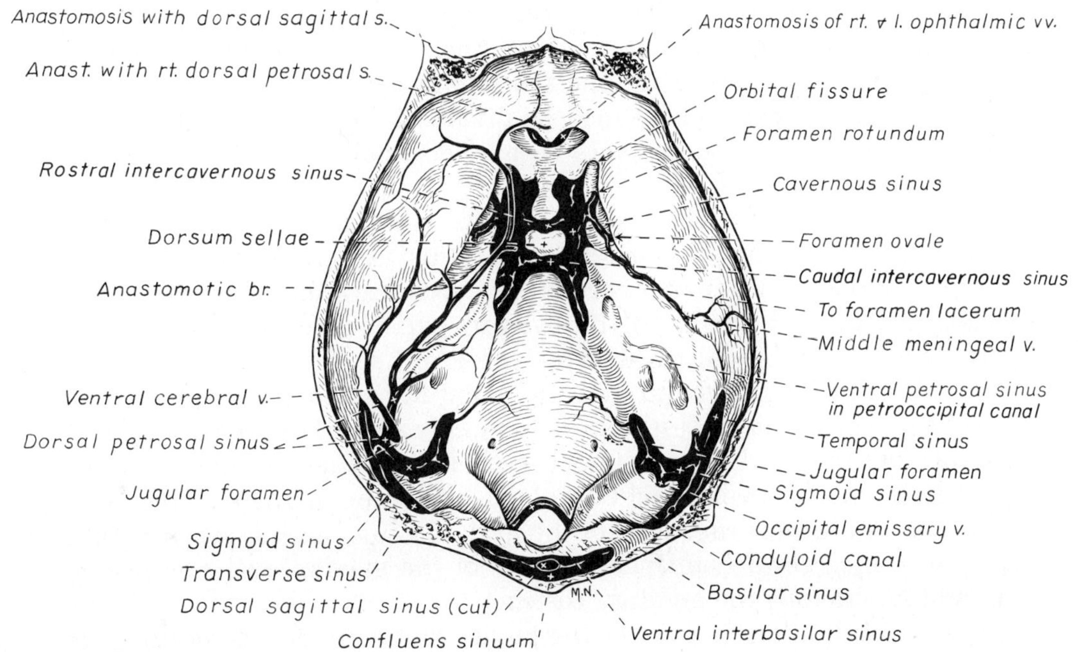

FIGURE 219. Cranial venous sinuses, dorsal aspect, calvaria removed (modified from Reinhard, Miller and Evans, 1962).

strated radiographically by compressing the external jugular veins and injecting a radiopaque solution into the angularis oculi vein on the side of the face near the medial angle of the eye. This procedure can be used clinically to diagnose space-occupying lesions that interfere with this pathway. It can also be used anatomically to inject the venous system with various materials.

BRAIN

The brain is composed of the embryologically segmented brain stem and two suprasegmental portions, the cerebrum (telencephalon) and the cerebellum (dorsal metencephalon). The **brain stem** includes the myelencephalon (medulla), the ventral metencephalon (pons), the mesencephalon (midbrain) and the diencephalon (interbrain—epithalamus, thalamus and hypothalamus).

Dissect and identify the following structures on the intact brain that has been provided.

Cerebrum—Surface Structures

The cerebrum is divided into two cerebral hemispheres by the **longitudinal fissure.** Each cerebral hemisphere has outward folds (convolutions) called **gyri** and inward folds called **sulci.** Identify the following gyri and sulci (Figs. 220, 221): the rostral and caudal parts of the lateral rhinal sulcus; the pseudosylvian fissure; the rostral and caudal sylvian gyri; the ectosylvian sulcus and gyrus; the suprasylvian sulcus and gyrus; the cruciate sulcus; the postcruciate and precruciate gyri; the coronal sulcus; the marginal gyrus and sulcus; and the ectomarginal gyrus.

Each cerebral hemisphere may be divided into lobes named for that portion of the calvaria that covers them. The relationship is not precise and varies among species.

The **frontal lobe** is that portion of each cerebral hemisphere rostral to the cruciate sulcus. The precruciate gyrus is part of this lobe and functions as part of the motor cortex. The **parietal lobe** is caudal to the cruciate sulcus and dorsal to the sylvian gyri. It extends caudally to approximately the caudal third of the cerebral hemisphere. The postcruciate and rostral suprasylvian gyri are found in this lobe and function as part of the motor and somesthetic sensory cerebral cortex. The **occipital lobe** includes the caudal third of the cerebral hemisphere. Caudal portions of this lobe on both medial and lateral sides function as the visual cortex. The **temporal lobe** is composed of the gyri and sulci on the ventrolateral aspect of the cerebral hemisphere. Parts of the sylvian gyri are located here and function as the auditory cortex.

The **rhinal sulcus** separates the phylogenetically new cerebrum, the neopallium, from the older olfactory cerebrum, the paleopallium, below. Portions of the paleopallium that are visible are the **olfactory bulb,**

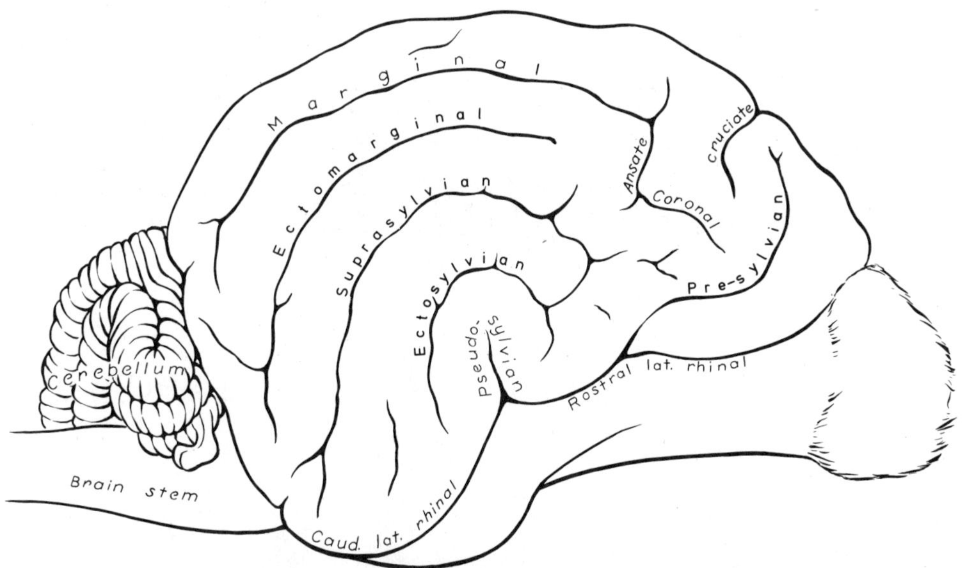

FIGURE 220. Sulci of brain, right lateral view.

which rests on the cribriform plate, and the **olfactory peduncle,** which joins the bulb to the cerebral hemisphere (Fig. 222). The **olfactory peduncle** courses caudally with a band of fibers on its ventral surface. Caudally this band divides into **lateral** and **medial olfactory tracts.** Observe the lateral olfactory tract passing caudally to the **piriform lobe,** which forms a ventral bulge just lateral to the pituitary gland and medial to the temporal lobe of the neopallium. The medial olfactory tract cannot be observed.

Each gyrus contains gray matter superficially and white matter in

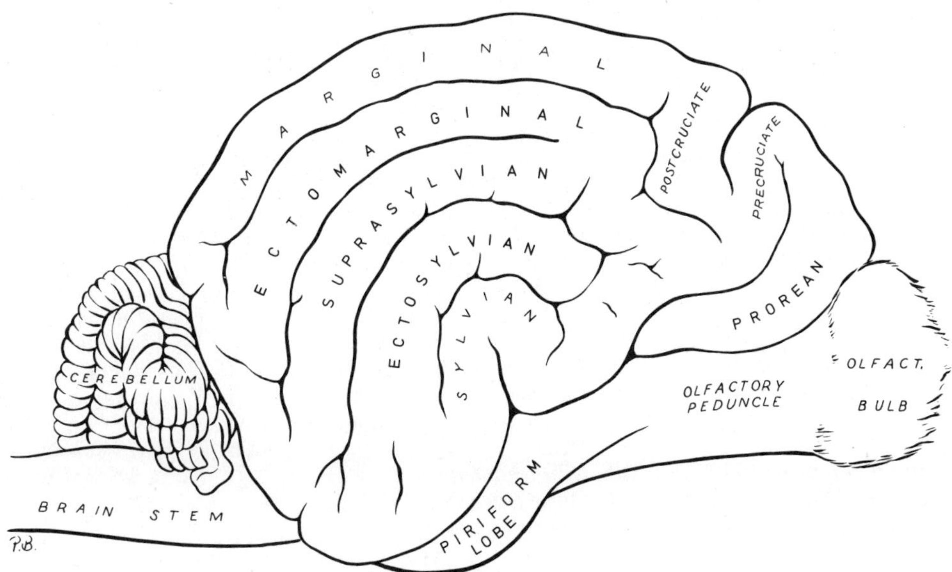

FIGURE 221. Gyri of brain, right lateral view.

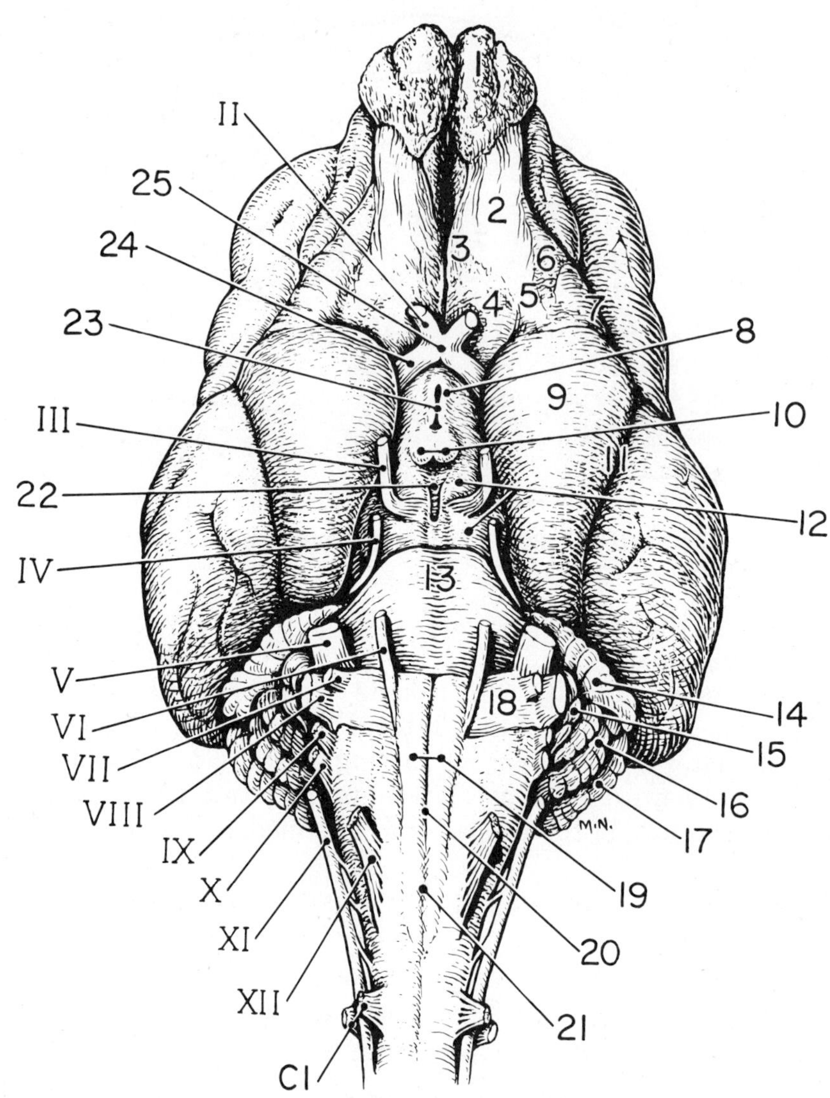

FIGURE 222. Ventral view of brain and cranial nerves.

1. *Olfactory bulb*
2. *Olfactory peduncle*
3. *Medial olfactory tract*
4. *Rostral perforated substance*
5. *Lateral olfactory tract*
6. *Lateral olfactory gyrus*
7. *Rostral rhinal sulcus*
8. *Tuber cinereum*
9. *Piriform lobe*
10. *Mamillary bodies*
11. *Caudal rhinal sulcus*
12. *Crus cerebri*
13. *Pons, transverse fibers*
14. *Ventral paraflocculus*
15. *Flocculus*
16. *Dorsal paraflocculus*
17. *Ansiform lobule*
18. *Trapezoid body*
19. *Pyramids*
20. *Ventral median fissure*
21. *Decussation of pyramids*
22. *Caudal perforated substance in interpeduncular fossa*
23. *Infundibulum*
24. *Optic tract*
25. *Optic chiasm*

II. Optic nerve
III. Oculomotor nerve
IV. Trochlear nerve
V. Trigeminal nerve
VI. Abducent nerve
VII. Facial nerve
VIII. Vestibulocochlear nerve
IX. Glossopharyngeal nerve
X. Vagus nerve
XI. Accessory nerve
XII. Hypoglossal nerve
C1. First cervical nerve.

its center. The gray matter, or **cerebral cortex** of the neopallium, is composed of six layers of neuronal cell bodies. The white matter, **corona radiata,** contains the processes of neurons coursing to and from the overlying cortex.

Cerebellum

The **cerebellum** is derived from the dorsal portion of the metencephalon and lies caudal to the cerebrum and dorsal to the fourth ventricle. The **transverse cerebral fissure** separates it from the cerebrum. The dural tentorium cerebelli is located in this fissure. The cerebellum is connected to the brain stem by three cerebellar peduncles on each side of the fourth ventricle and by portions of the roof of the fourth ventricle.

The **choroid plexus** is a compact mass of pia, blood vessels and ependyma. A choroid plexus develops where neural tube neuroepithelium did not proliferate to form parenchyma but remained as a single layer of the neuroepithelial cells, a roof plate. These areas are found in the medulla (roof plate of the fourth ventricle), the diencephalon (roof plate of the third ventricle), and telencephalon (roof plate of the lateral ventricle). At these sites the vessels in the pia covering the single layer of neuroepithelial cells proliferate to form a dense plexus of capillaries intimately related to the neuroepithelial cells. These cells and blood vessels are involved in passive and active secretion of cerebrospinal fluid into the ventricular system. The choroid plexus of the fourth ventricle protrudes into the lumen of the fourth ventricle and is visible caudolateral to the cerebellum on the dorsal surface of the medulla.

Identify the **transverse fibers of the pons** on the ventral surface of the brain stem. Follow these fibers laterally as they course dorsocaudally into the cerebellum on each side as the **middle cerebellar peduncle** (Figs. 222, 223, 230). At the point where they merge into the cerebellum, cut this peduncle with a scalpel. Continue the cut slightly rostral and caudal to the middle cerebellar peduncle and detach the cerebellum from the pons on that side. This will sever the **rostral cerebellar peduncle,** which attaches rostrally, and the **caudal cerebellar peduncle,** which attaches caudally. These can also be seen caudally by lifting the vermis of the cerebellum dorsally from the medulla. This exposes the caudal surface of the caudal cerebellar peduncle. Cut these peduncles on the opposite side and remove the cerebellum. The rostral cerebellar peduncle contains mainly efferent axons from the cerebellum to the brain stem. Afferent axons to the cerebellum from the brain stem and spinal cord pass primarily through the middle and caudal cerebellar peduncles.

The cerebellum is composed of lateral **cerebellar hemispheres** and a middle portion, the **vermis.** The gyri of the cerebellum are known as **folia.** These are grouped into three lobes and numerous cerebellar lobules that have specific names. The vermis comprises the entire middle portion of the cerebellum directly above the fourth ventricle. Some of its lobules are found on the ventral surface of the cerebellum facing the roof plate of

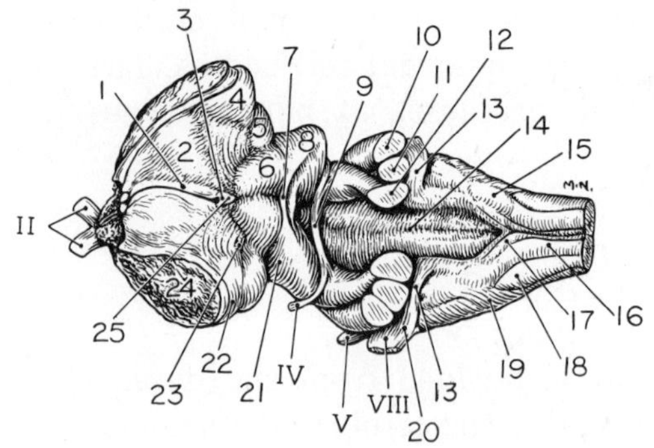

FIGURE 223. Dorsal view of brain stem.

1. *Stria habenularis thalami*
2. *Thalamus*
3. *Habenular commissure*
4. *Lateral geniculate nucleus*
5. *Medial geniculate nucleus*
6. *Rostral colliculus*
7. *Commissure of caudal colliculus*
8. *Caudal colliculus*
9. *Crossing of trochlear nerve fibers in rostral medullary velum*
10. *Middle cerebellar peduncle*
11. *Caudal cerebellar peduncle*
12. *Rostral cerebellar peduncle*
13. *Dorsal cochlear nucleus in acoustic stria*
14. *Median sulcus in fourth ventricle*
15. *Lateral cuneate nucleus*
16. *Fasciculus cuneatus*
17. *Fasciculus gracilis*
18. *Spinal tract of trigeminal nerve*
19. *Superficial arcuate fibers*
20. *Left ventral cochlear nucleus*
21. *Brachium of caudal colliculus*
22. *Optic tract*
23. *Brachium of rostral colliculus*
24. *Cut surface between cerebrum and brain stem*
25. *Pineal body*

II. *Optic nerves*
IV. *Trochlear nerve*
V. *Trigeminal nerve*
VIII. *Vestibulocochlear nerve*

the fourth ventricle. Each hemisphere projects over the cerebellar peduncles and the adjacent brain stem. A lateral component lies in the cerebellar fossa of the petrosal part of the temporal bone.

Make a median incision through the vermis, hemisectioning the cerebellum. Examine the cut surface. Note the pattern of white matter as it branches and arborizes from the medulla of the cerebellum in the folia (see Figs. 228, 230). The **medulla of the cerebellum** is the white matter in its central portion that contains nuclei and connects with all the folia and the cerebellar peduncles. Observe the laminae of foliate white matter and the cerebellar cortex. Make a transverse section of one-half of the cerebellum through its medulla to observe the lateral extent of the medullary white matter (see Fig. 228).

Brain Stem—Surface Structures

To expose the dorsal surface structures of the brain stem, the left cerebrum will be removed by the following dissection.

Separate the two cerebral hemispheres at the longitudinal fissure. Expose the band of fibers that course transversely from one hemisphere to the other in the depth of the fissure. This structure is the corpus callosum. Completely divide the corpus callosum longitudinally along the median plane in the depth of the longitudinal fissure. Cut deep enough to include the hippocampal commissure and body of the fornix but do not cut into the thalamus (Figs. 224, 225, 230). Continue the cut rostrally and ventrally through the rostral commissure just dorsal to the optic chiasm and rostral to the thalamus. On the ventral surface, follow the optic tract in a dorsocaudal direction from the optic chiasm and cut the fibers of the internal capsule rostral and medial to this tract. The fibers of the internal capsule attach the cerebral hemisphere to the brain stem. Gently lift the medial side of the cerebrum off of the thalamus and continue this separation over the dorsal aspect of the diencephalon. Cut any remaining attachments and remove the hemisphere from the diencephalon.

Examine the surface of the brain stem and locate the following structures (Figs. 222, 223).

Diencephalon

The **diencephalon** consists of a large, centrally located thalamus, a smaller hypothalamus below and a very small epithalamus on the dorsal midline.

The optic or second cranial nerves form the **optic chiasm** of the diencephalon rostral to the pituitary gland (Figs. 222, 224). The **optic tracts** course laterally and dorsocaudally from the chiasm, pass over the lateral surface of the diencephalon and enter the lateral geniculate nucleus of the thalamus. In this pathway, each tract curves around the caudal edge of the internal capsule.

Caudal to the optic chiasm on the median plane is the **hypophysis,** which is attached by the **infundibulum** to the **tuber cinereum** of the hypothalamus. If the gland is missing, the lumen of the infundibulum will be evident. This lumen communicates with the overlying third ventricle of the diencephalon.

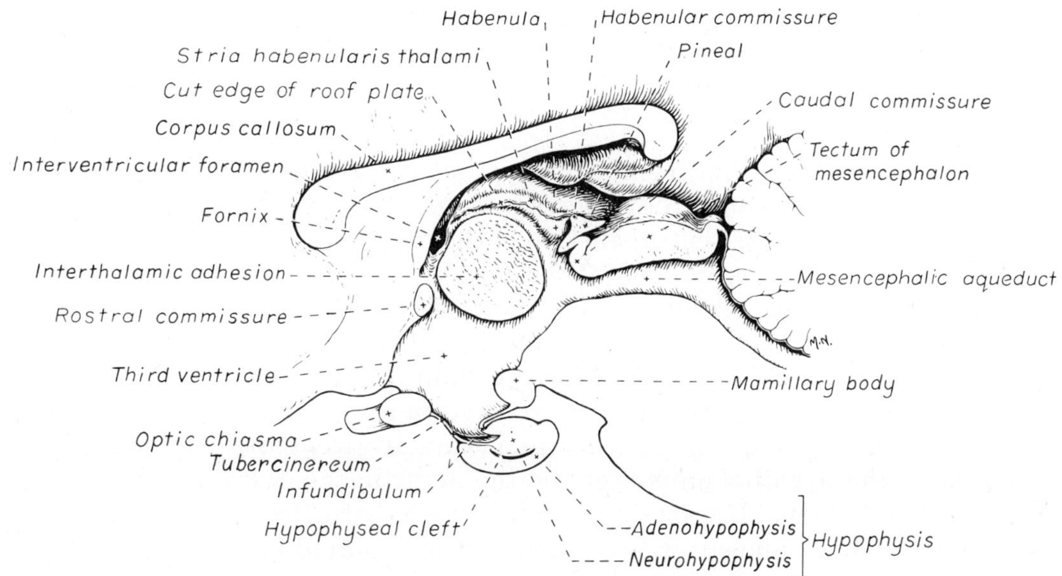

FIGURE 224. Diencephalon, median section.

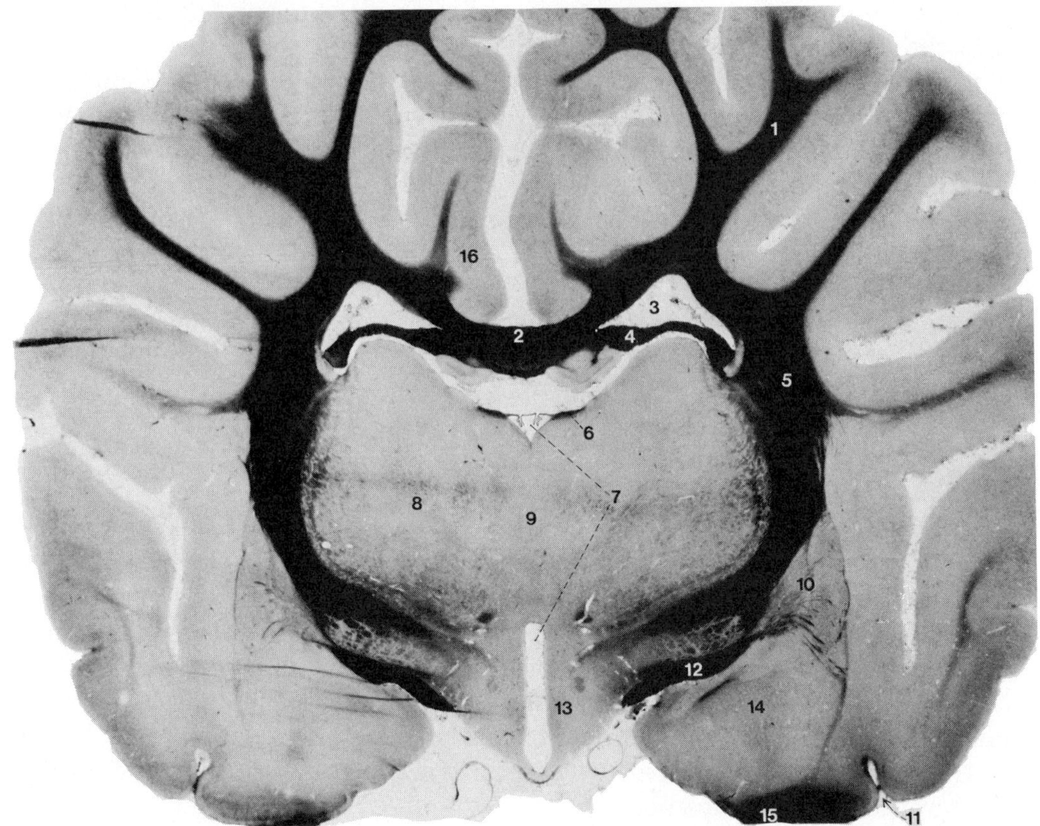

FIGURE 225. Diencephalon and cerebral hemispheres. (In this and the following transverse sections the white matter is stained with iron hematoxylin and appears black in the photographs).

1. *Corona radiata*
2. *Corpus callosum*
3. *Lateral ventricle*
4. *Crus of fornix*
5. *Internal capsule*
6. *Stria habenularis*
7. *Third ventricle (roof plate added)*
8. *Thalamus*
9. *Interthalamic adhesion*
10. *Lentiform nucleus*
11. *Lateral rhinal sulcus*
12. *Optic tract*
13. *Hypothalamus*
14. *Amygdala*
15. *Pyriform lobe*
16. *Cingulate gyrus*

The **mamillary bodies** of the hypothalamus bulge ventrally, caudal to the tuber cinereum. They demarcate the most caudal extent of the hypothalamus on the ventral surface of the diencephalon.

The internal capsule bounds the diencephalon laterally and was cut when the left cerebrum was removed. The thalamus and epithalamus can be seen on the dorsal aspect of the diencephalon (Figs. 223, 224).

Three structures compose the **epithalamus.** They all are located adjacent to the median plane. The **stria habenularis** lies on either side of the midline, coursing dorsally and caudally from the rostroventral aspect of the hypothalamus over the thalamus to the dorsocaudal aspect of the diencephalon. Here the stria enters the **habenular nucleus.** Caudal to the habenular nucleus is the small, unpaired **pineal body.** This caudal projection from the diencephalon is small in the dog but very prominent in larger domestic animals.

A space can usually be found between the stria habenularis of each side. This is the dorsal part of the **third ventricle** (Fig. 226). It is covered by a thin remnant of the roof plate of the neural tube, a layer of ependyma that extends from one stria habenularis to the other. Branches of the caudal cerebral artery course over the diencephalon and form the **choroid plexus of the third ventricle.** This is usually pulled free when the brain is removed from the cranial cavity. Rostrally the choroid plexus of the

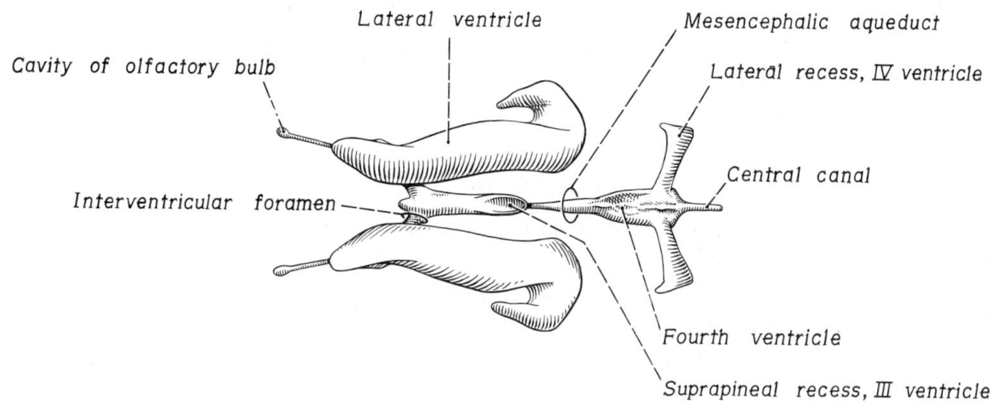

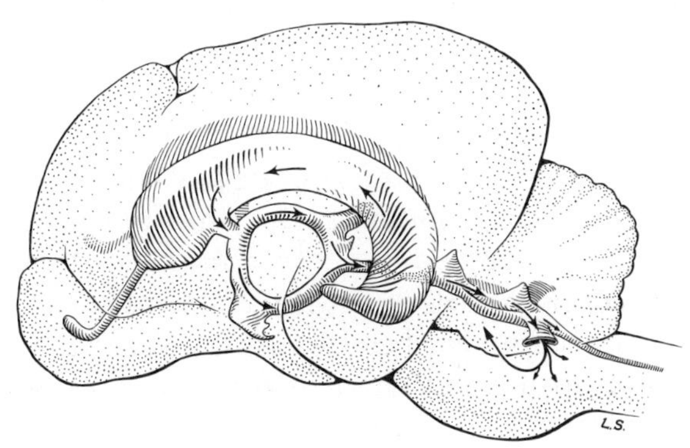

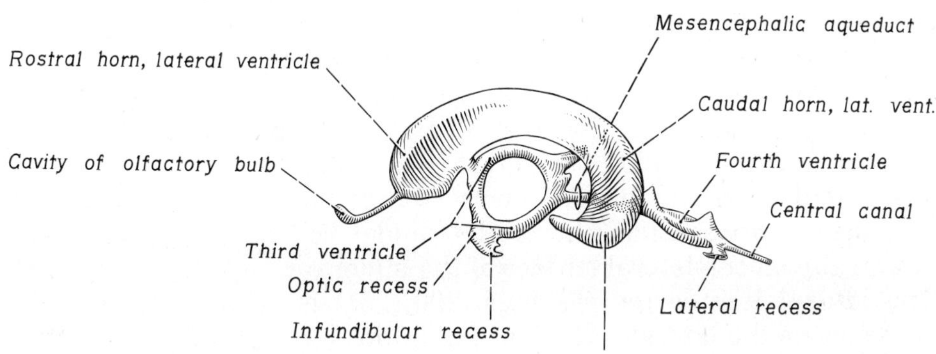

FIGURE 226. Ventricles of brain. Direction of arrows indicates flow of cerebrospinal fluid (from de Lahunta, 1983).

third ventricle is continuous with the choroid plexus of the lateral ventricle at the interventricular foramen. This foramen is caudal to the column of the fornix at the level of the rostral commissure. These structures will be seen in the dissection of the telencephalon.

The **thalamus** lies between the stria habenularis medially and the internal capsule laterally. A lateral eminence on the caudodorsal surface of the thalamus is the **lateral geniculate body,** which receives fibers of the optic tract and functions in the visual system. The lateral geniculate body is connected with the rostral colliculus of the midbrain. Caudoventral to the lateral geniculate body is the **medial geniculate body** of the thalamus. This nucleus functions in the auditory system and is connected to the caudal colliculus of the midbrain by the brachium of the caudal colliculus.

In the third ventricle, between the stria habenularis of each side, observe the **interthalamic adhesion** between the right and left sides of the thalamus. This area appears round on median section because the third ventricle encircles it. In transverse section the narrow, vertically oriented third ventricle appears as a perpendicular slit below the interthalamic adhesion. Its lateral and ventral walls are formed by the hypothalamus. The dorsal portion of the third ventricle is small and tubular. It passes over the interthalamic adhesion but its thin roof plate, which is attached on each side to the stria habenularis, cannot be observed grossly.

Mesencephalon

Between the mamillary bodies of the hypothalamus and the transverse fibers of the pons is the ventral surface of the **mesencephalon** (midbrain). The transverse fibers of the pons cover part of the mesencephalon ventrally. Descending tracts that connect portions of the cerebral cortex with lower brain stem centers and the spinal cord course on the ventral surface of the midbrain. These are grouped together on each side as the **crus cerebri** (Fig. 227). The **oculomotor** or third cranial nerve leaves the midbrain medial to the crus (Fig. 222).

The mesencephalic structures dorsal to the mesencephalic aqueduct compose the tectum of the midbrain (Fig. 224) The **mesencephalic aqueduct** is a short, narrow tube, derived from the neural canal in the midbrain, that connects the third ventricle rostrally with the fourth ventricle caudally. Four dorsal bulges, the **corpora quadrigemina,** are evident on the dorsal side. The rostral pair are the **rostral colliculi,** which function with the auditory system (Fig. 223).

The **trochlear** or fourth cranial nerve courses laterally out of the roof of the fourth ventricle adjacent to the caudal colliculus. It continues rostroventrally on the lateral surface of the midbrain.

The **lateral lemniscus** (see Figs. 230, 231) is a band of auditory system axons on the lateral side of the midbrain. It courses rostrodorsally from the cochlear nucleus to the caudal colliculus and originates medial to the middle cerebellar peduncle. Many of these fibers arise from the

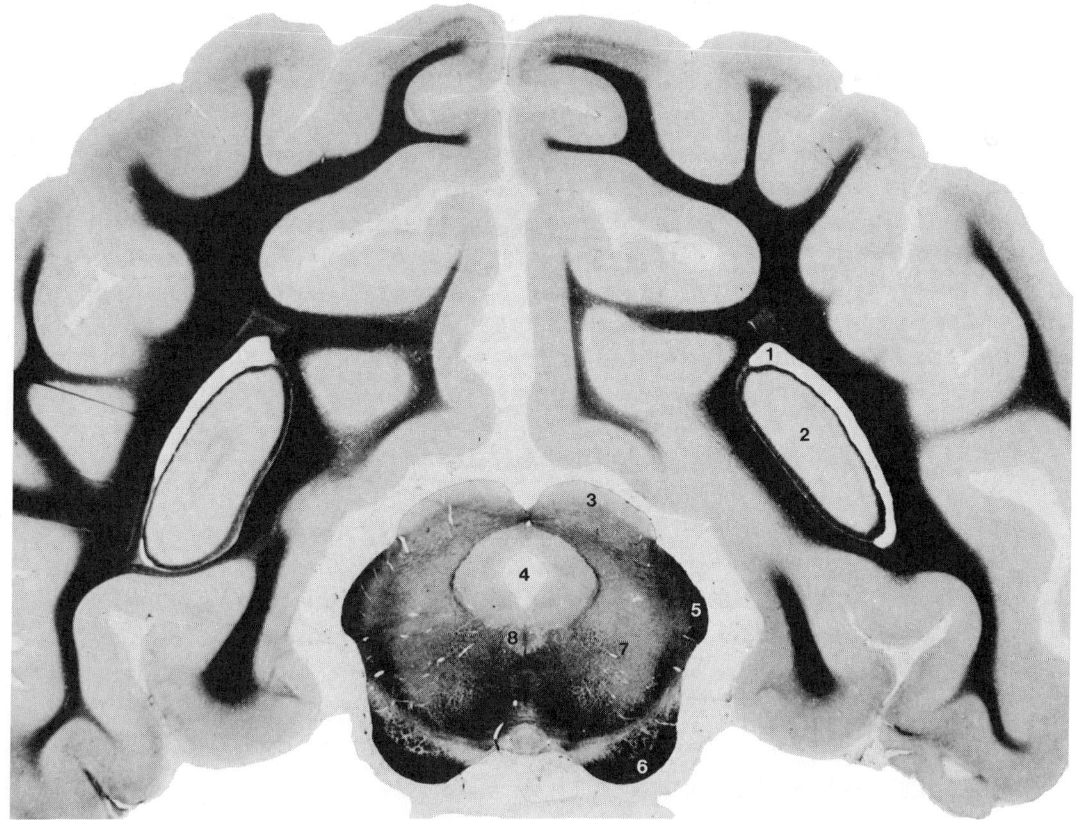

FIGURE 227. Mesencephalon and cerebral hemispheres. (The white matter is stained with iron hematoxylin and appears black in the photograph.)

1. *Lateral ventricle*
2. *Hippocampus*
3. *Rostral colliculus*
4. *Mesencephalic aqueduct*
5. *Brachium of caudal colliculus*
6. *Crus cerebri*
7. *Reticular formation*
8. *Oculomotor nucleus*

cochlear nucleus. The **brachium of the caudal colliculus** (Figs. 223, 230, 231) runs rostroventrally from the caudal colliculus to the medial geniculate body of the thalamus. On the dorsal surface the **commissure of the caudal colliculi** (Fig. 223) can be seen crossing between these two structures. The rostral colliculus is connected to the lateral geniculate body of the thalamus by a short **brachium of the rostral colliculus** (Fig. 223).

Ventral Metencephalon

The metencephalic portion of the rhombencephalon includes a segment of the brain stem, the **pons,** and the dorsal development, the **cerebellum.** The ventral surface of the pons includes the **transverse fibers of the pons,** which course laterally into the middle cerebellar peduncles. This large band of fibers borders the trapezoid body of the medulla caudally. Its rostral border covers part of the ventral surface of the midbrain. The **trigeminal nerve** is associated with the pons and can be found entering the pons along the caudolateral aspect of the transverse fibers (Figs. 222, 230). The descending fibers of the crus cerebri enter the pons dorsal to the transverse fibers, where they are called the **longitudinal fibers of the pons.** These longitudinal fibers are covered ventrally by the transverse fibers. The longitudinal fibers that do not terminate in pontine nuclei continue caudally on the ventral surface of the trapezoid body of the medulla as the pyramids. Many of the axons in the crus cerebri and the longitudinal fibers of the pons, and most of those in the transverse fibers of the pons, comprise a large cerebro-ponto-cerebellar pathway. Synapse occurs in the pontine nuclei that are covered by the transverse fibers and crossing occurs through the transverse fibers. Therefore, impulses that arise in the left cerebrum are projected to the right cerebellar hemisphere.

The **rostral medullary velum** forms the roof of the fourth ventricle between the caudal colliculi of the mesencephalon rostrally and the midventral surface of the cerebellum caudally. The crossing fibers of the trochlear nerves course through this velum (Fig. 223). The velum in the preserved specimen lies on the floor of the fourth ventricle and covers the caudal opening of the mesencephalic aqueduct. Insert a probe under the caudal cut edge and raise the velum to demonstrate its attachment and the continuity of the fourth ventricle with the aqueduct.

Myelencephalon (Medulla)

The myelencephalon, or medulla, extends from the transverse fibers of the pons to the level of the ventral rootlets of the first cervical nerve. The **trapezoid body** is the transverse band of fibers rostrally that course parallel but caudal to the transverse pontine fibers (Figs. 222, 230, 231). It is continuous with the vestibulocochlear nerve and cochlear nuclei laterally on the side of the medulla and functions in the auditory system. The **pyramids** are a pair of longitudinally coursing fiber bundles on either side of the ventral median plane. They emerge from the transverse fibers as the caudal continuations of axons from the longitudinal fibers of the pons that did not terminate in pontine nuclei. They course caudally across the trapezoid body to continue on the ventral surface of the medulla. They are separated by the **ventral median fissure.** This fissure can be followed caudally until it is obliterated over a short distance by the **decussation of the pyramids** located at the level of the emerging hypoglossal nerve fibers. The decussation itself is difficult to see, for it occurs as the pyramidal fibers are passing dorsally into the parenchyma of the medulla. Pyramidal axons continue in the spinal cord as the corticospinal tracts.

The **abducent** or sixth cranial nerve leaves the medulla through the trapezoid body on the lateral border of each pyramid.

Cranial nerves VII and VIII are located on the lateral side of the medulla. Cranial nerve VII, the **facial nerve** (Fig. 222), is smaller and leaves the lateral surface of the medulla through the trapezoid body caudal to the trigeminal nerve and rostroventral to the eighth cranial nerve.

Cranial nerve VIII, the **vestibulocochlear nerve** (Figs. 222, 223, 228), is on the lateral side of the medulla at the most lateral extent of the trapezoid body just lateral to the facial nerve. Part of the vestibulocochlear nerve contributes fibers to the trapezoid body. Part enters the medulla directly, and part courses over the dorsal surface of the medulla and caudal cerebellar peduncle in the **acoustic stria** (Fig. 223). This occurs just caudal to where the caudal cerebellar peduncle passes dorsally into the cerebellum. The **cochlear nuclei** are located in this nerve as it courses over the lateral and dorsal surfaces of the medulla (Figs. 223, 228, 230).

The **hypoglossal nerve,** cranial nerve XII, emerges as a number of fine rootlets from the ventrolateral side of the myelencephalon caudal to the trapezoid body. Observe that these fibers are in the same sagittal plane as the third and sixth cranial nerves rostrally and the ventral rootlets of the spinal cord caudally. All contain somatic efferent neurons. The junction of the myelencephalon and the spinal cord is between the

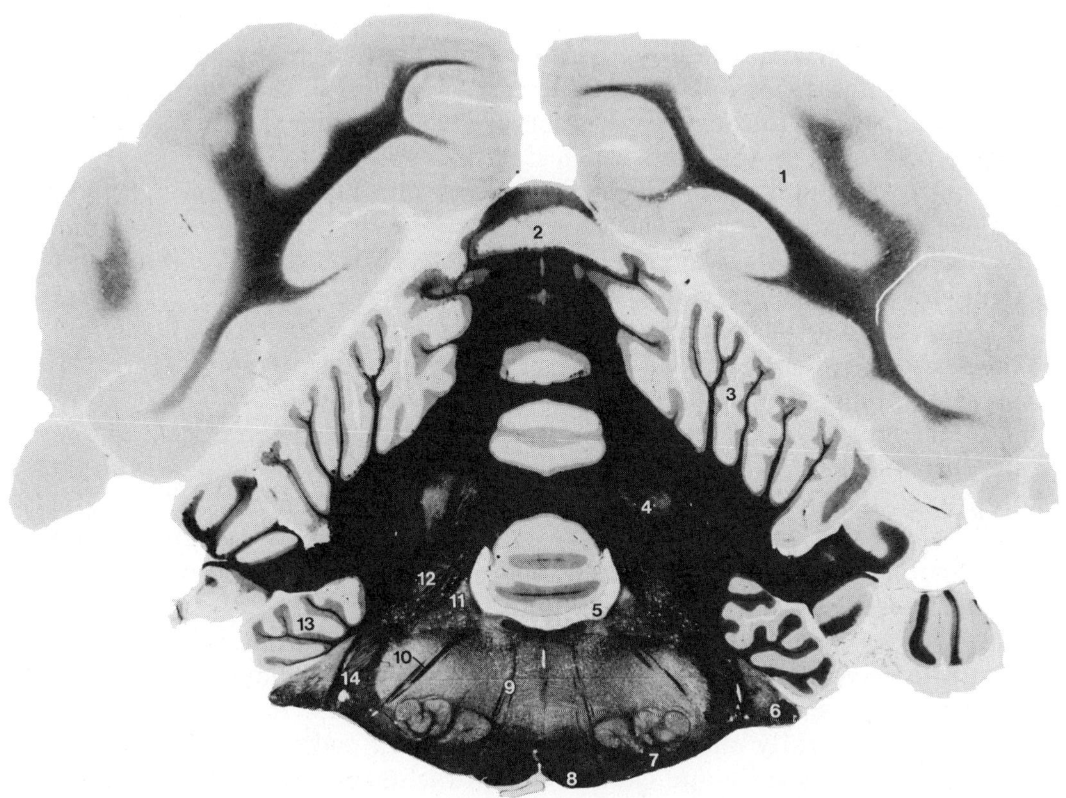

FIGURE 228. Cerebellum and myelencephalon. (The white matter is stained with iron hematoxylin and appears black in the photograph.)

1. *Occipital lobe*
2. *Cerebellar vermis*
3. *Cerebellar hemisphere*
4. *Cerebellar nucleus*
5. *Fourth ventricle*
6. *Cochlear nuclei and vestibulocochlear nerve*
7. *Trapezoid body*
8. *Pyramidal tract*
9. *Abducent nerve fibers*
10. *Descending facial nerve fibers*
11. *Vestibular nuclei*
12. *Caudal cerebellar peduncles*
13. *Flocculus*
14. *Spinal tract of trigeminal nerve*

hypoglossal fibers and the ventral rootlets of the first cervical spinal nerve. The hypoglossal nerve outside the skull is much larger owing to the addition of connective tissue components.

Dorsolateral to the emergence of the hypoglossal fibers and the fibers of the ventral root of the first cervical spinal nerve, a nerve runs lengthwise along the lateral surface of the spinal cord. This is the eleventh cranial nerve, the **accessory nerve** (Figs. 216, 222, 232). Its spinal rootlets emerge from the lateral surface of the spinal cord as far caudally as the seventh cervical segment. They emerge between the level of the dorsal and ventral rootlets of the cervical spinal nerves and course cranially through the vertebral canal and foramen magnum. A few cranial rootlets emerge from the lateral side of the medulla caudal to the tenth cranial nerve and join the accessory nerve as it courses by the medulla (Fig. 222). These are usually torn off the medulla in removing the brain for dissection. The accessory nerve leaves the cranial cavity through the jugular foramen and tympano-occipital fissure along with the ninth and tenth cranial nerves.

Cranial nerves IX and X, the **glossopharyngeal** and **vagal nerves,** leave the lateral side of the myelencephalon caudal to the eighth cranial nerve and rostral to the accessory nerve. These rootlets are small and are rarely preserved on the brain when it is removed.

Examine the dorsal surface of the pons and medulla (Fig. 223). On either side of the fourth ventricle are the cut ends of the three cerebellar peduncles. The **rostral cerebellar peduncle** is medial and courses rostrally into the mesencephalon; the **middle cerebellar peduncle** is lateral and arises from the transverse fibers on the lateral side of the pons; and the **caudal cerebellar peduncle** is in the middle, entering from the myelencephalon after passing beneath the acoustic stria.

The groove in the center of the floor of the fourth ventricle is the **median sulcus.** On the lateral wall the longitudinal groove is the **sulcus limitans.** Just lateral to this latter groove, at the level of the acoustic stria, there is a slight dorsal bulge of the medulla. This demarcates the location of the **vestibular nuclei.** Most of the vestibular neurons in the vestibulocochlear nerve terminate here.

The roof of the fourth ventricle caudal to the cerebellum is the **caudal medullary velum.** It is a thin layer composed of ependyma lining the ventricle and a supporting layer of vascularized pia. It attaches to the cerebellum rostrally and to the caudal cerebellar peduncle and the fasciculus gracilis laterally and caudally. Its attachment caudally at the apex is known as the **obex.** At this level the fourth ventricle is continuous with the central canal of the spinal cord.

At the level of the eighth cranial nerve, there is an opening in the caudal medullary velum known as the **lateral aperture** of the fourth ventricle (see Fig. 217). The cerebrospinal fluid produced in the ventricular system communicates with the subarachnoid space of the meninges via this aperture. It then courses through the subarachnoid space over the entire surface of the brain and spinal cord and is absorbed into the venous system. Most of this absorption occurs where the arachnoid is in close apposition to the cerebral venous sinuses and where it has formed specialized structures known as arachnoid villi. Cerebrospinal fluid is also absorbed from the subarachnoid space, where the spinal nerves leave the vertebral canal through the intervertebral foramina, and along the olfactory and optic nerves.

The choroid plexus of the fourth ventricle bulges into the lumen of the ventricle on each side of the dorsal midline. Each plexus extends outward through the lateral aperture, where it was seen caudal to the cerebellum before the latter was removed.

Examine the dorsal surface of the myelencephalon caudal to the fourth ventricle. The structures to be observed can be more easily recognized if the pia-arachnoid is removed and the medulla is examined under a dissecting microscope. The median groove is the **dorsal median sulcus.** The narrow longitudinal bulge flanking the sulcus is the **fasciculus gracilis** (Fig. 223). This longitudinal tract ascends the entire length of the spinal cord in this position. At the caudal end of the myelencephalon it ends at the **nucleus gracilis.** This nucleus is located at the caudal end of the fourth ventricle, where the fasciculus widens and ends. This fasciculus and nucleus function primarily in pelvic limb proprioception.

The groove lateral to the fasciculus gracilis is the **dorsal intermediate sulcus.** The longitudinal bulge lateral to this is the **fasciculus cuneatus.** This tract also ascends the dorsal aspect of the spinal cord, starting in the midthoracic region. The fasciculus cuneatus diverges laterally at the caudal end of the fourth ventricle and ends in a slight bulge. This bulge represents the **lateral cuneate nucleus** and is known as the **cuneate tubercle.** Rostrally the lateral cuneate nucleus is continuous with the caudal cerebellar peduncle. This fasciculus and nucleus primarily function in thoracic limb proprioception.

The groove on the caudodorsal surface of the myelencephalon lateral to the fasciculus cuneatus is the **dorsolateral sulcus.** The longitudinal bulge lateral to it is the **spinal tract of the trigeminal nerve.** The axons in this tract are from sensory cell bodies in the trigeminal ganglion that supply the head. The trigeminal tract extends caudally to the level of the first cervical segment of the spinal cord because of the large number of cell bodies serving its extensive peripheral distribution in the head. This tract emerges on the lateral surface of the myelencephalon caudal to a band of obliquely ascending fibers, the **superficial arcuate fibers.** The arcuate fibers connect structures in the medulla with the caudal cerebellar peduncle.

The dorsal rootlets of the spinal cord enter through the dorsolateral sulcus along the spinal cord.

Telencephalon (Cerebrum)

The left cerebrum was previously removed (see Fig. 230) by cutting the rostral commissure, corpus callosum and hippocampal commissure;

separating the two halves of the body of the fornix on the median plane; and sectioning the internal capsule, which attached the cerebrum to the thalamus of the brain stem.

There are three commissural pathways crossing between the hemispheres, one for each phylogenetic division of the cerebrum. The corpus callosum connects the neopallial portion of each hemisphere and is the largest of the three pathways. The rostral commissure connects the paleopallial or olfactory components of each hemisphere. The hippocampal commissure is small and is located just caudal to the junction of the crus of each fornix. This connects the archipallial components of each hemisphere. The internal capsule consists of projection fibers that course between the brain stem and cerebral hemisphere. Association fibers remain within the hemisphere, coursing between adjacent or distant gyri.

The **corpus callosum** (Figs. 224, 225, 229, 230) consists of a rostral

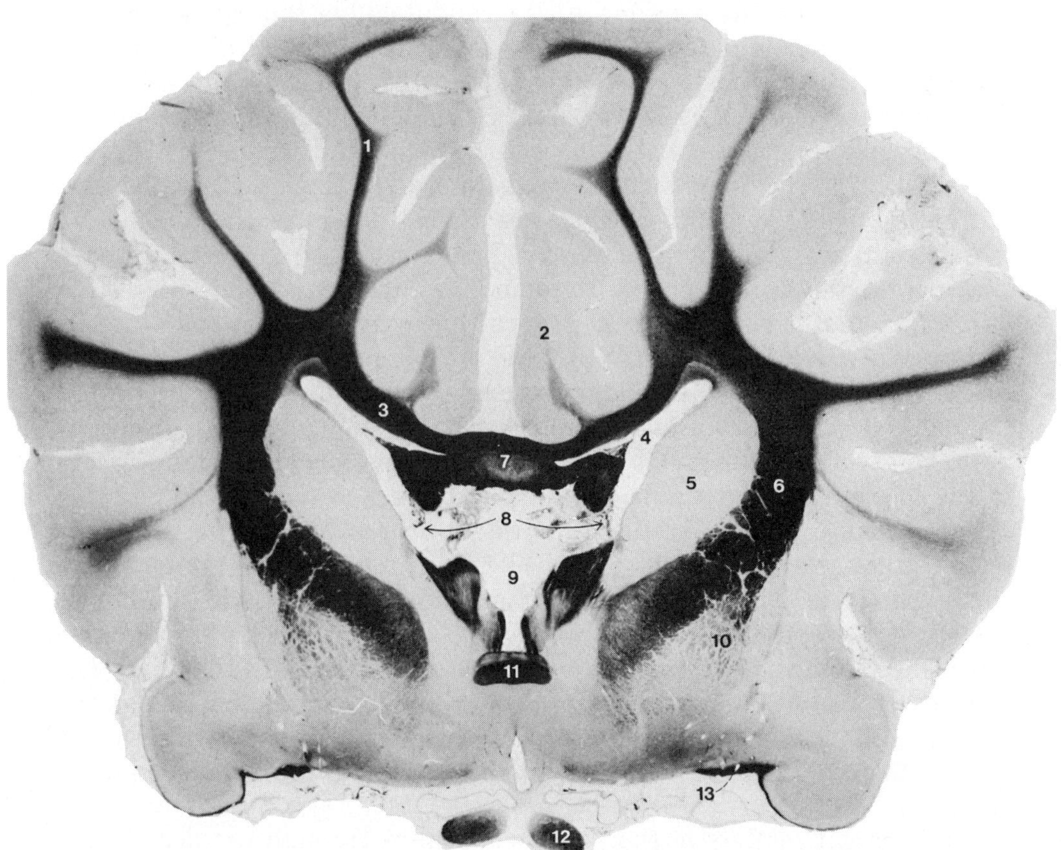

FIGURE 229. Telencephalon. (The white matter is stained with iron hematoxylin and appears black in the photograph.)

1. *Corona radiata*
2. *Cingulate gyrus*
3. *Corpus callosum*
4. *Lateral ventricle*
5. *Caudate nucleus*
6. *Internal capsule*
7. *Body of fornix*
8. *Interventricular foramen*
9. *Third ventricle*
10. *Lentiform nucleus*
11. *Rostral commissure*
12. *Optic nerve*
13. *Lateral olfactory tract*

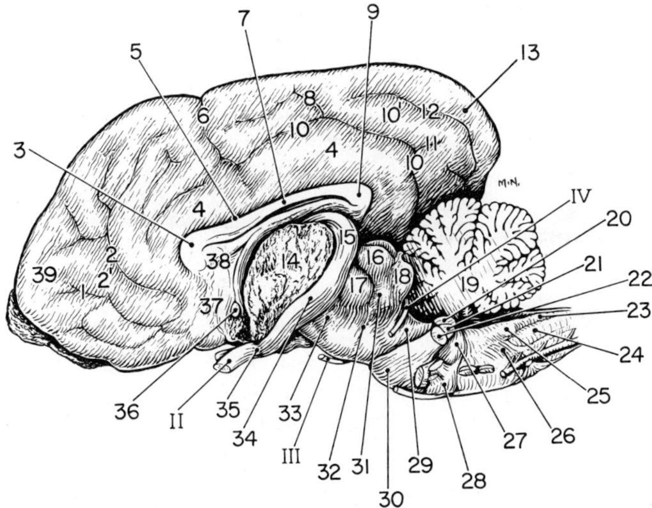

FIGURE 230. Medial surface of right cerebrum and lateral surface of brain stem.

1. *Ectogenual sulcus*
2. *Genual sulcus*
2'. *Genual gyrus*
3. *Genu of corpus callosum*
4. *Cingulate gyrus*
5. *Callosal sulcus*
6. *Cruciate sulcus*
7. *Body of corpus callosum*
8. *Lesser cruciate sulcus*
9. *Splenium of corpus callosum*
10. *Calcarine sulcus*
10'. *Splenial gyrus*
11. *Caudal horizontal ramus of splenial sulcus*
12. *Suprasplenial sulcus*
13. *Occipital gyrus*
14. *Cut surface between cerebrum and brain stem*
15. *Optic tract at lateral geniculate body*
16. *Rostral colliculus*
17. *Medial geniculate body*
18. *Caudal colliculus*
19. *Arbor vitae cerebelli*
20. *Rostral cerebellar peduncle*
21. *Caudal cerebellar peduncle*
22. *Middle cerebellar peduncle*
23. *Fasciculus cuneatus*
24. *Spinal tract of trigeminal nerve*
25. *Lateral cuneate nucleus*
26. *Superficial arcuate fibers*
27. *Cochlear nuclei*
28. *Trapezoid body*
29. *Lateral lemniscus*
30. *Transverse fibers of pons*
31. *Brachium of caudal colliculus*
32. *Transverse crural tract*
33. *Crus cerebri*
34. *Left optic tract*
35. *Optic chiasm*
36. *Rostral commissure*
37. *Paraterminal gyrus*
38. *Septum pellucidum*
39. *Prorean gyrus*
II. *Optic nerve*
III. *Oculomotor nerve*
IV. *Trochlear nerve*

genu, middle body and caudal splenium. It is the commissural pathway for axons crossing between the neopallium of each cerebrum. The **internal capsule** (Fig. 231) contains projection fibers that course between the telencephalon and the diencephalon and descend from the telencephalon to the brain stem and the spinal cord. The fibers that were previously seen in the mesencephalon as the crus cerebri descend through the internal capsule. The projection pathway of most sensory modalities to the cerebrum for conscious perception involves synapses in the thalamus. The axons of these thalamic cell bodies then enter the internal capsule to reach the cerebrum.

Examine the isolated left cerebral hemisphere. On its medial surface, locate the portions of the corpus callosum and note the thin vertical sheet of tissue ventral to the corpus callosum. This is the **septum pellucidum,** which is more developed rostrally where it extends from the genu to the rostral commissure. A thickening in the septum dorsal and rostral to the rostral commissure represents the **septal nuclei.** Caudal to the rostral commissure the septum attaches the corpus callosum to a column of fibers

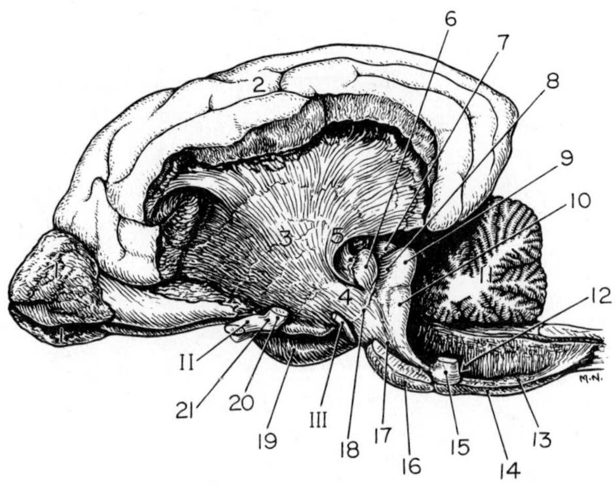

FIGURE 231. Lateral view of brain, internal capsule exposed.

1. *Olfactory bulbs*
2. *Left cerebral hemisphere*
3. *Internal capsule (lateral view)*
4. *Crus cerebri*
5. *Acoustic radiation*
6. *Medial geniculate nucleus*
7. *Rostral colliculus*
8. *Brachium of caudal colliculus*
9. *Caudal colliculus*
10. *Lateral lemniscus*
11. *Cerebellum*
12. *Location of dorsal nucleus of trapezoid body*
13. *Location of olivary nucleus*
14. *Pyramid*
15. *Trapezoid body*
16. *Transverse fibers of pons*
17. *Pyramidal and corticopontine tracts (longitudinal fibers of pons)*
18. *Transverse crural tract*
19. *Piriform lobe*
20. *Optic tract (cut to show internal capsule)*
21. *Optic chiasm*

II. *Optic nerve*
III. *Oculomotor nerve*

that courses rostrally and then descends in a rostroventral curve behind the rostral commissure. These fibers are part of the **fornix.** They connect the hippocampus with the diencephalon and rostral cerebrum. The hippocampus is a specialized region of cerebral cortex that will be seen shortly. Cut the septum pellucidum to separate the corpus callosum from the fornix. The fornix begins caudally by the accumulation of fibers on the lateral side of the hippocampus. These form the **crus** of the fornix. The crura join rostral to the hippocampus and dorsal to the thalamus to form the **body** of the fornix.

The body of the fornix courses rostrally and then descends rostroventrally as the **column** of the fornix. At the rostral commissure some fibers course dorsal and rostral to the rostral commissure, but most descend caudal to the commissure and continue caudoventrally, lateral to the third ventricle, to reach the mamillary body of the hypothalamus. This descending column may be more evident on the intact right cerebral hemisphere. Pierce the septum and separate the corpus callosum from the column and body of the fornix.

The curved cavity exposed lateral to the septum pellucidum and ventral to the corpus callosum is the **lateral ventricle.** It communicates with the third ventricle of the diencephalon by the **interventricular foramen** (Fig. 224), which is located caudal to the column of the fornix at the level of the rostral commissure. The caudodorsal wall of this foramen is the roof plate and choroid plexus of the third ventricle. This is usually pulled free and lost when the brain is removed.

Locate the caudal part of the lateral rhinal sulcus in the left temporal lobe. Cut through this sulcus into the lateral ventricle. The smooth, curved bulge exposed in the wall of the ventricle is the caudal surface of the hippocampus as it ascends in a dorsorostral curve. To remove the hippocampus intact from the ventricle, cut the column of the fornix dorsal to the rostral commissure. Grasp the fornix with the forceps and, with the blunt end of the scalpel, gently roll the hippocampus out of the lateral ventricle. Its attachment ventrally in the temporal horn may be cut to free the hippocampus completely.

Phylogenetically, the cerebral cortex can be divided into three regions: paleopallium, neopallium and archipallium. The paleopallium consists of the cortex of the olfactory peduncle. This is separated laterally from the neopallium by the rhinal sulcus. The neopallium includes all of the gyri on the external surface of the cerebrum. The **hippocampus** belongs to the archipallium and is an internal gyrus of the telencephalon that has been rolled into the lateral ventricle by the lateral expansion of the neopallium. Notice that the hippocampus begins ventrally in the temporal lobe and curves, first caudodorsally and then rostrodorsally, over the diencephalon to reach its caudodorsal aspect. At that point the hippocampus ends, and a **hippocampal commissure** connects the hippocampus of each cerebrum. The fibers of the hippocampus continue rostrally as the body and column of the fornix. Notice the crus of the fornix on its lateral surface. Place the previously removed hippocampus over the exposed left diencephalon to see its normal relationship with that structure.

Attached to the lateral free edge of the crus of the fornix is a network of blood vessels covered by meninges and ependyma. This is the **choroid plexus of the lateral ventricle.** Its anatomic structure is the same as the choroid plexus of the third and fourth ventricles. The surface of the choroid plexus that faces the lumen of the ventricle is the ependymal layer derived from the embryonic neural tube. This layer of ependyma is attached on one side to the free edge of the fornix. To complete the wall of the lumen of the lateral ventricle, it must be attached on its other side. This attachment is to a small tract, the stria terminalis, in the groove between the thalamus and the caudate nucleus. This layer of ependyma from the groove to the fornix forms part of the medial wall of the lateral ventricle. Branches of the middle and caudal cerebral arteries covered by pia push this layer of ependyma into the lumen of the lateral ventricle. The result is the formation of a choroid plexus that is continuous with the choroid plexus of the third ventricle at the interventricular foramen.

Examine the **rostral commissure** (Figs. 224, 229, 230). On each side this commissure connects rostrally to the olfactory peduncles and caudally to the piriform lobes.

In each cerebrum the neuronal cell bodies are located either on the surface in the **cerebral cortex** or deep to the surface in a **basal nucleus.** Examine the floor of the lateral ventricle of the removed left hemisphere. The bulge that enlarges rostrally is the **caudate nucleus.** This is one of the subcortical nuclei of the telencephalon, a part of the corpus striatum. Its rostral extremity is the head. Caudal to this the body rapidly narrows into a small tail, which courses over the internal capsule fibers in the floor of the ventricle. With the blunt end of a scalpel, free the caudate nucleus from the medial side of the internal capsule.

Dorsolateral to the caudate nucleus the internal capsule forms the lateral angle of the lateral ventricle. At the dorsolateral angle of the lateral ventricle the fibers of the internal capsule meet those of the corpus callosum. This interdigitation of fibers is known as the **corona radiata,** which radiates in all directions to reach the gray matter of the cerebral cortex. A longitudinal section of the cerebral hemisphere in the dorsal plane at the level of this interdigitation reveals a mass of white matter in the center of the hemisphere. This mass is referred to as the **centrum semiovale.**

Remove the pia and arachnoid from the surface of the right cerebral hemisphere. Expose the corona radiata by removing the gray matter, the cerebral cortex, with the scalpel handle. (Following fixation the brain specimen was frozen and then thawed. This procedure dissociates cell bodies from fiber tracts and makes it possible to remove the gray matter without damaging the corona radiata.)

Begin this removal of gray matter on the medial side of the hemisphere. The cingulate gyrus is located dorsal to the corpus callosum. Remove the gray matter of the cingulate gyrus to expose its fibers, which form the **cingulum.** Many of the fibers in the cingulum are long association fibers that course longitudinally from one end of the hemisphere to the other. Demonstrate this by freeing some fibers rostral to the genu of the corpus callosum and by stripping them caudally. Remove the cingulum and demonstrate the transverse course of the fibers of the corpus callosum by stripping these from their cut edge toward the hemisphere.

Remove the gray matter from the gyri on the lateral surface of the rostral half of the hemisphere. Examine the white matter of the gyri. Short association fibers, the arcuate fibers, course between adjacent gyri.

The internal capsule is situated lateral to the caudate nucleus. Expose the lateral surface of the internal capsule (Fig. 231).

When you have completed the dissection of the right cerebral hemisphere, all that remains are the corpus callosum, the internal capsule and part of the corona radiata.

Remove the corpus callosum of the right cerebral hemisphere to expose the lateral ventricle, with the caudate nucleus on the floor rostrally and the hippocampus on the floor caudally. The internal capsule forms the lateral wall (Fig. 231).

Make a median section of the brain stem from the optic chiasm through the medulla (Fig. 224). Observe the interthalamic adhesion and note the smooth surface of the third ventricle surrounding it. The ventricle is bounded dorsally by the roof plate between each stria habenularis. It connects rostrally with each lateral ventricle through the interventricular foramen, which is caudal to the column of the fornix and dorsal to the rostral commissure. Caudally it is continuous with the mesencephalic aqueduct. Rostroventral to the interventricular foramen the ventricle is bounded by the lamina terminalis. The **lamina terminalis** demarcates the most rostral extent of the embryonic neural tube on the median plane. (From this point each cerebral hemisphere develops laterally.) The third ventricle extends into the infundibulum of the pituitary gland as the **infundibular recess** (Fig. 226).

Follow the ventricular system caudally from the third ventricle into the mesencephalic aqueduct and into the fourth ventricle. Note that the roof of the fourth ventricle consists of the rostral medullary velum, the cerebellum and the caudal medullary velum. The choroid plexus of the fourth ventricle is formed in the caudal medullary velum. The fourth ventricle continues into the central canal of the spinal cord at the obex.

Functional and structural correlates of the nervous system are listed in Table 5.

SPINAL CORD

If a prepared dissection of the spinal cord is not available, remove all of the vertebral arches and the attached muscles from your specimen to expose the spinal cord. Observe the spinal cord with its dural covering. Between the dura of the spinal cord and the periosteum of the vertebrae is the **epidural space,** which contains loose connective tissue, fat and blood vessels.

Table 5. *FUNCTIONAL AND STRUCTURAL CORRELATES OF THE NERVOUS SYSTEM*

Neuroanatomical Structure	Function
Frontal lobe	Behavior
Frontoparietal lobe (Sensorimotor cortex)	Upper motor neuron (central motor system) Conscious perception of somatic, visceral and proprioceptive sensations
Temporal lobe	Behavior
Occipital lobe	Vision
Olfactory bulb, peduncle, tract, pyriform lobe	Olfaction—Smell
Pons—Transverse fibers Middle cerebellar peduncle	Pathway from cerebrum to cerebellum
Rostral cerebellar peduncle	Pathway from cerebellum to brain stem and cerebrum
Caudal cerebellar peduncle	Pathway from spinal cord and medulla with vestibular proprioceptive systems to cerebellum
Mamillary bodies	Limbic system (behavior)
Crus cerebri	Pathway from cerebrum to brain stem and spinal cord—Upper motor neuron
Internal capsule	Pathway between cerebrum and brain stem—Upper motor neuron and thalamocortical projections—Sensory and motor
Hypothalamus	Visceral functions Autonomic nervous system Limbic system Pituitary control

***Table 5.** FUNCTIONAL AND STRUCTURAL CORRELATES OF THE NERVOUS SYSTEM Continued*

Neuroanatomical Structure	Function
Lateral geniculate nucleus	Visual system (conscious perception)
Rostral colliculus	Visual system (reflexes)
Medial geniculate nucleus	Auditory system (conscious perception)
Caudal colliculus	
Brachium of caudal colliculus	
Lateral lemniscus	Auditory system (reflexes)
Trapezoid body	
Acoustic stria	
Cochlear nuclei	
Pyramid	Upper motor neuron
Corticospinal tracts	
Fasciculus gracilis	Pelvic limb—Conscious pathway of general proprioception
Nucleus gracilis	
Fasciculus cuneatus	Thoracic limb—General proprioception
Medial cuneate nucleus	Conscious perception pathway
Lateral cuneate nucleus	Cerebellar pathway
Spinal tract of trigeminal nerve	Projection pathway of somatic afferent sensation from head
Septal nuclei	Limbic system
Fornix–Hippocampus	Limbic system
Cingulum	
Caudate nucleus	Upper motor neuron
Lentiform nucleus	
Stria habenularis	Limbic system
Habenular nucleus	

The spinal cord is divided into segments (Fig. 232). A group of dorsal and ventral rootlets leave each spinal cord segment on each side and combine respectively to form the **spinal nerve** at the level of the intervertebral foramen. Note the **spinal ganglia** in the intervertebral foramina.

There are eight cervical spinal cord segments, thirteen thoracic, seven lumbar, three sacral and about five caudal. The roots of the first cervical spinal nerve leave the vertebral canal through the lateral vertebral foramen in the arch of the atlas. The second cervical roots leave caudal to the atlas. The cervical roots of segments 3 through 7 leave the vertebral canal through the intervertebral foramina cranial to the vertebra of the same number. The roots of the eighth cervical segment pass caudal to the seventh (last) cervical vertebra. The roots of all the remaining spinal cord segments pass through the intervertebral foramina caudal to the vertebra of the same number.

In the caudal cervical region over the fifth to seventh cervical vertebrae, there is an enlargement of the spinal cord that nearly fills the canal. This is the **cervical intumescence.** Its presence is due to an increase in white matter and cell bodies that are associated with the innervation of the thoracic limb. The intumescence occurs from the sixth cervical segment of the spinal cord through the first thoracic segment. Another enlargement occurs in the midlumar vertebral region for the innervation of the pelvic limb. The **lumbar intumescence** begins at about the fourth lumbar

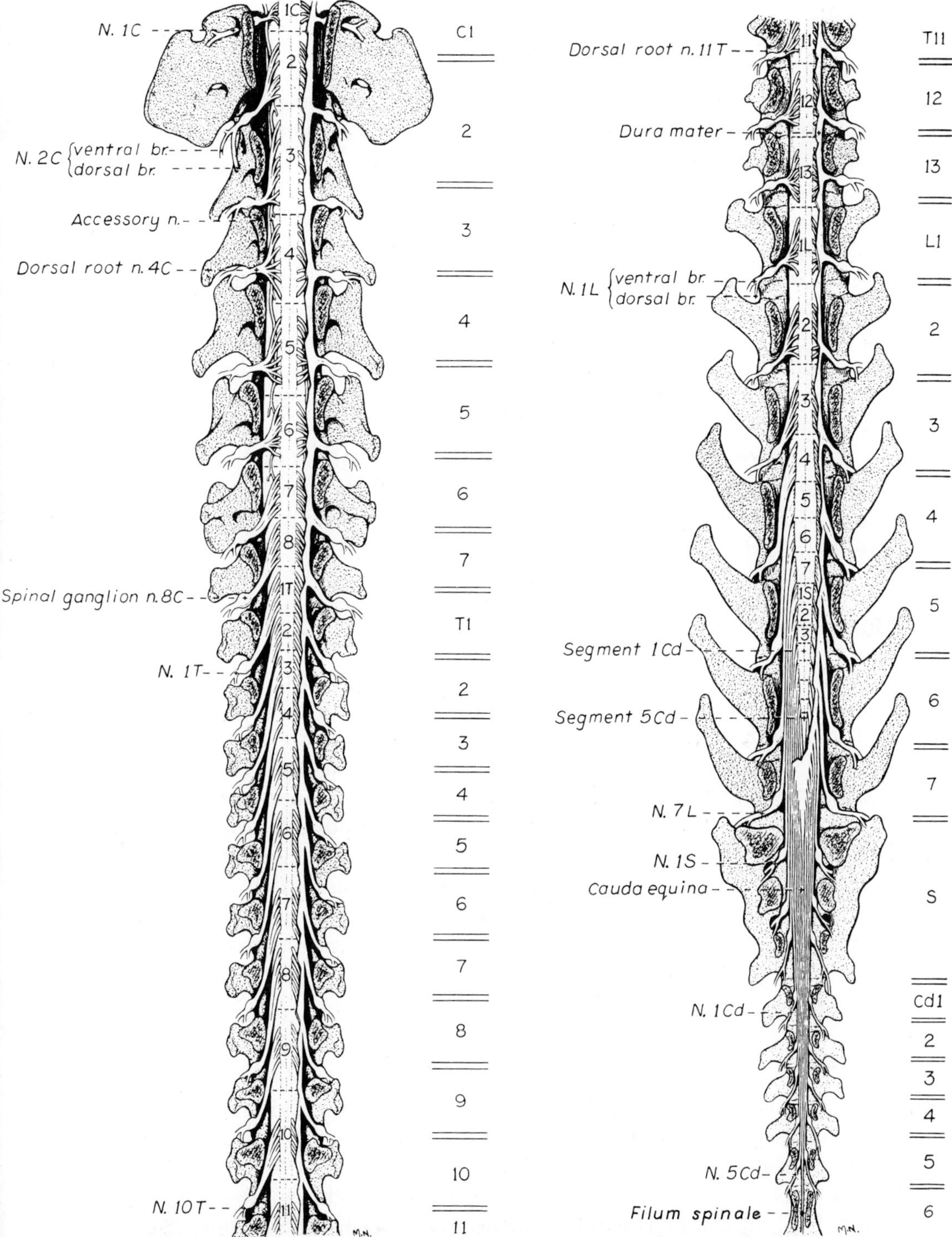

FIGURE 232. Dorsal roots of spinal nerves and spinal cord segments. Dorsal view, vertebral arches removed. (Figures on the right represent levels of vertebral bodies.)

segment and gradually narrows caudally as the spinal cord comes to an end near the intervertebral space between the sixth and seventh lumbar vertebrae. The narrow end of the parenchyma of the spinal cord is known as the **conus medullaris.** The spinal cord terminates in the **filum terminale,** which is a narrow cord of meninges that attaches the conus medullaris to the caudal vertebrae. The **cauda equina** includes the conus medullaris together with the adjacent lumbar and sacral roots that extend caudally in the vertebral canal.

Observe the relationship of the spinal cord segments to the corresponding vertebrae. The only spinal cord segments that are found entirely within their corresponding vertebrae are the last two thoracic and the first two (or occasionally three) lumbar segments. All other spinal cord segments reside in the vertebral canal cranial to the vertebra of the same number (Fig. 232). This is most pronounced in the caudal lumbar and sacrocaudal segments of the spinal cord. In general, the three sacral segments lie within the fifth lumbar vertebra and the five caudal segments within the sixth lumbar vertebra.

The nerve roots of the first ten thoracic segments and those caudal to the third lumbar segment are long because of the distance between their origin at the spinal cord and their passage through the intervertebral foramen.

Meninges

The thick, fibrous dura may be cut longitudinally along the entire length of the dorsal aspect of the spinal cord to reveal the dorsal rootlets and their length at different levels.

The thin arachnoid and pia are apposed to each other in the embalmed specimen and remain on the spinal cord. On the lateral surface of the spinal cord the pia thickens and forms a longitudinal cord of connective tissue called the **denticulate ligament.** This ligament attaches to the arachnoid and dura laterally, midway between the roots of adjacent spinal cord segments. At each intervertebral foramen the dura forms a strong attachment to the foramen, and the subdural and subarachnoid spaces end.

Vessels

There is a longitudinal **ventral spinal artery** and one or two dorsal spinal arteries on the spinal cord. They are formed by **spinal branches** of the paired vertebral arteries in the cervical region, intercostal arteries in the thoracic region and lumbar arteries in the lumbar region. The ventral spinal artery is continuous cranially with the basilar artery.

There are ventral internal vertebral venous plexuses on the floor of the vertebral canal in the epidural space. These were seen previously in the atlanto-occipital region to be continuous cranially with a branch of the sigmoid sinus. Anastomoses occur between the plexus of each side and the branches of the vertebral and azygos veins and caudal vena cava.

Transverse Sections

Study transverse sections of the spinal cord at segments C4, C8, T4, T12, L2, L6 and S1 under a dissecting microscope. Compare the shape of the gray matter of these segments and relate it to their areas of innervation (Fig. 233).

The gray matter of the spinal cord in transverse section is in the shape of a butterfly or the letter H. It consists primarily of neuronal cell bodies. The dorsal extremity on each side is the **dorsal horn,** which receives the entering dorsal (sensory) rootlets. The ventral extremity is the **ventral horn,** which sends axons out via the ventral (motor) rootlets. In the thoracolumbar region the **lateral horn** projects laterally from the gray matter midway between the dorsal and ventral gray horns. This lateral horn contains the cell bodies of the preganglionic sympathetic neurons. In the center of the gray matter of the spinal cord is the **central canal.** This remnant of the embryonic neural tube is continuous rostrally with the fourth ventricle.

The white matter of the spinal cord can be divided into three pairs of funiculi. Dorsally, a shallow, longitudinal groove extends the entire length of the spinal cord. This is the **dorsal median sulcus.** The longitudinal furrow along which the dorsal rootlets enter the cord is the **dorsolateral**

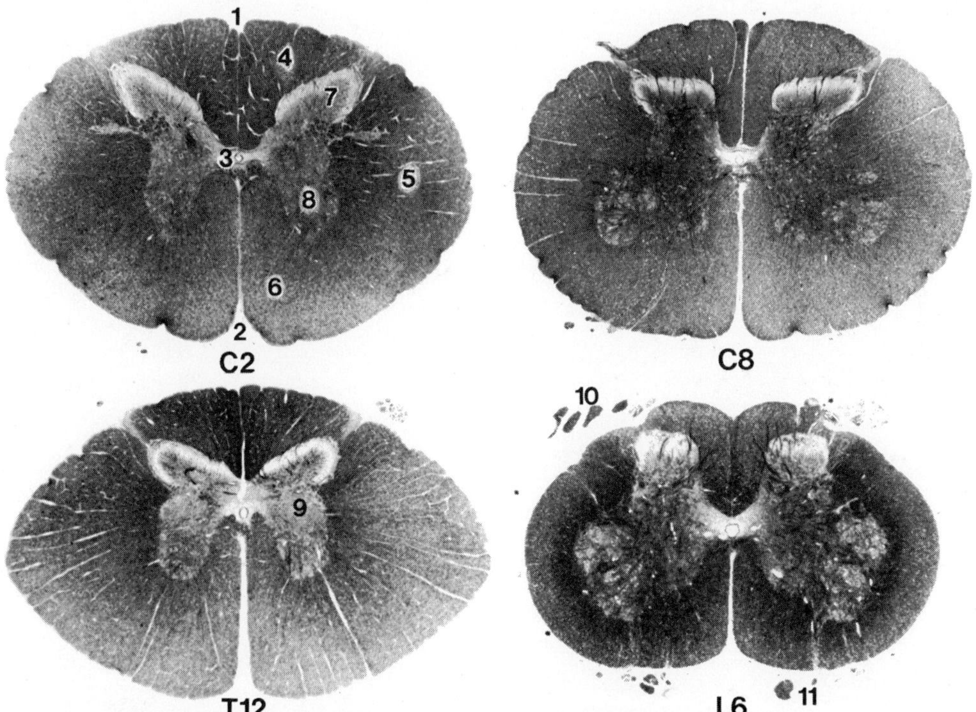

FIGURE 233. Transverse section of spinal cord: second cervical segment (C2), eighth cervical segment (C8), twelfth thoracic segment (T12) and sixth lumbar segment (L6).

1. Dorsal median septum
2. Ventral median fissure
3. Central canal
4. Dorsal funiculus
5. Lateral funiculus
6. Ventral funiculus
7. Dorsal gray column
8. Ventral gray column
9. Intermediate gray column
10. Dorsal rootlets
11. Ventral rootlets

sulcus. Between these two sulci is the **dorsal funiculus** of the spinal cord.

Between the dorsolateral sulcus and the line of exit of the ventral rootlets, the **ventrolateral sulcus,** is located the **lateral funiculus.** The **ventral funiculus** is the white matter between the line of exit of the ventral rootlets and the longitudinal groove on the ventral side of the spinal cord, the **ventral median fissure.** In some species the funiculi have been subdivided topographically into specific ascending and descending tracts. Such anatomical information in domestic animals is still incomplete.

REFERENCES

Adams, D. R., 1986. Canine Anatomy: A Systematic Study. Iowa State University Press, Ames.

Ammann, K., E. Seiferle and G. Pelloni, 1978. Atlas of Topographical Surgical Anatomy of the Dog. Verlag Paul Parey, Berlin.

Anderson, A. C., 1970. The Beagle as an Experimental Dog. Iowa State University Press, Ames.

Barone, R., 1976. Anatomie Comparée des Mammiferes Domestiques. Vol. I—Osteologie; Vol. II—Arthrologie et Myologie; Vol. III—Splanchnologie, Foetus et Ses Annexes. Editions Vigot Frères, Paris.

Baum, H., and O. Zietzschman, 1936. Handbuch der Anatomie des Hundes. 2nd Ed. Verlag Paul Parey, Berlin.

Bolk, L., et al., 1931–1938. Handbuch der Vergleichenden Anatomie der Wirbeltiere. 6 vols. Verlag Urban & Schwarzenberg, Berlin.

Bourdelle, E., and C. Bressou, 1953. Anatomie Régionale des Animaux Domestiques. Vol. IV—Carnivores: Chien et Chat. Editions J-B Ballière, Paris.

Bruni, A. C., and U. Zimmerl, 1950. Anatomia degli Animali Domestici. 2nd Ed. 2 vols. Vallardi, Milan.

Budras, K.-D., and W. Fricke, 1983. Atlas der Anatomie des Hundes. Schlutersche, Hannover.

de Lahunta, A., 1983. Veterinary Neuroanatomy and Clinical Neurology. 2nd Ed. W. B. Saunders Company, Philadelphia.

de Lahunta, A., and R. E., Habel, 1986. Applied Veterinary Anatomy. W. B. Saunders Company, Philadelphia.

Dyce, K. M., W. O. Sack and C. J. G. Wensing, 1987. Textbook of Veterinary Anatomy. W. B. Saunders Company, Philadelphia.

Ellenberger, W., and H. Baum, 1943. Handbuch der Vergleichenden Anatomie der Haustiere. 18th Ed. Springer-Verlag, Berlin.

Evans, H. E., and G. C. Christensen, 1979. Miller's Anatomy of the Dog. 2nd Ed. W. B. Saunders Company, Philadelphia.

Feher, G., 1980. A haziallatok funkcionalis anatomiaja. 3 vols. Mezogazdasagi Kiado, Budapest.

International Committee on Veterinary Anatomical Nomenclature. 1983. Nomina Anatomica Veterinaria 3rd Ed. World Association of Veterinary Anatomists, Ithaca, N. Y.

Jenkins, T. W., 1978. Functional Mammalian Neuroanatomy. 2nd Ed. Lea & Febiger, Philadelphia.

Nickel, R., A. Schummer and E. Seiferle, 1961–1967. Lehrbuch der Anatomie der Haustiere. Vol. I–IV. Verlag Paul Parey, Berlin. (Revisions with additional authors and translations into English, 1973–1984. Springer-Verlag, Berlin.)

Noden, D. M., and A. deLahunta, 1985. The Embryology of Domestic Animals: Developmental Mechanisms and Malformations. Williams & Wilkins, Baltimore.

Pierard, J., 1972. Anatomie Appliquée des Carnivores Domestiques, Chien et Chat. Somabec Lté, Quebec.

Popesko, P., 1977. Atlas of Topographical Anatomy of the Domestic Animals. 2nd Ed. W. B. Saunders Company, Philadelphia.

Romer, A. S., and T. S. Parsons, 1986. The Vertebrate Body. 6th Ed. W. B. Saunders Company, Philadelphia.

Schebitz, H., and H. Wilkens, 1986. Atlas of Radiographic Anatomy of the Dog. 4th Ed. Verlag Paul Parey, Berlin & W. B. Saunders Company, Philadelphia.

Singer, M., 1962. The Brain of the Dog in Section. W. B. Saunders Company, Philadelphia.

Stockard, C. R., 1941. The genetic and endocrinic basis for the differences in form and behavior. Amer. Anat. Memoirs: *19*. Wistar Institute, Philadelphia.

INDEX

Note: Numbers in *italics* refer to illustrations. Page numbers followed by t refer to tables.